Faraday's Experimental Researches in Electricity

Guide to a First Reading

Faraday's
Experimental Researches
in Electricity

Guide to a First Reading

Howard J. Fisher

Green Lion Press

Santa Fe, New Mexico

Manufactured in the United States of America.

Published by Green Lion Press,
1611 Camino Cruz Blanca, Santa Fe, New Mexico 87501 USA.

Telephone (505) 983-3675; FAX (505) 989-9314;

mail@greenlion.com
www.greenlion.com.

Green Lion Press books are printed on acid-free paper. Both softbound and cloth-bound editions have sewn bindings designed to lie flat and allow heavy use by students and researchers. Clothbound editions meet the guidelines for perma-nence and durability of the Committee on Production Guidelines for Book Longevity of the Council on Library Resources.

Printed and bound by Sheridan Books, Inc., Chelsea, Michigan.

Cover illustration based on "Michael Faraday in his Basement Laboratory, 1852" by Harriet Jane Moore (watercolor on paper), used by permission of The Bridgeman Art Library. Cover design by Dana Densmore with help from William H. Donahue and Nadine Shea.

Cataloging-in-Publication Data:

Fisher, Howard J.
Faraday's Experimental Researches in Electricity: Guide to a First Reading / by Howard J. Fisher

Includes abridged text of Michael Faraday's *Experimental Researches in Electricity*, index, bibliography, introductions, and notes.

ISBN 1-888009-14-4 (sewn softcover binding)
ISBN 1-888009-13-6 (cloth binding with dust jacket)

1. Faraday, Experimental Researches in Electricity. 2. History of Science.
3. Physics. 4. Electricity. 5. Magnetism. 6. Electrochemistry.

I. Michael Faraday (1791–1867). II. Fisher, Howard J. (1942–) III. Title.

QC517.F27 F57 2001

Library of Congress Card Number 00-112273

Table of Contents

The Green Lion's Preface

Michael Faraday's incomparable *Experimental Researches in Electricity* is a thrilling book, presenting scientific discoveries and arguments of the highest order in elegant—and mathematics-free—English narrative. Its accessibility has, however, been limited by two factors.

First, it has been out of print for many years and has become difficult to find. Green Lion Press has remedied this situation with a new complete facsimile edition. Still, the complete *Experimental Researches in Electricity* is a large book, consisting of three volumes and some 1500 pages, and even at the modest Green Lion pricing, not a casual purchase.

Second, it was written more than a century and a half ago, and makes use of some experimental apparatus, terms, and concepts that are unfamiliar to modern readers.

What seemed, distressingly, to be happening was that, for all but the few Faraday scholars and aficionados, Faraday's work was becoming subsumed under Maxwell, and Maxwell was becoming replaced by "Maxwell's equations" so that even the Faraday influence in Maxwell was increasingly obliterated.

What was needed, and what we longed to do, was to make Faraday accessible again to general historians of science, physicists, students of humanities, and interested nonspecialists outside the academic world with an interest in how present-day ideas of electricity evolved and how one very careful and creative thinker and very ingenious experimenter tried to sort out and envision the phenomena of magnetism and electricity.

To meet this aim, we of course had to start by bringing the complete *Experimental Researches in Electricity* back into print. But in addition, we would need to provide a single volume of selections which showed the breadth, the depth, the complexity, and the excitement of Faraday's investigations and Faraday's vision, with notes that would help readers understand his terms and follow his reasoning.

We immediately thought of Howard Fisher as the prospective author of such a guided text. We knew his reputation as a deep and lively student of Faraday's works, authentically recreating the experiments as well as entering imaginatively and energetically into Faraday's thinking, the thinking that is unfolded in the investigations Faraday chronicled in *Experimental Researches in Electricity*. We also knew Fisher's articles in *The College* and its successor, *The St. John's Review,* and thought they showed just the sort of approach that Faraday required:

one that would respect and embrace the full breadth of Faraday's vision and not try to reduce it to too-pat or too-succinct conclusions, or favor the passages that seemed to agree with what present-day electromagnetism favors for models. Happily, Fisher agreed to take on the project, and the result you see before you.

The selections were made in such a way as to avoid breaking the flow of Faraday's thoughts. *Experimental Researches in Electricity* is divided into twenty nine Series. The Series selected for this Guide (about a third of the total text of *Experimental Researches in Electricity*) are presented largely unabridged. The Series that have been omitted are those that either seemed too challenging for a "first reading" or, though in themselves interesting and worthy, could not be included without inappropriate inflation of the book's size and price.

The design of this Guide allows the reader to follow Faraday's text continuously, as long as all is flowing well without help. There are no interruptions of the text for comments, not even in the form of footnote references (the only footnote references are to Faraday's own footnotes). Comments are at the bottom of the pages, separated from Faraday's text by a rule, and are keyed to Faraday's numbered paragraphs and to target phrases within those paragraphs. The reader need only consult the comments when some confusion arises.

The aim of the notes is to help the reader understand the text. They do this in two primary ways. First, notes explain terms and concepts that are no longer current, describe experimental apparatus, and illustrate things which may be unclear with additional diagrams. Second, the notes provide guidance in avoiding misinterpretation (often based on anachronistic preconceptions). In reading Faraday, we not only need to familiarize ourselves with the practices of his day, but must also set aside what we "know" about electricity and magnetism, in order to appreciate fully Faraday's way of eliciting principles and concepts directly from the experiments themselves.

Fisher has written an introduction for each Faraday Series included in this Guide. These introductions clarify the issues and place the investigations and thinking of the Series in its context in the whole web of *Experimental Researches in Electricity*, and each is a little gem of an essay which makes the reader eager to plunge into the promised adventure of the coming Series. The whole ends with the thrilling Faraday paper "On the Physical Character of the Lines of Magnetic Force" which stands as a culmination of Faraday's vision.

Fisher has also provided an index of unusual detail and helpfulness. Reading through this index is like reading an insightful summary of the unfolding and development of Faraday's thought through these key selected investigations. One feels one is learning a lot about Faraday's work, or perhaps that one is getting a refresher of the whole project encompassed by the Guide, with the added bonus of page references for every topic, to use to delve more deeply into the most enticing ones.

In preparing this Guide, Fisher brought with him, in addition to his understanding derived from years of study, experimentation, discussion and reflection, a keen sense of typography and design. The format of the book is based on his original proposals, refined in consultation with Green Lion design staff. The book is typeset in New Baskerville, a font based on designs contemporary with the original text.

In accord with Green Lion Press standards, Faraday's footnotes and Fisher's comments both appear on the same page as the paragraphs to which they refer, and diagrams are likewise on the same page or, in a few cases where that proved impossible, at least on the same spread as the text that they illustrate. This allows the reader to consult the notes and diagrams easily, without unduly interrupting the flow of Faraday's narrative by annoying page flipping.

This book, conceived in love of Faraday, beautifully and engagingly prepared by Howard Fisher, and lovingly edited and produced by the Green Lion, will, we trust, go forth to open up Faraday's wonderful work to a much wider audience, doing justice to the magnificent text to which we have the honor of providing this introduction. May it encourage many to go on to read and delight in the complete *Experimental Researches in Electricity.*

Dana Densmore
William H. Donahue
for Green Lion Press

Acknowledgments

It is a pleasure to acknowledge the generous assistance I have received from many quarters during the preparation of this book. Kathryn Kinzer and Vicki Cone, and later Lisa Richmond, together with their staff at the Greenfield Library of St. John's College in Annapolis, have provided continuing and unfailingly cordial assistance.

Chester Burke (St. John's College, Annapolis), Erik Fisher (University of Colorado, Boulder), and Dietmar Hoettecke (Oldenburg University, Germany) read portions of the manuscript. Their discerning comments helped greatly to improve the commentary in both content and expression. I am additionally indebted to Dr. Hoettecke for his kindness in providing the photograph of Oldenburg University's splendid reproduction of Faraday's spherical condensers, constructed by the Research Group on Higher Education and History of Science. My very special thanks go to Ardis A. Welch of Baltimore for sharing her extensive knowledge of nineteenth-century typographical and design practices.

This book has benefited from numerous discussions with students. I particularly wish to thank Grant Edmonds, China Layne, and Iddrisu Tia—members of the Faraday preceptorial at St. John's College in the Fall of 1997. Eve Gibson and Todd Pytel, also at St. John's, took part in an extended and fruitful reading of Faraday's Eleventh Series.

For specific information and materials I am indebted to the following individuals: David Gross, former director of the National Aquarium, Washington, D.C., generously arranged an instructive and enjoyable visit with the Aquarium's venerable electric eel. Frank A. J. L. James offered me a most gracious welcome to the Royal Institution and has since supplied invaluable information and references concerning both Faraday's apparatus and his laboratory Diary. Susan J. McLean, National Geophysical Data Center, Boulder, Colorado, provided historical information about the magnetic dip at London. Neil Brown, Senior Curator for Classical Physics, The Science Museum, London, provided firsthand information about the Gowin Knight compound magnet and the Wheatstone rotating mirror apparatus presently under the Museum's care. Wendi Wobbe is the owner of the Wheatstone concertina pictured in the Twelfth Series' introduction.

Steven Turner and Roger Sherman, both of the National Museum of American History, located, identified, and photographed the Arago disk described in Faraday's First Series. For those generous efforts, and for the knowledgeable advice each has offered regarding historical instrumentation and apparatus, I here express my warm appreciation.

Thomas King Simpson, my former colleague at St. John's College, inoculated me with Faraday fire some thirty years ago through conversations, articles, and one memorable remark. Today it would be still be difficult to name a more deeply suggestive thinker, whether regarding Faraday, Maxwell, or a host of other writers or subjects. It has been my more recent pleasure to benefit from writings of and conversations with David Gooding (University of Bath, United Kingdom). Both individuals will find their remarks cited more than once in this book.

That this book exists at all is due to the vision and vigor of Dana Densmore and William Donahue, founders and keepers of the Green Lion Press. The project of a Faraday Guide was Dana Densmore's conception, and she has worked tirelessly and lovingly on its behalf, extending herself far beyond the nominal role of editor. Our numerous and energetic discussions of Faraday's text, and mine, have greatly benefited the latter. They have also, over time, revealed a generous colleague and valued friend, for which I am very grateful.

Howard Fisher
December, 2000

To the Reader

FARADAY'S *Experimental Researches in Electricity* is a three-volume collection of papers he had originally published in the *Philosophical Transactions* and other journals between 1821 and 1855. Paramount among them are the twenty-nine numbered research *Series*—reports of his experimentation and speculation not only about electricity but about magnetism too. Faraday reflects searchingly on those twin powers, their distinctive characters, and their relations with one another and with the materials in which they act. For this Guide, I have made selections from thirteen of Faraday's Series, together with one additional paper.

As Faraday himself was anxious to point out, the *Experimental Researches* was not originally conceived as a single work. The papers composing each of its volumes were written at diverse periods and about subjects distinguished by their newness, not by their systematic relations with one another. The individual volumes were not even compiled at the same time,* so that one may well question whether it makes sense to read Faraday's book as a *whole* at all. I think it does; and it will be part of my work to point out themes, resonances, and narrative structures that lend a perhaps unexpected coherence to this eclectic work. When Faraday published the first volume, he himself noted with surprised gratification how well the individual papers managed to sustain a level of mutual consistency and integrity.

By introducing itself as *Guide to a First Reading*, this book immediately invites questions: What *is* a "first" reading? What sort of guidance does it need? Certainly a first reading is likely to be an incomplete one. Inevitably we come across incidents and remarks we do not understand, or suspect we do not understand; but some following passage may shed retrospective light upon them. Some passages we frankly pass over, at least for the present.

A first reading is usually quite free of acquisitive impulses—we are not ordinarily digging for a particular nugget of information. Instead, perhaps, we are following the story or discussion for its inherent interest and its narrative excellence. In subsequent readings we are more likely to have adopted clear objectives which, while beneficial in many ways, may predominate over a book's intrinsic course. In the extreme, we may transform a *reading* into a mining operation.

---

* They were published in 1839, 1844, and 1855, respectively.

If the book is a really good one, we may let ourselves be carried away by it during a first reading. Events and passages may strike us when we don't expect to be struck, perhaps sounding resonances we did not even know we harbored within ourselves. Self-discovery is thus a frequent attendant to first readings. Rereadings, by contrast, are more likely to be shaped by our own and others' intentions and opinions, which may lead us to take control of the book rather than the other way around.

First readings, then, are to be greatly valued. What is more, there can be *many* "first" readings of the same book. I almost think that the better the book, the more numerous are the "first" readings it can support. So I hope that you will make this reading of Faraday *first* in the same sense as Mark Twain's "First Melon I Ever Stole": the first, that is, of many delectable and memorable ones!

But doesn't the very idea of a *guide* imply that a "first reading" is something that requires *correction*? Isn't a guide's function precisely to eliminate doubts and errors, to explain difficulties, to tell us what in our author may be sound and unsound, what is essential and what dispensable? Or, if the spontaneity and freedom of a "first" reading are such good things, does not the work of a guide necessarily undermine those very qualities?

Every discipline, indeed, produces commentaries and expositions whose aim is precisely to spare the reader the delays and frustrations attendant upon false starts, unresolved questions and obsolete concepts. With sufficient assistance of that sort, a reader may give to his first reading the character and benefits of a second or third reading, if that is desired. My intention is different. I think your first reading of *Experimental Researches in Electricity* deserves cultivation, not correction. Therefore I will not try to resolve preliminary and tentative interpretations; rather I will acknowledge them and encourage you to live with them—as indeed Faraday himself must live with them. I will occasionally suggest answers, but much more frequently point out questions. I will call attention to particulars of nineteenth-century vocabulary and usage that merely distract when they appear unfamiliar but become sources of pleasure and edification when they are found to illuminate contemporary idioms. From time to time I will supply information about apparatus, practices or techniques which Faraday might reasonably suppose his reader to know.

I will also try to help you guard against routinely adopting modern terminology, habitual associations, and conventional but potentially inappropriate concepts in electricity and magnetism. You will gain most from Faraday's discussion if you take it as much as possible on his own terms. At the same time, you are likely to enjoy a fuller and deeper appreciation of present-day theoretical concepts if you have a longer perspective from which to approach them.

Faraday's text, and this Guide, make use of a few formal devices that may need explanation. By far the most conspicuous is Faraday's remarkable system of numbered paragraphs. Readers are invariably astonished at this comprehensive scheme, which in the *Experimental Researches* (he utilized it in other publications too) extends throughout 21 years and 1114 pages of text. This single device already announces Faraday's researches as continuous and whole, the subject of an astoundingly synoptic vision. On a more prosaic level, the numerical curriculum helps to sustain that vision in practice, since it permits Faraday to provide ongoing yet unobtrusive cross-references within and among the individual Series. Thus, for example, the number 1749 set off with period and parentheses—(1749.)—directs the reader to consult paragraph 1749 for supporting remarks or other material related to the topic under immediate discussion.

This Guide distinguishes scrupulously between Faraday's narrative and my remarks. Within each Series, his text always occupies the top of the page. My comments appear in smaller type at the bottom of the page, beneath a separator line, and are keyed to each numbered paragraph—in many cases also to a particular phrase. Faraday's own footnotes appear beneath his text, but above the separator line. Each Series is preceded by an editor's introduction, clearly marked.

Unless captioned otherwise, illustrations in the text are Faraday's own; but I have sometimes relettered them for clarity. Illustrations in the comments and in the editor's introductions are from various sources.

Since the selections in this Guide represent only about a third of the complete *Experimental Researches*, I should say something about the principles I have followed in deciding on retentions and omissions. Believing strongly that serious readers should be offered whole acts of thought, I have tried to avoid the kind of targeted excerpting whose main purpose is to confirm a favored presupposition or to support a scholarly position. While probably no selection can be wholly free of such influences, I have done my best to guard against them. Similarly, I hope I have avoided triumphalism—a selective glorification of such of Faraday's concepts and theories as have succeeded in gaining conventional acceptance.

In keeping with my earlier remarks about the *incompleteness* of a first reading, I have omitted topics which I judge to be so challenging that they had better be put off for subsequent study. Much of Faraday's electrochemistry falls in this category, as does much of his thought about the elusive "electrotonic state." In both these cases, I believe that initial work with the selections here chosen will greatly benefit any future study of the omitted portions.

Inevitably, there remain readings and research topics that are important, exciting, fruitful, informative, accessible, and beautiful— which are nevertheless omitted because of space limitations. I have, for example, passed lightly over self-induction, over Faraday's critique of the supposed atomic constitution of matter, and over his efforts to unify electricity and gravity. Each of these writings could be fairly argued a better candidate for inclusion than some selection actually chosen—but had I made the implied substitution, the same argument could have been made just as forcefully in the other direction. Let us remember that one's first reading need not be one's last, and that all good things need not be done at once. It is my hope that readers of this Guide will be moved to consult the complete *Experimental Researches in Electricity* for the many treasures which could not be included here. An unabridged reprint edition of the work, the first in thirty years, is being published by Green Lion Press concurrently with this Guide.

First Series — Editor's Introduction

THROUGHOUT the *Experimental Researches,* Faraday contrives phenomena that permit, to a truly remarkable extent, the essential characters and forms of electric and magnetic action to reveal themselves quite directly. That, along with his extraordinary gift for prose narrative, helps to make his writings both accessible and rewarding to the nonspecializing thoughtful reader. Nevertheless, he regularly employs instruments, and alludes to theories (sometimes theories with which he is profoundly dissatisfied), to which many readers may desire some introduction. In these instances it is not so important for the reader to gain scientific or historical backgrounds it is to attain a measure of independence from present-day preconceptions and conventionalities about electricity and magnetism. As examples, we may point to the persistent notion of electricity as an active fluid, and to the image of electric *current* as the transport of that fluid. We acquire such ideas not only from our formal education but from the very artifacts of our culture. The idea of electric *flow* is seemingly confirmed every time we *plug in* a household appliance. The idea of electricity as a store of active substance is seemingly validated every time we replace a flashlight battery.

This makes electricity hard to think about, since to do so accurately requires us to remove a patina of insufficiently examined concepts, images, and habitual associations. We are after all surrounded with things we have always known to be "electrical."[*] Every one of us has grown up with electricity, the most ubiquitous of industrial age amenities—available, as used to be said, "at the touch of a button."

But when Faraday wrote, most of the things that were undoubtedly "electrical" in nature required at least some skillful effort, and often some specialized equipment, to witness; while those phenomena that were within everyday reach, and which were *perhaps* electrical in nature, were at the same time highly questionable. Was *lightning* "electrical"? Was the *spark* one produced when stroking a cat's fur on a dry day? Was the shock of the Mediterranean torpedo fish "electrical"?

For many years Faraday carried on a program of young people's lectures on scientific topics, which he offered annually during the Christmas holidays. Part of Faraday's success as a public lecturer was to

---

[*] Almost at a glance, it seems, we recognize the following as "electrical": light bulb, motor, spark, shock; but is it obvious that all these have anything in common, by virtue of which they are ranked together? And if so, what would that be? Note how much of our so-called "electrical" experience depends on *devices* which were in large measure shaped by theoretical conceptions.

bring natural phenomena, which in fact required practice, skill, and dexterity to produce, within the compass of experience and interpretive ability possessed by people of general background, even by children. Faraday's young audience had, otherwise, scant occasion to witness such wonderful phenomena. But they evidently needed little more than access to them—the wonder came of itself. We have the opposite problem of excessive familiarity with electricity, both in our experience and in our conventional discourse about it. The ubiquity of electrical and other natural powers has paradoxically distanced them from us and has deprived us of the ability to be "at home" among them. We recognize them as facts but not as the bearers of meanings. They seldom speak to us; they seldom occasion wonder.

Actually, even Faraday's audience seems to have experienced our problem, though not in connection with electricity. In the first of the 1859 young people's lectures[*] Faraday remarks on the difficulty of remembering to *wonder* (the special gift, he thinks, of children):

> Let us now consider, for a little while, how wonderfully we stand upon this world. Here it is we are born, bred, and live, and yet we view these things with an almost entire absence of wonder to ourselves respecting the way in which all this happens. So small, indeed is our wonder, that we are never taken by surprise; and I do think that, to a young person of ten, fifteen, or twenty years of age, perhaps the first sight of a cataract or a mountain would occasion him more surprise than he had ever felt concerning the means of his own existence—how he came here; how he lives; by what means he stands upright; and through what means he moves about from place to place. Hence, we come into this world, we live, and depart from it, without our thoughts being called specifically to consider how all this takes place; and were it not for the exertions of some few inquiring minds, who have looked *into* these things and ascertained the very beautiful laws and conditions by which we *do* live and stand upon the earth, we should hardly be aware that there was anything wonderful in it.

The purpose of the following remarks, then, is not to instruct readers in the fundamentals of electrical theory, nor is it to provide "historical background." It is rather to trace a path that springs not

---

[*] Michael Faraday, *On the Various Forces of Matter and their Relations to Each Other, a course of lectures delivered before a juvenile audience at the Royal Institution.* Ed. William Crookes. London, 1860. See the opening of Lecture I. The "1859 lectures," six in number, began in December 1859 and concluded in January 1860.

from our conventional representations of electricity but from a few seminal experiences, which you may at least imagine, and in some cases recreate for yourself. I hope that may help you to think about electric and magnetic phenomena independently of the conventionalities that currently frame them. Perhaps you will even be moved to *wonder* at them.

So our position is not so very different from that of Faraday's young Christmas Lecture audiences after all: They needed to see instances of electric and magnetic action; we need to see them with fresh eyes. For that reason the remainder of this introduction will incorporate some generous excerpts from those lectures. You could say, as one early reader did, that I've arranged to have *Faraday* write the introduction!

If, therefore, you read the remainder of these preliminary remarks before beginning the *Experimental Researches*, you will still be engaged in reading Faraday's words. But if you decide instead to turn right away to the *Researches* themselves—and there is much to be said in favor of that—you will find that I have supplied references to the topics discussed so that you may turn back and consult individual sections of the introduction as needed.

Frictional electricity

Almost the only one of our "electrical" experiences that does not depend on sophisticated industrial devices is *frictional electricity*—the electricity developed when walking across some kinds of carpeting on a dry day, or sliding along a sofa upholstered with certain fabrics— electricity developed, in general, by rubbing one material against another.

The earliest known instance of frictional electricity, and the one that gave electricity its name, is mentioned by Thales of Miletus (600 B.C.E.) as a peculiarity of the substance *amber* (Greek *elektron*). This fossilized resin, when rubbed with silk or flannel, acquires the power of attracting bits of thread or dust. Similar attractive powers are found in glass, especially when rubbed with silk, and also in rubber, especially when stroked with fur. The condition, properly called *electrification*, can often be transferred from one body to another, either directly by contact or through intermediate bodies; and such transferability suggested to many investigators the idea of electricity as a mobile or even fluid substance that becomes concentrated or accumulated in the electrified body. It is probably to that conception of a material fluid that we owe the electrical term most familiar to us, the term *charge*— from its archaic meaning of a material *load* or *weight*. As we will see in his writings, however, Faraday was consistently skeptical of this "fluid" image.

Faraday described several of the chief characteristics of frictional electricity in the Christmas Lectures of 1859. The lectures were transcribed verbatim and subsequently published with illustrations. The following excerpt is from Lecture V. The remarks in bracketed italics, which provide a running narrative of Faraday's manipulations, were supplied by William Crookes:

To-day we come to a kind of attraction even more curious than the last, namely, the attraction which we find to be of a double nature—of a curious and dual nature. And I want first of all to make the nature of this double-ness clear to you. Bodies are sometimes endowed with a wonderful attraction, which is not found in them in their ordinary state. For instance, here is a piece of shell-lac,* having the attraction of gravitation, having the attraction of cohesion; and if I set fire to it, it would have the attraction of chemical affinity to the oxygen in the atmosphere. Now all these powers we find in it as if they were parts of its substance; but there is another property which I will try and make evident by means of this ball, this bubble of air [*a light India-rubber ball, inflated and suspended by a thread*]. There is no attraction between this ball and this shell-lac at present: there may be a little wind in the room slightly moving the ball about, but there is no attraction. But if I rub the shell-lac with a piece of flannel [*rubbing the shell-lac, and then holding it near the ball*], look at the attraction which has arisen out of the shell-lac, simply by this friction, and which I may take away as easily by drawing it gently through my hand. [*The Lecturer repeated the experiment of exciting the shell-lac, and then removing the attractive power by drawing it through his hand.*] Again, you will see I can repeat this experiment with another substance; for if I take a glass rod and rub it with a piece of silk covered with what we call amalgam,** look at the attraction which it has, how it draws the ball towards it; and then, as before, by quietly rubbing it through the hand, the attraction will be all removed again, to come back by friction with this silk.

---

* *Shell-lac* is a resinous substance prepared from a secretion of certain insects. It can be cast into solid forms, as in Faraday's examples. Dissolved in alcohol, it becomes the "shellac" we know as a wood finish.

** *Amalgam*: from Greek *malassein*, to soften. Usually a soft metallic alloy with mercury; by extension, any combination with mercury. In his *History and Present State of Electricity* (London, 1767) Priestley described the benefits of impregnating oiled silk with "an amalgam of mercury and tin, with a very little chalk or whiting." A glass object rubbed with the treated silk "may be excited to a very great degree with very little friction."

But now we come to another fact. I will take this piece of shell-lac and make it attractive by friction; and remember that whenever we get an attraction of gravity, chemical affinity, adhesion, or electricity (as in this case), the body which attracts is attracted also; and just as much as that ball was attracted by the shell-lac, the shell-lac was attracted by the ball. Now, I will suspend this piece of excited shell-lac in a little paper stirrup, in this way [Fig. 33],* in order to make it move easily, and I will take another piece of shell-lac, and after rubbing it with flannel, will bring them near together. You will think that they ought to attract each other; but now what happens? It does not attract; on the contrary, it very strongly *repels*, and I can thus drive it round to any extent. These, therefore, repel each other, although they are so strongly attractive—repel each other to the extent of driving this heavy piece of shell-lac round and round in this way. But if I excite this piece of shell-lac, as before, and take this piece of glass and rub it with silk, and then bring

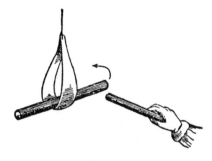

FIG. 33.

them near, what think you will happen? [*The Lecturer held the excited glass near the excited shell-lac, when they attracted each other strongly.*] You see, therefore, what a difference there is between these two attractions—they are actually two *kinds* of attraction concerned in this case, quite different to anything we have met with before; but the force is the same. We have here, then, a double attraction—a dual attraction or force—one attracting, and the other repelling.

Again, to shew you another experiment which will help to make this clear to you. Suppose I set up this rough indicator again [*the excited shell-lac suspended in the stirrup*]—it is rough, but delicate enough for my purpose; and suppose I take this other piece of shell-lac, and take away the power, which I can do by drawing it gently through the hand; and suppose I take a piece of flannel [Fig. 34], which I have shaped into a cap for it and made dry. I will put this shell-lac into the flannel, and here comes out a very beautiful result. I will rub

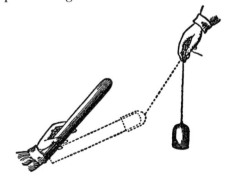

FIG. 34.

---

* Figure numbers are those of the published lecture.

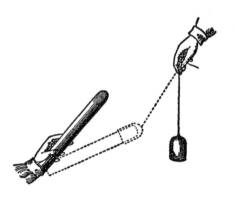

FIG. 34.

this shell-lac and the flannel together (which I can do by twisting the shell-lac round), and leave them in contact; and then, if I ask, by bringing them nearer our indicator—what is the attractive force?—it is nothing! But if I take them apart, and then ask what will they do when they are separated— why, the shell-lac is strongly repelled, as it was before, but the cap is strongly attractive; and yet if I bring them both together again, there is no attraction—it has all disappeared. [*The experiment was repeated.*] Those two bodies, therefore, still contain this attractive power: when they were parted, it was evident to your senses that they had it, though they do not attract when they are together.

This, then, is sufficient in the outset to give you an idea of the nature of the force which we call electricity. There is no end to the things from which you can evolve this power. When you go home, take a stick of sealing-wax—I have rather a large stick, but a smaller one will do—and make an indicator of this sort [Fig. 35].* Take a watch-glass (or your watch itself will do; you only want something which shall have a round face), and now, if you place a piece of flat glass upon that, you have a

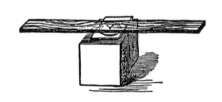

FIG. 35.

very easily moved centre. And if I take this lath and put it on the flat glass (you see I am searching for the centre of gravity of this lath—I want to balance it upon the watch-glass), it is very easily moved round; and if I take this piece of sealing-wax and rub it against my coat, and then try whether it is attractive [*holding it near the lath*], you see how strong the

---

* In Figure 35, a block of wood or other material serves as the base. Upon it rests the convex glass of his pocket-watch; then a piece of flat glass; finally the wooden strip ("lath") is balanced upon the whole. The curved and flat glass surfaces together make a low-friction pivot, which permits the lath to turn easily.

Because Faraday interrupts his thought in mid-sentence ("take a stick of sealing-wax ... and make an indicator of this sort"), you might think the sealing wax is part of the indicator. It is not, though. As Faraday soon explains, the function of the sealing wax is to be made attractive by rubbing, like the glass rod and the stick of shell-lac in his previous examples.

attraction is; I can even draw it about. Here, then, you have a very beautiful indicator, for I have, with a small piece of sealing-wax and my coat, pulled round a plank of that kind; so you need be in no want of indicators to discover the presence of this attraction. There is scarcely a substance which we may not use. Here are some indicators. I bend round a strip of paper into a hoop [Fig. 36], and we have as good an indica-tor as can be required. See how it rolls along, travelling after the sealing-wax. If I make them smaller, of course we have them running faster, and sometimes they are actually attracted up into the air. Here also is a little collodion* balloon. It is so electrical that it will scarcely leave my hand unless to go to the other. See, how curiously electrical it is: it is hardly possible for me to touch it without making it electrical; and here is a piece which clings to any-thing it is brought near, and which it is not easy to lay down. And here is another substance, gutta-percha,** in thin strips: it is astonishing how, by rubbing this in your hands, you make it electrical. But our time forbids us to go further into this subject at present. You see clearly there are two kinds of electricities which may be obtained by rubbing shell-lac with flannel, or glass with silk.

FIG. 36.

In the experiments illustrated in Figs. 33 and 34 of the preceding excerpt, Faraday showed that *the electric condition exists in two varieties,* which moreover are *antithetical*: that is, they are capable of nullifying or neutralizing one another. The conventional terms *positive* (for the electric condition exhibited by glass) and *negative* (for the condition exhibited by shell-lac) express perfectly that relation of opposition.*** It does not take too much more experimentation of the sort illustrated

---

* *Collodion* is a glutinous material used as a coating in photography and medicine, also in theatrical makeup.

** *Gutta-percha* is the tough plastic substance also called "hard rubber" because it contains more resin than true rubber. It is often used for pocket combs.

*** The nomenclature "positive" and "negative" was introduced by Benjamin Franklin. He interpreted the contrary electrical conditions as representing *excess* (+) and *deficiency* (−) of a single electrical fluid, whereas other theorists had postulated the existence of dual electric fluids—a "vitreous" fluid in glass and a "resinous" fluid in rubber, shell-lac, sealing wax, and similar materials. Faraday willingly employs the Franklin terminology in his *Experimental Researches* but, as I noted earlier, he is highly skeptical of *any* "fluid" imagery, single or dual.

in Fig. 33 to verify that *oppositely-electrified bodies attract one another,* and *similarly-electrified bodies repel one another.* We have, however, yet to explain how *nonelectrified* bodies are attracted to *electrified* bodies, whether positive or negative—as in Faraday's figure 36.

The distinctively *electrical* power, then—and for now our principal sign of the presence and degree of electrification—is *attraction and repulsion.* All of the "indicators" Faraday exhibited in the preceding excerpt made use of the attractive or repulsive power of electrified bodies.

The electroscope

The *electroscope* is a more refined indicator that employs repulsion to show the electrical condition of bodies near it or in contact with it. The

drawing shows a common *leaf electroscope,* which consists of two metal foil leaves suspended side by side from a metal support; this in turn is connected to a sensing plate. The foil leaves are protected from air currents by a glass enclosure.

If now an electrified rod is brought into the vicinity of the plate, we observe the leaves begin to diverge. Their separation increases as the rod approaches, and it decreases if the rod is again withdrawn. Is this indeed a case of mutual repulsion, as in Faraday's shell-lac indicator, which was repelled by a similarly-charged rod of shell-lac?

The question is clarified to some extent if we permit the electrified rod to *touch* the plate; for then the leaves separate—*and remain separated even when the rod is removed to a great distance.* It is reasonable to infer that through contact, the electrified rod has communicated a portion of its electrification to the electroscope, and that consequently the two leaves, being now in a similar electrical condition, repel one another and diverge. It follows then that the angle of their divergence will indicate (roughly) the degree of the electroscope's charge, and hence (even more roughly) the degree of electrification of the body that contacted it.

Thus it seems necessary to infer that even when the divergence is maintained by the approach, without contact, of an electrified rod, the diverging leaves have taken on similar electrical conditions. And yet electrification has not been permanently *transferred* from rod to leaves, as the divergence ceases as soon as the rod is withdrawn. Evidently an electrical state in the *leaves* has been somehow "induced"—that is the conventional name, but it explains nothing, as Faraday fully realizes—by the presence of the electrified *rod*. Faraday will allude to this kind of *electric induction* in the opening paragraph of the present Series; and it will become the prime subject of investigation of the Eleventh Series.

A form of electroscope much favored by Faraday uses only a *single* moving indicator, usually a dried straw lightly weighted with a pith or cork ball.* As the drawing shows, it is pivoted at one end. The principle is evidently the same as for the leaf electroscope: when the supporting post is electrified it communicates some of its condition to the straw, which is therefore repelled by the similarly-electrified post.

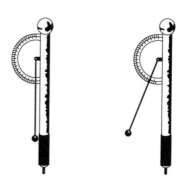

"Static" electricity and electric discharge

Loss of the electrified or charged condition is *discharge*. Discharge may occur when an electrified body is brought into contact with a much larger body. In the lecture excerpt, for example, Faraday called attention to the discharge (he did not use that term) of electrified rods that occurred when he passed them across his hands.** Discharge can also occur with a *spark*, as Faraday will demonstrate on page 11 below.

We are thus led to distinguish between the persistent or *static* condition of electrification, which is manifest primarily by the power to attract and repel, and the condition of *discharging*, which is a passing condition, usually too short-lived to be studied in itself. Thus frictional electricity was conceived as *essentially* static—literally, quiescent or unchanging. That conception survives in our present term, "static electricity." It is however an awkward nomenclature, because the phenomena we usually have in mind when we use that term—from the shock we sometimes experience in a carpeted room to the crashing

---

* The chief function of the ball, though, is not to add weight but to provide a blunt surface in place of the sharp straw end. As Faraday will discuss in the Third Series, a sharp or pointed body readily tends to discharge into the air. The terminating ball helps prevent such a discharge.

** For example: "look at the attraction which has arisen out of the shell-lac, simply by this friction, and which I may take away as easily by drawing it gently through my hand" on page 4 above. Another way to discharge a body, or even prevent it from acquiring an electrical charge, is by suitably connecting it to the *earth*—thereby *grounding* (or, as the British say, *earthing*) it.

noises that intrude in radio and telephone reception—are instances of *discharge*, not in any way quiescent. I can think of only one everyday instance of "static" electricity that really is static: that is *clinging*—as when a rubbed balloon clings to a wall, or when clothing clings together after removal from a clothes dryer.

For centuries after its discovery in amber, frictional electricity, as evidenced by attraction and repulsion, was the *only* electricity. Even Faraday calls it "ordinary" electricity (for example, paragraphs 24 and 25 in the First Series, below); and when he uses the term "excitation," he refers specifically to the raising of an electrical condition by rubbing or friction.* Up to now the only frictional processes we have considered have been *discontinuous*: In the lecture excerpt above, for example, Faraday described experiments that repeatedly cycled between frictional *excitation* of glass or shell-lac, and subsequent *discharge* of those materials. But later in the same talk he employs a mechanism that exhibits *continuous* excitation and discharge. This device, and a whole class of similar ones, Faraday calls electric "machines." Let us return to his account:

> ... And now we will return for a short time to the subject treated at the commencement of this lecture. You see here [Fig. 41] a large machine, arranged for the purpose of

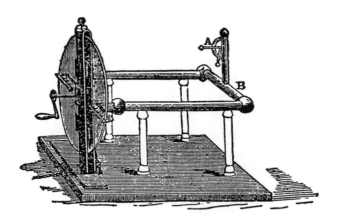

FIG. 41.

> rubbing glass with silk, and for obtaining the power called electricity; and the moment the handle of the machine is turned, a certain amount of electricity is evolved as you will

---

* For example: "I will suspend this piece of excited shell-lac..." on page 5 above.

see by the rise of the little straw indicator [*at* A].* Now, I know from the appearance of repulsion of the pith ball at the end of the straw, that electricity is present in those brass conductors and I want you to see the manner in which that electricity can pass away. [*Touching the conductor* B *with his finger, the Lecturer drew a spark from it, and the straw electrometer immediately fell.*] There, it has all gone; and that I have really taken it away, you shall see by an experiment of this sort. If I hold this cylinder of brass by the glass handle, and touch the conductor with it, I take away a little of the electricity. You see the spark in which it passes, and observe that the pith-ball indicator has fallen a little, which seems to imply that so much electricity is lost; but it is not lost: it is here in this brass; and I can take it away and carry it about, not because it has any substance of its own, but by some strange property which we have not before met with as belonging to any other force. Let us see whether we have it here or not. [*The Lecturer brought the charged cylinder to a jet from which gas was issuing; the spark was seen to pass from the cylinder to the jet, but the gas did not light.*] Ah! the gas did not light, but you saw the spark; there is, perhaps, some draught in the room which blew the gas on one side, or else it would light. We will try this experiment afterwards. You see from the spark that I can transfer the power from the machine to this cylinder, and then carry it away and give it to some other body....

But with regard to the travelling of electricity from place to place, its rapidity is astonishing. I will, first of all, take these pieces of glass and metal, and you will soon understand how it is that the glass does not lose the power which it acquired when it is rubbed by the silk. By one or two experiments I will shew you. If I take this piece of brass and bring it near the machine, you see how the electricity leaves the latter, and passes to the brass cylinder. And, again, if I take a rod of metal and touch the machine with it, I lower the indicator; but when I touch it with a rod of glass, no power is drawn away— shewing you that the electricity is conducted by the glass and the metal in a manner entirely different: and to make you see that more clearly, we will take one of our Leyden jars—

The Leyden jar

I interrupt Faraday's narrative because, astonishingly, he seems to feel no need to introduce the *Leyden jar* to his young audience. Perhaps those devices were as commonplace to them as flashlight batteries are

---

* Note that this indicator is an electroscope of the kind described on page 9 above.

to us. In this cutaway view of a typical Leyden jar, the glass is coated inside and out with metal foil. The central post is brass and extends down to the bottom of the jar, with a brass foot resting on the inner foil. The head of the post is supported by an insulating stopper.

A Leyden jar can be *electrified*, which is accomplished as follows: With the electric machine in operation, bring the knob of the Leyden jar into contact with the prime conductor (B in Faraday's Fig. 41). The outer surface of the jar is grasped in the hand or, alternatively, connected to "ground". At the moment of contact you may notice the indicator dip; then after a few moments it regains its previous position. It would thus appear that mutual contact *transferred electrification from the prime conductor to the jar*, but that continued operation of the machine restored the conductor to its previous degree of charge. As Faraday resumes the lecture, he verifies that electrification did indeed occur, by obtaining a *spark* from the jar:

If I take a piece of metal, and bring it against the knob at the top and the metallic coating at the bottom, you will see the electricity passing through the air as a brilliant spark.* It takes no sensible time to pass through this; and if I were to take a long metallic wire, no matter what the length—at least as far as we are concerned—and if I make one end of it touch the outside, and the other touch the knob at the top, see how the electricity passes!—it has flashed instantaneously through the whole length of this wire**...

Here is another experiment, for the purpose of shewing the conductibility of this power through some bodies, and not through others. Why do I have this arrangement made of brass? [*Pointing to the brass work of the electrical machine*, Fig. 41]. Because it conducts electricity. And why do I have these columns made of glass? Because they obstruct the passage of electricity.

---

* Faraday brings the piece of metal into contact with both the knob and the metallic coating of the jar. But the spark forms before contact is actually achieved and therefore passes "through the air."

** Despite his word "flashed," Faraday does not mean that a *spark* passed along the length of the wire. The verb here means *to move or proceed rapidly*. Thus the electricity passed swiftly through the whole length of the wire; but it formed a *spark* only between the points that were about to make contact, just as before.

And why do I put that paper tassel [Fig. 43] at the top of the pole, upon a glass rod, and connect it with this machine by means of a wire? You see at once that as soon as the handle of the machine is turned, the electricity which is evolved travels along this wire and up the wooden rod, goes to the tassel at the top, and you see the power of repulsion with which it has endowed these strips of paper, each spreading outwards to the ceiling and sides of the room. The outside of that wire is covered with gutta-percha. It would not serve to keep the force from you when touching it with your hands, because it would burst through; but it answers our purpose for the present. And so you perceive how easily I can manage to send this power of electricity from place to place, by choosing the materials which can conduct the power.

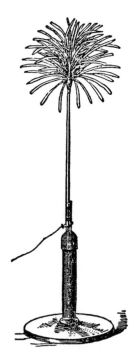

FIG. 43.

Suppose I want to fire a portion of gunpowder. I can readily do it by this transferable power of electricity. I will take a Leyden jar, or any other arrangement which gives us this power, and arrange wires so that they may carry the power to the place I wish;* and then placing a little gunpowder on the extremities of the wires, the moment I make the connection by this discharging rod, I shall fire the gunpowder. [*The connection was made, and the gunpowder ignited.*] And if I were to shew you a stool like this, and were to explain to you its construction, you could easily understand that we use glass legs because these are capable of preventing the electricity from going away to the earth. If, therefore, I were to stand on this stool, and receive the electricity through this conductor, I could give it to anything that I touched. [*The Lecturer stood upon the insulating stool, and placed himself in connection with the*

---

* Faraday runs a pair of wires from the charged Leyden jar to a small heap of gunpowder, where their ends are positioned a fraction of an inch apart. At the jar, he connects one wire to the metallic coating and supports the other wire a few inches away from the knob. He then joins knob and wire with a metal "discharging rod"; a spark passes between the wire ends at the gunpowder, and the powder ignites.

conductor of the machine.] Now, I am electrified—I can feel my hair rising up as the paper tassel did just now. Let us see whether I can succeed in lighting gas by touching the jet with my finger. [*The Lecturer brought his finger near a jet from which gas was issuing, when, after one or two attempts, the spark which came from his finger to the jet set fire to the gas.*] You now see how it is that this power of electricity can be transferred from the matter in which it is generated and conducted along wires and other bodies, and thus be made to serve new purposes utterly unattainable by the powers we have spoken of on previous days; and you will not now be at a loss to bring this power of electricity into comparison with those which we have previously examined; and to-morrow we shall be able to go further into the consideration of these transferable powers.

Conductors and insulators

In the paragraphs above, Faraday has shown that materials differ in their ability to transmit electrification from body to body. Thus arises the classification of materials into *conductors*, which are capable of conveying the electric condition, and *insulators* (from *insula*, island), which are not. The distinction is by no means absolute, but most *metals* are excellent conductors; and *air* is (usually) quite an effective insulator. That is why the basic framework of much elementary electrical apparatus is often a network of metal wires strung in air.

The signs of electrification

Notice that in the course of the demonstrations just completed Faraday has revealed *spark* as an additional indication of electrification, supplementing *attraction and repulsion*. When, on page 11 above, Faraday drew a spark from the electrical machine, the simultaneous fall of the straw indicator confirmed the spark as a manifestation of electric discharge.

One class of electrical manifestations that Faraday does not present in his young people's lectures is the physiological—particularly that convulsion of muscles we call *shock*.[*] Nevertheless, the physiological detection of electricity had already become, for him as for other investigators, an important research technique. Perhaps the earliest studies of animal tissue in an explicitly electrical context had been those of Luigi Galvani.

---

[*] Faraday's sensation of "hair rising up" under electrification (top of this page) is not an instance of muscular contraction, but one of mutual electrostatic repulsion of the hair fibers—analogous to the repulsion of the tassels in his figure 43.

Animal electricity

Galvani had long studied the effects of electricity on animal organs; he established that discharges from electrical machines or Leyden jars would contract the muscle of a frog's leg. Galvani had also studied the torpedo-fish; and he knew, either from his own findings or from reports published around 1770, that the torpedo's shock causes muscular contractions in the nearby fishes that experience it. The parallel physiological effects produced by ordinary electricity, on the one hand, and by the shock of the torpedo-fish, on the other, suggested that the torpedo's distinctive power is itself a species of electrical discharge—a discharge of "animal electricity," which must, by the similarity of effects, be at least analogous to frictional electricity. Of far greater importance, however, is Galvani's further conjecture: that even the *normal* muscular activity in animals is effected by "animal electricity." In 1791 Galvani published the account of an experiment which, he thought, fully confirmed that idea.

In Galvani's experiment two metal rods, one copper and one iron, are touched respectively to the main nerve center and the leg nerve of a recently dissected frog. When the free ends of the rods are brought together, the leg kicks! This happens only when there is a continuous metallic path from the nerve center to the leg nerve; and Galvani inferred that there was a *flow of animal electricity to the leg*, through the medium of the conductive rods. In his view, the metallic rods provided an artificial pathway, paralleling that which nature ordinarily provides for animal electricity in the animate frog at the moment of kick.

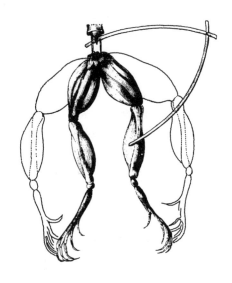

Galvani's interpretation was a reasonable one, but it depended entirely on his supposition that the copper and iron wires played the part only of passive conductors. Alessandro Volta, the professor of physics at the University of Pavia, learned of Galvani's experiment, entered into correspondence with him, and confirmed the observation himself. But he put forward a very different interpretation, perhaps suggested by his observation that the use of *dissimilar metals* appeared to be essential—which would not be expected if the rods were mere conductors. Volta asserted that the convulsion of the leg muscle was not due to conduction of "animal electricity" through the rods, but rather to electricity that was actually *generated* by the two metals as a result of their mutual contact. He proposed a comprehensive *contact theory*: that when dissimilar materials come into contact, one becomes positively and the other negatively charged. If then a conductive path is provided between the two materials (for instance, by the frog leg muscle), discharge will take place through that path.

Voltaic pile, cell and battery

In subsequent experiments Volta developed apparatus to exploit and enhance this "contact power" of dissimilar metals. One of them was a vertical column (the "voltaic pile") of alternating copper and zinc discs separated by pieces of cardboard moistened with brine or weak acid solution. Another voltaic device was a series of cups, each containing a pair of dissimilar metal plates immersed in saline or dilute acid solution. Clearly the elementary or unit arrangement of this sort is a *single pair of plates*. The series of cups was dubbed 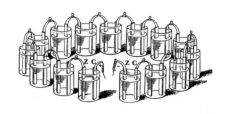 "crown of cups" because they were usually arranged in a circle, bringing the extremities of the series conveniently near to one another on the work table.

By Faraday's time, voltaic apparatus had undergone considerable advancement. Faraday had at his disposal a powerful device which 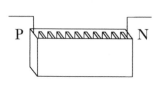 featured wide copper and zinc plates mounted vertically in a grooved box, sometimes called a "trough." The plates divide the box's interior into separate *cells* filled with dilute acid solution. The modern term "cell" derives from this design, which was originated by William Cruikshank, a British chemist. The aggregate of cells was called a "battery," invoking a military metaphor.*

---

* As an artillery battery is a coordinated assemblage of single guns, the voltaic battery is an interconnected series of single voltaic cells. There is perhaps a suggestion in the terminology that a voltaic cell is like an artillery piece (both may be seen as instruments of *power*, for example). Similarly, in paragraphs 270 and 275 in the Third Series, Faraday uses the term *Leyden battery* to denote an arrangement of individual Leyden jars wired in parallel. On the other hand, it is possible that the term "battery" has in these instances only the force of a generic collective.

Current electricity

The voltaic battery was distinguished by its capability to produce powerful electrical discharge *continuously*,* as Volta announced in a letter to the Royal Society in 1800. He believed that every voltaic arrangement produced a flow of electric fluid about the complete circuit; and he even referred to that circulation, or *current*, as a "*mouvement perpetuel.*" He inferred continuity of electric flow from physiological effects of the circuit: on the tongue and eyes,** on muscles and on the skin—effects which continued for as long as the voltaic connections were maintained. Later investigators succeeded in demonstrating continuous chemical activity (deposition of copper, decomposition of water) in other materials through which a "voltaic current" was directed. But Faraday will express increasing impatience with the entire Voltaic conceptual scheme. I have already noted his dissatisfaction with the "fluid flow" image of continuous electric discharge—although he cannot avoid *using* the word "current." Faraday will also find Volta's appeal to "contact" less and less adequate as an explanation of the current's ability to sustain itself.

Magnetism

The development of reliable, powerful sources of current electricity in turn made possible the discovery of still another indicator of electrical activity, namely, its *magnetic effects.* Magnetic manifestations of electricity had to await the appearance of these current sources because, seemingly, only *current* electricity had a magnetic dimension: *static* electricity was apparently devoid of it. But before we take up the intriguing phenomena which magnetism presents in relation to electricity, let us look at the elements of magnetic activity itself, without any electrical reference. This topic Faraday also discusses in his 1859 Christmas Lectures:***

---

* Frictional electric machines, as we saw, were also capable of producing a continuous discharge; but the voltaic discharge is of far greater quantity, as will become evident when the chemical and magnetic effects of the two electrical sources are tabulated. Faraday is deeply involved in this work; see paragraphs 371–376 in the Third Series.

** Sylvanus P. Thompson (*Elementary Lessons in Electricity and Magnetism*, 1894) reports: "A certain *taste* resembling green vitriol … is noticed if the two wires from the poles of a single voltaic cell are placed in contact with the tongue. Ritter discovered that a feeble current transmitted through the eyeball produces the sensation as of a bright *flash* of light…" Readers are cautioned not to imitate these demonstrations. Quite apart from the electrical effects, there is too much opportunity for injury from mechanical or chemical causes.

*** I am giving the following excerpt somewhat out of order. In Faraday's presentation it actually preceded some of the electrical topics we have already looked at.

Now, there are some curious bodies in nature (of which I have two specimens on the table) which are called *magnets* or *loadstones*—ores of iron, of which there is a great deal sent from Sweden. They have the attraction of gravitation, and attraction of cohesion, and certain chemical attraction; but they also have a great attractive power, for this little key is held up by this stone. Now, that is not chemical attraction—it is not the attraction of chemical affinity, or of aggregation of particles, or of cohesion, or of electricity (for it will not attract this ball if I bring it near it); but it is a separate and dual attraction—and, what is more, one which is not readily removed from the substance, for it has existed in it for ages and ages in the bowels of the earth.

Now, we can make artificial magnets (you will see me tomorrow make artificial magnets of extraordinary power). And let us take one of these artificial magnets, and examine it and see where the power is in the mass, and whether it is a dual power. You see it attracts these keys, two or three in succession, and it will attract a very large piece of iron. That, then, is a very different thing indeed to what you saw in the case of the shell-lac; for that only attracted a light ball, but here I have several ounces of iron held up. And if we come to examine this attraction a little more closely, we shall find it presents some other remarkable differences: first of all, one end of this bar [Fig. 37] attracts this key, but the middle does not attract. It is not, then, the *whole* of the substance which attracts. If I place this little key in the middle, it does not adhere; but if I place it there, a little nearer the end, it does, though feebly. Is it not, then, very curious to find that there is an attractive

FIG. 37.

power at the extremities which is not in the middle—to have thus in one bar two places in which this force of attraction resides? If I take this bar and balance it carefully on a point, so that it will be free to move round, I can try what action this piece of iron has on it. Well, it attracts one end, and it also attracts the other end, just as you saw the shell-lac and the glass did, with the exception of its not attracting in the middle. But if now, instead of a piece of iron, I take a *magnet,* and examine it in a similar way, you see that one of its ends *repels* the suspended magnet—the force then is no longer attraction, but repulsion; but if I take the other end of the magnet and bring it near, it shews attraction again.

You will see this better, perhaps, by another kind of experiment. Here [Fig. 38] is a little magnet and I have coloured the ends differently, so that you may distinguish one from the other.* Now this end [S] of the bar magnet attracts the *uncoloured* end of the little magnet. You see it pulls it towards it with great power; and as I carry it round, the uncoloured end still follows. But now, if I gradually bring the middle of the bar magnet opposite the uncoloured end of the needle, it has no effect upon it, either of attraction or repulsion, until, as

FIG. 38.

I come to the opposite extremity [N], you see that it is the *coloured* end of the needle which is pulled towards it. We are now therefore dealing with two kinds of power, attracting different ends of the magnet—a double power, already existing in these bodies, which takes up the form of attraction and repulsion. And now, when I put up this label with the word MAGNETISM, you will understand that it is to express this double power.

Now, with this loadstone you may make magnets artificially. Here is an artificial magnet [Fig. 39] in which both ends have been brought together in order to increase the attraction. This mass will lift that lump of iron; and, what is more, by placing this keeper [K], as it is called, on the top of the magnet, and taking hold of the handle, it will adhere sufficiently strongly to allow itself to be lifted up— so wonderful is its power of attraction. If you take a needle, and just draw one of its ends along one extremity of the magnet, and then draw the other end along the other extremity, and then gently

FIG. 39.

place it on the surface of some water (the needle will generally float on the surface, owing to the slight greasiness communicated to it by the fingers), you will be able to get all the phenomena of attraction and repulsion, by bringing another magnetised needle near to it.

I want you now to observe, that although I have shewn you in these magnets that this double power becomes evident principally at the extremities, yet the *whole* of the magnet is concerned in giving the power. That will at first seem rather strange; and I must therefore shew you an experiment to prove that this is not an accidental matter, but that the whole

* Labels N and S, supplied by Faraday's editor, indicate the north-seeking and south-seeking ends of the magnet, respectively. The next exercise will show that the N and S ends of a single magnet are *different in kind,* in that they attract opposite ends of the magnetized needle. We can also infer that between two magnets, *unlike ends attract* and *like ends repel* one another; but Faraday takes no notice of that maxim here.

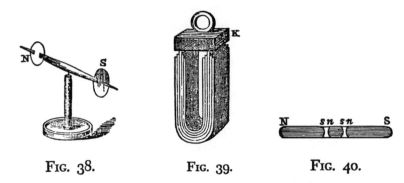

FIG. 38.   FIG. 39.   FIG. 40.

of the mass is really concerned in this force, just as in falling the whole of the mass is acted upon by the force of gravitation. I have here [Fig. 40] a steel bar, and I am going to make it a magnet by rubbing it on the large magnet [Fig. 39]. I have now made the two ends magnetic in opposite ways. I do not at present know one from the other, but we can soon find out. You see when I bring it near our magnetic needle [Fig. 38] one end repels and the other attracts; and the middle will neither attract nor repel—it *cannot,* because it is *half-way between the two ends.* But now, if I break out that piece [*n s*], and then examine it—see how strongly one end [*n*] pulls at this end [S, Fig. 38], and how it repels the other end [N]. And so it can be shewn that every part of the magnet contains this power of attraction and repulsion, but that the power is only rendered evident at the end of the mass. You will understand all this in a little while; but what you have now to consider is, that every part of this steel is in itself a magnet. Here is a little fragment which I have broken out of the very centre of the bar, and you will still see that one end is attractive and the other is repulsive. Now, is not this power a most wonderful thing—and very strange the means of taking it from one substance and bringing it to other matter? I cannot make a piece of iron or anything else heavier or lighter than it is. Its cohesive power it must and does have; but, as you have seen by these experiments, we can add or subtract this power of magnetism, and almost do as we like with it....

Notice that Faraday's demonstrations bring forth *attraction and repulsion* as the principal signs of magnetism—as was also the case for (static) electrification. And, in another way reminiscent of electrification, we can communicate magnetism from one body to another, as when Faraday made the bar (Fig. 40) magnetic by drawing its ends along the respective ends of the large magnet. On the other hand, there seems nothing in magnetism quite like the "conductors" of electricity. We cannot prompt a magnet's distinctive condition to leave it, migrate through another body, and take up residence in a third. Even when a body is "magnetized" by an existing magnet, it is far from clear that any of the magnetic condition is *removed* from the first magnet and *conveyed* to the second: perhaps a magnetic state is simply raised up in the second body without any diminution of the first. Thus the magnetic phenomena present far weaker images of a "fluid" than electrical phenomena do—and, as I have already indicated, Faraday found even the electrical intimations of a fluid imagery unconvincing.

We should take note of an additional point that Faraday does not call attention to in his example. When iron is magnetized by contact with—or even by approach to—an existing magnet, it develops powers specifically *contrary* to those of the dominating magnet. In the sketch, for example, an iron bar has been laid across the ends of a horseshoe magnet. As a result, the bar end adjacent to the magnet's north-seeking extremity develops the *south-seeking* character, while that adjacent to the magnet's south-seeking end takes on the *north-seeking* character. Faraday will rely on this fact in paragraph 38 of the First Series.

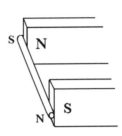

Magnetism's ability to raise up a *contrary* magnetism in an adjacent body is clearly analogous to the tendency of an electrified body to arouse an opposite electrical condition in surrounding surfaces—what I called *electric induction* on page 9 above; and other writers routinely denominated that magnetic action by the term "induction of magnetism." Curiously, however, Faraday seems reluctant to recognize a magnetic usage for the term *induction*. He avoids it even in paragraph 38, source of the very example I just cited. In fact no references to "magnetic" induction appear until the Second Series.

In other ways, too, Faraday's demonstrations exhibit an evident *duality* in magnetism, the north-seeking and south-seeking magnetic characters inviting comparison with the positive and negative varieties of electrification. And the rule, "unlikes attract; likes repel," which holds for electrification, appears to apply to magnetization as well.* On the other hand, the magnetic duality appears to be far profounder than the electric. The broken magnet demonstration implied that if a body is magnetic at all, it must possess both the N and the S characters *simultaneously*; whereas bodies appear to be capable of being electrified either positively *or* negatively, as wholes.

---

* We usually say, "*opposites* attract." But can we be so sure that "north-seeking" and "south-seeking" magnetic characters are *opposites*? Because electrification is so mobile, it was relatively easy to show that positive and negative electrifications can *nullify* one another; but when we bring N and S ends of different magnets together there is no mutual "discharge" such as we observe between oppositely electrified bodies.

The way in which Faraday articulates this doubleness in magnetism is significant. Many authors, especially textbook writers, point out that the N and S characters are "inseparable"—thereby invoking *independence* as the norm and *duality or relation* as the aberration that has to be explained. But Faraday describes the magnetic power as involving "the whole of the mass" (pages 19–20 above); and, later, declares that "every part of this steel" is itself a magnet. Evidently for him the dual magnetic power is *a whole*, not a conjunction of two inherently distinct powers.[*] In the *Experimental Researches* Faraday uses the term *polarity* to indicate just this kind of duality which is nevertheless a unity. So paradigmatic will such *polarity* become for Faraday, that in the Eleventh Series he will determine that it is rather the apparent separability of electric charge, not the relatedness of N and S magnetic characters, that needs explanation—and further, that the "explanation" consists in showing the apparent separability of charge to be *only* apparent. Electrification will be revealed as a profoundly polar condition, even as magnetism is.

"Magnetic curves"

If you place a bar-magnet beneath a sheet of paper, and over the paper sprinkle filings of iron or bits of steel wool, they will form patterns that suggest continuous *curves* passing from pole to pole, fanning out from one pole and concentrating again at the other. Here is an example from Faraday.

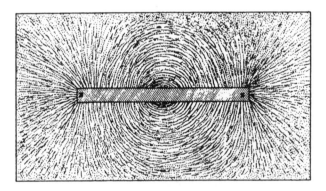

The suggestion of continuity can indeed be verified by using a single small magnetic needle to "map" individual curves by moving the needle always in the direction of its alignment. Certainly many of the curves

---

[*] How can a *whole* be *dual*? The question animates one of the many philosophically comical stories about the legendary town of Chelm, whose misguided inhabitants attempt to cut off the left end of a log—and discover to their dismay that when the operation is completed, the log still possesses a left as well as a right end!

evidently run from pole to pole, as shown in
this sketch of magnetic curves being
mapped about a horseshoe magnet.

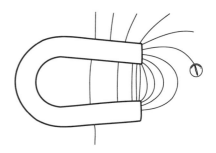

These pleasing curves, mere curiosities
at first, will under Faraday's experimental
investigation become powerful and fertile
articulations of the magnetic power. He will
eventually rename them "lines of force."

Magnetic action of a current

In 1819 Oersted discovered that a wire which carries a voltaic
current will affect the direction of a magnetic compass needle. The
existence of *some* connection between electric and magnetic action was
not, in itself, surprising. Such a connection had been suspected, in part
on the strength of recurring reports of lightning having magnetized
steel articles. But the *direction* of the magnetic action was surprising. In
the last of the 1859 Christmas Lectures, Faraday describes the
magnetic effect:

> Now, observe this: here is a piece of wire which I am about to
> make into a bridge of force—that is to say, a communicator
> between the two ends of the battery.* It is copper wire only, and
> is therefore not magnetic of itself. We will examine this wire with

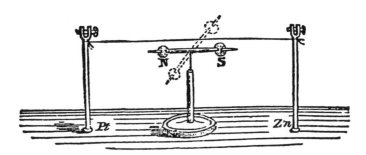

FIG. 51.

> our magnetic needle [Fig. 51]; and, though connected with one
> extreme end of the battery, you see that before the circuit is
> completed it has no power over the magnet [solid position]. But

---

* Faraday's colorful term "bridge of force" removes the connotation of *progress* from
what would conventionally be called a "path of current"; he does not wish to infect his
young audience's mind with *fluid flow* imagery! But in the published lecture, Figure 51
undercut Faraday's caution by showing *arrows* above the wire, which must inevitably
suggest *flow*. I have removed them.

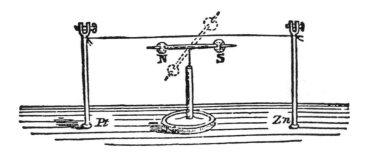

FIG. 51.

observe it when I make contact; watch the needle—see how it is swung round [*dotted position*], and notice how indifferent it becomes if I break contact again....

Before Faraday completes the battery connections the magnetic needle points (as it must) in the north-south direction; this is the position drawn in solid lines in the sketch and, as you can see, it is also the direction along which Faraday has strung the wire. But when he makes connections to the platinum (Pt) and zinc (Zn) plates of the battery, respectively, the magnetized needle turns aside (the drawing is a bit unclear, but the needle in fact remains horizontal). This evidently means that either some or all portions of the needle were urged in the east-west direction.

Such a direction is not only at right angles to the line of the current, it is also at right angles to an imaginary line from any point on the needle (in its initial position) directly to the wire. This was surprising to many investigators, because it was *neither an attraction to, nor a repulsion from, the wire*. When Faraday refers to "magnetic action at right angles to the current" (in paragraph 3 below), this is what he means.

One additional remark will become important later: If the needle is held above the wire instead of below it, all else remaining the same, its deflection will be in the opposite direction—that is, the end which went *east* when below the wire will go *west* when above it. Indeed, through a painstaking survey of *all* positions around a current-carrying wire, Faraday was able to confirm that the magnetic action of a current is disposed in *circles* perpendicular to and concentric with the wire.*

The galvanometer

As the attractive and repulsive powers of frictional electricity were developed and employed in the electroscope and other "indicators" of

---

* Faraday originally published that investigation some ten years prior to the First Series. It was subsequently included in the *second* volume of *Experimental Researches*.

electrification, so the magnetic influence of current electricity was raised up from a phenomenon in itself, to become an *instrument* for detecting the presence of electric current and even measuring its quantity. The instrument—or rather a whole class of instruments—was called "galvanometer," in honor of Luigi Galvani's discovery, however misinterpreted by him (see pages 15-16 above).*

It will be clear that any magnetic needle placed either above or below a wire running north-and-south will be capable of detecting current in the wire. The east-west influence of the current will then conspire with the north-south influence of the earth's magnetism to cause the needle to point in some oblique direction, When this happens, the angle of

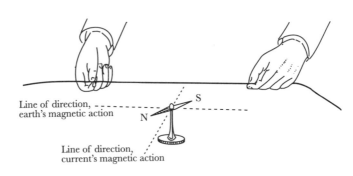

Line of direction, earth's magnetic action

Line of direction, current's magnetic action

deviation from the north-south line serves as a rough indicator of the magnetic effect of the current, in relation to the earth's magnetic influence at the location of the needle (see the sketch).

The greater the angle of deviation for a given current, the greater will be the galvanometer's *sensitivity*. Clearly, we can increase this angle either by (*i*) enhancing the action of the current, or (*ii*) diminishing the action of the earth. Faraday uses instruments that do both.

i. Wind the galvanometer wire parallel to itself several times about the needle. Each winding multiplies the magnetic effect; and therefore the whole coil is called a *multiplier*. But if many windings are used, the longer path arising from the increased length of wire tends to reduce the very current whose measurement is sought; so this method must be used with judgment.

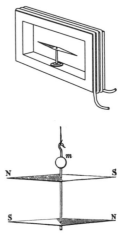

ii. Join together a *pair* of magnetic needles, oppositely directed; the second needle opposes the earth's directive action on the first. But you can deduce from the "additional remark" on page 24 that if the galvanometer wire is routed

---

* Volta, of course, has by no means been neglected. Honorific nomenclature includes the *volt* and *voltage*; and Faraday developed an instrument called *voltameter*, which he makes use of in the Seventh Series. In any case, Volta's own *contact theory* is, in Faraday's view, nearly as much a misinterpretation as Galvani's.

between the oppositely-directed needles, it will deflect them both in the *same direction*. Thus the current's effect will be enhanced, while the terrestrial effect is reduced; sensitivity is thereby increased.[*]

The "ballistic" galvanometer

I have described the galvanometer as it responds to steady currents; but in fact Faraday more often has to deal with *brief discharges*. A momentary pulse of current does not afford the galvanometer needle enough time to take up a steady position; instead, it throws the needle into *oscillation* about its rest position—inciting a large angle of swing if the pulse is strong, a small angle if weak. Faraday has a clever technique to detect weak discharges: by repeating the impulse at intervals corresponding to the vibration period of the needle, he builds up the needle's response just as we might, by judicious pushing, build up the motion of a child's swing.[**] A very weak galvanometer indication, therefore, which might be indiscernible or highly dubious in itself, can often be augmented in this way by repetition.

Perhaps deriving from an image of being "thrown" to a maximum deflection, the galvanometer is today said to be in a *ballistic* mode when used to detect brief current pulses. Faraday does not use the term, but we will find it useful—especially in the Third Series, where he will establish an important result concerning what, exactly, the ballistic galvanometer "measures."

As Faraday begins the *Experimental Researches*, then, it has long been known that an electric current, whatever it may be in itself, includes among its powers the ability to affect a magnet. At the opening of the First Series, Faraday will announce his discovery of the *reciprocal* effect—the capability of a magnet to affect an electric current.

The "direction" of current

If, as Volta taught, electric current is a circulation of electric fluid, then certainly that circulation must have a *direction* and therefore

---

[*] Nobili, whom Faraday frequently cites, carried the dual-needle idea to its logical completion by making the needles, as far as possible, *equally* magnetic. Such an instrument is entirely immune not only to the earth's magnetic influence but to any other extraneous magnetic influences. When the current does deflect the needle pair, therefore, its only opposition is the force—usually very small—that is required to twist the suspension thread; a very sensitive instrument results.

[**] The analogy with a swing, or indeed with any *pendulum*, is quite accurate. If you apply steady horizontal pressure to a child's swing it will take up a fixed position at some angle from the vertical. But a single, abrupt *blow* will hurl the swing to some maximum deflection, whence it will swing back and forth in regular but decaying oscillation.

current must be a directional phenomenon. The direction of fluid transport must—it would seem—be the direction of the electric current.

Now fluids flow from regions of excess to regions of deficient accumulation (it is the business of *pumps* to create such excesses and deficiencies as will achieve flow in the directions we desire); so if the fluid is electrically *positive* in nature, current flow will be from *positive to negative*. But what if the fluid is electrically *negative*?* Then it must flow from regions of excess to regions of deficient accumulation of *negative* fluid—that is, from *negative to positive*. And if there are both positive *and* negative electrical fluids in the same conductor, then both must flow *simultaneously in opposite directions*! It being then impossible to establish conclusively the number and identity of electrical fluid or fluids, partisans of electric fluid imagery had little choice but to assign a current direction arbitrarily, by convention. The convention adopted was from *positive to negative*, a convention that remains in force today. Thus in a conductor joining the poles of a voltaic battery, current is conventionally said to flow *from* the positive pole (the one whose electrification is similar to that of rubbed glass) *to* the negative pole.

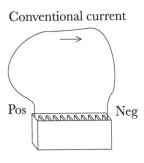

Conventional current

Pos   Neg

The fluid-based convention is easy to apply to a conductor; but when we try to include the battery in the picture, troublesome complications emerge. Does electrical "fluid" pass internally through the battery also? If so, it must flow from negative to positive! (Alternatively, perhaps the fluid is *created* at the battery's positive pole and *destroyed* at the negative pole.) Furthermore, imagine the conductor suddenly detached from the battery's negative pole. Why doesn't fluid continue to spill out from the detached end, at least momentarily? And if it *did* spill out, what would emerge would be (by convention) *positive* fluid; thus we would obtain *positive* fluid from what had a moment before been the *negative* end of the conductor!

I do not mean to suggest these as insoluble puzzles, only as complexities that lurk within the conventional notions. Faraday, as I

---

* As I remarked in a footnote on page 7, the term "positive" was adopted by Franklin to *mean* "excess of electrical fluid"; hence for him, current flow had to be from positive to negative *by definition*. That however is no guarantee that what we habitually call positive (for example, glass rubbed with silk), really contains an excess, and not a deficiency, of the theoretical fluid. In that sense, it is possible for "the fluid" to turn out to be electrically negative.

have by now repeatedly asserted, viewed any version of fluid theory with constant suspicion.* Nevertheless—and especially in the early Series—he has little choice but to employ much conventional electrical nomenclature; and it will therefore require a good deal of discovery and investigation to elucidate the meaning, on his terms, of *current* and *direction* of current. That is work that we, as readers, must do; but our task is not to figure out what Faraday already knows. Even more than we, *Faraday* has to discover meaning in those conventional terms, if meaning there is.

As in electricity, so in all areas of natural science. Faraday does not begin with principles, or even with a hypothesis, but with a willing hand and a ready eye. His experimental art endeavors to let the phenomena articulate themselves. His *narrative* art aims to recreate and transmit the presence and intelligibility that natural powers inherently possess. Thus arises the work that we readers have indeed to do—but we are permitted to do it along with Faraday himself, as at once our colleague and our guide.

---

* Ten years before publication of the First Series he had written the following: "Those who consider electricity as a fluid, or as two fluids, conceive that a current or currents of electricity are passing through the wire during the whole time it forms the connection between the poles of an active apparatus. There are many arguments in favour of the materiality of electricity, and but few against it; but still it is only a supposition; and it will be as well to remember ... that we have no proof of the materiality of electricity, or of the existence of any current through the wire."

EXPERIMENTAL RESEARCHES
IN
ELECTRICITY.

FIRST SERIES.

§ 1. *On the Induction of Electric Currents.* § 2. *On the Evolution of Electricity from Magnetism.* § 3. *On a new Electrical Condition of Matter.*
§ 4. *On Arago's Magnetic Phenomena.*

[Read November 24, 1831.]

1. THE power which electricity of tension possesses of causing an opposite electrical state in its vicinity has been expressed by the general term Induction; which, as it has been received into scientific language, may also, with propriety, be used in the same general sense to express the power which electrical currents may possess of inducing any particular state upon matter in their immediate neighbourhood, otherwise indifferent. It is with this meaning that I purpose using it in the present paper.

2. Certain effects of the induction of electrical currents have already been recognised and described: as those of magnetization; Ampère's experiments of bringing a copper disc near to a flat spiral; his repetition with electro-magnets of Arago's extraordinary experiments, and perhaps a few others. Still it appeared unlikely that these could be all the effects which induction by currents could produce; especially as, upon dispensing with iron, almost the whole of them disappear, whilst yet an infinity of bodies, exhibiting definite phenomena of induction with electricity of tension, still remain to be acted upon by the induction of electricity in motion.

---

1. *electricity of tension*: that is, *frictional* or *static electricity* (see discussion starting on page 3 of the editor's introduction). The reason for the epithet "tension" will be clearer in the Third Series. An example of static *induction* is the following: if a positively electrified body approaches other, originally unelectrified bodies, we find that those neighboring bodies develop negative electrification on their surfaces—usually, but not always, on areas facing the intruding body. Faraday will undertake a detailed investigation of static electric induction in the Eleventh Series. Since it appears that electrical *currents* are also capable of arousing determinate states in neighboring bodies, Faraday proposes to extend the term "induction" to these cases too.

2. Faraday will describe Arago's and Ampère's experiments in paragraphs 78, 81, and 129, below.

3. Further: Whether Ampère's beautiful theory were adopted, or any other, or whatever reservation were mentally made, still it appeared very extraordinary, that as every electric current was accompanied by a corresponding intensity of magnetic action at right angles to the current, good conductors of electricity, when placed within the sphere of this action, should not have any current induced through them, or some sensible effect produced equivalent in force to such a current.

4. These considerations, with their consequence, the hope of obtaining electricity from ordinary magnetism, have stimulated me at various times to investigate experimentally the inductive effect of electric currents. I lately arrived at positive results; and not only had my hopes fulfilled, but obtained a key which appeared to me to open out a full explanation of Arago's magnetic phenomena, and also to discover a new state, which may probably have great influence in some of the most important effects of electric currents.

5. These results I purpose describing, not as they were obtained, but in such a manner as to give the most concise view of the whole.

§ 1. Induction of Electric Currents.

6. About twenty-six feet of copper wire one twentieth of an inch in diameter were wound round a cylinder of wood as a helix, the different spires of which were prevented from touching by a thin interposed twine. This helix was covered with calico, and then a second wire applied in the same manner. In this way twelve helices were superposed, each containing an average length of wire of twenty-seven feet, and all in the same direction. The first, third, fifth, seventh, ninth, and eleventh of these helices were connected at their extremities end to end, so as to form one helix; the others were connected in a similar manner; and thus two principal helices were produced, closely

---

3. Current electricity is accompanied by magnetic action at "right angles," as discussed in the editor's introduction. Ampère's "beautiful" theory had declared magnets to be *essentially* aggregations of microscopic electrical currents in matter. Why, then, asks Faraday, on it or on any other theory that purports to explain the magnetic effect of currents, do we not observe a *reciprocal* effect: capability by a *magnet* to affect or produce a *current*?

6. *spires*: that is, *windings* (Latin *spira*, coil).

The six odd-numbered helices are connected *in series* ("so as to form one helix"); similarly the six even-numbered helices. If the average helix is, like the first, about twenty-*six* feet long (notwithstanding Faraday's subsequent figure twenty-*seven*), each principal helix would indeed be about "one hundred and fifty-five feet in length."

interposed, having the same direction, not touching anywhere, and each containing one hundred and fifty-five feet in length of wire.

7. One of these helices was connected with a galvanometer, the other with a voltaic battery of ten pairs of plates four inches square, with double coppers and well charged; yet not the slightest sensible deflection of the galvanometer needle could be observed.

8. A similar compound helix, consisting of six lengths of copper and six of soft iron wire, was constructed. The resulting iron helix contained two hundred and fourteen feet of wire, the resulting copper helix two hundred and eight feet; but whether the current from the trough was passed through the copper or the iron helix, no effect upon the other could be perceived at the galvanometer.

9. In these and many similar experiments no difference in action of any kind appeared between iron and other metals.

10. Two hundred and three feet of copper wire in one length were coiled round a large block of wood; other two hundred and three feet of similar wire were interposed as a spiral between the turns of the first coil, and metallic contact everywhere prevented by twine. One of these helices was connected with a galvanometer, and the other with a battery of one hundred pairs of plates four inches square, with double coppers, and well charged. When the contact was made, there was a sudden and very slight effect at the galvanometer, and there was also a similar slight effect when the contact with the battery was broken. But whilst the voltaic current was continuing to pass through the one helix, no galvanometrical appearances nor any effect like induction upon the other helix could be perceived, although the active power of the

---

7. The diagram shows how the battery and the galvanometer are connected to their respective helices. In paragraphs 16 and 17 Faraday will call the battery's helix the "inducing wire," and the other helix the "wire under induction."

Faraday's galvanometer consists of a pair of oppositely-directed magnetic needles, with a coil (or a pair of coils) wound around the lower needle—thus combining the two principles described on page 25 of the editor's introduction. Faraday will describe his galvanometers in detail beginning at paragraph 87 of the present Series and again at paragraph 205 of the Second Series.

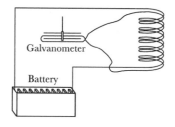

Faraday's voltaic battery is described on page 16 of the editor's introduction. "Double coppers" means that each copper plate is folded to face both sides of the corresponding zinc plate; this reduces the formation of gaseous hydrogen films on the copper surfaces, which would hinder the current.

battery was proved to be great, by its heating the whole of its own helix, and by the brilliancy of the discharge when made through charcoal.

11. Repetition of the experiments with a battery of one hundred and twenty pairs of plates produced no other effects; but it was ascertained, both at this and the former time, that the slight deflection of the needle occurring at the moment of completing the connexion, was always in one direction, and that the equally slight deflection produced when the contact was broken, was in the other direction; and also, that these effects occurred when the first helices were used (6. 8.).

12. The results which I had by this time obtained with magnets led me to believe that the battery current through one wire, did, in reality, induce a similar current through the other wire, but that it continued for an instant only, and partook more of the nature of the electrical wave passed through from the shock of a common Leyden jar than of the current from a voltaic battery, and therefore might magnetise a steel needle, although it scarcely affected the galvanometer.

13. This expectation was confirmed; for on substituting a small hollow helix, formed round a glass tube, for the galvanometer, introducing a steel needle, making contact as before between the battery and the inducing wire (7. 10.), and then removing the needle before the battery contact was broken, it was found magnetised.

---

10. *discharge ... made through charcoal*: a glowing discharge in air between carbon points or rods. It is the "arc" in arc welding and in theatrical arc lamps.

12. *...it continued for an instant only*: Faraday is here speaking of the induced current.

results ...obtained with magnets: Faraday does not tell us exactly what results he has in mind; but the fleeting galvanometer deflections noticed in paragraphs 10 and 11 are enough to suggest that any induced current can only be a transient discharge, like that of a Leyden jar (see the editor's introduction), rather than continuous like that of a voltaic battery. It might therefore act for too short a time to cause an evident deflection of the galvanometer here. But since current electricity produces magnetism (see the introduction), it ought to be able to magnetize a small needle—perhaps, he thinks, such magnetization will prove a more sensitive indicator of induced current than the galvanometer.

13. He introduces the needle before *making* the battery connections to the inducing wire, and removes it before *breaking* the connections. Thus the needle is exposed to the effect, in the wire under induction, of a sudden *commencement* of current in the inducing wire, but not to the effect of a *cessation* of that current. The result is magnetization of the needle.

14. When the battery contact was first made, then an unmagnetised needle introduced into the small indicating helix (13.), and lastly the battery contact broken, the needle was found magnetised to an equal degree apparently as before; but the poles were of the contrary kind.

15. The same effects took place on using the large compound helices first described (6. 8.).

16. When the unmagnetised needle was put into the indicating helix, before contact of the inducing wire with the battery, and remained there until the contact was broken, it exhibited little or no magnetism; the first effect having been nearly neutralised by the second (13. 14.). The force of the induced current upon making contact was found always to exceed that of the induced current at breaking of contact; and if therefore the contact was made and broken many times in succession, whilst the needle remained in the indicating helix, it at last came out not unmagnetised, but a needle magnetised as if the induced current upon making contact had acted alone on it. This effect may be due to the accumulation (as it is called) at the poles of the unconnected pile, rendering the current upon first making contact more powerful than what it is afterwards, at the moment of breaking contact.

17. If the circuit between the helix or wire under induction and the galvanometer or indicating spiral was not rendered complete *before* the connexion between the battery and the inducing wire was completed or broken, then no effects were perceived at the galvanometer. Thus, if the battery communications were first made, and then the wire under induction connected with the indicating helix, no magnetising power was there exhibited. But still retaining the latter communications, when those with the battery were broken, a magnet was formed in the helix, but of the second kind (14.), i. e. with poles indicating a current in the same direction to that belonging to the battery current, or to that always induced by that current at its cessation.

---

14. Now he introduces the needle *after* making the battery connections and removes it after breaking the connections. This time the needle will be exposed to the effect, in the wire under induction, of a sudden *cessation* of current in the inducing wire, but not to the effect of its *commencement*. The result is equal magnetization of the needle, but in the opposite direction.

16. Here the needle is exposed to the effects, in the wire under induction, *first* of commencement and *then* of cessation of current in the inducing wire. The second effect evidently cancels the first. Faraday notices a residual magnetism after many cycles of this sort, which indicates that the cancellation is not perfect. He blames a peculiarity of the battery (the very one that "double coppers" helps to alleviate); but other explanations are possible.

18. In the preceding experiments the wires were placed near to each other, and the contact of the inducing one with the battery made when the inductive effect was required; but as the particular action might be supposed to be exerted only at the moments of making and breaking contact, the induction was produced in another way. Several feet of copper wire were stretched in wide zigzag forms, representing the letter W, on one surface of a broad board; a second wire was stretched in precisely similar forms on a second board, so that when brought near the first, the wires should everywhere touch, except that a sheet of thick paper was interposed. One of these wires was connected with the galvanometer, and the other with a voltaic battery. The first wire was then moved towards the second, and as it approached, the needle was deflected. Being then removed, the needle was deflected in the opposite direction. By first making the wires approach and then recede, simultaneously with the vibrations of the needle, the latter soon became very extensive; but when the wires ceased to move from or towards each other, the galvanometer needle soon came to its usual position.

---

18. Is the induction transient by nature, or only because commencement and cessation of current in the inducing wire are each momentary events? With the inducing wire carrying a *continuous* current, he moves it physically towards and away from the wire under induction. Induction is observed, and it is *not* momentary: it persists as long as the motion of the inducing wire lasts! The induction is therefore associated with *motion* or *change* of the inducing agent. Note how he builds up resonant vibration in the galvanometer needle so as to magnify its response (see page 26 of the editor's introduction).

19. To see what he means by "same" and "contrary" directions of induced current, imagine a voltaic cell inserted in the galvanometer circuit (Faraday actually describes this in the next paragraph, though for a different purpose). With connections as shown, the inducing wire and the wire under induction must experience steady currents in the *same direction* because their parallel ends are attached to similar voltaic plates in their respective circuits. Mark the direction of galvanometer deflection. If then the voltaic cell is removed and the wires are made to approach and recede as Faraday describes, galvanometer deflections in the *marked* direction will indicate "induced current … in the same direction as the inducing current," while deflections in the *unmarked* direction will indicate "induced current … in the contrary direction to the inducing current." Once "calibrated" in this way, the galvanometer can be used routinely to indicate, by the direction of its deflection, the direction of any other currents routed through it.

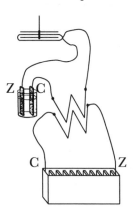

19. As the wires approximated, the induced current was in the *contrary* direction to the inducing current. As the wires receded, the induced current was in the same direction as the inducing current. When the wires remained stationary, there was no induced current (54.).

20. When a small voltaic arrangement was introduced into the circuit between the galvanometer (10.) and its helix or wire, so as to cause a permanent deflection of 30° or 40°, and then the battery of one hundred pairs of plates connected with the inducing wire, there was an instantaneous action as before (11.); but the galvanometer needle immediately resumed and retained its place unaltered, notwithstanding the continued contact of the inducing wire with the trough: such was the case in whichever way the contacts were made (33.).

21. Hence it would appear that collateral currents, either in the same or in opposite directions, exert no permanent inducing power on each other, affecting their quantity or tension.

22. I could obtain no evidence by the tongue, by spark, or by heating fine wire or charcoal, of the electricity passing through the wire under induction; neither could I obtain any chemical effects, though the contacts with metallic and other solutions were made and broken alternately with those of the battery, so that the second effect of induction should not oppose or neutralize the first (13. 16.).

23. This deficiency of effect is not because the induced current of electricity cannot pass fluids, but probably because of its brief duration and feeble intensity; for on introducing two large copper plates into the circuit on the induced side (20.), the plates being immersed in brine, but prevented from touching each other by an interposed cloth, the effect at the indicating galvanometer, or helix, occurred as before. The induced electricity could also pass through a voltaic trough (20.). When, however, the quantity of interposed fluid was reduced to a drop, the galvanometer gave no indication.

---

20. *a small voltaic arrangement*: that is, a pair of plates, as described in the previous comment. Induction was not affected by a constant current in the wire.

22. Sensation on the tongue, chemical change, spark, heating, etc. are all additional signs of electrical action (see editor's introduction); but evidently none of them is as responsive to the brief and weak induced current as either the galvanometer or the steel needle described in paragraphs 13 and 16.

23. From the lack of chemical effects or sensation on the tongue, one might suspect that induced currents cannot pass through *fluids* as ordinary currents can; but this is ruled out since the induced current must have passed through the voltaic cell added in paragraph 20. He also confirms that the induced current can pass through *saltwater*, provided there is more than a drop present.

24. Attempts to obtain similar effects by the use of wires conveying ordinary electricity were doubtful in the results. A compound helix similar to that already described, containing eight elementary helices (6.), was used. Four of the helices had their similar ends bound together by wire, and the two general terminations thus produced connected with the small magnetising helix containing an unmagnetised needle (13.). The other four helices were similarly arranged, but their ends connected with a Leyden jar. On passing the discharge, the needle was found to be a magnet; but it appeared probable that a part of the electricity of the jar had passed off to the small helix, and so magnetised the needle. There was indeed no reason to expect that the electricity of a jar, possessing as it does great tension, would not diffuse itself through all the metallic matter interposed between the coatings.

25. Still it does not follow that the discharge of ordinary electricity through a wire does not produce analogous phenomena to those arising from voltaic electricity; but as it appears impossible to separate the effects produced at the moment when the discharge begins to pass, from the equal and contrary effects produced when it ceases to pass (16.), inasmuch as with ordinary electricity these periods are simultaneous, so there can be scarcely any hope that in this form of the experiment they can be perceived.

26. Hence it is evident that currents of voltaic electricity present phenomena of induction somewhat analogous to those produced by electricity of tension, although, as will be seen hereafter, many differences exist between them. The result is the production of other

---

24. Can discharge of "ordinary electricity" (static electricity—see pages 9 and 10 of the introduction) through wires similarly induce currents? Instead of a voltaic battery he discharges a Leyden jar (see pages 11 and 12 of the introduction) through the inducing helix. Although a needle became magnetized as in paragraph 13, he suspects that the Leyden discharge was powerful enough to overcome the insulation between the wires and so pass directly to the helix under induction (remember he called this form of electricity "electricity of *tension*" in paragraph 1); so the result is suspect. If the experiment had been certain, it would have indicated that "ordinary" electricity and voltaic electricity have similar powers—an important result, as it is not *obvious* that both of them are "electricity" in the same sense. Faraday will investigate that question thoroughly in the Third Series.

25. Because of the suddenness of static discharge, commencement and cessation of the inducing current take place almost together; so there is no chance to isolate the effect of one from the contrary effect of the other. If a current *is* induced, then, this form of experiment cannot show it—and therefore cannot rule it out, either.

currents, (but which are only momentary,) parallel, or tending to parallelism, with the inducing current. By reference to the poles of the needle formed in the indicating helix (13. 14.) and to the deflections of the galvanometer-needle (11.), it was found in all cases that the induced current, produced by the first action of the inducing current, was in the contrary direction to the latter, but that the current produced by the cessation of the inducing current was in the same direction (19.). For the purpose of avoiding periphrasis, I propose to call this action of the current from the voltaic battery, *volta-electric induction.* The properties of the second wire, after induction has developed the first current, and whilst the electricity from the battery continues to flow through its inducing neighbour (10. 18.), constitute a peculiar electric condition, the consideration of which will be resumed hereafter (60.). All these results have been obtained with a voltaic apparatus consisting of a single pair of plates.

§ 2. Evolution of Electricity from Magnetism.

27. A welded ring was made of soft round bar-iron, the metal being seven eighths of an inch in thickness, and the ring six inches in external diameter. Three helices were put round one part of this ring, each containing about twenty-four feet of copper wire one twentieth of an inch thick; they were insulated from the iron and each other, and superposed in the manner before described (6.), occupying about nine inches in length upon the ring. They could

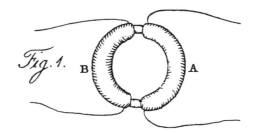

Fig. 1. B A

---

26. This paragraph is sometimes misunderstood. Faraday is *not* saying it is "evident" that induction by voltaic currents is analogous to induction by *discharges* of static electricity—to say so would ignore the difficulty identified in the previous paragraph! Rather he alludes to the original meaning of "induction" set forth in paragraph 1: he has shown that voltaic currents induce *currents,* as static electrification induces *electrification.*

Note Faraday's summary of direction relations: commencement of a current induces a current in the *contrary* direction, while cessation induces a current in the *same* direction as the current that ceased. Recall that in paragraph 19, an induced current in the contrary direction was obtained when the inducing wire approached the wire under induction; thus both the *commencement of a current,* and the *approach of a constant current,* produce the same effect on the wire under induction.

Finally, induction effects which were at first observed only with powerful batteries were subsequently detected even when only a single cell was used.

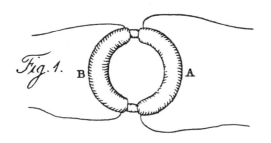

be used separately or conjointly; the group may be distinguished by the letter A (Pl. I. fig. 1.). On the other part of the ring about sixty feet of similar copper wire in two pieces were applied in the same manner, forming a helix B, which had the same common direction with the helices of A, but being separated from it at each extremity by about half an inch of the uncovered iron.

28. The helix B was connected by copper wires with a galvanometer three feet from the ring. The helices of A were connected end to end so as to form one common helix, the extremities of which were connected with a battery of ten pairs of plates four inches square. The galvanometer was immediately affected, and to a degree far beyond what has been described when with a battery of tenfold power helices *without iron* were used (10.); but though the contact was continued, the effect was not permanent, for the needle soon came to rest in its natural position, as if quite indifferent to the attached electro-magnetic arrangement. Upon breaking the contact with the battery, the needle was again powerfully deflected, but in the contrary direction to that induced in the first instance.

29. Upon arranging the apparatus so that B should be out of use, the galvanometer be connected with one of the three wires of A (27.), and the other two made into a helix through which the current from the trough (28.) was passed, similar but rather more powerful effects were produced.

30. When the battery contact was made in one direction, the galvanometer needle was deflected on the one side; if made in the

---

27. To visualize helices having "the same common direction," imagine them wound on a straight iron rod, which is afterwards bent into the ring shown in Fig. 1.

28. The connections are the same as those diagrammed in the note to paragraph 7 above, except that the coils are linked by the iron ring instead of being wound coaxially with one another. Faraday finds the same transient induction as before, but *greatly* magnified. (Could this be because of the presence of iron? Recall his remark in paragraph 2 noting that in the *absence* of iron, many of the reported magnetic effects of electric current become indiscernible.)

other direction, the deflection was on the other side. The deflection on breaking the battery contact was always the reverse of that produced by completing it. The deflection on making a battery contact always indicated an induced current in the opposite direction to that from the battery; but on breaking the contact the deflection indicated an induced current in the same direction as that of the battery. No making or breaking of the contact at B side, or in any part of the galvanometer circuit, produced any effect at the galvanometer. No continuance of the battery current caused any deflection of the galvanometer-needle. As the above results are common to all these experiments, and to similar ones with ordinary magnets to be hereafter detailed, they need not be again particularly described.

31. Upon using the power of one hundred pairs of plates (10.) with this ring, the impulse at the galvanometer, when contact was completed or broken, was so great as to make the needle spin round rapidly four or five times, before the air and terrestrial magnetism could reduce its motion to mere oscillation.

32. By using charcoal at the ends of the B helix, a minute *spark* could be perceived when the contact of the battery with A was completed. This spark could not be due to any diversion of a part of the current of the battery through the iron to the helix B; for when the battery contact was continued, the galvanometer still resumed its perfectly indifferent state (28.). The spark was rarely seen on breaking contact. A small platina wire could not be ignited by this induced current; but there seems every reason to believe that the effect would

---

30. Note that from these observations it follows that *commencement* of an inducing current of one direction has the same direction of effect as does *cessation* of an inducing current of the contrary direction. Moreover, with these very strong effects it is even more evident than before, that a *continuing* current in the inducing wire fails to cause deflection of the galvanometer.

32. A *spark* between the ends of the coil under induction, which had been sought unsuccessfully in paragraph 22, is now obtained. Since the spark appears only on making (and sometimes on breaking) battery contact, we need not fear that it is the result of leakage from the inducing wire to the wire under induction—the concern raised in paragraph 24. For such leakage, if present, would continue for as long as the battery remained connected, and we should then have observed the spark continuously.

The spark appears on commencing, but seldom on terminating, current in the inducing wire. Recall that also in paragraph 16, induction associated with commencement of the battery current (evidenced by magnetizing a small needle) predominated over induction associated with the current's cessation.

ignited: here meaning *heated to a glow*, not *set aflame.*

be obtained by using a stronger original current or a more powerful arrangement of helices.

33. A feeble voltaic current was sent through the helix B and the galvanometer, so as to deflect the needle of the latter 30° or 40°, and then the battery of one hundred pairs of plates connected with A; but after the first effect was over, the galvanometer needle resumed exactly the position due to the feeble current transmitted by its own wire. This took place in whichever way the battery contacts were made, and shows that there again (20.) no permanent influence of the currents upon each other, as to their quantity and tension, exists.

34. Another arrangement was then employed connecting the former experiments on volta-electric induction (6–26.) with the present. A combination of helices like that already described (6.) was constructed upon a hollow cylinder of pasteboard: there were eight lengths of copper wire, containing altogether 220 feet; four of these helices were connected end to end, and then with the galvanometer (7.); the other intervening four were also connected end to end, and the battery of one hundred pairs discharged through them. In this form the effect of the galvanometer was hardly sensible (11.), though magnets could be made by the induced current (13.). But when a soft iron cylinder seven eighths of an inch thick, and twelve inches long, was introduced into the pasteboard tube, surrounded by the helices, then the induced current affected the galvanometer powerfully, and with all the phenomena just described (30.). It possessed also the power of making magnets with more energy, apparently, than when no iron cylinder was present.

35. When the iron cylinder was replaced by an equal cylinder of copper, no effect beyond that of the helices alone was produced. The iron cylinder arrangement was not so powerful as the ring arrangement already described (27.).

36. Similar effects were then produced by *ordinary magnets:* thus the hollow helix just described (34.) had all its elementary helices connected with the galvanometer by two copper wires, each five feet in

---

34. A straight helix allows insertion and removal of an iron core, and confirms that the presence of iron greatly magnifies induction.

36. The fact that iron enhances the induction (paragraph 34) while copper has no effect (paragraph 35) suggests that the *magnetic* influence of the inducing current is inherently involved. He confirms this by dispensing with the inducing current entirely, substituting for it an ordinary magnet with the clever arrangement of Figure 2. As the following paragraphs will report, similar induced currents are obtained.

length; the soft iron cylinder was introduced into its axis; a couple of bar magnets, each twenty-four inches long, were arranged with their opposite poles at one end in contact, so as to resemble a horse-shoe magnet, and then contact made between the other poles and the ends of the iron cylinder, so as to convert it for the time into a magnet (fig. 2.): by breaking the magnetic contacts, or reversing them, the magnetism of the iron cylinder could be destroyed or reversed at pleasure.

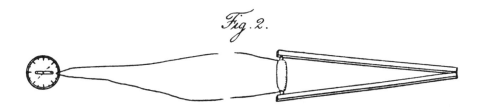

Fig. 2.

37. Upon making magnetic contact, the needle was deflected; continuing the contact, the needle became indifferent, and resumed its first position; on breaking the contact, it was again deflected, but in the opposite direction to the first effect, and then it again became indifferent. When the magnetic contacts were reversed the deflections were reversed.

38. When the magnetic contact was made, the deflection was such as to indicate an induced current of electricity in the opposite direction to that fitted to form a magnet, having the same polarity as

---

37. *Upon making magnetic contact, the needle was deflected...*: Faraday first brings the twin magnets in contact with the iron core of the helix, then retracts them. Both actions cause the galvanometer needle to deflect briefly, though in opposite directions. Evidently a momentary current is induced in the helix, first when the helix and its core are subjected to magnetic influence, then in the opposite direction when that influence is removed.

When the magnetic contacts were reversed the deflections were reversed: Clearly the direction of the induced current depends on the direction of the magnetic influence, as well as whether that influence is being brought to bear on, or being removed from, the wire under induction. In the next paragraph Faraday will pay specific attention to the direction of magnetic action.

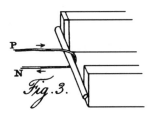

Fig. 3.

that really produced by contact with the bar magnets. Thus when the marked and unmarked poles were placed as in fig. 3, the current in the helix was in the direction represented, P being supposed to be the end of the wire going to the positive pole of the battery, or that end towards which the zinc plates face, and N the negative wire. Such a current would have converted the cylinder into a magnet of the opposite kind to that formed by contact with the poles A and B; and such a current moves in the opposite direction to the currents which in M. Ampère's beautiful theory are considered as constituting a magnet in the position figured.[1]

---

[1] The relative position of an electric current and a magnet is by most persons found very difficult to remember, and three or four helps to the memory have been devised by M. Ampère and others. I venture to suggest the following as a very simple and effectual assistance in these and similar latitudes. Let the experimenter think he is looking down upon a dipping needle, or upon the pole of the earth, and then let him think upon the direction of the motion of the hands of a watch, or of a screw moving direct; currents in that direction round a needle would make it into such a magnet as the dipping needle, or would themselves constitute an electro-magnet of similar qualities; or if brought near a magnet would tend to make it take that direction; or would themselves be moved into that position by a magnet so placed; or in M. Ampère's theory are considered as moving in that direction in the magnet. These two points of the position of the dipping-needle and the motion of the watch hands being remembered, any other relation of the current and magnet can be at once deduced from it.

---

38. *Thus when the marked and unmarked poles were placed as in fig. 3…*: Faraday's Figure 3 is an interpretive simplification of Figure 2. Its aim is to express the direction of the induced current in relation to the direction of the inducing magnetic action, specifically. The verb "placed" indicates that Faraday is referring to the current that develops while the twin magnets are being brought into contact with the iron cylinder, not while they are being retracted from it. Moreover, since the "marked pole" of a magnet is our *north* (north-seeking) pole, we know that when the magnetic contact is made, the cylinder end adjacent to that pole—away from the viewer—will become a south pole and the end towards the viewer a north pole (see introduction, page 21).

…the current in the helix was in the direction represented…: Faraday can infer the direction of the induced current by observing the direction of the galvanometer's deflection, as discussed earlier in the comment to paragraph 19. He expresses that direction by a convenient fiction: Imagine a wire connected

39. But as it might be supposed that in all the preceding experiments of this section, it was by some peculiar effect taking place during the formation of the magnet, and not by its mere virtual approximation, that the momentary induced current was excited, the following

---

to positive (P) and negative (N) battery plates and looped about the iron cylinder as shown in Figure 3; the current that is induced in the actual helix has the same direction as the current that would develop in an imaginary wire so connected.

Such a current would have converted the cylinder into a magnet of the opposite kind to that formed by contact with the poles...: See Faraday's note to the present paragraph, which reviews the well-known relation (as to direction) between an electric current and its magnetic effects. Whether expressed in Faraday's own image of the dipping-needle or as the more modern "right-hand rule," that relation implies that if a voltaic battery really *were* connected to the wire loop of Figure 3—and if the twin magnets were absent—the iron cylinder would become so magnetized as to have its north end *away* from the viewer and its south end *facing* the viewer. Thus the induced current's direction is such as to produce magnetism that is *opposed* to the magnetism which the cylinder actually acquires under the action of the twin magnets—that is, opposed to the very magnetic action that induces the current! Subsequently, the Estonian scientist Heinrich Lenz will propose that in *all* cases of induction, the induced currents have direction such as to oppose the change that produces them.

Comment on **Faraday's footnote 1**: Although Faraday is obviously quite pleased with his artifice for expressing the direction relations between a current and the magnetism it produces, it is hard to think of anything handier than the modern *right-hand rule*, which in its *electromagnetic* form stems from a "corkscrew" image devised by Maxwell: Suppose a current-carrying wire be coiled about an iron rod (which will thus become magnetized). And let the rod be grasped in the right hand, with fingers parallel to the coil windings and pointing in the direction from positive to negative. Then if the thumb is extended along the rod, the end it points to is the *north* (north-seeking) end. Furthermore, since we have seen that iron only *heightens* the magnetic effect, the same rule must apply to a coil, even without an iron core.

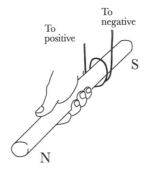

In the light of this rule, a current having the direction indicated in Figure 3 would cause the iron cylinder to develop a north pole away from the reader, a south pole towards the reader—thereby *opposing* the effect which the twin magnets actually produce in the cylinder.

experiment was made. All the similar ends of the compound hollow helix (34.) were bound together by copper wire, forming two general terminations, and these were connected with the galvanometer. The soft iron cylinder (34.) was removed, and a cylindrical magnet, three quarters of an inch in diameter and eight inches and a half in length, used instead. One end of this magnet was introduced into the axis of the helix (fig. 4.), and then, the galvanometer-needle being stationary,

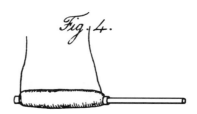

Fig. 4.

the magnet was suddenly thrust in; immediately the needle was deflected in the same direction as if the magnet had been formed by either of the two preceding processes (34. 36.). Being left in, the needle resumed its first position, and then the magnet being withdrawn the needle was deflected in the opposite direction. These effects were not great; but by introducing and withdrawing the magnet, so that the impulse each time should be added to those previously communicated to the needle, the latter could be made to vibrate through an arc of 180° or more.

40. In this experiment the magnet must not be passed entirely through the helix, for then a second action occurs. When the magnet

---

39. Previously (Figure 2), an iron core *already in the helix* was alternately magnetized and demagnetized by contact with exterior magnets. Now (Figure 4) a *permanent* magnet is alternately inserted into and removed from the helix. Notice that Faraday introduces the magnet in two steps: *First* he inserts, say, the *N* end of the magnet into the right-hand end of the helix: this act deflects the galvanometer momentarily. *Then* he thrusts the magnet fully into the coil (but not so far as to exit from the left end; see next paragraph): the galvanometer deflects in the same direction as if an iron core had been *magnetized in place*, developing an *N* end at the left and an *S* end at the right. The reverse deflection occurs when the magnet begins to be withdrawn; hence by moving the magnet back and forth within the helix he can, as before, cause a resonant buildup of the galvanometer's response so as to make the effect more evident.

What do *magnetization of the iron in place* (Figure 2) and *approach of already-magnetized iron to a place* have in common, whereby they produce the same effect on the galvanometer?

is introduced, the needle at the galvanometer is deflected in a certain direction; but being in, whether it be pushed quite through or withdrawn, the needle is deflected in a direction the reverse of that previously produced. When the magnet is passed in and through at one continuous motion, the needle moves one way, is then suddenly stopped, and finally moves the other way.

41. If such a hollow helix as that described (34.) be laid east and west (or in any other constant position), and a magnet be retained east and west, its marked pole always being one way; then whichever end of the helix the magnet goes in at, and consequently whichever pole of the magnet enters first, still the needle is deflected the same way: on the other hand, whichever direction is followed in withdrawing the magnet, the deflection is constant, but contrary to that due to its entrance.

42. These effects are simple consequences of the *law* hereafter to be described (114).

43. When the eight elementary helices were made one long helix, the effect was not so great as in the arrangement described. When only one of the eight helices was used, the effect was also much diminished. All care was taken to guard against any direct action of the inducing magnet upon the galvanometer, and it was found that by moving the magnet in the same direction, and to the same degree on the outside of the helix no effect on the needle was produced.

---

40. It may be helpful to imagine a magnet that is *very much longer* than the helix, so that, say, the *N* end of the magnet can approach, enter, and even exit from the other end of the helix, while the (trailing) *S* end of the magnet remains outside and distant. Whether the *N* end is approaching, within, or exiting the helix, the galvanometer will deflect in a constant direction so long as motion continues; and it will deflect in the reverse direction if the motion is reversed (that is, if the magnet is withdrawn). The galvanometer will also reverse its direction if, instead of being withdrawn, the magnet continues forward until the trailing *S* end approaches and enters the helix. Thus the *N* and *S* ends produce opposite deflections when they move through the helix in the same direction, and the deflection produced by each reverses if the direction of its motion reverses.

43. Since the galvanometer is basically a magnetic needle, it is important to make sure that its deflections are indeed produced by *currents in the galvanometer wire*, and not by direct magnetic attractions between it and the moving magnet. By making sure that there is no deflection when the magnet is moved *outside* the helix, Faraday shows that any such direct effect must be insignificant. (As explained in the editor's introduction, the double-needle galvanometer design reduces its susceptibility to such extraneous magnetic influences.)

44. The Royal Society are in possession of a large compound magnet formerly belonging to Dr. Gowin Knight, which, by permission of the President and Council, I was allowed to use in the prosecution of these experiments: it is at present in the charge of Mr. Christie, at his house at Woolwich, where, by Mr. Christie's kindness I was at liberty to work; and I have to acknowledge my obligations to him for his assistance in all the experiments and observations made with it. This magnet is composed of about 450 bar magnets, each fifteen inches long, one inch wide, and half an inch thick, arranged in a box so as to present at one of its extremities two external poles (fig. 5.). These poles projected horizontally six inches from the box, were each twelve inches high and three inches wide. They were nine inches apart; and when a soft iron cylinder, three quarters of an inch in diameter and twelve inches long, was put across from one to the other, it required a force of nearly one hundred pounds to break the contact. The pole to the left in the figure is the marked pole.[2]

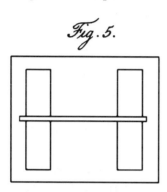

Fig. 5.

[2] To avoid any confusion as to the poles of the magnet, I shall designate the pole pointing to the north as the marked pole; I may occasionally speak of the north and south ends of the needle, but do not mean thereby north and south poles. That is by many considered the true north pole of a needle which points to the south; but in this country it is often called the south pole.

---

44. *The Royal Society are in possession of a large compound magnet...*: This mammoth magnet still exists and has been housed at The Science Museum, London, since 1899. The sketch shows its construction.

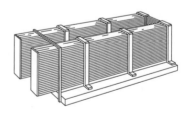

Comment on **Faraday's footnote 2**: If two magnetic poles attract one another, they must be opposite in kind. The south-seeking pole of a compass needle must therefore be pointing to a *north-seeking pole* in the earth's southern regions; and similarly there must be a *south-seeking pole* located in the earth's northern parts. If "north pole" means the north-seeking pole of a needle, there is no problem; but if "north pole" means "a pole like the one that is situated in the earth's northern regions," then the true "north pole" would be the one which points south! Happily, the latter nomenclature has fallen out of use.

45. The indicating galvanometer, in all experiments made with this magnet, was about eight feet from it, not directly in front of the poles, but about 16° or 17° on one side. It was found that on making or breaking the connexion of the poles by soft iron, the instrument was slightly affected; but all error of observation arising from this cause was easily and carefully avoided.

46. The electrical effects exhibited by this magnet were very striking. When a soft iron cylinder thirteen inches long was put through the compound hollow helix, with its ends arranged as two general terminations (39.), these connected with the galvanometer, and the iron cylinder brought in contact with the two poles of the magnet (fig. 5.), so powerful a rush of electricity took place that the needle whirled round many times in succession.

47. Notwithstanding this great power, if the contact was continued, the needle resumed its natural position, being entirely uninfluenced by the position of the helix (30.). But on breaking the magnetic contact, the needle was whirled round in the opposite direction with a force equal to the former.

48. A piece of copper plate wrapped *once* round the iron cylinder like a socket, but with interposed paper to prevent contact, had its edges connected with the wires of the galvanometer. When the iron was brought in contact with the poles the galvanometer was strongly affected.

49. Dismissing the helices and sockets, the galvanometer wire was passed over, and consequently only half round the iron cylinder (fig. 6.); but even then a strong effect upon the needle was exhibited, when the magnetic contact was made or broken.

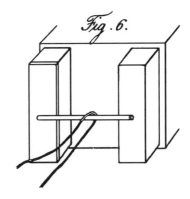

Fig. 6.

---

45. As in paragraph 43, Faraday here too tests for any direct influence of this very large magnet on the galvanometer. A slight effect is detected, but it is easily taken into account. As we see in the next paragraph, the extraneous direct effect is very small compared with the induction.

48. The copper "socket" is formed about the iron cylinder as shown in the sketch. It amounts to a single turn of what might be considered a *wide, flat wire.*

To galvanometer

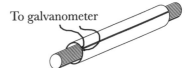

50. As the helix with its iron cylinder was brought towards the magnetic poles, *but without making contact,* still powerful effects were produced. When the helix, without the iron cylinder, and consequently containing no metal but copper, was approached to, or placed between the poles (44.), the needle was thrown 80°, 90°, or more, from its natural position. The inductive force was of course greater, the nearer the helix, either with or without its iron cylinder, was brought to the poles; but otherwise the same effects were produced, whether the helix, &c. was or was not brought into contact with the magnet; i.e. no permanent effect on the galvanometer was produced; and the effects of approximation and removal were the reverse of each other (30.).

51. When a bolt of copper corresponding to the iron cylinder was introduced, no greater effect was produced by the helix than without it. But when a thick iron wire was substituted, the magneto-electric induction was rendered sensibly greater.

52. The direction of the electric current produced in all these experiments with the helix, was the same as that already described (38.) as obtained with the weaker bar magnets.

53. A spiral containing fourteen feet of copper wire, being connected with the galvanometer, and approximated directly towards the marked pole in the line of its axis, affected the instrument strongly; the current induced in it was in the reverse direction to the current theoretically considered by M. Ampère as existing in the magnet (38.), or as the current in an electro-magnet of similar polarity. As the spiral was withdrawn, the induced current was reversed.

54. A similar spiral had the current of eighty pairs of 4-inch plates sent through it so as to form an electro-magnet, and then the other spiral connected with the galvanometer (53.) approximated to it; the needle vibrated, indicating a current in the galvanometer spiral the reverse of that in the battery spiral (18. 26.). On withdrawing the latter spiral, the needle passed in the opposite direction.

55. Single wires, approximated in certain directions towards the magnetic pole, had currents induced in them. On their removal, the

---

53. Now Faraday substitutes a spiral in place of a helix. Unlike a helix, the spiral "explores" a single plane.

the current induced in it was in the reverse direction to the current theoretically considered by M. Ampère…: This remark does not contradict Ampère's theory. Like a similar statement in paragraph 38, it signifies that the induced current's direction is such as to produce magnetism in a direction *opposed* to that of the inducing magnet.

55. Even *single wires* show reversal of induction when approach turns to recession, or vice versa.

currents were inverted. In such experiments the wires should not be removed in directions different to those in which they were approximated; for then occasionally complicated and irregular effects are produced, the causes of which will be very evident in the fourth part of this paper.

56. All attempts to obtain chemical effects by the induced current of electricity failed, though the precautions before described (22.), and all others that could be thought of, were employed. Neither was any sensation on the tongue, or any convulsive effect upon the limbs of a frog, produced. Nor could charcoal or fine wire be ignited (133.). But upon repeating the experiments more at leisure at the Royal Institution, with an armed loadstone belonging to Professor Daniell and capable of lifting about thirty pounds, a frog was *very powerfully convulsed* each time magnetic contact was made. At first the convulsions could not be obtained on breaking magnetic contact; but conceiving the deficiency of effect was because of the comparative slowness of separation, the latter act was effected by a blow, and then the frog was convulsed strongly. The more instantaneous the union or disunion is effected, the more powerful the convulsion. I thought also I could perceive the *sensation* upon the tongue and the *flash* before the eyes; but I could obtain no evidence of chemical decomposition.

57. The various experiments of this section prove, I think, most completely the production of electricity from ordinary magnetism.

---

56. *ignited*: here, "heated to a glow," as in paragraph 32 above.

armed loadstone: one that has been equipped with iron caps covering its polar regions. Such adornment had long been recognized as effective in increasing the lifting power of a loadstone, albeit for reasons imperfectly understood. Faraday will lay the groundwork for understanding this and similar magnetic effects in the Twenty-sixth Series and will outline a comprehensive vision of them in his essay, "On the Physical Character of the Lines of Magnetic Force."

Notice the variety of additional signs of electric current which Faraday looks for, besides the already-mentioned galvanometer deflection and crystal magnetization. Undetectable at first, several of these signs become apparent when he employs the armed loadstone. Again, as in paragraph 32 above, the inductive effect is at first obtained on *making* but not on *breaking* the primary circuit. But now he notices that *when the break is made more abruptly*, the effect is achieved when breaking the circuit as well. Might this shed light on the earlier residual magnetization (paragraph 16), which Faraday had attributed to "accumulation" at the battery poles? Since it now appears that the inductive effect is intensified when the connection or disconnection is rapid, and diminished when it is slow, perhaps for some reason *separations* tend to be consistently slower in completion than *connections* are.

tongue ... eyes: See the footnote on page 17 of the editor's introduction.

That its intensity should be very feeble and quantity small, cannot be considered wonderful, when it is remembered that like thermo-electricity it is evolved entirely within the substance of metals retaining all their conducting power. But an agent which is conducted along metallic wires in the manner described; which, whilst so passing possesses the peculiar magnetic actions and force of a current of electricity; which can agitate and convulse the limbs of a frog; and which, finally, can produce a spark[3] by its discharge through charcoal (32.), can only be electricity. As all the effects can be produced by ferruginous electro-magnets (34.), there is no doubt that arrangements like the magnets of Professors Moll, Henry, Ten Eyke, and others, in which as many as two thousand pounds have been lifted, may be used for these experiments; in which case not only a brighter spark may be obtained, but wires also ignited, and, as the current can pass liquids (23.), chemical action be produced. These effects are still more likely to be obtained when the magneto-electric arrangements to be explained in the fourth section are excited by the powers of such apparatus.

58. The similarity of action, almost amounting to identity, between common magnets and either electro-magnets or volta-electric currents, is strikingly in accordance with and confirmatory of M. Ampère's theory, and furnishes powerful reasons for believing that the action is the same in both cases; but, as a distinction in language is still necessary, I propose to call the agency thus exerted by ordinary magnets, *magneto-electric* or *magnelectric* induction (26.).

59. The only difference which powerfully strikes the attention as existing between volta-electric and magneto-electric induction, is the suddenness of the former, and the sensible time required by the latter; but even in this early state of investigation there are circumstances

---

[3] For a mode of obtaining the spark from the common magnet which I have found effectual, see the Philosophical Magazine for June 1832, p. 5. In the same Journal for November 1834, vol. v. p. 349, will be found a method of obtaining the magneto-electric spark, still simpler in its principle, the use of soft iron being dispensed with altogether —*Dec.* 1838.

---

57. Note that it is not entirely a closed question whether the power which has been shown to be produced from magnetism *is* electricity! The question of identity of the several supposed forms of electricity is the topic of the Third Series.

thermo-electricity: a form of electricity observed by Seebeck in 1821. He found that if bismuth and antimony wires were soldered together and their free ends connected to a galvanometer, a current passed when the junction was warmed to a temperature higher than the rest of the circuit.

ferruginous: of or containing iron.

which seem to indicate, that upon further inquiry this difference will, as a philosophical distinction, disappear (68.).

§ 3. *New Electrical State or Condition of Matter.*[4]

60. Whilst the wire is subject to either volta-electric or magneto-electric induction, it appears to be in a peculiar state; for it resists the formation of an electrical current in it, whereas, if in its common condition, such a current would be produced; and when left uninfluenced it has the power of originating a current, a power which the wire does not possess under common circumstances. This electrical condition of matter has not hitherto been recognised, but it probably exerts a very important influence in many if not most of the phenomena produced by currents of electricity. For reasons which will immediately appear (71.), I have, after advising with several learned friends, ventured to designate it as the *electro-tonic* state.

61. This peculiar condition shows no known electrical effects whilst it continues; nor have I yet been able to discover any peculiar powers exerted, or properties possessed, by matter whilst retained in this state.

62. It shows no reaction by attractive or repulsive powers. The various experiments which have been made with powerful magnets upon such metals as copper, silver, and generally those substances not magnetic, prove this point; for the substances experimented upon, if

---

[4] This section having been read at the Royal Society and reported upon, and having also, in consequence of a letter from myself to M. Hachette, been noticed at the French Institute, I feel bound to let it stand as part of the paper; but later investigations (intimated 73. 76. 77.) of the laws governing these phenomena, induce me to think that the latter can be fully explained without admitting the electro-tonic state. My views on this point will appear in the second series of these researches.

---

59. A *philosophical* distinction is a distinction in essence or nature. Faraday anticipates that there will be found no essential difference between induction as produced by currents (volta-electric induction) and induction as produced by ordinary magnets (magneto-electric induction).

60. Faraday thinks the lack of continuance of induced current needs to be explained by some opposing condition *in the wire* (rather than simple cessation of the cause). But as his note 4 indicates, his confidence in that opinion is giving way to second thoughts; and even in paragraph 62 he acknowledges that there is no *observable* evidence for the condition. I have decided to omit most of his discussion of this puzzling notion, which still troubles commentators. Yet it has a remarkable fascination for him, and Faraday will repeatedly propose, withdraw, and again propose the idea of an "electro-tonic state" throughout the *Researches.*

electrical conductors, must have acquired this state; and yet no evidence of attractive or repulsive powers has been observed...

*   *   *

71. This peculiar state appears to be a state of tension, and may be considered as *equivalent* to a current of electricity, at least equal to that produced either when the condition is induced or destroyed. The current evolved, however, first or last, is not to be considered a measure of the degree of tension to which the electro-tonic state has risen; for as the metal retains its conducting powers unimpaired (65.), and as the electricity evolved is but for a moment, (the peculiar state being instantly assumed and lost (68.),) the electricity which may be led away by long wire conductors, offering obstruction in their substance proportionate to their small lateral and extensive linear dimensions, can be but a very small portion of that really evolved within the mass at the moment it assumes this condition. Insulated helices and portions of metal instantly assumed the state; and no traces of electricity could be discovered in them, however quickly the contact with the electrometer was made, after they were put under induction, either by the current from the battery or the magnet. A single drop of water or a small piece of moistened paper (23. 56.) was obstacle sufficient to stop the current through the conductors, the electricity evolved returning to a state of equilibrium through the metal itself, and consequently in an unobserved manner.

*   *   *

73. All the results favour the notion that the electro-tonic state relates to the particles, and not to the mass, of the wire or substance under induction, being in that respect different to the induction exerted by electricity of tension. If so, the state may be assumed in liquids when no electrical current is sensible, and even in non-conductors; the current itself, when it occurs, being as it were a contingency due to the existence of conducting power, and the momentary propulsive force exerted by the particles during their arrangement. Even when conducting power is equal, the currents of electricity, which as yet are the only indicators of this state, may be unequal, because of differences as to number, size, electrical condition, &c. &c. in the particles themselves. It will only be after the laws which govern this new state are ascertained, that we shall be able

---

73. *Even when conducting power is equal, the currents of electricity ... may be unequal*: A strange statement to modern readers, for whom equal conducting power (under identical circumstances) must imply equal currents *by definition*! But Faraday will explore this matter more thoroughly in the next Series.

to predict what is the true condition of, and what are the electrical results obtainable from, any particular substance.

*   *   *

79. The momentary existence of the phenomena of induction now described is sufficient to furnish abundant reasons for the uncertainty or failure of the experiments, hitherto made to obtain electricity from magnets, or to effect chemical decomposition or arrangement by their means.

80. It also appears capable of explaining fully the remarkable phenomena observed by M. Arago between metals and magnets when either are moving (120.), as well as most of the results obtained by Sir John Herschel, Messrs. Babbage, Harris, and others, in repeating his experiments; accounting at the same time perfectly for what at first appeared inexplicable; namely, the non-action of the same metals and magnets when at rest. These results, which also afford the readiest means of obtaining electricity from magnetism, I shall now proceed to describe.

§ 4. Explication of Arago's Magnetic Phenomena.

81. If a plate of copper be revolved close to a magnetic needle, or magnet, suspended in such a way that the latter may rotate in a plane parallel to that of the former, the magnet tends to follow the motion of the plate; or if the magnet be revolved, the plate tends to follow its motion; and the effect is so powerful, that magnets or plates of many pounds weight may be thus carried round. If the magnet and plate be at rest relative to each other, not the slightest effect, attractive or repulsive, or of any kind, can be observed between them (62.). This is the phenomenon discovered by M. Arago; and he states that the effect takes place not only with all metals, but with solids, liquids, and even gases, i. e. with all substances (130.).

---

79. *the momentary existence of the phenomena of induction*: As Faraday noted in paragraph 30 and elsewhere, an induced current continues only upon the *commencement* or the *cessation* of the current that induces it. Since these are usually momentary events, the induced current is correspondingly momentary.

I have omitted from this paragraph a footnote that Faraday added in 1832, addressing a claim of historical priority.

81. The sketch shows one form of what had become known as "Arago's wheel." When the copper disc is rotated, the suspended needle endeavors to rotate in the same direction. Similarly if the needle is rotated, the disc follows it. It is this mutual "dragging" effect that requires explanation. A typical nineteenth-century Arago device is shown on page 71 below.

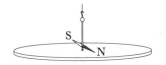

82. Mr. Babbage and Sir John Herschel, on conjointly repeating the experiments in this country,[5] could obtain the effects only with the metals, and with carbon in a peculiar state (from gas retorts), i. e. only with excellent conductors of electricity. They refer the effect to magnetism induced in the plate by the magnet; the pole of the latter causing an opposite pole in the nearest part of the plate, and round this a more diffuse polarity of its own kind (120.). The essential circumstance in producing the rotation of the suspended magnet is, that the substance revolving below it shall acquire and lose its magnetism in sensible time, and not instantly (124.). This theory refers the effect to an attractive force, and is not agreed to by the discoverer, M. Arago, nor by M. Ampère, who quote against it the absence of all attraction when the magnet and metal are at rest (62. 126.), although the induced magnetism should still remain; and who, from experiments made with a long dipping needle, conceive the action to be always repulsive (125.).

83. Upon obtaining electricity from magnets by the means already described (36. 46.), I hoped to make the experiment of M. Arago a new source of electricity; and did not despair, by reference to terrestrial magneto-electric induction, of being able to construct a new electrical machine. Thus stimulated, numerous experiments were made with the magnet of the Royal Society at Mr. Christie's house, in

[5] Philosophical Transactions, 1825, p. 467.

---

82. Babbage and Herschel hypothesized that each pole of the magnetic needle induces delayed and temporary magnetism of the opposite kind— hence attractive—in the metal plate, so that when the needle rotates, each magnetic pole would tend to drag along with itself the oppositely-magnetized regions of the disc that lie immediately behind it. But the hypothesis fails to explain why such magnetism is not detected when the apparatus is stationary. Faraday does not mention another objection—that it implies the existence of magnetism in *copper*, a metal never previously found to be magnetic. But he will soon describe a way in which copper *can*, in a sense, sustain magnetism (paragraphs 120, 138, 215).

83. In this wonderful moment, Faraday evidently sees that the Arago apparatus incorporates what appear to be the essential elements in the production of electricity from magnetism: magnet, electrical conductor, and their relative motion. Since the motion of a wheel is *continuous*, he anticipates that Arago's wheel might be made the basis for a new kind of electric machine—one capable of continuous production of electricity from magnetism rather than by friction. He proceeds to construct variants on the Arago device, each designed to capture and thus bring to light any induced electrical currents that may pervade the copper disc.

all of which I had the advantage of his assistance. As many of these were in the course of the investigation superseded by more perfect arrangements, I shall consider myself at liberty to rearrange them in a manner calculated to convey most readily what appears to me to be a correct view of the nature of the phenomena.

84. The magnet has been already described (44.). To concentrate the poles, and bring them nearer to each other, two iron or steel bars,

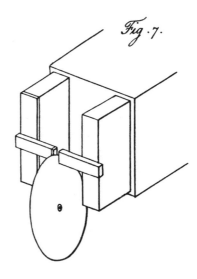

Fig. 7.

each about six or seven inches long, one inch wide, and half an inch thick, were put across the poles as in fig. 7, and being supported by twine from slipping, could be placed as near to or far from each other as was required. Occasionally two bars of soft iron were employed, so bent that when applied, one to each pole, the two smaller resulting poles were vertically over each other, either being uppermost at pleasure.

85. A disc of copper, twelve inches in diameter, and about one fifth of an inch in thickness, fixed upon a brass axis, was mounted in frames so as to allow of revolution either vertically or horizontally, its edge being at the same time introduced more or less between the magnetic poles (fig. 7.). The edge of the plate was well amalgamated for the purpose of obtaining a good but moveable contact, and a part round the axis was also prepared in a similar manner.

---

84. By mounting suitably shaped iron bars on the poles, Faraday is able to use the magnet with either a vertical or a horizontal disc; the sketch shows both

disc positions. As noted in the introduction (page 21), a bar that is mounted on, say, the north pole of the magnet develops the contrary (south) magnetic character at its end adjacent to the pole and the similar (north) magnetic character at its end adjacent to the disc. Thus whether straight or bent, the iron bars present a strong magnetic action

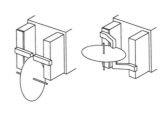

that is concentrated in a small area of the disc. Faraday refers to the ends of the bars adjacent to the disc as "smaller resulting poles."

85. *amalgamated*: here, *coated with mercury*. Faraday's intention is to improve contact with the sliding "collectors" mentioned in the next paragraph.

86. Conductors or electric collectors of copper and lead were constructed so as to come in contact with the edge of the copper disc (85.), or with other forms of plates hereafter to be described (101.). These conductors were about four inches long, one third of an inch wide, and one fifth of an inch thick; one end of each was slightly grooved, to allow of more exact adaptation to the somewhat convex edge of the plates, and then amalgamated. Copper wires, one sixteenth of an inch in thickness, attached, in the ordinary manner, by convolutions to the other ends of these conductors, passed away to the galvanometer.

87. The galvanometer was roughly made, yet sufficiently delicate in its indications. The wire was of copper covered with silk, and made sixteen or eighteen convolutions. Two sewing-needles were magnetized and fixed on to a stem of dried grass parallel to each other, but in opposite directions, and about half an inch apart; this system was suspended by a fibre of unspun silk, so that the lower needle should be between the convolutions of the multiplier, and the upper above them. The latter was by much the most powerful magnet,

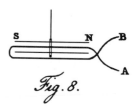

Fig. 8.

---

86. *Conductors or electric collectors*: Faraday makes electrical connections to the disc edge and its supporting axis. Note that the connections must permit sliding, so that the disc may rotate beneath them. Faraday will give additional details of the arrangement in paragraph 88.

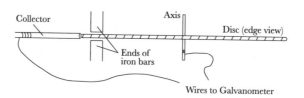

87. *The galvanometer*: See also the general description of the double-needle galvanometer and multiplier in the introduction. Note the two needles are *not equally magnetized*, in contrast to a later model; the stronger needle therefore gives "terrestrial direction to the whole." *Unspun* silk is desirable for suspending the needle assembly, since spun silk has an inherent twist which might either retard or assist the magnetic deflection, either case constituting interference. When the coil is placed in the "magnetic meridian" as shown in Fig. 8, a current through it will deflect a designated end of the needle assembly *east* or *west*, depending on the direction of the current.

and gave terrestrial direction to the whole; fig. 8. represents the direction of the wire and of the needles when the instrument was placed in the magnetic meridian: the ends of the wires are marked A and B for convenient reference hereafter. The letters S and N designate the south and north ends of the needle when affected merely by terrestrial magnetism; the end N is therefore the marked pole (44.). The whole instrument was protected by a glass jar, and stood, as to position and distance relative to the large magnet, under the same circumstances as before (45.).

88. All these arrangements being made, the copper disc was adjusted as in fig. 7, the small magnetic poles being about half an inch apart, and the edge of the plate inserted about half their width between them. One of the galvanometer wires was passed twice or thrice loosely round the brass axis of the plate, and the other attached to a conductor (86.), which itself was retained by the hand in contact with the amalgamated edge of the disc at the part immediately between the magnetic poles. Under these circumstances all was quiescent, and the galvanometer exhibited no effect. But the instant the plate moved, the galvanometer was influenced, and by revolving the plate quickly the needle could be deflected 90° or more.

89. It was difficult under the circumstances to make the contact between the conductor and the edge of the revolving disc uniformly good and extensive; it was also difficult in the first experiments to obtain a regular velocity of rotation: both these causes tended to retain the needle in a continual state of vibration; but no difficulty existed in ascertaining to which side it was deflected, or generally, about what line it vibrated. Afterwards, when the experiments were made more carefully, a permanent deflection of the needle of nearly 45° could be sustained.

90. Here therefore was demonstrated the production of a permanent current of electricity by ordinary magnets (57.).

91. When the motion of the disc was reversed, every other circumstance remaining the same, the galvanometer needle was deflected with equal power as before; but the deflection was on the opposite side, and the current of electricity evolved, therefore, the reverse of the former.

---

88. Connections are made to the disc's center and circumference so that the line joining them is *a radius passing between the poles*. When—and only when—the disc is rotated, current passes through the galvanometer.

92. When the conductor was placed on the edge of the disc a little to the right or left, as in the dotted positions fig. 9, the current of electricity was still evolved, and in the same direction as at first (88. 91.). This occurred to a considerable distance, i. e. 50° or 60° on each side of the place of the magnetic poles. The current gathered by the conductor and conveyed to the galvanometer was of the same kind on both sides of the place of greatest intensity, but gradually diminished in force from that

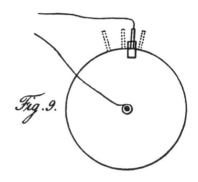

Fig. 9.

place. It appeared to be equally powerful at equal distances from the place of the magnetic poles, not being affected in that respect by the direction of the rotation. When the rotation of the disc was reversed, the direction of the current of electricity was reversed also; but the other circumstances were not affected.

93. On raising the plate, so that the magnetic poles were entirely hidden from each other by its intervention, (*a.* fig. 10,) the same effects were produced in the same order, and with equal intensity as before. On raising it still higher, so as to bring the place of the poles to *c*, still the effects were produced, and apparently with as much power as at first.

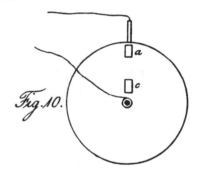

Fig. 10.

94. When the conductor was held against the edge as if fixed to it, and with it moved between the poles, even though but for a few degrees, the galvanometer needle moved and indicated a current of electricity, the same as that which would have been produced if the wheel had revolved in the same direction, the conductor remaining stationary.

---

92. *When the conductor was placed on the edge of the disc a little to the right or left*: that is, right or left of the pole. Compare Figure 9 with Figure 7. The magnet's poles are placed, one behind the page, one in front of the page, at the position of the small rectangle in Figure 9. The current is greatest when the radius defined by the collectors passes directly between the poles; it diminishes in intensity when the radius skirts them.

95. When the galvanometer connexion with the axis was broken, and its wires made fast to two conductors, both applied to the edge of the copper disc, then currents of electricity were produced, presenting more complicated appearances, but in perfect harmony with the above results. Thus, if applied as in fig. 11, a current of electricity through the galvanometer was produced; but if their place was a little shifted, as in fig. 12, a current in the contrary direction resulted; the fact being, that in the first instance the galvanometer indicated the difference between a strong current through A and a weak one through B, and in the second, of a weak current through A and a strong one through B (92.), and therefore produced opposite deflections.

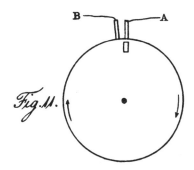

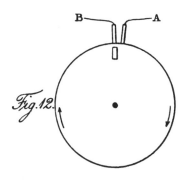

96. So also when the two conductors were equidistant from the magnetic poles, as in fig. 13, no current at the galvanometer was perceived, whichever way the disc was rotated, beyond what was momentarily produced by irregularity of contact; because equal currents in the same direction tended to pass into both. But when the two conductors were connected with one wire, and the axis with the

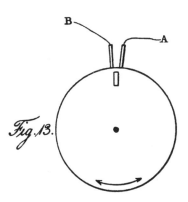

95. Note that when a current is obtained from collectors that are both placed on the circumference, as in Figures 11 and 12, Faraday does not suppose that current flows directly between A and B in the disc. Rather he interprets the result in accordance with paragraph 92—as the *difference* between (in Figure 11, for example) a strong current from center to A and a weaker current from center to B.

other wire, (fig. 14,) then the galvanometer showed a current according with the direction of rotation (91.); both conductors now acting consentaneously, and as a single conductor did before (88.).

97. All these effects could be obtained when only one of the poles of the magnet was brought near to the plate; they were of the same kind as to direction, &c., but by no means so powerful.

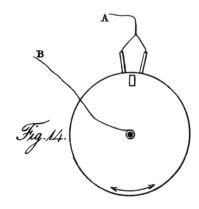

Fig. 14.

98. All care was taken to render these results independent of the earth's magnetism, or of the mutual magnetism of the magnet and galvanometer needles. The contacts were made in the magnetic equator of the plate, and at other parts; the plate was placed horizontally, and the poles vertically; and other precautions were taken. But the absence of any interference of the kind referred to, was readily shown by the want of all effect when the disc was removed from the poles, or the poles from the disc; every other circumstance remaining the same.

---

96. *Consentaneously*:   that is, *in agreement*, or, as he says, acting "as a single conductor."

98. *All care was taken to render these results independent of the earth's magnetism...*: This is important because the relation between the magnetic action, the motion of the plate (disc), and the induced current is one of mutual *direction*—Faraday will attempt to express that relation in the following paragraph. The direction of the magnet's action is clear; but if the earth's magnetism acted in some other direction the overall magnetic direction would be indeterminate. That is why Faraday lists making the plate *horizontal* as one of several "precautions" to be taken in this connection. At the latitude of London the earth's magnetism acts in a nearly vertical direction; hence with the plate horizontal, and the poles vertical, the earth's magnetic action on the plate is essentially parallel to the magnet's, and no uncertainty of direction is introduced.

But the strongest indication that Faraday's results are independent of the earth's magnetism is that when the magnet is withdrawn and the rotating disc is left to the action of earth's magnetism alone, *no current* is detected. The earth, then, exercises either no influence, or a negligible one, in these experiments. In the next Series, however, Faraday will employ more sensitive instruments to exhibit the earth's magnetic action directly.

99. The relation of the current of electricity produced, to the magnetic pole, to the direction of rotation of the plate, &c. &c., may be expressed by saying, that when the unmarked pole (44. 84.) is beneath the edge of the plate, and the latter revolves horizontally, screw-fashion, the electricity which can be collected at the edge of the plate nearest to the pole is positive. As the pole of the earth may mentally be considered the unmarked pole, this relation of the rotation, the pole, and the electricity evolved, is not difficult to remember. Or if, in fig. 15, the circle represent the copper disc revolving in the direction of the arrows, and *a* the outline of the unmarked pole placed beneath the plate, then the electricity collected at *b* and the neighbouring parts is positive, whilst that collected at the centre *c* and other parts is negative (88.). The currents in the plate are therefore from the centre by the magnetic poles towards the circumference.

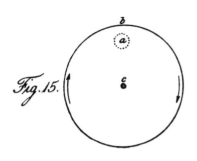

Fig. 15.

100. If the marked pole be placed above, all other things remaining the same, the electricity at *b*, fig., 15, is still positive. If the marked pole

---

99. As I noted in the introduction (pages 27 and 28), Faraday regards the prevailing rule for assigning *direction* to electric current as a linguistic formality only. While far from accepting the idea of current as a literal transport of substance, he has little choice but to adopt the conventional phraseology, that the direction of a current is "from" positive "to" negative. Therefore the last two sentences in this paragraph confuse some readers, who reason that if "the electricity collected at *b* ... is positive," while "that collected at the centre ... is negative," then the direction of current in the plate would appear to be *from circumference to center*—inasmuch as the conventional direction of electric current is from positive to negative; yet Faraday states that the currents are directed "from the centre ... towards the circumference." Think of it rather this way: It is the *galvanometer* that detects the current direction (see the comment to paragraph 19 above). If then the current in the galvanometer indicates that the disc's circumference is positive and its center negative, then the current *through the galvanometer* flows from disc circumference to disc center. The current *in the disc* then, in order to complete the circuit, must have the direction from center to circumference—as Faraday states.

As the pole of the earth may mentally be considered the unmarked pole...: Faraday is speaking of the pole in the northern hemisphere. Recall his note to paragraph 44, and its associated comment. The "unmarked" pole of a magnet is the south-seeking pole, and the magnetic pole in the northernmost region of the earth must be a south-seeking pole—for it cannot seek itself!

be placed below, or the unmarked pole above, the electricity is reversed. If the direction of revolution in any case is reversed, the electricity is also reversed.

101. It is now evident that the rotating plate is merely another form of the simpler experiment of passing a piece of metal between the magnetic poles in a rectilinear direction, and that in such cases currents of electricity are produced at right angles to the direction of the motion, and crossing it in the place of the magnetic pole or poles.

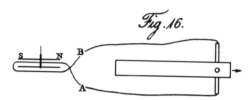

Fig. 16.

This was sufficiently shown by the following simple experiment: A piece of copper plate one-fifth of an inch thick, one inch and a half wide, and twelve inches long, being amalgamated at the edges, was placed between the magnetic poles, whilst the two conductors from the galvanometer were held in contact with its edges; it was then drawn through between the poles of the conductors in the direction of the arrow, fig. 16; immediately the galvanometer needle was deflected, its north or marked end passed eastward, indicating that the wire A

---

101. In Figure 16 the magnet is not shown, but the small circle indicates where its poles are placed in relation to the moving plate and the collectors; similarly for Figures 17–23. Faraday's galvanometer consists of a pair of oppositely-directed magnetic needles which are aligned north and south when they are not deflected by current in the coil. The letters N and S respectively indicate the north- and south-seeking ends of the *predominating* needle; thus in the absence of current, the extremity labeled N points north.

The galvanometer needle was deflected, its north or marked end passed eastward, indi-cating that the wire A received negative and the wire B positive electricity: How does Faraday infer the direction of this current? He can, of course, simply compare the observed direction of deflection with the deflection produced by a voltaic cell connected as described—its positive plate to B and its negative plate to A; a similar calibration technique was described in the comment to paragraph 19. But his specific mention of the *eastward* motion of the *north-seeking* end of the needle suggests that he has a mnemonic device in mind, like the one proposed in his footnote to paragraph 38 above. Alternatively, we may look to the electro-magnetic right-hand rule, discussed in my comment to that footnote: Suppose that a current develops in the galvanometer coil of Figure 16; and let its direction be from B to A—that is, *clockwise* about the lower galvanometer needle. Then the galvanometer coil will become an electro-magnet. Now hold

received negative and the wire B positive electricity; and as the marked pole was above, the result is in perfect accordance with the effect obtained by the rotatory plate (99.).

102. On reversing the motion of the plate, the needle at the galvanometer was deflected in the opposite direction, showing an opposite current.

103. To render evident the character of the electrical current existing in various parts of the moving copper plate, differing in their relation to the inducing poles, one collector (86.) only was applied at the part to be examined near to the pole, the other being connected

---

your right hand with fingers curving clockwise and parallel to the page. Since the thumb points into the page, the end of the coil that lies beneath the page will be north-seeking and the end that lies above the page will be south-seeking. What then will be its effect upon a magnetic needle, initially situated above and outside the coil, like the one marked NS in the figure? Clearly the extremity labeled N will tend to move *away* from the north-seeking end of the galvanometer coil—that is, it will tend *towards* the reader. But since that extremity originally pointed north, its displacement towards the reader represents an *eastward* motion, just as Faraday describes. Thus an eastward displacement of the north-seeking end of the needle indicates a current having direction *from* B (which must therefore play the role of the positive terminal of the source of the current) *towards* A (which plays the role of the negative terminal). Finally, the unlabeled needle introduces no complications since, as the introduction stated (page 25), a current routed *between* two oppositely-directed magnetic needles tends to turn them both in the same direction.

...*and as the marked pole was above*...: Faraday refers to the marked pole *of the magnet*; it is situated above the *page*. Thus the north-seeking pole is situated towards the reader at the location marked by the small circle. The south-seeking pole is situated below the page—away from the reader—at the same location.

As the direction relations shown in Figure 3 could be epitomized in an "electro-magnetic right-hand rule" (see my comment following Faraday's footnote to paragraph 38), so too the direction relations in Figure 16 can be summarized in the "magneto-electric right-hand rule" sketched here. Let the index finger point away from the north-seeking pole, or toward the south-seeking pole (Y)—where "toward" and "away" are reckoned through the air, not through the iron magnet or its pole pieces. Hold the third finger parallel to an adjacent conductor. Then if the conductor is moved, parallel to itself, in the direction of the thumb (X), the third finger will indicate the (conventional) direction of current induced in the moving conductor (Z).

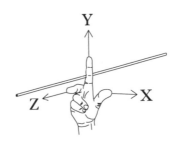

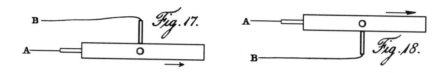

with the end of the plate as the most neutral place: the results are given at fig. 17–20, the marked pole being above the plate. In fig. 17, B received positive electricity; but the plate moving in the same direction, it received on the opposite side, fig. 18, negative electricity; reversing the motion of the latter, as in fig. 20, B received positive electricity; or reversing the motion of the first arrangement, that of fig. 17 to fig. 19, B received negative electricity.

104. When the plates were previously removed sideways from between the magnets, as in fig. 21, so as to be quite out of the polar axis, still the same effects were produced, though not so strongly.

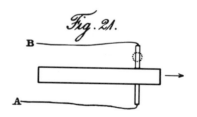

105. When the magnetic poles were in contact, and the copper plate was drawn between the conductors near to the place, there was but very little effect produced. When the poles were opened by the width of a card, the effect was somewhat more, but still very small.

106. When an amalgamated copper wire, one eighth of an inch thick, was drawn through between the conductors and poles (101.), it produced a very considerable effect, though not so much as the plates.

---

105. *When the magnetic poles were in contact:* that is, in contact with one another.

107. If the conductors were held permanently against any particular parts of the copper plates, and carried between the magnetic poles with them, effects the same as those described were produced, in accordance with the results obtained with the revolving disc (94.).

108. On the conductors being held against the ends of the plates, and the latter then passed between the magnetic poles, in a direction transverse to their length, the same effects were produced (fig. 22.). The parts of the plates towards the end may be considered either as mere conductors, or as portions of metal in which the electrical current is excited, according to their distance and the strength of the magnet; but the results were in perfect harmony with those before obtained. The effect was as strong as when the conductors were held against the sides of the plate (101.).

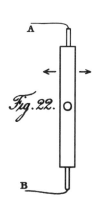

109. When a mere wire, connected with the galvanometer so as to form a complete circuit, was passed through between the poles, the galvanometer was affected; and upon moving the wire to and fro, so as to make the alternate impulses produced correspond with the vibrations of the needle, the latter could be increased to 20° or 30° on each side the magnetic meridian.

110. Upon connecting the ends of a plate of metal with the galvanometer wires, and then carrying it between the poles from end to end (as in fig. 23.), in either direction, no effect whatever was produced upon the galvanometer. But the moment the motion became transverse, the needle was deflected.

111. These effects were also obtained from *electro-magnetic poles*, resulting from the use of copper helices or spirals, either alone or with iron cores (34. 54.). The directions of the motions were precisely the same; but the action was much greater when the iron cores were used, than without.

---

110. Note the null result obtained from the arrangement depicted in Figure 23 makes clear that the induced current cannot have direction parallel to the motion of the conductor.

112. When a flat spiral was passed through edgewise between the poles, a curious action at the galvanometer resulted; the needle first went strongly one way, but then suddenly stopped, as if it struck against some solid obstacle, and immediately returned. If the spiral were passed through from above downwards, or from below upwards, still the motion of the needle was in the same direction, then suddenly stopped, and then was reversed. But on turning the spiral halfway round, i. e. edge for edge, then the directions of the motions were reversed, but still were suddenly interrupted and inverted as before. This double action depends upon the halves of the spiral (divided by a line passing thorough its centre perpendicular to the direction of its motion) acting in opposite directions; and the reason why the needle went to the same side, whether the spiral passed by the poles in the one or the other direction, was the circumstance, that upon changing the motion, the direction of the wires in the approaching half of the spiral was changed also. The effects, curious as they appear when witnessed, are immediately referable to the action of single wires (40. 109.).

113. Although the experiments with the revolving plate, wires, and plates of metal, were first successfully made with the large magnet belonging to the Royal Society, yet they were all ultimately repeated with a couple of bar magnets two feet long, one inch and a half wide, and half an inch thick; and, by rendering the galvanometer (87.) a little more delicate, with the most striking results. Ferro-electro-magnets, as those of Moll, Henry, &c. (57.), are very powerful. It is very essential, when making experiments on different substances, that thermo-electric effects (produced by contact of the fingers, &c.) be avoided, or at least appreciated and accounted for; they are easily distinguished by their permanency, and their independence of the magnets, or of the direction of the motion.

114. The relation which holds between the magnetic pole, the moving wire or metal, and the direction of the current evolved, i. e. *the law* which

---

114. *The law* (Faraday's italics) here depicted and stated is consistent with the magneto-electric right-hand rule proposed in the note to paragraph 101. For refer to Figure 24. Let your index finger point through the air, away from the marked pole; and hold the third finger parallel to conductor PN. Then if the thumb is to point in the direction of the arrows which correspond to that conductor, the third finger must point from P to N, as Faraday states.

But in Faraday's present formulation he makes explicit reference to the magnetic curves and even employs a locution of "cutting" them. Is this simply metaphor—the curves mere landmarks to aid his expression of the direction relations? Or is Faraday here raising the possibility that the curves have material or physical significance in the magnetic production of electricity?

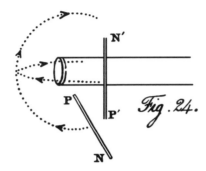

Fig. 24.

governs the evolution of electricity by magneto-electric induction, is very simple, although rather difficult to express. If in fig. 24. P N represent a horizontal wire passing by a marked magnetic pole, so that the direction of its motion shall coincide with the curved line proceeding from below upwards; or if its motion parallel to itself be in a line tangential to the curved line, but in the general direction of the arrows; or if it pass the pole in other directions, but so as to cut the magnetic curves[6] in the same general direction, or on the same side as they would be cut by the wire if moving along the dotted curved line;—then the current of electricity in the wire is from P to N. If it be carried in the reverse directions, the electric current will be from N to P. Or if the wire be in the vertical position, figured P′ N′, and it be carried in similar directions, coinciding with the dotted horizontal curve so far, as to cut the magnetic curves on the same side with it, the current will be from P′ to N′. If the wire be considered a tangent to the curved surface of the cylindrical magnet, and it be carried round that surface into any other position, or if the magnet itself be revolved on its axis, so as to bring any part opposite to the tangential wire,—still, if afterwards the wire be moved in the directions indicated, the current of electricity will be from P to N; or if it be moved in the opposite direction, from N to P; so that as regards the motions of the wire past the pole, they may be reduced to two, directly opposite to each other, one of which produces a current from P to N, and the other from N to P.

115. The same holds true of the unmarked pole of the magnet, except that if it be substituted for the one in the figure, then, as the wires are moved in the direction of the arrows, the current of electricity would be from N to P, and when they move in the reverse direction, from P to N.

---

[6] By magnetic curves, I mean the lines of magnetic forces, however modified by the juxtaposition of poles, which would be depicted by iron filings; or those to which a very small magnetic needle would form a tangent.

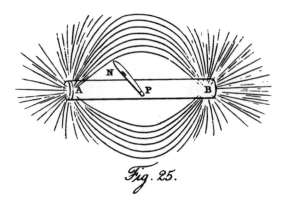

Fig. 25.

116. Hence the current of electricity which is excited in metal when moving in the neighbourhood of a magnet, depends for its direction altogether upon the relation of the metal to the resultant of magnetic action, or to the magnetic curves, and may be expressed in a popular way thus; Let AB (fig. 25.) represent a cylinder magnet, A being the marked pole, and B the unmarked pole; let PN be a silver knife-blade resting across the magnet with its edge upward, and with its marked or notched side towards the pole A; then in whatever direction or position this knife be moved edge foremost, either about the marked or the unmarked pole, the current of electricity produced will be from P to N, provided the intersected curves proceeding from A abut upon

---

116. Figure 25 does not satisfactorily convey the following: The knife-blade rests on its back, with cutting edge facing the top of the page. The shank P points towards the reader, the tip N points away. The "notch" is the thumbnail groove that is always cut into one side of a folding pocket-knife blade. The blade is "silver," hence nonmagnetic (but are pocket-knives ever made of silver?). In the drawing, the "notched side" of the blade faces the marked pole, A; and the magnetic curves proceeding from A "abut" (that is, touch) "upon the notched surface of the knife." Note that the knife-blade described—having tip and shank, edge and back, and sides notched and unnotched—defines *directionality* along three orthogonal axes. A relation framed in terms of it thus exhibits "handedness" as much as does the "right-hand rule." Does Faraday's representation in terms of a *blade* derive from the image of "cutting" magnetic curves, which he voiced in paragraph 114?

Faraday suggests construction of a wooden model. The invitation is charming, but puzzling: Was not the knife-blade device itself a conceptual model—and are we then to make a *model of a model*? I do not think that is Faraday's meaning. Rather, the suggestion should be seen as a small example of Faraday's characteristically evolutionary representations of phenomena—passing in this case from a rule or "*law*" (paragraph 114) to a diagram (Figure 24), to the knife-blade *image*, to an actual, tangible device. Notice the increasing *visible presence* of these successive representations.

the notched surface of the knife, and those from B upon the unnotched side. Or if the knife be moved with its back foremost, the current will be from N to P in every possible position and direction, provided the intersected curves abut on the same surfaces as before. A little model is easily constructed, by using a cylinder of wood for a magnet, a flat piece for the blade, and a piece of thread connecting one end of the cylinder with the other, and passing through a hole in the blade, for the magnetic curves: this readily gives the result of any possible direction.

117. When the wire under induction is passing by an electro-magnetic pole, as for instance one end of a copper helix traversed by the electric current (34.), the direction of the current in the approaching wire is the same with that of the current in the parts or sides of the spirals nearest to it, and in the receding wire the reverse of that in the parts nearest to it.

118. All these results show that the power of inducing electric currents is circumferentially exerted by a magnetic resultant or axis of power, just as circumferential magnetism is dependent upon and is exhibited by an electric current.

---

117. The helix is connected to the + and − terminals of a voltaic cell and thus carries current having the direction indicated. By the "electro-magnetic" right-hand rule (paragraph 38) the north and south ends are as shown. Then if a portion of the wire descends past the north end, the *law* depicted in Figure 24 (or, alternatively, the "magneto-electric" right-hand rule described in my comment to paragraph 101) assigns current in the wire as shown by the arrow. The sketch should clarify to which parts of the helix the induced current direction is "same" and to which parts it is "reverse" as the wire approaches, passes, and recedes beyond the helix.

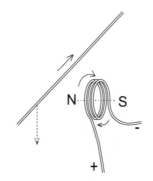

118. *circumferentially exerted*: In paragraph 114 Faraday declared that electric current is induced in a wire when it moves in such a way as to cross, or "cut," magnetic curves. But as his Figure 25 suggests, the curves about a magnetized bar spread out in roughly *radial* patterns from the polar regions; the directions of a wire which is to "cut" them must therefore be tangential or *circumferential.* He also reminds us that the magnetic action about a *straight current-carrying wire* is similarly disposed "in circles" (see the introduction). Thus magnetism and electricity are seen to display twofold symmetry: not only is each capable of evolving the other, but both powers share a similar geometry.

119. The experiments described combine to prove that when a piece of metal (and the same may be true of all conducting matter (213.)) is passed either before a single pole, or between the opposite poles of a magnet, or near electro-magnetic poles, whether ferruginous or not, electrical currents are produced across the metal transverse to the direction of motion; and which therefore, in Arago's experiments, will approximate towards the direction of radii. If a single wire be moved like the spoke of a wheel near a magnetic pole, a current of electricity is determined through it from one end towards the other. If a wheel be imagined, constructed of a great number of these radii, and this revolved near the pole, in the manner of the copper disc (85.), each radius will have a current produced in it as it passes by the pole. If the radii be supposed to be in contact laterally, a copper disc results, in which the directions of the currents will be generally the same, being modified only by the coaction which can take place between the particles, now that they are in metallic contact.

120. Now that the existence of these currents is known, Arago's phenomena may be accounted for without considering them as due to the formation in the copper of a pole of the opposite kind to that approximated, surrounded by a diffuse polarity of the same kind (82.); neither is it essential that the plate should acquire and lose its state in a finite time; nor on the other hand does it seem necessary that any repulsive force should be admitted as the cause of the rotation (82.).

---

119. *electrical currents are produced ... transverse to the direction of motion*: Faraday might well have added, "and perpendicular to the intersected magnetic curves." Here then is another instance of the Law invoked in paragraph 114 above; and since in the Arago experiments the metal disc necessarily rotates in the *circumferential* direction, the induced currents must be in the transverse or *radial* direction. Thus Arago's disc might be compared to a bicycle wheel strung with a huge number of radial spokes, joined at hub and rim. In an actual disc, the "spokes" would not be insulated from one another; but since the current direction is radial anyway, any resulting interaction ("coaction") should not make much difference.

120. *Arago's phenomena may be accounted for...*: During the lengthy succession of experiments from paragraph 84 to the present, a reader might easily forget that it was precisely Arago's rotation phenomena, previously described in paragraph 81, which had suggested those experiments. At their commencement Faraday said that he hoped to "make the experiment of M. Arago a new source of electricity" (paragraph 83). Now he is in a position to explain Arago's results without having to invoke hypothetical *magnetic poles in copper,* conjectural *time delays,* or a mysterious *repulsive force.*

Arago's wheel, by James W. Queen, Philadelphia, late 19th century. Smithsonian Institution, National Museum of American History, Electricity Collection, catalog no. 325373. The pivoting bar is not original. Photo courtesy of National Museum of American History.

Arago's wheel. Illustration from Ganot: Éleménts de Physique, translated by E. Atkinson. New York, William Wood & Co., 1886.

121. The effect is precisely of the same kind as the electromagnetic rotations which I had the good fortune to discover some years ago.[7] According to the experiments then made which have since been abundantly confirmed, if a wire (P N fig. 26.) be connected with the positive and negative ends of a voltaic battery, so that the positive electricity shall pass from P to N, and a marked magnetic pole N be placed near the wire between it and the spectator, the pole will move in a direction tangential to the wire, i. e. towards the right, and the wire will move tangentially towards the left, according to the directions of the arrows. This is exactly what takes place in the rotation of a plate beneath a

[7] Quarterly Journal of Science, vol. xii. pp. 74, 186, 416, 283.

---

121. *[Arago's] effect is precisely of the same kind as the electromagnetic rotations which I had the good fortune to discover some years ago*: In the course of experiments carried out in 1821, Faraday discovered that a current-carrying wire and the pole of a magnet tend to revolve about one another. (Sadly, his "good fortune" was soon marred by nasty charges that he had filched an idea of Wollaston's.) Those early experiments are relevant here because they suggest that a current, having been induced in a rotating disc under the influence of a magnet, should then act reciprocally upon the inducing magnet to set its pole into rotation (or attempted rotation) about the current.

the pole will move ... towards the right, and the wire will move tangentially towards the left...: Note that Faraday is reporting these motions as *independent facts*, not as one and the same motion described from different viewpoints. In the experiments of 1821, each of the motions indicated in Figure 26 was obtained as a *separate* result. But why does he not now refer the rotation of the pole, at least, to his mnemonic device of paragraph 38, which like the electromagnetic right-hand rule specifies the locomotive tendencies of a magnet in the vicinity of an electric current? First, perhaps, because those expedients involve currents in *circular* coils; so there can be no question of their direct application to currents in a straight wire. But there may also be a more fundamental reason, which I will suggest in the introduction to the Second Series.

This is exactly what takes place in the rotation of a plate beneath a magnetic pole: He points out parallels between the 1821 experiments and the present rotating disc experiments. According to the "bicycle wheel" image of paragraph 119, we recognize the existence of a radial current in the disc, beneath pole N. The radial current bears "exactly ... the same relation" to the pole in Figure 27 as the wire-borne current bears to the pole in Figure 26. But Figure 27 has a dual significance. Insofar as it represents Faraday's experiments with the rotating disc, the pole N belongs to a huge compound magnet and is fixed in position. But the figure also represents the Arago experiments, in which case N is one pole of a delicate magnetic needle suspended over the axis of the horizontal disc (paragraph 81) and therefore free to move. The pole in Figure 27 will in that case move to the right, just as the pole in Figure 26 does—in other words, Arago's needle will turn clockwise.

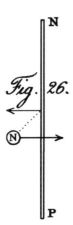

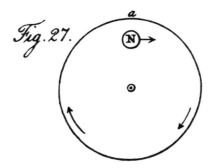

magnetic pole; for let N (fig. 27.) be a marked pole above the circular plate, the latter being rotated in the direction of the arrow: immediately currents of positive electricity set from the central parts in the general direction of the radii by the pole to the parts of the circumference *a* on the other side of that pole (99. 119.), and are therefore exactly in the same relation to it as the current in the wire (P N, fig. 26.), and therefore the pole in the same manner moves to the right hand.

122. If the rotation of the disc be reversed, the electric currents are reversed (91.), and the pole therefore moves to the left hand. If the contrary pole be employed, the effects are the same, i. e. in the same direction, because currents of electricity, the reverse of those described, are produced, and by reversing both poles and currents, the visible effects remain unchanged. In whatever position the axis of the magnet be placed, provided the same pole be applied to the same side of the plate, the electric current produced is in the same direction, in consistency with the law already stated (114, &c.); and thus every circumstance regarding the direction of the motion may be explained.

123. These currents are *discharged* or *return* in the parts of the plate on each side of and more distant from the place of the pole, where, of course, the magnetic induction is weaker: and when the collectors are applied, and a current of electricity is carried away to the galvanometer (88.), the deflection there is merely a repetition, by the same current or part of it, of the effect of rotation in the magnet over the plate itself.

---

123. The "bicycle wheel" image of paragraph 119, with its radial spokes, may also help to clarify the *return paths* of induced currents, which are represented in the accompanying sketch. The sketch is suggestive only; it does not reflect direct measurements.

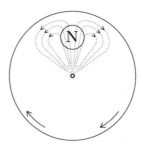

the effect of rotation: Do not construe this phrase as though it read "rotation's effect"—denoting an *action* attributed to rotation as a *subject* (subjective genitive). Compare it instead to phrases like "the city of London" or "the virtue of courage"—in which the second noun names or specifies the more general first noun (appositive genitive). Thus Faraday's phrase denotes a *specific effect*, namely, the *rotation* that is observed when a magnet is suspended over a rotating plate—that is to say, the Arago rotations.

the deflection there is merely a repetition...: Deflection of the galvanometer needle by currents induced in the spinning disc is a "repetition" of the rotation of Arago's magnetic needle in the sense that both phenomena exemplify the same principles of action. We know already that galvanometer deflections obey the electromagnetic right-hand rule. To see that Arago rotations conform to the same rule, consider the induced currents represented in the sketch above; the right-hand rule holds that the current circulating clockwise will act like a south magnetic pole facing the reader, while the counterclockwise current will act like a north magnetic pole. Thus the north pole N of a magnetic needle suspended above the plate will tend towards the right—that is, towards the direction of rotation of the plate—as Arago's needle is indeed observed to do.

Notice that if Faraday had been unable to infer the "return" paths of the currents in the disc, it would not be permissible to apply the electromagnetic right-hand rule to them and so demonstrate the identicalness of Arago rotations and galvanometer deflections. For the rule presupposes a *circular* current path, as I mentioned in the comment to paragraph 121.

124. It is under the point of view just put forth that I have ventured to say it is not necessary that the plate should acquire and lose its state in a finite time (120.); for if it were possible for the current to be fully developed the instant *before* it arrived at its state of nearest approximation to the vertical pole of the magnet, instead of opposite to or a little beyond it, still the relative motion of the pole and plate would be the same, the resulting force being in fact tangential instead of direct.

125. But it is possible (though not necessary for the rotation) that *time* may be required for the development of the maximum current in the plate, in which case the resultant of all the forces would be in advance of the magnet when the plate is rotated, or in the rear of the magnet when the latter is rotated, and many of the effects with pure electro-magnetic poles tend to prove this is the case. Then, the tangential force may be resolved into two others, one parallel to the plane of rotation, and the other perpendicular to it; the former would be the force exerted in making the plate revolve with the magnet, or the magnet with the plate; the latter would be a repulsive force, and is probably that, the effects of which M. Arago has also discovered (82.).

126. The extraordinary circumstance accompanying this action, which has seemed so inexplicable, namely, the cessation of all phenomena when the magnet and metal are brought to rest, now

---

124. The question of *time delay* becomes inessential on Faraday's account, as any delay in induction would merely shift the place of greatest concentration of induced current—to a location *near* the poles instead of *directly between* them; but the explanation of Arago's rotations would still apply.

125. *the tangential force*: Radial currents form in the rotating disc beneath the magnetic pole; but if their development requires *time*, the radius which conveys the greatest current will lie *beyond* the pole in the direction of plate rotation, as drawn here. But the magnetic action of a current is disposed in circles, as Faraday's "electromagnetic rotations" revealed (paragraph 121; see also the introduction). Thus the pole will experience a force *tangential* to the circle of action—it is represented by the solid arrow in the sketch. Only the component of force "parallel to the plane of rotation" is effective in moving the suspended needle—it is represented by the small dotted arrow.

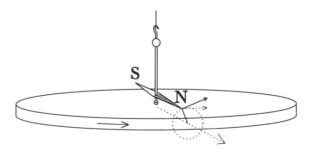

receives a full explanation (82.); for then the electrical currents which cause the motion cease altogether.

127. All the effects of solution of metallic continuity, and the consequent diminution of power described by Messrs. Babbage and Herschel,[8] now receive their natural explanation, as well also as the resumption of power when the cuts were filled up by metallic substances, which, though conductors of electricity, were themselves very deficient in the power of influencing magnets. And new modes of cutting the plate may be devised, which shall almost entirely destroy its power. Thus, if a copper plate (81.) be cut through at about a fifth or sixth of its diameter from the edge, so as to separate a ring from it, and this ring be again fastened on, but with a thickness of paper intervening (fig. 29.), and if Arago's experiment be made with this compound plate so adjusted that the section shall continually travel opposite the pole, it is evident that the magnetic currents will be greatly interfered with, and the plate probably lose much of its effect.[9]

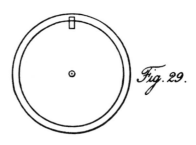

Fig. 29.

An elementary result of this kind was obtained by using two pieces of thick copper, shaped as in fig. 28. When the two neighbouring edges were amalgamated and put together, and the arrangement passed between the poles of the magnet, in a direction parallel to these edges, a current was urged through the wires attached to the outer angles, and the galvanometer became strongly affected; but when a single film of paper was interposed, and the experiment repeated, no sensible effect could be produced.

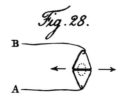

Fig. 28.

B

A

8 Philosophical Transactions, 1825, p. 481.

9 This experiment has actually been made by Mr. Christie, with the results here described, and is recorded in the Philosophical Transactions for 1827, p. 82.

---

127. *solution of metallic continuity*: In this odd phrase, "solution" bears the antique meaning of *coming into a state of discontinuity*—Faraday simply means that Babbage and Herschel made *cuts* in the rotating metal plate (thus impairing its continuity) and found that its ability to generate rotation in the suspended magnet was diminished. When those cuts were filled up again by materials that were electrically conductive but not magnetic, the transmitted rotations regained their original vigor.

128. A section of this kind could not interfere much with the induction of magnetism, supposed to be of the nature ordinarily received by iron.

*   *   *

130. The cause which has now been assigned for the rotation in Arago's experiment, namely, the production of electrical currents, seems abundantly sufficient in all cases where the metals, or perhaps even other conductors, are concerned; but with regard to such bodies as glass, resins, and, above all, gases, it seems impossible that currents of electricity, capable of producing these effects, should be generated in them. Yet Arago found that the effects in question were produced by these and by all bodies tried (81.). Messrs. Babbage and Herschel, it is true, did not observe them with any substance not metallic, except carbon, in a highly conductive state (82.). Mr. Harris has ascertained their occurrence with wood, marble, freestone and annealed glass, but obtained no effect with sulphuric acid and saturated solution of sulphate of iron, although these are better conductors of electricity than the former substances.

131. Future investigations will no doubt explain these difficulties, and decide the point whether the retarding or dragging action spoken of is always simultaneous with electric currents.[10] The existence of the action in metals, only whilst the currents exist, i.e. whilst motion is given (82. 88.), and the explication of the repulsive action observed by M. Arago (82. 125), are powerful reasons for referring it to this cause; but it may be combined with others which occasionally act alone.

[10] Experiments which I have since made convince me that this particular action is always due to the electrical currents formed; and they supply a test by which it may be distinguished from the action of ordinary magnetism, or any other cause, including those which are mechanical or irregular, producing similar effects (254.).

---

128. A section or cut such as Faraday has just described could not interfere with ordinary magnetism, which passes freely through air, paper, and other electrically nonconductive materials. In the 26th Series, though, Faraday will begin to explore a notion that magnetic conduction is not categorical but may, like electric conduction, admit of degrees.

130. *Yet Arago found that the effects in question were produced ... by all bodies tried*: On Faraday's account, nonconductive plates should be incapable of producing rotations in the suspended magnet. Yet Arago claims to have observed such rotations when using plates of *all* materials, and Harris makes a similar claim for several nonconductors. Faraday does not dispute these assertions outright, but he notes that Babbage and Herschel observed the phenomena *only* when conductive plates were used.

132. Copper, iron, tin, zinc, lead, mercury, and all the metals tried, produced electrical currents when passed between the magnetic poles: the mercury was put into a glass tube for the purpose. The dense carbon deposited in coal gas retorts, also produced the current, but ordinary charcoal did not. Neither could I obtain any sensible effects with brine, sulphuric acid, saline solutions, &c., whether rotated in basins, or inclosed in tubes and passed between the poles.

133. I have never been able to produce any sensation upon the tongue by the wires connected with the conductors applied to the edges of the revolving plate (88.) or slips of metal (101.). Nor have I been able to heat a fine platina wire, or produce a spark, or convulse the limbs of a frog. I have failed also to produce any chemical effects by electricity thus evolved (22. 56.).

134. As the electric current in the revolving copper plate occupies but a small space, proceeding by the poles and being discharged right and left at very small distances comparatively (123.); and as it exists in a thick mass of metal possessing almost the highest conducting power of any, and consequently offering extraordinary facility for its production and discharge; and as, notwithstanding this, considerable currents may be drawn off which can pass through narrow wires, forty, fifty, sixty, or even one hundred feet long; it is evident that the current existing in the plate itself must be a very powerful one, when the rotation is rapid and the magnet strong. This is also abundantly proved

---

132. Faraday confirms again that induced currents are obtained in metals, which are conductive solids. But conductive *liquids* fail to develop such currents. Notice, here as elsewhere, Faraday's readiness to acknowledge apparently unfavorable results.

133. Sensation on the tongue, heat, spark, and the other physiological and chemical effects here mentioned are recognized indicators of the presence and strength of electric discharge. But he obtains *none of them*, except deflection of the galvanometer needle, with the electricity evolved by the rotating disc or by other moving conductors. Since the galvanometer was found earlier to be the most sensitive among these indications (paragraph 22, *comment*), the current evolved in these experiments would seem to be rather feeble.

134. *it is evident that the current existing in the plate itself must be a very powerful one*: This remark does not contradict the implications of feebleness in the previous paragraph. It is the "current existing in the plate itself" that Faraday calls "powerful," whereas the current that passes through the galvanometer is but a small fraction of that overall current which is developed in the plate. Perhaps this recognition contributes to Faraday's thought, expressed in the next paragraph, that magneto-electric induction might become a significant new source of electricity.

by the obedience and readiness with which a magnet ten or twelve pounds in weight follows the motion of the plate and will strongly twist up the cord by which it is suspended.

135. Two rough trials were made with the intention of constructing *magneto-electric machines*. In one, a ring one inch and a half broad and twelve inches external diameter, cut from a thick copper plate, was mounted so as to revolve between the poles of the magnet and represent a plate similar to those formerly used (101.), but of interminable length; the inner and outer edges were amalgamated, and the conductors applied one to each edge, at the place of the magnetic poles. The current of electricity evolved did not appear by the galvanometer to be stronger, if so strong, as that from the circular plate (88.).

136. In the other, small thick discs of copper or other metal, half an inch in diameter, were revolved rapidly near to the poles, but with the axis of rotation out of the polar axis; the electricity evolved was collected by conductors applied as before to the edges (86.). Currents were procured, but of strength much inferior to that produced by the circular plate.

*   *   *

138. The remark which has already been made respecting iron (66.), and the independence of the ordinary magnetical phenomena of that substance and the phenomena now described of magneto-electric induction in that and other metals, was fully confirmed by many results of the kind detailed in this section. When an iron plate similar to the copper one formerly described (101.) was passed between the magnetic poles, it gave a current of electricity like the copper plate, but decidedly of less power; and in the experiments upon the induction of electric currents (9.), no difference in the kind of action between iron and other metals could be perceived. The power therefore of an iron plate to drag a magnet after it, or to intercept magnetic action, should be carefully distinguished from the similar power of such metals as silver, copper, &c. &c., inasmuch as in the iron by far the greater part of the effect is due to what may be called ordinary magnetic action.

135–136. *magneto-electric machines*: Standard "electric machines" are *frictional*, as described in the introduction. Now Faraday foresees a new kind of machine based on the "magneto-electric" principle—evolution of electricity from magnetism. The hope is to produce a stronger current than he had obtained from the disc, but the designs reported here were evidently disappointing in their performance.

There can be no doubt that the cause assigned by Messrs. Babbage and Herschel in explication of Arago's phenomena is the true one, where iron is the metal used.

*   *   *

Royal Institution, November 1831.

---

138. The importance of *induced currents* in the Arago rotation experiments has by now been fully established, for copper and other nonmagnetic rotating discs. But what about magnetic materials like iron? The currents induced in iron, which is a merely moderate conductor of electricity, are not especially powerful, and therefore Faraday is willing to conclude that in the case of iron, the "dragging" phenomena noticed by Arago are due more to *ordinary magnetic attraction* than to the magnetic action of induced currents. Is that a sound inference? Could you suggest an experiment to test it? Consider Faraday's remark in paragraph 128 above. What would be the expected behavior of an *iron* disc, divided into concentric rings as in Faraday's Figure 29 above?

Second Series — Editor's Introduction

Terrestrial magnetic curves

In the First Series Faraday referred to so-called "magnetic curves," which can be mapped about common magnets by means of a small compass. A similar mapping can be carried out over large areas of the earth's surface; the resulting charts being invaluable for compass navigation.* The chart shows sample magnetic curves for the British Isles circa 1900.** As you can see, magnetic compasses point to different "norths" at different locations, and it is the business of navigational charts to compare "compass north" to geographical north throughout their territories or districts.

The magnetic meridian

Magnetic north at any location is determined by the direction of a compass needle at that location—correcting of course for any extraneous interfering objects. The magnetic north-south line for a given compass location, together with the zenith point (directly overhead), determine the *magnetic meridian*—so named by analogy to the *celestial meridian* that is defined by the north and south celestial poles, together with the zenith. The magnetic meridian is significant for experiments described in the Second Series, since a flat surface located in the plane of the magnetic meridian is *not intersected by any terrestrial magnetic curves.*

Properly speaking, the meridian is not the plane just described, but rather the intersection of that plane with the earth's surface—that is, a great circle running through both terrestrial magnetic poles and the observer's location. But Faraday, like other writers, occasionally equivocates—sometimes by "meridian" referring to the meridian circle

---

* Practical navigation charts show, not actual magnetic curves, but regions over which the curves agree in direction within 1°. However each kind of chart is in principle convertible into the other.

** After S. P. Thompson (1908).

(for example, at paragraph 183), other times to the plane in which that circle lies (paragraph 181). Ordinarily, though, no confusion results.

Magnetic dip

Compass needles are usually restricted to turning in a roughly horizontal plane; but the earth's magnetic curves are not generally  horizontal, as we may quickly discover with a magnetized needle mounted between horizontal pivots, and thus free to turn in a vertical plane. At most latitudes such a needle will exhibit significant inclination, or "dip," from the horizontal. When the vertical plane of the "dip needle" coincides with the magnetic meridian, we can be sure that the dip needle (unlike the horizontal compass) reveals the true direction of terrestrial magnetic action at that point.* The drawing shows a laboratory dip needle.

In the northern hemisphere, the north-seeking end of the needle dips; in the southern hemisphere it is the south-seeking end that dips. The magnetic dip at London when Faraday wrote was roughly 70 degrees from the horizontal.

Magnetic curves about a current

When Oersted discovered the magnetic action of an electric current he inferred that it must be disposed in the form of circles, situated coaxially about the wire. For, he reasoned,

> without this condition, it seems impossible that the one part of the ... wire, when placed below the magnetic pole, should drive it towards the east, and when placed above it towards the west; for it is the nature of a circle that the motions in opposite parts should have an opposite direction.**

Faraday, however, believed at first that what Oersted described was consistent with simple attractions and repulsions between the needle and the wire; and in a series of experiments*** made some ten years

---

* This of course assumes that the needle has been *balanced.* That is best done before magnetization, but could you detect an out-of-balance condition in a previously-magnetized needle?

** J. C. Oersted, "Experiments on the Effect of a Current of Electricity on the Magnetic Needle," *Ann. Phil., 16* (1820), 276. (English translation of Latin original.)

*** "On some new Electro-Magnetical Motions, and on the Theory of Magnetism," *Quarterly Journal of Science, 12,* 74 (1821). It was reprinted in volume II of the Experimental Researches, but since it is not included in this edition I offer the present brief account.

before the appearance of the First Series, he set out to chronicle these expected attractions—only to find himself led by the phenomena to a similar conclusion of circularity, but with this difference: Oersted had proposed only that *some* essential attribute of current surrounds the wire and displays a circular structure.

Faraday's experiments made it clear that the "circles" are in all respects to be identified with the *magnetic curves* associated with permanent magnets.* Current electricity does not merely have an "Effect ... on the Magnetic Needle" (as Oersted's title put it)— it actually *produces magnetism*! The drawing shown here is Faraday's and depicts circular magnetic curves surrounding a current-carrying wire, as revealed by iron filings.**

The magnetic curves encircling a straight conducting wire have *direction*, just as the magnetic curves which envelop a coil or helix did. The *electro-magnetic right-hand rule* (paragraph 38, *note* and *comment*) served to represent direction in the helical case. And since a straight wire is just a coil that has been "unwound," one might expect the rule for a coil to apply to the straight wire also, simply by making appropriate changes. Let us see what happens when we try.

Grasp a coil as shown in the sketch (*a*), with fingers pointing in the (conventional) direction of current from positive to negative. The thumb points towards the reader; and according to the electromagnetic right-hand rule this indicates that the side of the coil facing the reader acts like the north pole of a magnet. For example, the

to battery negative

to battery positive

(a)

marked end of a compass needle will always point *away from* (and the south-seeking end *towards*) a more or less definable region on that side of the coil.

Now unwind the coil, keeping the fingers in position along the wire.

---

* Ampère went even further, declaring that *magnets themselves* were essentially bundles of circulating electric currents; but this "beautiful" theory (Faraday's word) was nevertheless viewed by Faraday with constant skepticism. While Faraday demands more from science than mere agreement with the facts, he also demands more than beauty!

** The figure shown actually belongs to the 29th Series; but Faraday makes so many references in the Second Series to the circular magnetic curves, it seems appropriate to give it here.

to battery negative

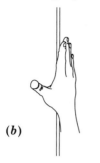

(b)

to battery positive

to battery negative

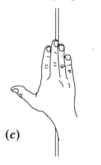

(c)

to battery positive

to battery negative

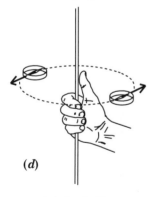

(d)

to battery positive

If the rule of paragraph 38 still applies, the thumb ought to continue to point to a region that resembles a magnetic north pole. Thus with the hand in the position shown (*b*), there should be such a region somewhere towards the reader. But clearly the hand could equally well take other positions while still retaining the fingers in the direction of the current—as, for example, in (*c*), where the thumb points to the left. It could even revolve completely about the wire, in which case the thumb would point successively in *all tangential directions*. But this contradicts the very idea of a "pole-like" region: if a compass needle always points tangentially, there can be no common intersection towards which its pointing directions converge. The sketch of iron filings on the previous page shows clearly that the circular magnetic curves about a straight current have no common intersection, and hence there can be no "pole" associated with them.*

The right-hand rule for the helix is not, therefore, equally useful for representing the straight conductor. Neither is Faraday's own mnemonic device which he proposed in his note to paragraph 38; both schemes presuppose a circular current and are therefore limited in the same way.

Although the right-hand rule of paragraph 38 does not very aptly represent the straight wire case, modern authors describe another "right-hand rule" which applies more directly:** Let the right hand grasp the conductor, with thumb pointing in the direction from positive to negative as depicted in sketch (*d*). Then the fingers show which way the north-seeking poles of compass needles would point if they were placed at various positions around the curve. Thus the directionality of the curves is expressed.

* How then do magnetic polar regions arise at the ends of a wire helix, since they did not exist when the wire was straight? In the 26th Series Faraday will begin to build a new vision of magnetic polarity, one that will prove capable of addressing this question.

** It can be vexing that there are so many "right-hand rules"! Henceforth whenever I refer to one of them, I will endeavor to identify it unambiguously.

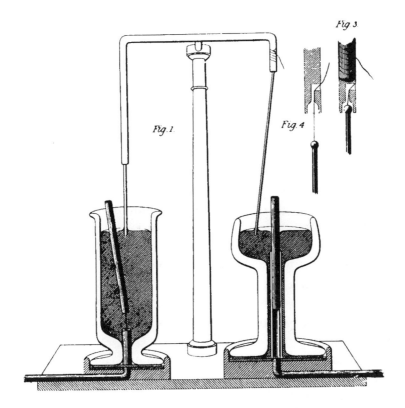

Now if every straight current is indeed encircled by closed magnetic curves, then one end of a magnet, if it could be sufficiently isolated from its opposite end, ought to be capable of following those curves in the appropriate directions, and so performing *rotations* about the current. These are the "electromagnetic rotations" which Faraday discovered in 1821, and to which he referred in the First Series (paragraph 121). He had built for him the device pictured above, which simultaneously displays (*on the left*) one end of a magnet circulating about an electric current, and (*on the right*) a current-carrying conductor circulating about one pole of a magnet. Note that according to Newton's Third Law, if either example of rotation is possible, the other must be possible too. On the other hand, the "circular" magnetic action of electric current is so unlike any of the forces which Newton considered, one might well question whether the action *ought* to conform to Newton's laws!

"Conducting power"

As described in the introduction to the First Series, materials may be classified as either *conductors* of electricity or *insulators*. But surely each of these classes admits of degrees. It is a commonplace, for us at least, that glass is a "better" insulator than wood, and copper a "better" electrical conductor than iron. But what, specifically, constitutes a conductor's excellence or deficiency?

Some years earlier, Humphrey Davy had experimentally ranked the metals for conducting power.* Davy's procedure was ingenious, if questionable in some respects. He connected a test wire AB and a trough V of acidulated water so as to present *parallel conducting paths* to the current from one or more batteries, as diagrammed here. With fewer than a certain number of batteries, discharge took place solely through the wire; there was no sign of discharge through the water. But when additional batteries were connected, current began to develop in the water too—as indicated by bubbles of oxygen and hydrogen forming at the positive and negative surfaces, respectively.**

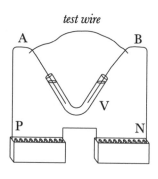

Davy thought that when current began to form in trough V it represented *overflow*, as it were, of the current in wire AB. He viewed the wire as having a limited capacity for conduction, which could be measured by the maximum number of voltaic cells whose discharge it could sustain, without producing bubbling in the trough. Davy therefore made two series of trials. First, comparing wires of identical material but different dimensions, he found that the maximum number of cells dischargeable without initiating bubbling in the trough was inversely proportional to the lengths and directly proportional to the cross-sectional areas of the wires.*** Second, comparing wires of different materials but identical dimensions, he determined that the maximum number of cells was highest for silver, lowest for lead. Davy was thus able to rank the metals for conduction capacity, as follows: silver (best), followed by copper, zinc, iron, tin, and lead (worst).

Davy's ranking of the metals was widely accepted, but his interpretation of conducting power in terms of "overflow" was not. Faraday

---

* Humphrey Davy was Faraday's predecessor and mentor at the Royal Institution. Faraday cites Davy's results in the present Series; and the experiments have additional interest for us inasmuch as Faraday himself, at that time Davy's laboratory assistant, participated in them.

** These bubbles indicate the electrical decomposition, or *electrolysis*, of water. Faraday will discuss electrolysis briefly in the Third Series, and extensively in the Seventh Series.

*** In 1773 Henry Cavendish had obtained similar results for conduction of static electric discharges, using his own body as the instrument to gauge the relative severity of electrical shocks! Davy pointed out that a dependence on cross-section showed that electrical conduction took place *throughout the body of the conductor* and not, as some believed, exclusively on the surface.

exhibits no more than a benign neglect towards the "overflow" image in the present Series,* though he cites Davy's order of the metals several times. In fact Davy's observations—that the conducting power of a metal wire increases if the wire grows in *cross-section area* but decreases if the wire grows in *length*—also suggest a very different image of conducting power. For it is easy to see that if each portion of the conducting material presents *obstruction* to the current—though, indeed, much less so than the surrounding air—then increased wire length would represent a longer succession of obstacles, thus increasing the resistance of the whole wire. Similarly, increased cross-section would provide the equivalent of additional discharge paths, thereby reducing the resistance presented by the whole wire.** Thus what Davy regarded as the measure of a wire's conducting power increases as the wire's resistance or overall obstruction decreases, and it decreases as the wire's resistance increases. It is therefore possible to view "conducting power" as *reciprocal resistance*, rather than through Davy's image of limited discharge capacity. As Faraday declares in the present Series (paragraph 213), "the obstruction is inversely as the conducting power."

But if Faraday has discounted the "overflow" image, he maintains a vigilant attitude toward the image of *obstruction*, which is itself difficult to visualize except in terms of the fluid-flow analogy. To be sure, Faraday employs images of both *flow* and *obstruction* in his thinking— he has no choice! But we never find him deducing consequences from those images as though from firm theoretical principles. He seems to regard them rather as disposable aids for the temporary anchoring of one's thought, destined—it is hoped—to evolve into a more luminous, more intelligible vision.

Divided currents—series and parallel connections

The "simple voltaic circle," consisting of a voltaic cell and a conductor connected between its poles, has only one configuration. But circuits that involve more than two elements can be connected in various ways, which creates multiple possibilities for distribution of the current. It will be helpful to distinguish two basic circuit forms.

---

* Besides having other doubts about the "overflow" idea Faraday knows, or will soon learn, that even when the water trough is the *only* discharge path, it takes a certain *minimum* number of voltaic cells (usually two or three) to initiate observable electrolysis. Thus the commencement of discharge through the water reflects properties of the *water*, as much as of the test wire.

** You can probably find a number of analogies to this relation in road traffic, or in the passage of water through pipes.

Since wires have two endpoints there are two ways in which a pair of wires may be joined together. If only one endpoint of each wire is joined to the other, the two wires together make up a single discharge path. Conductors so arranged are said to be connected *in series*, or as Faraday often expresses it, "in a single circuit."

If wires A and B as sketched here are identical, the circuit reduces to the unambiguous "voltaic circle," in which one and the same current must pass both through A and B. But what if the wires differ in their materials, lengths, cross-sections, or in any other characteristics relating to their respective conducting powers? It might seem obvious that, even so, any current that exists in either wire must exist in both wires. But it is only *obvious* if one assumes the "flow" image of elec-

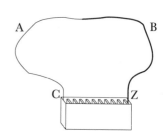

Series configuration

tric current. According to that image, any fluid that passes through wire A in the series configuration above must also pass through wire B; consequently it declares that *when conductors are connected in series, identical current must pass through both of them.*

The *parallel* configuration arises when corresponding endpoints of both wires are connected together, as this sketch shows. In this case the

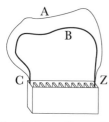

Parallel configuration

wires present alternative paths for current discharge; and again, the fluid-flow image makes definite pronouncements about the current disposition. According to the flow image, any fluid that passes out of the battery pole that does not flow through wire A must flow through wire B; and any that does not flow through B must flow through A. Consequently *when conductors are connected in parallel, the current that flows to them jointly must be the sum of the currents that flow in them individually.*

Both series and parallel cases, then, fall under a single principle, according to the fluid image: the total quantity of electric fluid approaching any point or junction must equal the total quantity of fluid departing from it.

At several places in the Second Series, you will find Faraday constructing *mixed* or multiple conducting configurations like the ones diagrammed here. Sometimes he makes suppositions about them that, strictly viewed, depend upon the highly questionable fluid-flow image (see paragraph 204, for example). More often, though, he tries to design his apparatus in such a way as to

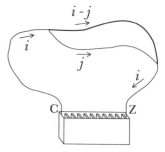

Mixed configuration

avoid the question of multiple conductors by keeping individual currents separate from one another. One striking example of this endeavor results in his construction of the differential galvanometer in paragraph 205.

Just as conductors can be connected in series or parallel configurations, other devices can be, too. A voltaic battery consists of individual cells connected in series, while a Leyden battery is a group of separate jars connected in parallel. And voltaic batteries themselves may be connected in series or in parallel, as shown in the diagram below.

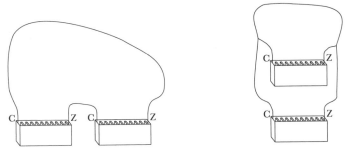

Voltaic batteries in series (left) and in parallel (right)

A note on "Ohm's Law"

This note is for readers who have already undertaken some electrical study, for they are likely to experience a problem that others escape. Modern readers who have even minimal familiarity with conventional electrical doctrines will no doubt have been introduced to Ohm's Law, which states that *current strength* is proportional to *conductance* and the so-called *electromotive force,** jointly. Ohm's Law has shaped our customary concepts of electric current and conductance (or *resistance*, the reciprocal of conductance); and if you are familiar with modern electrical thinking, you may, perhaps without even realizing it, expect to fit Faraday's thought and experimentation into its ambit. But Ohm's Law is *no part* of Faraday's vision in the Second Series!

The reason for its absence is only partly chronological. It is true that Ohm's work would not achieve general dissemination until more than a decade after the publication of Faraday's Second Series;** so it is

---

* *Electromotive force* is the hypothetical power that was at one time supposed to move electricity around the circuit.

** Georg Simon Ohm published his Law in essentially its final form in 1826 and, with full mathematical treatment, in 1827. His work was at first ignored, then subjected to a variety of attacks, including—from one critic—a charge of "heresy". Not until the 1840's would Ohm's theory gain general scientific acceptance—first in England and America; then, finally, in his own country, Germany.

unlikely that Faraday would have had occasion to consider his ideas—which were, besides, highly mathematical in their expression. But there is also a more fundamental barrier. Ohm modeled his account of current conduction on the theory of conduction and distribution of *heat*, which had been previously worked out by Fourier. Fourier, however, attributed to heat the mathematical properties of an *indestructible fluid*—so that, inevitably, the fluid imagery was carried over into Ohm's treatment of electric current.

The same imagery shows its influence in the definition of *electromotive force* as a power which supposedly "moves" electricity around. But Faraday sees no reason to assume that electricity advances or progresses by *pushes and pulls*, like ordinary bodies; such a "force" is therefore merely an extension of the fluid concepts.

Thus Ohm's Law implicitly involves the same fluid imagery that Faraday has continually doubted. We will do well, therefore, to suspend any customary Ohmic tenets as we read the Experimental Researches. While Ohm's thinking is not fundamentally *incompatible* with Faraday's,* it is settled and formalized where Faraday's is exploratory and evolving.

---

* We will find a few disparities between Faraday's thinking and Ohm's in this very Series—for example the argument at paragraph 213. But Maxwell, who had the gift of harmonizing even the most divergent concepts through mathematical inventiveness, shows in his *Treatise* how many of Faraday's mature ideas may be combined with Ohm's treatment of conduction. In any case the question will arise again, for us, in the 29th Series.

SECOND SERIES.

THE BAKERIAN LECTURE.

§ 5. Terrestrial Magneto-electric Induction. § 6. Force and Direction of Magneto-electric Induction generally.

Read January 12,1832.

§ 5. *Terrestrial Magneto-electric Induction.*

140. WHEN the general facts described in the former paper were discovered, and the *law* of magneto-electric induction relative to direction was ascertained (114.), it was not difficult to perceive that the earth would produce the same effect as a magnet, and to an extent that would, perhaps, render it available in the construction of new electrical machines. The following are some of the results obtained in pursuance of this view.

141. The hollow helix already described (6.) was connected with a galvanometer by wire eight feet long; and the soft iron cylinder (34.) after being heated red hot and slowly cooled, to remove all traces of magnetism, was put into the helix so as to project equally at both ends, and fixed there. The combined helix and bar were held in the magnetic direction or line of dip, and (the galvanometer needle being motionless) were then inverted, so that the lower end should become the upper, but the whole still correspond to the magnetic direction; the needle was immediately deflected. As the latter returned to its first position, the helix and bar were again inverted; and by doing this two or three times, making the inversions and vibrations to coincide, the needle swung through an arc of 150° or 160°.

142. When one end of the helix, which may be called A, was uppermost at first (B end consequently being below), then it mattered not in which direction it proceeded during the inversion, whether to the right hand or left hand, or through any other course; still the galvanometer needle passed in the same direction. Again, when B end

---

141. The "line of dip," and the *dip needle* that reveals it, are discussed in the introduction to the present Series. Note here again and in most of the following experiments the technique, by now routine, of building up resonant vibration in the galvanometer so as to make its response more evident to the eye.

was uppermost, the inversion of the helix and bar in any direction always caused the needle to be deflected one way; that way being the opposite to the course of the deflection in the former case.

143. When the helix with its iron core in any given position was inverted, the effect was as if a magnet with its marked pole downwards had been introduced from above into the inverted helix. Thus, if the end B were upwards, such a magnet introduced from above would make the marked end of the galvanometer needle pass west. Or the end B being downwards, and the soft iron in its place, inversion of the whole produced the same effect.

144. When the soft iron bar was taken out of the helix and inverted in various directions within four feet of the galvanometer, not the slightest effect upon it was produced.

145. These phenomena are the necessary consequence of the inductive magnetic power of the earth, rendering the soft iron cylinder a magnet with its marked pole downwards. The experiment is analogous

---

143. The coil with iron core is initially held B end down, and parallel to the "dip" (see the introduction to the present Series). When rotated 180° to leave B up, the galvanometer deflects to the same side as if the coil had been held *stationary* with B up, while a magnet descended towards it with north-seeking pole down. Faraday shows in paragraph 145 how this result is to be interpreted.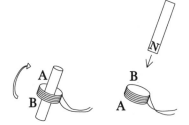

144. *not the slightest effect upon it was produced*: Thus the galvanometer deflection is wholly due to current induced in the galvanometer wire; movement of the iron bar among the earth's magnetic curves has no *direct* effect on the galvanometer needle.

145. *rendering the soft iron cylinder a magnet with its marked pole downwards*: When held parallel or nearly parallel to the earth's magnetic curves, the iron core becomes temporarily magnetized in conformity with the earth—thus in the northern hemisphere, whichever end is downward *faces north* and becomes a *north-seeking* end. Therefore when the coil and core are inverted together, the magnetism of the core reverses. It makes sense that the effect of such an inversion should resemble the effect of a *permanent* magnet descending, N-end first, into the coil; for in both of the cases contemplated, the coil would ultimately attain the same condition—from an *opposite* condition if by inversion (the actual case), from a lesser degree of the *same* condition if by insertion of a magnet (the imagined case). The alterations differ only in degree—that is the significance of the similar galvanometer deflections.

the inductive magnetic power of the earth: that is, the earth's ability to raise up a magnetic condition in the iron cylinder, as just described. In the introduction to the First Series, I pointed out that Faraday declines to use the term

to that in which two bar magnets were used to magnetize the same cylinder in the same helix (36.), and the inversion of position in the present experiment is equivalent to a change of the poles in that arrangement. But the result is not less an instance of the evolution of electricity by means of the magnetism of the globe.

146. The helix alone was then held permanently in the magnetic direction, and the soft iron cylinder afterwards introduced; the galvanometer needle was instantly deflected; by withdrawing the cylinder as the needle returned, and continuing the two actions simultaneously, the vibrations soon extended through an arc of 180°. The effect was precisely the same as that obtained by using a cylinder magnet with its marked pole downwards; and the direction of motion, &c. was perfectly in accordance with the results of former experiments obtained with such a magnet (39.). A magnet in that position being used, gave the same deflection, but stronger. When the helix was put at right angles to the magnetic direction or dip, then the introduction or removal of the soft iron cylinder produced no effect at the needle. Any inclination to the dip gave results of the same kind as those already described, but increasing in strength as the helix approximated to the direction of the dip.

147. A cylinder magnet, although it has great power of affecting the galvanometer when moving into or out of the helix, has no power of continuing the deflection (39.); and therefore though left in, still the magnetic needle comes to its usual place of rest. But upon repeating (with the magnet) the experiment of inversion in the direction of the dip (141.), the needle was affected as powerfully as before; the disturbance of the magnetism in the steel magnet, by the earth's inductive force upon it, being thus shown to be nearly, if not quite, equal in amount and rapidity to that occurring in soft iron. It is probable that in this way magneto-electrical arrangements may become very useful in

---

"induction" to denote this *magnetic* action until the present paragraph, even though the terminology was widely accepted and the effect itself would appear to be clearly analogous to static *electric* induction. Perhaps he is reluctant to let terminology depend too freely on *mere* analogy. If so, he will remove some of that dependence in the Eleventh Series, which reinterprets electrical induction in a remarkable way—and in the Twenty-eighth Series, which does the same for magnetic induction.

147. *disturbance of the magnetism*: The cylinder magnet has permanent magnetism of its own, but insofar as it is an iron bar it is still susceptible to influence of the earth. Thus it must experience a *disturbance of magnetism* when inverted, corresponding to the *complete reversal of magnetism* suffered by the iron bar in paragraph 143. It is this increment in magnetic condition, not the total condition itself, that corresponds to the induced current.

indicating the disturbance of magnetic forces, where other means will not apply; for it is not the whole magnetic power which produce the visible effect, but only the difference due to the disturbing causes.

148. These favourable results led me to hope that the direct magneto-electric induction of the earth might be rendered sensible; and I ultimately succeeded in obtaining the effect in several ways. When the helix just referred to (141. 6.) was placed in the magnetic dip, but without any cylinder of iron or steel, and was then inverted, a feeble action at the needle was observed. Inverting the helix ten or twelve times, and at such periods that the deflecting forces exerted by the currents of electricity produced in it should be added to the momentum of the needle (39.), the latter was soon made to vibrate through an arc of 80° or 90°. Here, therefore, currents of electricity were produced by the direct inductive power of the earth's magnetism, without the use of any ferruginous matter, and upon a metal not capable of exhibiting any of the ordinary magnetic phenomena. The experiment in everything represents the effects produced by bringing the same helix to one or both poles of any powerful magnet (50.).

149. Guided by the law already expressed (114.), I expected that all the electric phenomena of the revolving metal plate could now be produced without any other magnet than the earth. The plate so often referred to (85.) was therefore fixed so as to rotate in a horizontal plane. The magnetic curves of the earth (114. *note*), *i. e.* the dip, passes though this plane at angles of about 70°, which it was expected would be an approximation to perpendicularity, quite enough to allow of magneto-electric induction sufficiently powerful to produce a current of electricity.

150. Upon rotation of the plate, the currents ought, according to the law (114. 121.), to tend to pass in the direction of the radii, through *all* parts of the plate, either from the centre to the circumference, or from the circumference to the centre, as the direction of the rotation of the

---

148. *direct magneto-electric induction of the earth*: In the foregoing experiments, the earth induced magnetism in iron; *that* magnetism in turn (during its increase or decrease) caused an induced current in the helix. Now he shows that the earth's magnetism can induce currents in a conductor *directly*, without the mediation of iron.

ferruginous: of or containing iron.

149. *an approximation to perpendicularity*: Faraday will show in paragraph 153 below that a rotating plate develops maximum current when it cuts magnetic curves at right angles.

150. *the currents ought ... to pass in the direction of the radii, through all parts of the plate*: When as in paragraph 84 the copper disc was placed between the poles

plate was one way or the other. One of the wires of the galvanometer was therefore brought in contact with the axis of the plate, and the other attached to a leaden collector or conductor (86.), which itself was placed against the amalgamated edge of the disc. On rotating the plate there was a distinct effect at the galvanometer needle; on reversing the rotation, the needle went in the opposite direction; and by making the action of the plate coincide with the vibrations of the needle, the arc through which the latter passed soon extended to half a circle.

151. Whatever part of the edge of the plate was touched by the conductor, the electricity was the same, provided the direction of rotation continued unaltered.

152. When the plate revolved *screw-fashion*, or as the hands of a watch, the current of electricity (150.) was from the centre to the circumference; when the direction of rotation was *un-screw*, the current was from the circumference to the centre. These directions are the same with those obtained when the unmarked pole of a magnet was placed beneath the revolving plate (99.).

---

of the Royal Society's magnet, there would have been a high concentration of magnetic curves intersecting the part of the disc near the poles, and few intersecting more distant parts. But here the earth's poles are so far away that *all* parts of the disc will be intersected equally by the terrestrial magnetic curves. There will be no region of concentration, and therefore *all* of the disc's radii will tend equally to develop induced currents when the disc is rotated in the horizontal plane.

151. Faraday obtains equal currents everywhere on the circumference, confirming that currents are passing equally along all radii of the disc. But where are the *return* currents in this case? When the inducing magnet acted on only a limited region of the disc, as in paragraph 123, he was able to infer that the return currents pass along the radii that intersect few or no magnetic curves; but here there are no such radii! Thus there is *no* return path through the disc; *all* of the current must complete its circuit through the galvanometer. That circumstance makes this arrangement far more efficient, as a *magneto-electric generator*—Faraday will call it a "new electrical machine" in paragraph 154—than the versions reported in the First Series. In those experiments the huge magnet exerted a much more intense influence than did the earth's magnetism. But it acted on only *part* of the disc, and only *part* of the induced current passed to the galvanometer to be detected (paragraph 134).

152. *These directions are the same with those obtained when the unmarked pole of a magnet was placed beneath the revolving plate*: This must of course be the case since, the plate being (nearly) perpendicular to the terrestrial magnetic curves, and being in the northern hemisphere, there *is* always an "unmarked" (that is, a south-seeking) pole beneath the plate—namely, the magnetic pole at the earth's arctic region.

153. When the plate was in the magnetic meridian, or in any other plane *coinciding* with the magnetic dip, then its rotation produced no effect upon the galvanometer. When inclined to the dip but a few degrees, electricity began to appear upon rotation. Thus when standing upright in a plane perpendicular to the magnetic meridian, and when consequently its own plane was inclined only about 20° to the dip, revolution of the plate evolved electricity. As the inclination was increased, the electricity became more powerful until the angle formed by the plane of the plate with the dip was 90°, when the electricity for a given velocity of the plate was a maximum.

154. It is a striking thing to observe the revolving copper plate become thus a *new electrical machine*; and curious results arise on comparing it with the common machine. In the one, the plate is of the best non-conducting substance that can be applied; in the other, it is the most perfect conductor: in the one, insulation is essential; in the other it is fatal. In comparison of the quantities of electricity produced, the metal machine does not at all fall below the glass one; for it can produce a constant current capable of deflecting the galvanometer needle, whereas the latter cannot. It is quite true that the force of the current thus evolved has not as yet been increased so as to render it available in any of our ordinary applications of this power; but there appears every reasonable expectation that this may hereafter be effected; and probably by several arrangements. Weak as the current may seem to be, it is as strong as, if not stronger than, any thermo-electric current; for it can pass fluids (23.), agitate the animal system, and in the case of an electro-magnet has produced sparks (32.).

155. A disc of copper, one fifth of an inch thick and only one inch and a half in diameter, was amalgamated at the edge; a square piece of sheet lead (copper would have been better) of equal thickness had a circular hole cut in it, into which the disc loosely fitted; a little mercury completed the metallic communication of the disc and its surrounding ring; the latter was attached to one of the galvanometer wires, and the

---

154. *In the one ... in the other*: These phrases refer to the *common electrostatic plate machine*, and the new *magneto-electric generator*, respectively. The plate machine evolves electricity by friction; its rotating plate is glass or other insulating material. Insulation is "essential" because frictional electricity does not arise when the rubbing surfaces are conductive. The magneto-electric machine evolves electricity by induction; its rotating disc must be copper or other conducting material in order to sustain the magnetically induced currents. Insulation would be "fatal," as was evident in paragraph 127: when slits or gaps were made in the disc, currents could not be induced and no electricity was evolved.

other wire dipped into a little metallic cup containing mercury, fixed upon the top of the copper axis of the small disc. Upon rotating the disc in a horizontal plane, the galvanometer needle could be affected, although the earth was the only magnet employed, and the radius of the disc but three quarters of an inch; in which space only the current was excited.

156. On putting the pole of a magnet under the revolving disc, the galvanometer needle could be permanently deflected.

157. On using copper wires one sixth of an inch in thickness instead of the smaller wires (86.) hitherto constantly employed, far more powerful effects were obtained. Perhaps if the galvanometer had consisted of fewer turns of thick wire instead of many convolutions of thinner, more striking effects would have been produced.

158. One form of apparatus which I purpose having arranged, is to have several discs superposed; the discs are to be metallically connected, alternately at the edges and at the centres, by means of mercury; and are then to be revolved alternately in opposite directions, *i. e.* the first, third, fifth, &c. to the right hand, and the second, fourth, sixth, &c. to the left hand; the whole being placed so that the discs are perpendicular to the dip, or interact most directly the magnetic curves of powerful magnets. The electricity will be from

---

155. *in which space only the current was excited*: This little device illustrates the magneto-electric principle's promise as a source of *power*, suggested in the previous paragraph. For observe how much current can be produced in such a tiny space!

157. *Perhaps if the galvanometer had consisted of fewer turns of thick wire…*: The induced current has to overcome the resistance—or "obstruction," to use Faraday's word in paragraph 213 below—of the galvanometer coil (the "multiplier") and connecting wires. When thicker connecting wires were used the current was greater. Similarly if the wire forming the galvanometer coil had likewise been shorter in length and greater in thickness, it too would have presented less resistance and permitted a greater current to develop. Why then does Faraday say only that *perhaps* more striking effects would have been produced? Probably because a coil of fewer turns would have been less effective as a "multiplier," so that the overall result would depend on which change—current increase or multiplier decrease—had predominated. ("Obstruction" is discussed in the introduction to the present Series; the galvanometer "multiplier" in the introduction to the First Series.)

158. *I purpose*: that is, *I intend*. He "purposes" a *battery* of generators! Such a combination might prove capable of developing great power; but in the next paragraph he declares that the enlargement of generating capacity is not his main interest.

the centre to the circumference in one set of discs; and from the circumference to the centre in those on each side of them; thus the action of the whole will conjoin to produce one combined and more powerful current.

159. I have rather, however, been desirous of discovering new facts and new relations dependent on magneto-electric induction, than of exalting the force of those already obtained; being assured that the latter would find their full development hereafter.

*   *   *

170. These results, in conjunction with the general law, before stated (114.) suggested an experiment of extreme simplicity, which yet, on trial, was found to answer perfectly. The exclusion of all extraneous circumstances and complexity of arrangement, and the distinct character of the indications afforded, render this single experiment an epitome of nearly all the facts of magneto-electric induction.

171. A piece of common copper wire, about eight feet long, and one twentieth of an inch in thickness, had one of its ends fastened to one of the terminations of the galvanometer wire, and the other end to the other termination; thus it formed an endless continuation of the galvanometer wire: it was then roughly adjusted into the shape of a rectangle, or rather of a loop, the upper part of which could be carried to and fro over the galvanometer, whilst the lower part, and the

---

170. *These results*: He refers to experiments recounted in paragraphs 160–169 above. I have omitted them since, as Faraday himself states, they are nicely epitomized in the far simpler experiment he is about to describe.

171. *a rectangle, or rather of a loop*: Faraday's Figure 30 depicts the rectangular wire in perspective view. One leg is held stationary upon the work table and oriented north and south, the other legs are free to swing east and west, as indicated by the dotted arrow. Although Faraday actually employs a galvanometer, his figure omits the needle and multiplier coil. He will explain in the next paragraph that he intends this artifice as a simplification, and that we are to imagine the galvanometer needle situated above the wire; it is depicted in the accompanying sketch.

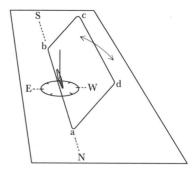

upper part … lower part: The "lower" part of the wire is the stationary leg, labeled *ab* in the accompanying sketch; the "upper" part is *bcda*.

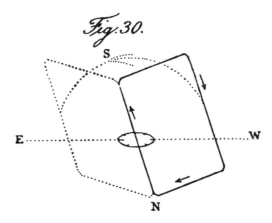

Fig. 30.

galvanometer attached to it, remained steady (Plate II. fig. 30.). Upon moving this loop over the galvanometer from right to left, the magnetic needle was immediately deflected; upon passing the loop back again, the needle passed in the contrary direction to what it did before; upon repeating these motions of the loop in accordance with the vibrations of the needle (39.), the latter soon swung through 90° or more.

172. The relation of the current of electricity produced in the wire, to its motion, may be understood by supposing the convolutions of the galvanometer away, and the wire arranged as a rectangle, with its lower edge horizontal and in the plane of the magnetic meridian, and a magnetic needle suspended above and over the middle part of this edge, and directed by the earth (fig. 30.). On passing the upper part of the rectangle from west to east in the position represented by the dotted line, the marked pole of the magnetic needle went west; the

---

172. *The relation ... may be understood by supposing the convolutions of the galvanometer away*: that is, by ignoring the galvanometer multiplier coil and supposing instead that the needle is suspended above a single straight wire.

a magnetic needle ... directed by the earth: that is, a needle pointing north and south.

On passing the upper part of the rectangle from west to east ... the marked pole ... went west: Refer to the sketch opposite. From the westward deflection of the marked (north-seeking) end of the galvanometer needle Faraday infers that current in the stationary leg of the wire must have direction from *a* to *b*. You can check this by the "electromagnetic" right-hand rule (paragraph 101, *comment*).

Since the wire is a closed loop, if the current direction is *a* to *b* in the lower leg, its direction must be *c* to *d* in the upper leg, whose eastward motion across the earth's magnetic curves induces the current. Satisfy yourself that this direction agrees with the "magneto-electric" right-hand rule (paragraph 38, *comment*). Remember that terrestrial magnetic curves in England and the northerly latitudes have a *downward* direction because there must be a *south-seeking pole* located in the earth's northern parts (paragraph 44, *note* and *comment*).

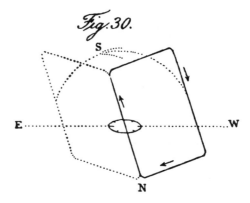

Fig.30.

electric current was therefore from north to south in the part of the wire passing under the needle, and from south to north in the moving or upper part of the parallelogram. On passing the upper part of the rectangle from east to west over the galvanometer, the marked pole of the needle went east, and the current of electricity was therefore the reverse of the former.

173. When the rectangle was arranged in a plane east and west, and the magnetic needle made parallel to it, either by the torsion of its suspension thread or the action of a magnet, still the general effects were the same. On moving the upper part of the rectangle from north to south, the marked pole of the needle went north; when the wire was moved in the opposite direction, the marked pole went south. The same effect took place when the motion of the wire was in any other azimuth of the line of dip; the direction of the current always being

---

173. The apparatus is rotated 90°, so that the stationary wire runs east-and-west and the upper part of the rectangle moves north-and-south. Hence the magnetic needle, too, must be turned out of its normal north-south orientation and aligned east-and-west in order to remain parallel to the stationary leg of the wire. To accomplish this, Faraday either twists the needle's suspension ("torsion of its suspension thread") or introduces an external magnet to attract the needle into the east-west direction ("action of a magnet"). But in the next paragraph he acknowledges that it is not really necessary to turn the whole assembly. Bending the upper part of the rectangle into the new position, while keeping the stationary leg and the galvanometer in place, proves sufficient.

still the general effects are the same: When the upper part of the wire is moved north and south, currents are induced just as before.

when the motion of the wire was in any other azimuth of the line of dip: that is, when the motion was in any other direction with respect to the north-south line (the "line of dip"). Currents are induced in all these cases, since the terrestrial magnetic curves will be "cut" by a horizontal wire moving in any horizontal direction perpendicular to its length. This will happen everywhere except in the equatorial latitudes, where the magnetic curves of the earth are essentially horizontal.

conformable to the law formerly expressed (114.), and also to the directions obtained with the rotating ball (164.).

174. In these experiments it is not necessary to move the galvanometer or needle from its first position. It is quite sufficient if the wire of the rectangle is distorted where it leaves the instrument, and bent so as to allow the moving upper part to travel in the desired direction.

175. The moveable part of the wire was then arranged *below* the galvanometer, but so as to be carried across the dip. It affected the instrument as before, and in the same direction; *i. e.* when carried from west to east under the instrument, the marked end of the needle went west, as before. This should, of course be the case; for when the wire is cutting the magnetic dip in a certain direction, an electric current also in a certain direction should be induced in it.

176. If in fig. 31. *dp* be parallel to the dip, and BA be considered as the upper part of the rectangle (171.), with an arrow *c* attached to it, both these being retained in a plane perpendicular to the dip,—then, however BA with its attached arrow is moved upon *dp* as an axis, if it afterwards proceed in the direction of the arrow, a current of electricity will move along it from B towards A.

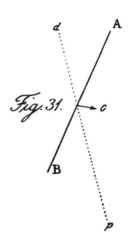

Fig. 31.

---

175. *across the dip*: that is, across the terrestrial magnetic curves.

176. *If in fig. 31. dp be parallel to the dip…*: As noted in the comment to paragraph 172, the direction of the terrestrial magnetic curves defining the dip is *downward*, from *d* to *p*.

…and BA be considered as the upper part of the rectangle … perpendicular to the dip: Think of arrow *c* and line *dp* as being perpendicular to one another in the plane of the page, and the wire BA as perpendicular to the page with end B towards the reader. In the actual trials described in paragraphs 172 and 173 the line of dip was not quite vertical, and hence the horizontal wire was not quite perpendicular to it. In the present case the single wire BA is strictly perpendicular to *dp*.

…if it afterwards proceed in the direction of the arrow, a current of electricity will move along it from B towards A: The mutually perpendicular wire BA and arrow *c* may together be given any orientation about *dp* as axis; but no matter what orientation is chosen, when BA moves in the direction of *c* a current will arise from B to A in accordance with the magneto-electric right-hand rule (paragraph 101, *comment*). For hold the thumb and forefinger parallel to the page, with thumb pointing in the direction of arrow *c* and forefinger pointing in the direction of dip from *d* to *p*. Then the middle finger, held perpendicular to both of them, will point into the page—that is, from B to A.

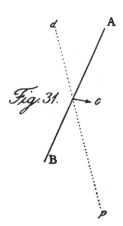

Fig. 31.

177. When the moving part of the wire was carried up or down parallel to the dip, no effect was produced on the galvanometer. When the direction of motion was a little inclined to the dip, electricity manifested itself; and was at a maximum when the motion was perpendicular to the magnetic direction.

178. When the wire was bent into other forms and moved, equally strong effects were obtained, especially when instead of a rectangle a double catenarian curve was formed of it on one side of the galvanometer, and the two single curves or halves were swung in opposite directions at the same time; their action then combined to affect the galvanometer: but all the results were reducible to those above described.

179. The longer the extent of the moving wire, and the greater the space through which it moves, the greater is the effect upon the galvanometer.

180. The facility with which electric currents are produced in metals when moving under the influence of magnets, suggests that henceforth precautions should always be taken, in experiments upon metals and magnets, to guard against such effects. Considering the universality of the magnetic influence of the earth, it is a consequence which

---

177. Now the moving wire *c* (in figure 31) is permitted to take *any* angle with respect to *dp*. If that angle is *zero*, no current is produced; and the current increases to a maximum when the angle between *c* and *dp* increases to a right angle.

178. A *catenarian curve* (L. *catena*, chain) is the curve assumed by a uniform, flexible, inextensible cord when hanging freely from two fixed points.

appears very extraordinary to the mind, that scarcely any piece of metal can be moved in contact with others, either at rest, or in motion with different velocities or in varying directions, without an electric current existing within them. It is probable that amongst arrangements of steam-engines and metal machinery, some curious accidental magneto-electric combinations may be found, producing effects which have never been observed, or, if noticed, have never as yet been understood.

---

181. Upon considering the effects of terrestrial magneto-electric induction which have now been described, it is almost impossible to resist the impression that similar effects, but infinitely greater in force, may be produced by the action of the globe, as a magnet, upon its own mass, in consequence of its diurnal rotation. It would seem that if a bar of metal be laid in these latitudes on the surface of the earth parallel to the magnetic meridian, a current of electricity tends to pass through it from south to north, in consequence of the travelling of the bar from west to east (172.), by the rotation of the earth; that if another bar in the same direction be connected with the first by wires, it cannot discharge the current of the first, because it has an equal tendency to have a current in the same direction induced within itself: but that if the latter be carried from east to west, which is equivalent to a diminution of the motion communicated to it from the earth (172.), then the electric current from south to north is rendered evident in the first bar, in consequence of its discharge, at the same time, by means of the second.

---

181. *if a bar of metal be laid in these latitudes on the surface of the earth parallel to the magnetic meridian...*: A metal bar lying in the north-south direction on the earth's surface is continually being carried west to east, like the upper wire in Figure 30. But is such a bar *cutting magnetic curves*? If it is, then the rotating earth does not carry the terrestrial magnetic curves around with itself! Can we put the question to test by observing the *current* which, if the bar is really cutting magnetic curves, should tend to form in it from south to north? Evidently not; for supposing the earth to be indeed spinning among its own stationary magnetic curves, a galvanometer connected between the bar's extremities will *itself* cut the same magnetic curves as the bar; an equivalent current will then form in it, and so nullify the current in the bar. (Faraday points out that if the intended "discharging wire" were *moved* relative to the bar, a current *would* form. But that would occur *whether or not* the terrestrial magnetic curves rotate along with the earth; so it cannot serve as a test for such rotation.)

182. Upon the supposition that the rotation of the earth tended, by magneto-electric induction, to cause currents in its own mass, these would, according to the law (114.) and the experiments, be, upon the surface at least, from the parts in the neighbourhood of or towards the plane of the equator, in opposite directions to the poles; and if collectors could be supplied at the equator and at the poles of the globe, as has been done with the revolving copper plate (150.), and also with magnets (220.), then negative electricity would be collected at the equator, and positive electricity at both poles (222.). But without the conductors, or something equivalent to them, it is evident these currents could not exist, as they could not be discharged.

183. I did not think it impossible that some natural difference might occur between bodies, relative to the intensity of the current produced or tending to be produced in them by magneto-electric induction, which might be shown by opposing them to each other; especially as Messrs. Arago, Babbage, Herschel, and Harris, have all found great differences, not only between the metals and other substances, but between the metals themselves, in their power of receiving motion from or giving it to a magnet in trials by revolution (130.). I therefore took two wires, each one hundred and twenty feet long, one of iron and the other of copper. These were connected with each other at their ends, and then extended in the direction of the magnetic meridian, so as to form two nearly parallel lines, nowhere in contact except at the extremities. The copper wire was then divided in the middle, and examined by a delicate galvanometer, but no evidence of an electrical current was obtained.

---

182. Recall that in the northern hemisphere, an "unmarked" (south-seeking) pole ultimately lies beneath all horizontal surfaces, while in the southern hemisphere a "marked" (north-seeking) pole does the same. If the rotating earth indeed *cuts* its own stationary magnetic curves, then bodies carried by the earth from west to east should tend to form currents—from the equator to the nearer pole in each hemisphere. If we could devise equatorial and polar collectors and connect them with a *stationary* wire (so as to avoid the problem identified in paragraph 181), it might be possible to detect a current. (In passing, note that *at* the equator the terrestrial magnetic curves are essentially horizontal; so a body placed on the earth's equatorial surface should develop no currents, as it will intersect no curves.)

183. *Arago, Babbage, Herschel, and Harris, have all found great differences…*: Recall that in Arago's original "rotation" experiments (paragraph 81) a spinning copper disk communicated its rotary motion to a bar magnet pivoted above it. The three later investigators substituted other materials in place of copper

184. By favour of His Royal Highness the President of the Society, I obtained the permission of His Majesty to make experiments at the lake in the gardens of Kensington-palace, for the purpose of comparing, in a similar manner, water and metal. The basin of this lake is artificial; the water is supplied by the Chelsea Company; no springs run into it, and it presented what I required, namely, a uniform mass of still pure water, with banks ranging nearly from east to west, and from north to south.

---

and found that different materials produced rotations having very different strengths. Although it is not clear that Babbage, Herschel, and Harris understood it, Faraday realizes that according to their results *the best electrical conductors* produced the strongest rotations, while the worst conductors—except for iron—produced the weakest rotations. (He will mention this in paragraphs 202 and 211 below.)

Now since Faraday has attributed the Arago rotations to currents induced by the magnet in the spinning disk, forcible or feeble rotations should on his view indicate strong or weak induced currents, respectively. Thus in the rotation experiments, *the strongest currents are induced in the best conductors*; and Faraday proposes to test whether that correlation is generally true. His purpose may appear obscure to modern readers, for whom currents must *by definition* be stronger in "better" conductors and weaker in "poorer" ones, under otherwise comparable circumstances—compare the comment on paragraph 73 above. But remember that Faraday is here investigating *induced* currents, not ordinary conduction currents. Conduction involves discharge and has an inverse relation to *obstruction*. Induction, on the other hand, like voltaic action, is *generative*. It is a wholly different process from discharge; and it is not obvious what relevance, if any, "obstruction" has to it. In paragraph 213 below we may find additional signs of this distinction between generation and discharge, in Faraday's evolving view.

Now it is clear from paragraph 181 that if *identical* stationary wires (joined at their ends) are indeed cutting terrestrial magnetic curves, they will form identical currents which cancel one another; overall, then, no current will pass. But if the strength of the induced current depends on the conducting power of the material, then wires of copper and iron should develop unequal currents under terrestrial magnetic induction since they differ in conducting power. If so, Faraday thinks, a galvanometer placed in one of the wires should indicate the excess of the greater current over the less—just as if two batteries having unequal numbers of cells were connected in opposite directions in a single circuit. Nevertheless, no observable deflection was obtained. If currents *are* indeed being developed in both wires, those currents must be virtually equal.

184. *His Royal Highness the President of the Society*: Augustus Frederick Hanover, Duke of Sussex, son of George III; he became President of the Royal Society in 1830. *His Majesty* is King William IV; he reigned 1830-1837, preceding Queen Victoria.

185. Two perfectly clean bright copper plates, each exposing four square feet of surface, were soldered to the extremities of a copper wire; the plates were immersed in the water, north and south of each other, the wire which connected them being arranged upon the grass of the bank. The plates were about four hundred and eighty feet from each other, in a right line; the wire was probably six hundred feet long. This wire was then divided in the middle, and connected by two cups of mercury with a delicate galvanometer.

186. At first, indications of electric currents were obtained; but when these were tested by inverting the direction of contact, and in other ways, they were found to be due to other causes than the one sought for. A little difference in temperature; a minute portion of the nitrate of mercury used to amalgamate the wires, entering into the water employed to reduce the two cups of mercury to the same temperature; was sufficient to produce currents of electricity, which affected the galvanometer, notwithstanding they had to pass through nearly five hundred feet of water. When these and other interfering causes were guarded against, no effect was obtained; and it appeared that even such dissimilar substances as water and copper, when cutting the magnetic curves of the earth with equal velocity, perfectly neutralized each other's action.

*   *   *

§6. *General remarks and illustrations of the Force and Direction of Magneto-electric induction.*

193. In the repetition and variation of Arago's experiment by Messrs. Babbage, Herschel, and Harris, these philosophers directed

---

185–186. Undeterred by the null results of the previous trial with copper and iron wires, he now pairs copper with *water*, for a greater difference in conducting power—and thus, he hopes, a greater inequality between the induced currents—than between copper and iron. But after removing several extraneous influences, he still detects no excess of one current over the other.

193. *the repetition and variation of Arago's experiment…*: that is, the experiments reviewed in the comment to paragraph 183 above, which had suggested that a material's conducting power governs the strength of the currents induced in it. Faraday could not confirm that connection in his experiments with long wires and at Kensington Lake, but he is evidently determined to make another trial with different apparatus. Or rather, not just "another" trial—over the next six paragraphs he describes *ten* variations on the original long-wire experiment! Not one of them shows the expected inequality between currents in different materials. Is Faraday displaying an admirable persistence or an unreasonable obstinacy?

their attention to the differences of force observed amongst the metals and other substances in their action on the magnet. These differences were very great,[1] and led me to hope that by mechanical combinations of various metals important results might be obtained (183). The following experiments were therefore made, with a view to obtain, if possible, any such difference of the action of two metals.

194. A piece of soft iron bonnet-wire covered with cotton was laid bare and cleaned at one extremity, and there fastened by metallic contact with the clean end of a copper wire. Both wires were then twisted together like the strands of a rope, for eighteen or twenty inches; and the remaining parts being made to diverge, their extremities were connected with the wires of the galvanometer. The iron wire was about two feet long, the continuation to the galvanometer being copper.

195. The twisted copper and iron (touching each other nowhere but at the extremity) were then passed between the poles of a powerful magnet arranged horse-shoe fashion (fig. 32.); but not the slightest effect was observed at the galvanometer, although the arrangement seemed fitted to show any electrical difference between the two metals relative to the action of the magnet.

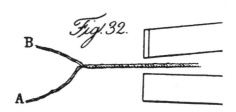

[1] [Philosophical Transactions,] 1825, p. 472; 1831, p. 78.

---

194. *bonnet-wire*: A soft, malleable wire used by milliners. An article on "The Early Spring Bonnets" in the March, 1894 *Ladies' Home Journal* instructs: "The lace is stiffened for these brims … with the very finest milliner's wire. … The lace may be made so that it will bend as desired. When the bonnet has served its time, the wire may be removed, and behold, the lace is as good as new."

195. A miniature equivalent of the experiment described in paragraphs 183–186. Induced currents in the two legs, which must have the *same* direction with respect to the magnet, will *oppose* one another in the circuit (see the diagram). If both currents are equal, they will nullify each other—the observed result. If one predominates, the galvanometer will indicate their difference.

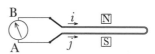

196. A soft iron cylinder was then covered with paper at the middle part, and the twisted portion of the above compound wire coiled as a spiral around it, the connexion with the galvanometer still being made at the ends A B. The iron cylinder was then brought in contact with the poles of a powerful magnet capable of raising thirty pounds; yet no signs of electricity appeared at the galvanometer. Every precaution was applied in making and breaking contact to accumulate effect, but no indications of a current could be obtained.

197. Copper and tin, copper and zinc, tin and zinc, tin and iron, and zinc and iron, were tried against each other in a similar manner (194), but not the slightest sign of electric currents could be procured.

198. Two flat spirals, one of copper and the other of iron, containing each eighteen inches of wire were connected with each other and with the galvanometer, and then put face to face so as to be in contrary directions. When brought up to the magnetic pole (53.), no electrical indications at the galvanometer were observed. When one was turned round so that both were in the same direction, the effect at the galvanometer was very powerful.

199. The compound helix of copper and iron wire formerly described (8.) was arranged as a double helix, one of the helices being all iron and containing two hundred and fourteen feet, the other all copper and containing two hundred and eight feet. The two similar ends A A of the copper and iron helix were connected together, and the other ends B B of each helix connected with the galvanometer; so that when a magnet was introduced into the centre of the arrangement, the induced currents in the iron and copper would tend to proceed in contrary directions. Yet when a magnet was inserted, or a soft iron bar within made a magnet by contact with poles, no effect at the needle was produced.

200. A glass tube about fourteen inches long was filled with strong sulphuric acid. Twelve inches of the end of a clean copper wire were bent up into a bundle and inserted into the tube, so as to make good superficial contact with the acid, and the rest of the wire passed along

---

196. *to accumulate effect*: that is, to build up resonant vibration in the galvanometer, as has often been noted.

200. *good superficial contact*: that is, good contact with the surface of the wire. By bundling up the wire he maximizes the surface immersed, thus making the best possible junction with the liquid acid.

the outside of the tube and away to the galvanometer. A wire similarly bent up at the extremity was immersed in the other end of the sulphuric acid, and also connected with the galvanometer, so that the acid and copper wire were in the same parallel relation to each other in this experiment as iron and copper were in the first (194). When this arrangement was passed in a similar manner between the poles of the magnet, not the slightest effect at the galvanometer could be perceived.

201. From these experiments it would appear, that when metals of different kinds connected in one circuit are equally subject in every circumstance to magneto-electric induction, they exhibit exactly equal powers with respect to the currents which either are formed, or tend to form, in them. The same even appears to be the case with regard to fluids, and probably all other substances.

202. Still it seemed impossible that these results could indicate the relative inductive power of the magnet upon the different metals; for that the effect should be in some relation to the conducting power seemed a necessary consequence (139), and the influence of rotating plates upon magnets had been found to bear a general relation to the conducting power of the substance used.

203. In the experiments of rotation (81.), the electric current is excited and discharged in the same substance, be it a good or bad conductor; but in the experiments just described the current excited in iron could not be transmitted but through the copper, and that excited in copper had to pass through iron; *i. e.*, supposing currents of dissimilar strength to be formed in the metals proportionate to

---

the same parallel relation: Faraday refers to the *geometrical* arrangement of the conductors. But electrically, the trough, wire, and galvanometer are all connected in series, not parallel, configuration.

201. *equally subject … to magneto-electric induction*: Note that by "induction" Faraday here means the magnetic influence that originates the current. At other places, though, "the induction" seems to refer to *the induced current itself.*

After exhaustive attempts to show otherwise, he finally acknowledges that conductors, even nonmetallic ones, develop (or tend to develop) equal induced currents when they are "equally subject … to magneto-electric induction" *and also* connected "in one circuit". But note his reservation in the next paragraph. He is still convinced that the evidence of the rotation experiments—which implied that the magnet has *unequal* inductive effects on materials having unequal conducting powers—is the general rule, and only some special circumstance in the "one circuit" (series) configuration has prevented its corroboration so far.

their conducting power, the stronger current had to pass through the worst conductor, and the weaker current through the best.

204. Experiments were therefore made in which different metals insulated from each other were passed between the poles of the magnet, their opposite ends being connected with the same end of the galvanometer wire, so that the currents formed and led away to the galvanometer should oppose each other; and when considerable lengths of different wires were used, feeble deflections were obtained.

---

203. *the stronger current had to pass through the worst conductor, and the weaker current through the best*: Faraday realizes that such is a consequence of the conductors being arranged in series. Even if currents *do* tend to form in their own conductors "proportionate to [the] conducting power" of their respective materials, obstruction by the other conductor would obscure the original proportionality. He thus rescues the proportionality in theory, by invoking an image of *obstruction* as the contrary of conducting power. But in practice, he will undertake to *remove* the external obstruction, by replacing the series configuration with a parallel (or rather, *anti*-parallel) design, described in his next paragraph.

204. The two wires formerly connected in series are now connected in *anti-parallel* arrangement, that is, "their opposite ends [are] connected with the same end of the galvanometer wire." In the upper diagram the wire ends are labeled with their corresponding galvanometer terminals. The lower sketch shows the configuration of connections. Suppose currents form unequally in the two wires, $i+u$ the larger and i the smaller; then the galvanometer will indicate u, the excess of one over the other—assuming, that is, that the currents divide at the junctions in the way indicated. But *that* assumption

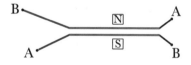

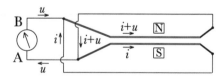

derives from the fluid analogy—see the introduction to the present Series. I think that is why Faraday will proceed to set aside this arrangement also, to employ instead a new dual coil galvanometer, described in his next paragraph. It seems that he does not trust *any* experimental design whose validity depends on the "fluid-flow" image.

Though Faraday has saved the proportionality between current strength and conducting power (paragraph 203) by acknowledging a new element of *obstruction* in the image of a conductor, his reconfigured circuits make no attempt to confirm that image; indeed they sidestep the matter altogether. Faraday, I think, is more interested in exhibiting the *effect* than in saving a verbal paradigm. But he will make use of the "obstruction" image again in paragraph 213.

205. To obtain perfectly satisfactory results a new galvanometer was constructed, consisting of two independent coils, each containing eighteen feet of silked copper wire. These coils were exactly alike in shape and number of turns, and were fixed side by side with a small interval between them, in which a double needle could he hung by a fibre of silk exactly as in the former instrument (87.). The coils may be distinguished by the letters KL, and when electrical currents were sent through them in the same direction, acted upon the needle with the sum of their powers; when in opposite directions, with the difference of their powers.

206. The compound helix (199. 8.) was now connected, the ends A and B of the iron with A and B ends of galvanometer coil K, and the ends A and B of the copper with B and A ends of galvanometer coil L, so that the currents excited in the two helices should pass in opposite directions through the coils K and L. On introducing a small cylinder magnet within the helices, the galvanometer needle was powerfully deflected. On disuniting the iron helix, the magnet caused with the copper helix alone still stronger deflection in the same direction. On reuniting the iron helix, and unconnecting the copper helix, the magnet caused a moderate deflection in the contrary direction. Thus it was evident that the electric current induced by a magnet in a copper wire was far more powerful than the current induced by the same magnet in an equal iron wire.

---

205. *a new galvanometer*: This new *differential* galvanometer has double windings, so arranged that appropriately-directed currents in the respective windings have opposing effects on the needle, which nullify one another completely if the currents are equal. Thus the instrument will faithfully disclose any inequality between the currents. It has the important theoretical advantage of keeping the two currents separate—thus relieving him of the need to make any assumptions about how they may or may not affect one another. As we saw, the discarded experimental designs of paragraphs 195 and 204 appeared to beg this question of mutual interaction of currents; so their validity could not be assured.

206. *the electric current induced by a magnet in a copper wire was far more powerful than the current induced by the same magnet in an equal iron wire*: The copper and iron windings of the compound helix develop far stronger currents than the short lengths of twisted wire previously illustrated in Figure 32. When the currents are sent in opposite senses, singly and together, to the respective galvanometer windings, the results show that the copper helix produced greater current than the iron one, under induction by the identical agent. This however cannot be viewed as a comparison of copper and iron *per se*, since the windings differ in length as well as material (see paragraph 8.); it is only a test of the *differential principle*—which is amply confirmed. He will fabricate a set of identical test windings in paragraph 208.

207. To prevent any error that might arise from the greater influence, from vicinity or other circumstances, of one coil on the needle beyond that of the other, the iron and copper terminations were changed relative to the galvanometer coils KL, so that the one which before carried the current from the copper now conveyed that from the iron, and vice versâ. But the same striking superiority of the copper was manifested as before. This precaution was taken in the rest of the experiments with other metals to be described.

208. I then had wires of iron, zinc, copper, tin, and lead, drawn to the same diameter (very nearly one twentieth of an inch), and I compared exactly equal lengths, namely sixteen feet, of each in pairs in the following manner: The ends of the copper wire were connected with the ends A and B of galvanometer coil K, and the ends of the zinc with the terminations A and B of the galvanometer coil L. The middle part of each wire was then coiled six times round a cylinder of soft iron covered with paper, long enough to connect the poles of Daniell's horse-shoe magnet (56.) (fig. 33.), so that similar helices of copper and zinc, each of six turns, surrounded the bar at two places equidistant from each other and from the poles of the magnet; but these helices were purposely arranged so as to be in contrary directions, and therefore send contrary currents through the galvanometer coils K and L.

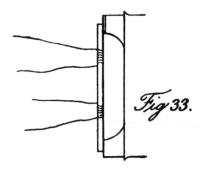

Fig 33.

209. On making and breaking contact between the soft iron bar and the poles of the magnet, the galvanometer was strongly affected; on detaching the zinc it was still more strongly affected in the same direction. On taking all the precautions before alluded to (207.), with

---

207. As a precaution, each measurement is made twice, exchanging the galvanometer coils between measurements. This would disclose any systematic discrepancy between the two galvanometer windings in their effects on the needle.

209. With copper and zinc windings connected to their respective galvanometer coils, there are strong deflections. This indicates that the copper and zinc windings have developed unequal currents, since deflection of a "differential" galvanometer indicates *difference* between two currents. Moreover, since the deflections produced using the copper winding *alone* are in the same directions, and stronger, it is clear that greater current was developed in the copper than in the zinc winding.

others, it was abundantly proved that the current induced by the magnet in copper was far more powerful than in zinc.

210. The copper was then compared in a similar manner with tin, lead, and iron, and surpassed them all, even more than it did zinc. The zinc was then compared experimentally with the tin, lead, and iron, and found to produce a more powerful current than any of them. Iron in the same manner proved superior to tin and lead. Tin came next, and lead the last.

211. Thus the order of these metals is copper, zinc, iron, tin, and lead. It is exactly their order with respect to conducting power for electricity, and, with the exception of iron, is the order presented by the magneto-rotation experiments of Messrs. Babbage, Herschel, Harris, &c. The iron has additional power in the latter kind of experiments, because of its ordinary magnetic relations, and its place relative to magneto-electric action of the kind now under investigation cannot be ascertained by such trials. In the manner above described it may be correctly ascertained.[2]

212. It must still be observed that in these experiments the whole effect between different metals is not obtained; for of the thirty-four feet of wire included in each circuit, eighteen feet are copper in both, being the wire of the galvanometer coils; and as the whole circuit is concerned in the resulting force of the current, this circumstance must tend to diminish the difference which would appear between the metals

---

[2] Mr. Christie, who being appointed reporter upon this paper, had it in his hands before it was complete, felt the difficulty (202.); and to satisfy his mind, made experiments upon iron and copper with the large magnet (44.), and came to the same conclusions as I have arrived at. The two sets of experiments were perfectly independent of each other, neither of us being aware of the other's proceedings.

---

211. *the order of these metals*: The metals had long been ranked in order of their electrical conducting power (see the introduction). When by means of the present experiment they are ranked according to *the current they develop under identical induction*, the order turns out to be the same! As mentioned already (paragraph 183, *comment*), Babbage, Herschel, and Harris's variations of Arago's rotation experiments gave a similar ranking—except for iron. Now Faraday sees that the iron disc in their experiments would have become magnetized, thus increasing the induction; and that for this reason iron would have taken a deceptively exalted position in their ranking of materials. (Of course Faraday's iron wire must have become magnetized too, but the magnetic influence of six turns of thin iron wire would have represented an inconsiderable addition to that of the massive iron core. Babbage and his colleagues had no comparable moderating mass of iron in their experiments.)

if the circuits were of the same substances throughout. In the present case the difference obtained is probably not more than a half of that which would be given if the whole of each circuit were of one metal.

213. These results tend to prove that the currents produced by magneto-electric induction in bodies [are] proportional to their conducting power. That they are *exactly* proportional to and altogether dependent upon the conducting power, is, I think, proved by the perfect neutrality displayed when two metals or other substances, as acid, water, &c. &c. (201. 186.), are opposed to each other in their action. The feeble current which tends to be produced in the worse conductor, has its transmission favoured in the better conductor, and the stronger current which tends to form in the latter has its intensity diminished by the obstruction of the former; and the forces of generation and obstruction are so perfectly balanced as to neutralize each other exactly. Now as the obstruction is inversely as the conducting power, the tendency to generate a current must be directly as that power to produce this perfect equilibrium.

---

212. *the difference obtained is probably not more than half of that which would be given…*: Faraday here suggests a rough correction for the copper's presence. Why "no more than half"? Perhaps because copper, the best conductor, makes up a little more than half ($18/34$) of each circuit; for the "obstruction" to current flow presented by a wire of constant cross-section would be proportional to that wire's *length* (see the introduction).

213. *the currents … are exactly proportional to and altogether dependent upon the conducting power*: Faraday showed in paragraphs 208–211 that "the currents produced by magneto-electric induction" display the same *order* as the conducting powers of the bodies in which they are produced. Now he maintains that the null results of paragraph 201 show that the induced currents are developed *in strict proportion* to the conducting powers. His prose argument is breathtakingly concise; it may be helpful to consider its elements individually:

the obstruction is inversely as the conducting power: In the circuit of paragraph 201 diagrammed here, let conductors 1 and 2 have conducting powers S_1 and S_2. Then they will present *obstructions* as R_1 and R_2, respectively, where

$$S_1 : S_2 :: R_2 : R_1 . \qquad (1)$$

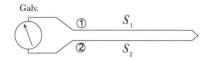

the forces of generation: This is a new phrase but Faraday clarifies it in the next sentence by substituting "the tendency to generate a current." Thus let conductors 1 and 2 experience different *generative tendencies* under magnetic induction, represented in this diagram as G_1 and G_2, respectively. If the current arising from G_1 in conductor 1 could discharge in that same

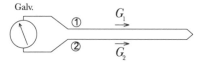

214. The cause of the equality of action under the various circumstances described, where great extent of wire (183.) or wire and water (184.) were connected together, which yet produced such different effects upon the magnet, is now evident and simple.

215. The effects of a rotating substance upon a needle or magnet ought, where ordinary magnetism has no influence, to be directly as the conducting power of the substance; and I venture now to predict that such will be found to be the case; and that in all those instances where non-conductors have been supposed to exhibit this peculiar influence, the motion has been due to some interfering cause of an ordinary kind; as mechanical communication of motion through the parts of the apparatus, or otherwise (as in the case Mr. Harris has pointed out[3]); or else to ordinary magnetic attractions. To distinguish the effects of the latter from those of the induced electric currents, I have been able to devise a most perfect test, which shall be almost immediately described (243.).

[3] Philosophical Transactions, 1831, p. 68.

---

conductor (as is the case in the rotation experiments), it would be simply proportional to G_1. But in the present circuit, the current induced in each wire has to discharge through the *other* wire, suffering diminution commensurate with the obstruction presented by that wire. Thus the current arising from G_1 will here be directly as G_1 but inversely as R_2; that is, $i_1 = G_1/R_2$. Similarly, the current arising from G_2 will be $i_2 = G_2/R_1$.

  so perfectly balanced as to neutralize each other exactly: The two currents cancel each other and must therefore be equal. Thus we have $G_1/R_2 = G_2/R_1$, or $G_1 : G_2 :: R_2 : R_1$ and therefore, from proportion (1) above,
$$G_1 : G_2 :: S_1 : S_2.$$
That is, "the tendency to generate a current must be directly as [the conducting] power..." Q.E.D.

  Note that the expression $i_1 = G_1/R_2$ assumes that current (i_1) is diminished by obstruction in its discharge path (R_2), but not by obstruction in its path of generation (R_1)! Faraday is maintaining, in effect, that "obstruction" is a concept that applies *only* to the act of discharge. That idea would constitute rank inconsistency on the fluid-flow image—since according to that image the electric fluid must "flow" through a battery or other active site just as it does through the passive conductors that complete the circuit; *obstruction* then ought to have its proper effect, wheresoever found. For Faraday, though, the idea may be part of an evolving vision in which *conduction, discharge,* and *obstruction* are all seen as basically identical—and specifically different from *generation*. In the Twelfth Series, Faraday will support and extend that vision. But in the 28th Series he will develop principles of induction that reinterpret the present exercise completely.

214. It is now "evident and simple" that the long-wire experiments of paragraphs 183 and 184 were simply large-scale versions of the ones described in paragraphs 194–200 and must give the same null results.

216. There is every reason to believe that the magnet or magnetic needle will become an excellent measurer of the conducting power of substances rotated near it; for I have found by careful experiment, that when a constant current of electricity was sent successively through a series of wires of copper, platina, zinc, silver, lead, and tin, drawn to the same diameter; the deflection of the needle was exactly equal by them all. It must be remembered that when bodies are rotated in a horizontal plane, the magnetism of the earth is active upon them. As the effect is general to the whole of the plate, it may not interfere in these cases; but in some experiments and calculations may be of important consequence.

---

217. Another point which I endeavoured to ascertain, was, whether it was essential or not that the moving part of the wire should, in cutting the magnetic curves, pass into positions of greater or lesser magnetic force;

---

216. As the argument in this paragraph is unusually telescopic, I will try to restate it step by step.

a constant current of electricity was sent successively through a series of wires: This is somewhat ambiguous; did he send current through the wires one at a time, or through all of them at once? But an entry in Faraday's laboratory *Diary*, describing an essentially identical experiment, clarifies the procedure: a single current is sent through the several wires connected *in series*, "so that [one and the same] current passed through all" (March 1, 1832).

the deflection of the needle was exactly equal by them all: A compass needle was placed as in the manner of Oersted on each conductor in turn, and showed equal deflections. Unstated inference: the needle deflection consistently indicates the strength of a current, no matter in what material that current appears. Therefore…

the … magnetic needle will become an excellent measurer of the conducting power of substances rotated near it: The magnetic needle that is suspended above an Arago-style rotating disc responds to currents induced in the disc, just as a compass needle responds to current sent through a wire. Therefore in Arago rotation experiments, as in Oersted experiments, the needle's response is independent of the conducting power of the current-carrying material. The needle's reaction in rotation experiments is therefore an accurate indicator of the induced currents; and since the current induced in a rotating disc is now known to be as the disc's conducting power, the needle will be a faithful indicator of that conducting power.

the effect is general to the whole of the plate: Since the earth's magnetic curves do not induce *currents* in an isolated rotating plate, there being no path of discharge, terrestrial magnetism is probably not a source of error in such conductivity measurements.

217. *whether it was essential or not that the moving part of the wire should … pass into positions of greater or lesser magnetic force…*: The image of "cutting" magnetic

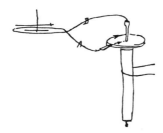

[A sketch from Faraday's Diary illustrating the experiment described in paragraph 218; it was not published with the Experimental Researches.]

or whether, always intersecting curves of equal magnetic intensity, the mere motion was sufficient for the production of the current. That the latter is true, has been proved already in several of the experiments on terrestrial magneto-electric induction. Thus the electricity evolved from the copper plate (149.), the currents produced in the rotating globe (161, &c.), and those passing through the moving wire (171.), are all produced under circumstances in which the magnetic force could not but be the same during the whole experiments.

218. To prove the point with an ordinary magnet, a copper disc was cemented upon the end of a cylinder magnet, with paper intervening; the magnet and disc were rotated together, and collectors (attached to the galvanometer) brought in contact with the circumference and the central part of the copper plate. The galvanometer needle moved as in former cases, and the direction of motion was the same as that which

---

curves is revealed, by the terrestrial magneto-electric induction experiments, as an image of motion simply, not necessarily motion along paths of *changing magnetic intensity*. Since the devices and paths of motion in those experiments were so tiny, compared to the earth, the magnetic force must have been virtually the same in all parts of the apparatus—yet currents were nevertheless evolved. Therefore change of magnetic intensity is not required.

But why would one even suppose that such change might be required? Probably because it is very hard to understand how "mere motion" could *produce* anything; whereas if the magnetic intensity were to vary along the moving wire's path, induction might turn out to be a kind of *conversion* of magnetic force into electrical force. However that possibility now appears to be ruled out. Experiments described in the following paragraphs explore the "cutting" image in the case of an ordinary magnet, and they too confirm it as expressing mere *relative motion* between magnetic curves and conductor. Later on, however, the magnetic curves will acquire more substantiality (paragraph 238). Will this also give a correspondingly physical cast to the image of cutting?

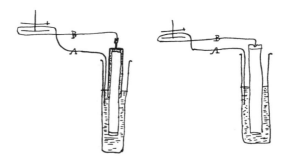

[Sketches from Faraday's Diary illustrating the experiments described in paragraphs 219 (left) and 220 (right); they were not published with the Experimental Researches.]

would have resulted, if the copper only had revolved, and the magnet been fixed. Neither was there any apparent difference in the quantity of deflection. Hence, rotating the magnet causes no difference in the results; for a rotatory and a stationary magnet produce the same effect upon the moving copper.

219. A copper cylinder, closed at one extremity, was then put over the magnet, one half of which it inclosed like a cap; it was firmly fixed, and prevented from touching the magnet anywhere by interposed paper. The arrangement was then floated in a narrow jar of mercury, so that the lower edge of the copper cylinder touched the fluid metal; one wire of the galvanometer dipped into this mercury, and the other into a little cavity in the centre of the end of the copper cap. Upon rotating the magnet and its attached cylinder, abundance of electricity passed through the galvanometer, and in the same direction as if the cylinder had rotated only, the magnet being still. The results therefore were the same as those with the disc (218.).

---

218. *a rotatory and a stationary magnet produce the same effect upon the moving copper*: This cautious statement will grow a little stronger in paragraph 220, where he will point out a "singular independence" between the magnet and its magnetic curves. To state it baldly—which he will not do until the Twenty-eighth Series—Faraday suspects that the magnetic curves remain *stationary* even when the magnet is rotated!

219. *one wire of the galvanometer dipped into this mercury, and the other into a little cavity in the centre of the end of the copper cap*: The mercury bath contacts the lower edge of the copper "cap," while a drop of mercury occupies the upper center of the cap. The bath and drop are connected to the galvanometer and therefore function as *collectors* for any current that may be induced in the cap.

220. That the metal of the magnet itself might be substituted for the moving cylinder, disc, or wire, seemed an inevitable consequence, and yet one which would exhibit the effects of magneto-electric induction in a striking form. A cylinder magnet had therefore a little hole made in the centre of each end to receive a drop of mercury, and was then floated pole upwards in the same metal contained in a narrow jar. One wire from the galvanometer dipped into the mercury of the jar, and the other into the drop contained in the hole at the upper extremity of the axis. The magnet was then revolved by a piece of string passed round it, and the galvanometer-needle immediately indicated a powerful current of electricity. On reversing the order of rotation, the electrical current was reversed. The direction of the electricity was the same as if the copper cylinder (219.) or a copper wire had revolved round the fixed magnet in the same direction as that which the magnet itself had followed. Thus a *singular independence* of the magnetism and the bar in which it resides is rendered evident.

---

220. *the metal of the magnet itself might be substituted for the moving cylinder, disc, or wire*: That portion of the magnet which floats above the mercury—more strictly, its outer surface—now corresponds to the copper disc of paragraph 218 and the "cap" of paragraph 219. The mercury bath and the axial wire serve as collectors, just as in paragraph 219. The galvanometer will therefore indicate any current induced in the floating portion of the magnet.

The magnet was then revolved ... and the galvanometer-needle immediately indicated a powerful current of electricity: As Faraday had anticipated, the exposed surface of the magnet itself evidently cuts magnetic curves and develops induced currents when revolved.

But what of the submerged portion of the magnet? Must not its surface too cut magnetic curves? Induced currents ought therefore to arise in that part of the magnet as well. Faraday does not mention such currents, probably because the mercury bath provides the equivalent of collectors and a discharge path for them. Thus any currents that may be induced in the submerged part of the magnet discharge through the mercury. They never reach the galvanometer and do not affect the observations.

a singular independence of the magnetism and the bar ... is rendered evident: Since a rotating magnet cuts its own magnetic curves, it is "evident" that those curves do not rotate with the magnet to which they belong. The question that was asked in vain for the *earth* (paragraph 181, *comment*) is here answered definitively for a small magnet; and it suggests a more fundamental question: What is the relation between a power and the body to which it belongs? That is a question we beg every time we ascribe power to matter: *gravity* to a body, *electricity* to a hypothetical fluid—even *vital powers* to living beings. The question is close to the surface when he investigates chemical powers in the Seventh Series. It stands behind the scene as he studies the electric eel in the Fifteenth Series. In the several concluding Series of *Experimental Researches*, he will make a heroic attempt to determine the question for magnetism, at least.

221. In the above experiment the mercury reached about half way up the magnet; but when its quantity was increased until within one eighth of an inch of the top, or diminished until equally near the bottom, still the same effects and the *same direction* of electrical current was obtained. But in those extreme proportions the effects did not appear so strong as when the surface of the mercury was about the middle, or between that and an inch from each end. The magnet was eight inches and a half long, and three quarters of an inch in diameter.

222. Upon inversion of the magnet, and causing rotation in the same direction, *i. e.* always screw or always unscrew, then a contrary current of electricity was produced. But when the motion of the magnet was continued in a direction constant in relation to its *own axis,*

---

221. Since the induced current varies when different extents of cylindrical surface are exposed, the lateral surfaces of the rotating magnet, as well as the ends, are evidently cutting magnetic curves.

But in those extreme proportions the effects did not appear so strong as when the surface of the mercury was about the middle...: If the lateral surfaces of the magnet cut magnetic curves, it makes sense that the induced current should decrease when less than half of the magnet is exposed; for with a smaller area between the collectors, fewer curves will be cut. But why should the induced current decrease again when *more* than half the magnet is exposed and more curves are cut? Clearly the curves cut by the "south" half of the magnet must have direction opposite to the curves that are cut by the "north" half, so that their inclusion reduces rather than augments the galvanometer current.

222. *inversion of the magnet, and causing rotation in the same direction*: "same," that is, with respect to the investigator. Suppose the magnet is rotating clockwise ("screw"); it is then stopped, turned top for bottom, and rotated once more in the clockwise direction—and similarly for counterclockwise ("unscrew") rotations. Thus nothing is altered except that the polarity of the magnet is reversed; and Faraday reports that the induced current also reverses.

a direction constant in relation to its own axis: Imagine the spinning magnet inverted without stopping its rotation. The resulting direction of rotation will be *constant* "in relation to its own axis"; but it will be *reversed* in relation to the observer. Thus both the polarity of the magnet and its direction of rotation are reversed; the two reversals evidently nullify one another and the direction of the induced current remains unchanged. As Faraday puts it, "electricity of the same kind was collected at both poles." Since he has now surveyed each end of the magnet separately, Faraday infers that if the spinning magnet could be supported in air (with suitable collectors at the equator and each pole), induced currents having the same direction would arise simultaneously between each pole and the equator. The currents are indicated in the accompanying sketch.

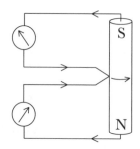

then electricity of the same kind was collected at both poles, and the opposite electricity at the equator, or in its neighbourhood, or in the parts corresponding to it. If the magnet be held parallel to the axis of the earth, with its unmarked pole directed to the pole star, and then rotated so that the parts at its southern side pass from west to east in conformity to the motion of the earth; then positive electricity may be collected at the extremities of the magnet, and negative electricity at or about the middle of its mass.

223. When the galvanometer was very sensible, the mere spinning of the magnet in the air, whilst one of the galvanometer wires touched the extremity, and the other the equatorial parts, was sufficient to evolve a current of electricity and deflect the needle.

*   *   *

231. The law under which the induced electric current excited in bodies moving relatively to magnets, is made dependent on the intersection of the magnetic curves by the metal (114.) being thus rendered more precise and definite (217. 220. 224.), seem now even to apply to the cause in the first section of the former paper (26.); and by rendering a perfect reason for the effects produced, take away any for supposing that peculiar condition, which I ventured to call the electro-tonic state (60.).

232. When an electrical current is passed through a wire, that wire is surrounded at every part by magnetic curves, diminishing in intensity according to their distance from the wire, and which in idea may be likened to rings situated in planes perpendicular to the wire or rather to the electric current within it. These curves, although different in form, are perfectly analogous to those existing between two contrary

---

with its unmarked pole directed to the pole star: Held as described, the rotated magnet becomes a magnetic model of the earth, similar and similarly situated (remember that the pole beneath the earth's arctic region must be homologous with an "unmarked" pole). The induced currents collected from the magnet fully correspond to those which he had contemplated for the earth in paragraph 182.

223. *sensible:* Today we would say *sensitive.*

231. *by rendering a perfect reason for the effects produced, take away any for supposing that peculiar condition, which I ventured to call the electro-tonic state:* Although he concedes here that there is no longer any need to suppose an "electrotonic state," Faraday still harbors a reservation. He will voice it at paragraph 242 below.

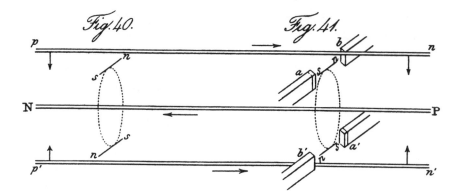

magnetic poles opposed to each other; and when a second wire, parallel to that which carries the current, is made to approach the latter (18.), it passes through magnetic curves exactly of the same kind as those it would intersect when carried between opposite magnetic poles (109.) in one direction; and as it recedes from the inducing wire, it cuts the curves around it in the same manner that it would do those between the same poles if moved in the other direction.

233. If the wire N P (fig. 40.) have an electric current passed through it in the direction from P to N, then the dotted ring may represent a magnetic curve round it, and it is in such a direction that if small magnetic needles be placed as tangents to it, they will become arranged as in the figure, n and s indicating north and south ends (44. *note*).

234. But if the current of electricity were made to cease for a while, and magnetic poles were used instead to give direction to the needles, and make them take the same position as when under the influence of the current, then they must be arranged as at fig. 41; the marked and unmarked poles ab above the wire, being in opposite directions to those $a'b'$ below. In such a position therefore the magnetic curves between the poles ab and $a'b'$ have the same general direction with

---

233. *n and s indicating north and south ends*: The labels n and s represent *north-seeking* and *south-seeking* poles, equivalent to *marked* and *unmarked* poles, respectively.

When consulting Faraday's figures 40 and 41 in connection with this and subsequent paragraphs, think of the dotted circles, the bar magnets, and the little lines labeled sn as directed perpendicularly into the paper. They lie within what Faraday described as "planes perpendicular to the wire" in paragraph 232.

234. *But if the current of electricity were made to cease for a while, and magnetic poles were used instead to give direction to the needles...*: By the device of imaginary magnetic poles, Faraday can visualize determinate *arcs* of the magnetic circles about a current as equivalent to the curves running between the poles of an ordinary magnet.

the corresponding parts of the ring magnetic curve surrounding the wire N P carrying an electric current.

235. If the second wire pn (fig. 40.) be now brought towards the principal wire, carrying a current, it will cut an infinity of magnetic curves, similar in direction to that figured, and consequently similar in direction to those between the poles ab of the magnets (fig. 41.), and it will intersect these current curves in the same manner as it would the magnet curves, if it passed from above between the poles downwards. Now, such an intersection would, with the magnets, induce an electric current in the wire from p to n (114.); and therefore as the curves are alike in arrangement, the same effect ought to result from the inter-section of the magnetic curves dependent on the current in the wire N P; and such is the case, for on approximation the induced current is in the opposite direction to the principal current (19.).

236. If the wire $p'n'$ be carried up from below, it will pass in the opposite direction between the magnetic poles; but then also the magnetic poles themselves are reversed (fig. 41.), and the induced current is therefore (114.) still in the same direction as before. It is also, for equally sufficient and evident reasons, in the same direction, if produced by the influence of the curves dependent upon the wire.

237. When the second wire is retained at rest in the vicinity of the principal wire, no current is induced through it, for it is intersecting no magnetic curves. When it is removed from the principal wire, it intersects the curves in the opposite direction to what it did before (235.); and a current in the opposite direction is induced, which therefore corresponds with the direction of the principal current (19.). The same effect would take place if by inverting the direction of motion of the wire in passing between either set of poles (fig. 41.), it

---

235. *it will intersect these current curves in the same manner as it would the magnet curves…*: The ordinary magnet becomes an interpretive device, by whose means Faraday can extend the "cutting" image from ordinary magnetic curves to the endless circles that surround a straight current-carrying wire. The beautiful imaginative reconstruction thereby shows why a wire which approaches a parallel current-carrying wire will, during its approach, experience a current in the direction *opposite* to the current it approaches.

237. *intersecting no magnetic curves*: that is, *moving through* no magnetic curves. Although the resting wire may be said to "intersect" curves in a purely geometrical sense, the "cutting" image implies *relative motion* (paragraph 217); thus a wire which rests in the vicinity of a constant current experiences *no* induced current. And similarly it is clear why a wire which *recedes* from a parallel current-carrying wire will, during its recession, experience a current in the *same* direction as the current from which it recedes.

were made to intersect the curves there existing in the opposite direction to what it did before.

238. In the first experiments (10. 13.), the inducing wire and that under induction were arranged at a fixed distance from each other, and then an electric current sent through the former. In such cases the magnetic curves themselves must be considered as moving (if I may use the expression) across the wire under induction, from the moment at which they begin to be developed until the magnetic force of the current is at its utmost; expanding as it were from the wire outwards, and consequently being in the same relation to the fixed wire under induction, as if *it* had moved in the opposite direction across them, or towards the wire carrying the current. Hence the first current induced in such cases was in the contrary direction to the principal current (17. 235.). On breaking the battery contact, the magnetic curves (which are mere expressions for arranged magnetic forces) may be conceived as contracting upon and returning towards the failing electrical current, and therefore move in the opposite direction across the wire, and cause an opposite induced current to the first.

239. When, in experiments with ordinary magnets, the latter, in place of being moved past the wires, were actually made near them (27. 36.), then a similar progressive development of the magnetic curves may be considered as having taken place, producing the effects which

---

238. *the magnetic curves themselves must be considered as moving … across the wire*: He observed in paragraph 10 that when the primary current *commences*, an adjacent stationary wire experiences induced current in the opposite direction—as though the wire under induction were moving *towards* the magnetic curves that encircle the primary current. But a stationary wire cannot be moving towards the magnetic curves; therefore the curves must be moving towards the wire! Faraday thus associates current growth with expansion of the encircling magnetic curves outwards, and current decay with contraction of the magnetic curves inwards.

Faraday describes the magnetic curves as "mere expressions for arranged magnetic forces". But in conceiving them as capable of *motion*, is he not granting the curves more substantiality than when they first appeared as simple traces of the pointing behavior of compass needles or iron filings?

239. *magnets … actually made*: When an iron rod is magnetized *in place* (as in Faraday's Figure 2 on page 41), we similarly infer that it propagates circular magnetic curves into the space surrounding itself; and that the curves contract again when it loses its magnetism. We would probably like to ask whether curves are *generated and destroyed* during these expansions and contractions—but note that such a question would portray the curves as having a physical existence fully and literally. Faraday does not seem ready to go so far, as yet.

would have occurred by motion of the wires in one direction; the destruction of the magnetic power corresponds to the motion of the wire in the opposite direction.

240. If, instead of intersecting the magnetic curves of a straight wire carrying a current, by approximating or removing a second wire (235.), a revolving plate be used, being placed for that purpose near the wire, and, as it were, amongst the magnetic curves, then it ought to have continuous electric currents induced within it; and if a line joining the wire with the centre of the plate were perpendicular to both, then the induced current ought to be, according to the law (114.), directly across the plate, from one side to the other, and at right angles to the direction of the inducing current.

241. A single metallic wire one twentieth of an inch in diameter had an electric current passed through it, and a small copper disc one inch and a half in diameter revolved near to and under, but not in actual con-

tact with it (fig. 39.). Collectors were then applied at the oppo-site edges of the disc, and wires from them connected with the galvanometer. As the disc revolved in one direction, the needle was deflected on one

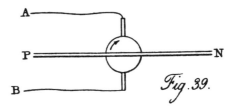

side; and when the direction of revolution was reversed, the needle was inclined on the other side, in accordance with the results anticipated.

242. Thus the reasons which induce me to suppose a particular state in the wire (60.) have disappeared; and though it still seems to me unlikely that a wire at rest in the neighbourhood of another carrying a powerful electric current is entirely indifferent to it, yet I am not aware of any distinct facts which authorize the conclusion that it is in a particular state.

---

241. As an exercise, try to predict the direction of the current this device should produce. Compare the wire PN in Figure 39 with the similarly-labeled wire in Figures 40 and 41 (page 122) to find the direction of the magnetic curves; then use the "law" (paragraph 114) or the magneto-electric right-hand rule to deduce the current direction in the upper and lower semicircles. For the direction of rotation shown, the induced current should run through the disk from A to B.

242. *the reasons which induce me to suppose a particular state in the wire have disappeared*: The imagery of movable magnetic curves capable of being cut and so stimulating electric currents, dispenses with any need for supposing the "electrotonic state." But the same imagery appears to imply that a wire is indeed "entirely indifferent" to magnetic curves, so long as it is not actively cutting them. Why does he regard such indifference as improbable?

243. In considering the nature of the cause assigned in these papers to account for the mutual influence of magnets and moving metals (120.), and comparing it with that heretofore admitted, namely, the induction of a feeble magnetism like that produced in iron, it occurred to me that a most decisive experimental test of the two views could be applied (215.).

244. No other known power has like direction with that exerted between an electric current and a magnetic pole; it is tangential, while all other forces, acting at a distance, are direct. Hence, if a magnetic pole on one side of a revolving plate follow its course by reason of its obedience to the tangential force exerted upon it by the very current of electricity which it has itself caused, a similar pole on the opposite side of the plate should immediately set it free from this force; for the currents which tend to be formed by the action of the two poles are in opposite directions; or rather no current tends to be formed, [f]or no magnetic curves are intersected (114.); and therefore the magnet should remain at rest. On the contrary, if the action of a north magnetic pole were to produce a southness in the nearest part of the copper plate, and a diffuse northness elsewhere (82.), as is really the case with iron; then the use of another north pole on the opposite side of the same part of the plate should double the effect instead of destroying it, and double the tendency of the first magnet to move with the plate.

---

243. *induction of a … magnetism like that produced in iron*: Faraday refers to the magnetic state raised up in iron when brought under the influence of an existing magnet. Note that this represents the same magnetic usage of "induction" he had previously avoided until paragraph 145.

comparing [Faraday's explanation] with that heretofore admitted…: Faraday understands the Arago rotations as manifestations of currents, magneto-electrically induced. Babbage and his colleagues had supposed the rotations to be caused by a magnetic condition *induced* according to the sense just described—albeit in materials that are normally nonmagnetic. We are now in a position to distinguish experimentally between the two accounts.

244. A rotating disk placed between north and south magnetic poles will intersect magnetic curves and develop induced currents. But magnetic curves do not run between like poles; so a disk placed between, say, two north poles should intersect no curves and develop no induced current.

On the other hand, if the rotating disk experiences magnetization like that of iron, paired external north poles will elicit a strong *south* pole in the disk; while north and south poles together will produce opposite poles, canceling one another within the disk. Thus if the dragging phenomenon is due to currents induced in the disk, it should be enhanced between opposite poles and diminished between like poles; while if due to a magnetization like that of iron it should be enhanced between like poles and diminished between opposite poles.

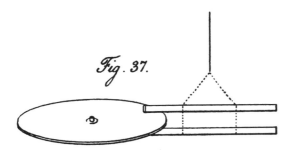

Fig. 37.

245. A thick copper plate (85.) was therefore fixed on a vertical axis, a bar magnet was suspended by a plaited silk cord, so that its marked pole hung over the edge of the plate, and a sheet of paper being interposed, the plate was revolved; immediately the magnetic pole obeyed its motion and passed off in the same direction. A second magnet of equal size and strength was then attached to the first, so that its marked pole should hang beneath the edge of the copper plate in a corresponding position to that above, and at an equal distance (fig. 37.). Then a paper sheath or screen being interposed as before, and the plate revolved, the poles were found entirely indifferent to its motion, although either of them alone would have followed the course of rotation.

246. On turning one magnet round, so that opposite poles were on each side of the plate, then the mutual action of the poles and the moving metal was a maximum.

247. On suspending one magnet so that its axis was level with the plate, and either pole opposite its edge, the revolution of the plate caused no motion of the magnet. The electrical currents dependent upon induction would now tend to be produced in a vertical direction across the thickness of the plate, but could not be so discharged, or at least only to so slight a degree as to leave all effects insensible; but

---

245–246. The results are consistent with those expected from induced currents, and they contradict what would be expected if the rotating disk sustained magnetization comparable to that of iron.

247. *suspending one magnet so that its axis was level with the plate*: One pole of the suspended magnet faces the circumferential edge of the plate, as sketched here. Only *horizontal* magnetic curves are now intersected by the moving plate; therefore currents will tend to be induced *vertically*—or, as Faraday puts it, "across the thickness of the plate." But since there is no effective discharge path for such currents, they cannot convey rotation to the magnet.

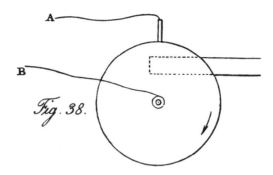

Fig. 38.

ordinary magnetic induction, or that on an iron plate, would be equally if not more powerfully developed in such a position (251.).

248. Then, with regard to the production of electricity in these cases:—whenever motion was communicated by the plate to the magnets, currents existed; when it was not communicated, they ceased. A marked pole of a large bar magnet was put under the edge of the plate; collectors (86.) applied at the axis and edge of the plate as on former occasions (fig. 38.), and these connected with the galvano-meter; when the plate was revolved, abundance of electricity passed to the instrument. The unmarked pole of a similar magnet was then put over the place of the former pole, so that contrary poles were above and below; on revolving the plate, the electricity was more powerful than before. The latter magnet was then turned end for end, so that marked poles were both above and below the plate, and then, upon revolving it, scarcely any electricity was procured. By adjusting the distance of the poles so as to correspond with their relative force, they at last were brought so perfectly to neutralize each other's inductive action upon the plate, that no electricity could be obtained with the most rapid motion.

247, continued. *ordinary magnetic induction*: that is, the raising up of a magnetic condition in a body by influence of a neighboring magnet, as described in paragraph 145 above. If the single magnet rotations observed in paragraph 245 were due to ordinary magnetic induction in the moving plate, then a magnet placed edgewise should produce an equal or even stronger condition in the plate. But evidently no such magnetic condition is developed, as no rotation is observed.

248. *A marked pole of a large bar magnet was put under the edge of the plate*: Note that the magnets here are fixed in place, in contrast to the suspended magnets of paragraphs 245–247. The results confirm that those configurations for which motion is "communicated by the plate to the magnets" are identical with those magnetic arrangements which induce currents in the plate.

249. I now proceeded to compare the effect of similar and dissimilar poles upon iron and copper, adopting for the purpose Mr. Sturgeon's very useful form of Arago's experiment. This consists in a circular plate of metal supported in a vertical plane by a horizontal axis, and weighted a little at one edge or rendered excentric so as to vibrate like a pendulum. The poles of the magnets are applied near the side and edges of these plates, and then the number of vibrations, required to reduce the vibrating arc a certain constant quantity, noted. In the first description of this instrument[4] it is said that opposite poles produced the greatest retarding effect, and similar poles none; and yet within a page of the place the effect is considered as of the same kind with that produced in iron.

250. I had two such plates mounted, one of copper, one of iron. The copper plate alone gave sixty vibrations, in the average of several experiments, before the arc of vibration was reduced from one constant mark to another. On placing opposite magnetic poles near to, and on each side of, the same place, the vibrations were reduced to fifteen. On putting similar poles on each side of it, they rose to fifty; and on placing two pieces of wood of equal size with the poles equally near, they became fifty-two. So that, when similar poles were used, the magnetic effect was little or none (the obstruction being due to the confinement of the air, rather), whilst with opposite poles it was the greatest possible. When a pole was presented to the edge of the plate, no retardation occurred.

[4] Edin. Phil. Journal, 1821, p. 124.

---

249. *Mr. Sturgeon's very useful form of Arago's experiment*: Sturgeon's disk is mounted vertically and weighted to swing like a pendulum, as described. Instead of a movable magnet, a fixed magnet is held near the plate. Then the force which in Arago's wheel had *accelerated* the movable magnet, acts here to *retard*, or "damp," the oscillations of the disk. The degree of retardation can be ascertained by noting how rapidly the oscillations decay to a specified fraction of their initial amplitude.

250. *one of copper, one of iron*: It is appropriate to compare *iron* and *copper* plates, inasmuch as iron was the only substance whose dragging response had been found disproportionate to its electrical conducting power (paragraph 211).

copper plate: When dummy poles are mounted on either side, the amplitude of plate oscillation decays by a specified amount in 52 vibrations. When the dummy poles are replaced by *like* magnetic poles, the same decay requires 50 vibrations—no significant change. But when *opposite* poles are presented to the plate, the same decay requires only 15 vibrations! Thus maximum retardation for copper is obtained between opposite poles, as would be expected if the retardation is caused by induced currents. Sturgeon had himself obtained similar results (paragraph 249), but he evidently failed to see that they contradict what would be expected from ordinary magnetic induction.

251. The iron plate alone made thirty-two vibrations, whilst the arc of vibration diminished a certain quantity. On presenting a magnetic pole to the edge of the plate (247.), the vibrations were diminished to eleven; and when the pole was about half an inch from the edge, to five.

252. When the marked pole was put at the side of the iron plate at a certain distance, the number of vibrations was only five. When the marked pole of the second bar was put on the opposite side of the plate at the same distance (250.), the vibrations were reduced to two. But when the second pole was an unmarked one, yet occupying exactly the same position, the vibrations rose to twenty-two. By removing the stronger of these two opposite poles a little way from the plate, the vibrations increased to thirty-one, or nearly the original number. But on removing it altogether, they fell to between five and six.

253. Nothing can be more clear, therefore, than that with iron, and bodies admitting of ordinary magnetic induction, opposite poles on opposite sides of the edge of the plate neutralize each other's effect, whilst similar poles exalt the action; a single pole end on is also sufficient. But with copper, and substances not sensible to ordinary magnetic impressions, similar poles on opposite sides of the plate neutralize each other; opposite poles exalt the action; and a single pole at the edge or end on does nothing.

254. Nothing can more completely show the thorough independence of the effects obtained with the metals by Arago, and those due to ordinary magnetic forces; and henceforth, therefore, the application of two poles to various moving substances will, if they appear at all magnetically affected, afford a proof of the nature of that affection. If

---

251–252. *iron plate*: The retardation was augmented by the approach of one magnetic pole, and increased still further between two like poles. But retardation was minimal between *unlike* poles adjusted for cancellation. In iron, then, the dominant effect is that of ordinary magnetic induction. Doubtless magneto-electric effects are also produced in iron, it being a conductor; but evidently they are overbalanced by the ordinary magnetic effects. The two kinds of induction are completely independent of one another; and it is clear why iron occupies an anomalistic position among materials in the Arago experiments (paragraph 211).

254. *the application of two poles to various moving substances will, if they appear at all magnetically affected, afford a proof of the nature of that affection*: Here the word "proof" carries its antique sense and means a *test*, not a confirmation. As Faraday goes on to recount, the effects of like and unlike poles on a material passed between them can indicate whether that material develops induced electrical currents, or exhibits magnetization like that of iron. Nevertheless, where materials other than iron are concerned, he thinks magnetization is improbable, even when the test results seem to favor it. He will reiterate that view in the next paragraph.

opposite poles produce a greater effect than one pole, the result will be due to electric currents. If similar poles produce more effect than one, then the power is not electrical; it is not like that active in the metals and carbon when they are moving, and in most cases will probably be found to be not even magnetical, but the result of irregular causes not anticipated and consequently not guarded against.

255. The result of these investigations tends to show that there are really but very few bodies that are magnetic in the manner of iron. I have often sought for indications of this power in the common metals and other substances; and once in illustration of Arago's objection (82.), and in hopes of ascertaining the existence of currents in metals by the momentary approach of a magnet, suspended a disc of copper by a single fibre of silk in an excellent vacuum, and approximated powerful magnets on the outside of the jar, making them approach and recede in unison with a pendulum that vibrated as the disc would do: but no motion could be obtained; not merely, no indication of ordinary magnetic powers, but none of any electric current occasioned in the metal by the approximation and recession of the magnet. I therefore venture to arrange substances in three classes as regards their relation to magnets; first, those which are affected when at rest, like iron, nickel, &c., being such as possess ordinary magnetic properties; then, those which are affected when in motion, being conductors of electricity in which are produced electric currents by the inductive force of the magnet; and, lastly, those which are perfectly indifferent to the magnet, whether at rest or in motion.

---

256. Although it will require further research, and probably close investigation, both experimental and mathematical, before the exact mode of action between a magnet and metal moving relatively to each other is ascertained; yet many of the results appear

---

255. *I therefore venture to arrange substances in three classes as regards their relation to magnets*: Notice that of his three magnetic classes, only substances of the first class ("like iron, nickel, &c.") are magnetic in themselves. Those of the second class are magnetic only through the electric currents they may conduct, while those of the third class exhibit neither magnetic nor electrically conductive properties. In later researches, Faraday will classify the magnetic materials differently. He will find reason to distinguish the magnetism of *iron* from other members of the first class; and he will demonstrate the existence of still another species of magnetic material which, it would appear, does not exhibit *poles* at all! See, for example, paragraph 2820 in the 26th Series.

sufficiently clear and simple to allow of expression in a somewhat general manner: If a terminated wire move so as to cut a magnetic curve, a power is called into action which tends to urge an electric current through it; but this current cannot be brought into existence unless provision be made at the ends of the wire for its discharge and renewal.

257. If a second wire move in the same direction as the first, the same power is exerted upon it, and it is therefore unable to alter the condition of the first: for there appear to be no natural differences among substances when connected in a series, by which, when moving under the same circumstances relative to the magnet, one tends to produce a more powerful electric current in the whole circuit than another (201. 214.).

258. But if the second wire move with a different velocity, or in some other direction, then variations in the force exerted take place; and if connected at their extremities, an electric current passes through them.

256–264. In these paragraphs Faraday will summarize the conditions under which currents can be induced as the result of *motion.*

257. *If a second wire move in the same direction as the first…*: that is, in the same direction as the wire specified in the previous paragraph. Additionally, Faraday's citation of paragraph 201 implies that the second wire is adjacent and parallel to the first, and that the two wires are joined at their ends to form a loop—in Faraday's phrase, "connected in a series"—since the experiments described in the eight paragraphs leading up to 201 displayed that configuration.

the same power is exerted upon it: Since the second wire moves through the same magnetic curves as the first, the same power (of induction) will act upon both wires.

it is therefore unable to alter the condition of the first: The "condition" Faraday means is the *electrical* condition of the first wire—thus the second moving wire does not alter the current which develops in the first wire. The currents in both wires (when they are connected in series) are known to be equal, since the experiments recounted in the cited paragraphs showed that they counteract one another exactly. But Faraday's use of the adverb "therefore" is puzzling: Does he mean that the currents are equal *simply* because the two wires move through the same magnetic curves? That would represent a major departure from his approach in paragraph 213, where he concluded that the current which a wire tends to develop under induction depends on the wire's *conductivity* as much as it depends on the magnetic landscape through which the wire moves. If Faraday has really altered his thinking to such an extent, what are his reasons? And why does he not call attention to the change in his views?

259. Taking, then, a mass of metal or an endless wire, and referring to the pole of the magnet as a centre of action, (which though perhaps not strictly correct may be allowed for facility of expression, at present,) if all parts move in the same direction, and with the same angular velocity, and through magnetic curves of constant intensity, then no electric currents are produced. This point is easily observed with masses subject to the earth's magnetism, and may be proved with regard to small magnets; by rotating them, and leaving the metallic arrangements stationary, no current is produced.

260. If one part of the wire or metal cut the magnetic curves, whilst the other is stationary, then currents are produced. All the results obtained with the galvanometer are more or less of this nature, the galvanometer extremity being the fixed part. Even those with the wire, galvanometer, and earth (170.), may be considered as without any error in the result.

261. If the motion of the metal be in the same direction, but the angular velocity of its parts relative to the pole of the magnet different, then currents are produced. This is the case in Arago's experiment, and also in the wire subject to the earth's induction (172.), when it was moved from west to east.

262. If the magnet moves not directly to or from the arrangement, but laterally, then the case is similar to the last.

263. If different parts move in opposite directions across the magnetic curves, then the effect is a maximum for equal velocities.

264. All these in fact are variations of one simple condition, namely, that all parts of the mass shall not move in the same direction across the curves, and with the same angular velocity. But they are forms of expression which, being retained in the mind, I have found useful when comparing the consistency of particular phenomena with general results.

Royal Institution,
December 21, 1831.

---

259. *referring to the pole of the magnet as a centre of action, (which though perhaps not strictly correct…*: Note his reservation about regarding the magnetic *pole* as a "center of action," that is, a place endowed with determinate *power.* That reservation is consistent with, if it does not actually arise from, the growing substantiality which the magnetic curves are beginning to display—perhaps it is actually the *curves,* and not the *poles,* that "own" the magnetic power. Such indeed *must* be the case for the circular magnetic curves that surround currents, as they clearly do not originate in magnetic "poles!"

Third Series — Editor's Introduction

Identity of electricities

Is electricity one or many? From the various names that are given to it, it would appear to be many: There is *frictional electricity* (also called "ordinary" or "common" electricity—our "static" electricity); there is also *voltaic electricity, thermoelectricity,* and even *animal electricity,* once called "galvanism." For the most part, the various electricities take their names from the sources or processes by which they are obtained. This makes sense, as there are evident differences among the powers exhibited by electricity from different sources. For example, frictional (static) electricity is distinguished in displaying attraction and repulsion; voltaic currents, on the other hand, excel in showing the magnetic deflections discovered by Oersted. There are also numerous other electrical effects, including chemical action, heating, shock and other physiological effects, and distinctive forms of discharge in air, including the spark. Every variety of electricity exhibits some of these characteristic phenomena, but as Faraday enters upon the Third Series of experimental researches, *no* variety of electricity exhibits them all. Moreover several of the powers claimed for various "electricities" are subjects of doubt and controversy.

Faraday is confident that all the electricities are essentially identical; but their identity is not obvious and must be demonstrated. He will therefore undertake a general electrical survey, hoping to show that the disparities observed among electrical powers are differences only in degree and circumstance—and that, therefore, the fundamental identity of all electricities may be unequivocally affirmed.

Faraday has introduced us to several of the characteristic signs of electrical activity already: the phenomena of *attraction and repulsion* and *magnetic action* are familiar from his discussions in the first two Series. Similarly, he has worked extensively with electricity from voltaic as well as frictional sources. But his survey will now extend to other electrical sources and phenomena that he has not previously reviewed so attentively. Here are some of them:

Discharge in air

Spark. Generally, air is an effective insulator that does not permit electrical discharge. But a typical plate electrical machine easily produces *sparks* across air gaps up to several inches long. Sparks of comparable length can also be obtained from an electrified Leyden jar. If an electroscope or other indicator is applied to the jar both before and after, it will be seen that the jar loses its electrified condition when the spark passes. Spark discharge is then clearly a case of sudden electrical discharge through air.

When an electrified Leyden jar once discharges by spark, no further discharge can take place until its previous condition of electrification is restored by external means.* A voltaic battery, though, can recover from a very great and sudden discharge almost immediately after the discharge ceases. The voltaic battery, therefore, can in principle produce a *succession* of discharges which follow, one closely upon another, so as to present the appearance of a sustained spark. Nevertheless, true spark requires very high electrical *tensions*, such as few voltaic devices are built to achieve.** Most people, therefore, have probably never seen a spark produced *voltaically*. Many light flashes commonly called "sparks" are really *arcs*, which are not quite the same thing.

Arc discharge. A form of discharge that both appears and is continuous is the electric arc, obtained in this way. Let two pointed carbon rods be connected to the terminals of a powerful voltaic battery. No current will pass until the rods touch. However, once the points have been brought into contact and current commences, the current will continue to flow *even when the rods are drawn apart to a short distance*. Evidently the electric conduction is taking place through the *air*, now very intensely heated. A kind of flame called the "arc" appears between the separated points; and the carbon points themselves, heated to white-hot incandescence, emit an intense and brilliant light—"the most beautiful light that man can produce by art," Faraday will declare in paragraph 280. The illustration is from Faraday's Christmas Lectures of 1859–60.

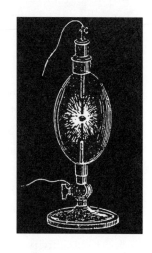

A difference between *arc* and *spark* is therefore evident. Sparks are sudden, brief discharges through significant lengths of air or other (ordinarily insulating) materials; when Faraday uses the verb "strike" he usually has this sudden discharge in mind. Arcs, on the other hand, are continuous discharges that begin with a *strike* through minimal distance or even with *contact*, and are only subsequently drawn out to significant lengths.

---

* An apparent exception is the phenomenon of "return," in which a Leyden jar, suddenly discharged, appears gradually and spontaneously to recover its electrified condition. Faraday will investigate the "return" charge in the Eleventh Series.

** See the following section on *quantity, intensity, and tension*.

Discharge through rarefied air. The discharge apparatus depicted on the opposite page had become a staple of electrical demonstrations under its perhaps inevitable sobriquet, the "electric egg." Faraday's drawing shows the stopcock in the open position so that the glass vessel is vented and the arc burns under normal atmospheric conditions. But the apparatus could also be sealed and partially exhausted for studies of electric discharge in rarefied air. Both "strike" and "arc" occur at increased distances when the air is moderately rarefied, as Faraday will recount in paragraph 274 below. As the degree of exhaustion is further increased, other, quite beautiful phenomena begin to appear: arcs become nebulous glows, which spread out in various tints and hues to occupy the whole vessel. Different colors appear in the presence of different gases; purple and rose are characteristic for discharges in air. In paragraph 306 Faraday describes some discharges produced in an "exhausted receiver" (an impromptu predecessor of the electric egg) as resembling *aurora borealis*—the "northern lights."

True *aurora* may be seen almost nightly within the Arctic circle, less frequently elsewhere in the northern hemisphere. In middle latitudes it usually takes the form of lightly-tinted streaks and streamers, which radiate across the sky from the direction of magnetic north on the horizon and may even form an arch, as depicted in the drawing above.* Comparable lights in the earth's south polar regions are called *aurora australis*. Like other investigators, Faraday continually finds suggestive similarities between electric discharges in rarefied air and the spectacular *aurora*.

---

* After S. P. Thompson

Discharge by points. An electrified body will discharge in air more readily if it displays an *edge* or a *point* than if it is everywhere smooth and blunt. Consider, for example, an electrical machine, energetically worked so as to produce sparks between spherical terminals placed a few inches apart. Let the machine be then stopped and discharged, the blunt terminals replaced with a pair of sharp metal points at the same separation, and the machine set into operation again. *No sparks now pass*, even though the machine is worked just as actively as before. Instead, one hears a distinctive hissing noise, which may be accompanied by a fine stream or brush of pale blue light originating at one of the points. Since both light and sound are indications of discharge in air, we may infer that a steady discharge is taking place between the points. This sustained discharge has evidently pre-empted sparking across a distance that previously permitted spark. It seems, therefore, that sharp surfaces facilitate discharge, while blunt surfaces can be increasingly electrified until they attain the degree required to initiate sparking.

Discharge by flame. Faraday gives an example of flame discharge below (paragraphs 271–272) when he shows that an electrical current, which had been blocked by an air gap at room temperature, will pass in detectible measure if the gap is exposed to flame. In the passage cited he seems to credit the air's increased conductivity to heat alone; but in the Twelfth Series he will express doubt about the efficacy of mere heating; rather, he there will suggest, discharge is effected by charged carbonaceous particles in the flame, which carry off electrification from the terminals of the air gap.

Animal electricity

In Mediterranean waters there dwells the curious *torpedo-fish*, which has the ability to stun its prey without having to make contact—apparently by some kind of electrical action. The animal was immortalized in Plato's dialogue *Meno*, and its distinctive powers have continued to attract the investigative attention of writers and researchers. Cavendish in 1776 pronounced the torpedo-fish electrical; and the young Faraday assisted Sir Humphrey Davy, then on a European tour, in experiments which, while confirming the electrical nature of that animal's peculiar action, failed to exhibit several of the usual signs of either common or voltaic electricity. Davy concluded that in this instance, at least, animal electricity must be specifically distinguished from other electricities.*

Other fishes are similarly capable of electrifying the water about them. The Nile boasts an electric catfish, and the South American river

---

* Faraday will cite Davy's pronouncement in paragraph 265 below in order to show that the identity of all electricities is viewed in many quarters as highly questionable.

Orinoco is home to the *gymnotus** or electric eel. Faraday has had some experience with gymnotus, and he will devote the entire Fifteenth Series of these researches to a study of that remarkable animal.

Thermoelectricity

In 1821 Seebeck discovered that a current can be produced by heating the junction between two different metals. If wires of bismuth and antimony, for example, are joined and their free ends connected to a galvanometer, the galvanometer will indicate a steady current if the bimetallic junction is maintained at either a higher or a lower temperature than the rest of the circuit. The current depends, both in magnitude and direction, on the relative temperature of the junction. If warmer than the rest of the circuit, the current flows across the junction from bismuth to antimony; if cooler, the direction is reversed. In either direction, the galvanometer deflection increases or decreases with the relative temperature difference and it is constant so long as that difference remains constant.**

Chemical action

When a current passes in ordinary conductors, it does not alter the conducting material. The typical conductor is a solid wire of metal, although liquid mercury is a very serviceable conductor. Many other liquids will conduct voltaic currents, but they generally undergo some kind of chemical change during conduction. Water *decomposes* if an electric current is led through it—a chemical action which can be most easily recognized if the wires leading into and out of the liquid are platinum, a material which does not itself participate readily in chemical reactions: bubbles of *oxygen gas* will be evolved at the positive wire, and bubbles of *hydrogen gas* at the negative, so long as current passes.

When current is led through a water solution of copper sulfate ("blue vitriol"), several reactions occur. Again using platinum wires, we find that (*i*) bubbles of *oxygen gas* are released at the positive wire; (*ii*) an ever-thickening deposit of *copper* forms on the negative wire; and (*iii*) the solution exhibits ever-increasing concentrations of *sulfuric acid*. The copper can only have come from the copper sulfate; so it is pretty clear that copper sulfate is being continuously decomposed and

---

* The animal called *gymnotus* by Faraday and his contemporaries is *Gymnotus electricus* or, especially in America, *Electrophorus*. It should not be confused with *Gymnotus carapo*, a weakly electric member of the knifefish family.

** A circuit containing two junctions, immersed respectively in boiling water and freezing water (the two calibration temperatures of the thermometer), can in principle, therefore, serve as a reliable standard of constant current.

that, in the presence of water, sulfuric acid is continuously being formed in its place.

Notice that in both examples, the products of decomposition appear *separately* at the positive and negative wires, respectively. In paragraph 309 of the present Series, Faraday will identify this separation as an *essential characteristic* of "true electro-chemical decomposition." Moreover, the chemical changes proceed so long *and only so long* as current continues to pass. Thus the chemical action of electricity can serve as an effective *current indicator*, even as the galvanometer is a current indicator based on magnetic action.

Quantity, intensity, tension

If the distinguishing characteristics exhibited by electricities from different sources are indeed only differences in degree, and not essential or "philosophical" differences, then it is important to identify the aspect or aspects in which electricity is capable of such variation. Two characteristics which had long been conventionally attributed to electricity, and with respect to which electricity could be said to be "more" or "less," were *quantity* and *intensity*. But what did these terms mean? In a note to paragraph 360 below, Faraday will make a rather strange remark:

> The term *quantity* in electricity is perhaps sufficiently definite as to sense; the term *intensity* is more difficult to define strictly. I am using the terms in their ordinary and accepted meaning.

But of course the "ordinary and accepted" meanings are *very much in question.* Or, to the extent that they appear settled, their definiteness may disguise questionable assumptions. That appears to be the case especially for "quantity of electricity," which may be a perfectly clear notion on the image of *electricity as a fluid,* but is much more difficult to define independent of such imagery. So it is more hindrance than help that everybody has *some* favorite notion as to what *quantity* and *intensity* should be taken to mean.

Faraday will go a long way towards defining *intensity* (and *tension,* its closely related measure) in the Eleventh Series. He will help clarify the measure of *quantity* in the present Series—see, for example, paragraphs 362–366. But at the outset those concepts reflect little more than a conventional and somewhat arbitrary selection of electrical phenomena and behaviors, which it may be helpful to review in advance. First, then, some conventional signs which Faraday invokes as indicating *intensity* or *tension* include the following:

Attraction and repulsion (paragraph 270). These are most characteristic of frictional electricity. Voltaic devices do not cause readily detectible attractions or repulsions unless they involve about one hundred cells (or two hundred plates) in a series. We therefore generally associate frictional and other forms of "common" electricity with *high* tension, voltaic electricity with *low* tension. But evidently the tension of voltaic sources is capable of increase by aggregation; and the number of voltaic plates in a battery can serve as a rough token of the electric tension it is able to produce.

Spark (paragraphs 269, 340). Increased length of spark is commonly taken as a sign of greater tension. For common electricity, the electric machine provides a supporting example: When its discharge terminals are separated to a greater distance, the machine has to be turned a greater number of times before a spark will strike. Similarly, it requires a voltaic battery of at least several hundred cells to strike even the shortest spark in air, while it took a prodigious battery of 11,000 zinc-silver cells (constructed by Warren de la Rue) to spark through a space of about 2/3 inch.

Leakage and dissipation. These and other forms of slow discharge are generally supposed to proceed more rapidly at high tensions than at low. In paragraph 365 of the present Series, Faraday cites an instance of such reasoning concerning leakage between the adjacent windings of a galvanometer coil.

Electrometer measurements. Although writers often cite the electrometer* as indicating electrical tension, inter-pretation of electrometer readings is not a simple matter. Consider Henley's electrometer, for example, which Faraday mentioned in the First Series. To begin with, it displays an *angle of divergence* between fixed and pivoted arms. Then since the pivoted arm has weight, one may calculate the *mechanical force* required to effect this divergence; and that force may in turn be attributed to the mutual repulsion of the electrified post and ball—presumably, the greater the *degree of electrification*, the  stronger the repulsion. Thus the instrument may be seen as indicating *angle, force,* or *degree of electrification*—each interpretation a bit more hypothetical than the previous one. But none of them is electric *tension!* Clearly, much more has to be established before a relation between electrometer readings and the concept of tension can be asserted.**

---

* Ordinary electrical indicators or *electroscopes* could in many cases be equipped with numerical scales. They were then dignified with the name "electrometers"—the distinguishing suffix "-meter" being intended to signify a *measuring* instrument.

** With the aid of a superior electrometer developed by Coulomb, Faraday will carry out much of the work necessary for this project in the Eleventh Series.

Faraday also invokes some conventional signs of *quantity of electricity*, which include the following:

Identical Leyden jars charged to equal degrees (according to the electrometer) are presumed to represent equal quantities of electricity (paragraphs 363–364).

Plate electrical machine. Each turn of the machine is presumed to develop an equal quantity of electricity (paragraphs 363–364).

Chemical decomposition. Any cumulative effect will indicate electrical quantity in relation to time, and chemical decomposition is probably the most widely recognized such effect.

Time. The time required to charge a given body to a given condition (according to the electrometer), or to discharge it from that condition, is another cumulative effect: Other factors being equal, the longer the time, the greater the quantity.

Measurements with the "ballistic" galvanometer (paragraph 366). In paragraphs 361–366 of the present Series, Faraday will show that the long-period or "ballistic" galvanometer gives readings that parallel *chemical* indications of quantity as well as quantity measured by *standard Leyden jars*—even when other factors such as tension and rate of discharge are widely varied. The long-period galvanometer is thus established as an instrument capable of measuring quantity of electricity.

THIRD SERIES.

§ 7. *Identity of Electricities derived from different sources.*
§ 8. *Relation by measure of common and voltaic Electricity.*

[Read January 10th and 17th, 1833.]

§ 7. *Identity of Electricities derived from different sources.*

265. THE progress of the electrical researches which I have had the honour to present to the Royal Society, brought me to a point at which it was essential for the further prosecution of my inquiries that no doubt should remain of the identity or distinction of electricities excited by different means. It is perfectly true that Cavendish,[1] Wollaston,[2] Colladon,[3] and others, have in succession removed some of the greatest objections to the acknowledgement of the identity of common, animal, and voltaic electricity, and I believe that most philosophers consider these electricities as really the same. But on the other hand it is also true, that the accuracy of Wollaston's experiments have been denied;[4] and also that one of them, which really is no proper proof of chemical decomposition by common electricity (309. 327.), has been that selected by several experimenters as the test of chemical action (336. 346.). It is a fact, too, [that] many philosophers are still drawing distinctions between the electricities of different sources; or at least doubting whether their identity is proved. Sir Humphry Davy, for instance, in his paper on the Torpedo,[5] thought it probable that animal electricity would be found of a peculiar kind; and referring to it, to common electricity, voltaic electricity and magnetism, has said,

---

[1] Phil. Trans. 1776, p. 196.

[2] Ibid. 1801, p. 434.

[3] Annales de Chimie, 1826, p. 62, &c.

[4] Phil. Trans. 1832, p. 282, note.

[5] Phil. Trans. 1829, p. 17. "Common electricity is excited upon non-conductors, and is readily carried off by conductors and imperfect conductors. Voltaic electricity is excited upon combinations of perfect and imperfect conductors, and is only transmitted by perfect conductors or imperfect conductors of the best kind. Magnetism, if it be a form of electricity, belongs only to perfect conductors; and, in its modifications, to a peculiar class of them.* Animal electricity resides only in the imperfect conductors forming the organs of living animals, &c."

. . . .

*Dr. Ritchie has shown this is not the case, Phil. Trans. 1832, p. 294.

"Distinctions might be established in pursuing the various modifications or properties of electricity in these different forms, &c." Indeed I need only refer to the last volume of the Philosophical Transactions to show that the question is by no means considered as settled.[6]

266. Notwithstanding, therefore, the general impression of the identity of electricities, it is evident that the proofs have not been sufficiently clear and distinct to obtain the assent of all those who were competent to consider the subject; and the question seemed to me very much in the condition of that which Sir H. Davy solved so beautifully,— namely, whether voltaic electricity in all cases merely eliminated, or did not in some actually produce, the acid and alkali found after its action upon water. The same necessity that urged him to decide the doubtful point, which interfered with the extension of his views, and destroyed

---

[6] Phil. Trans. 1832, p. 259. Dr. Davy, in making experiments on the torpedo, obtains effects the same as those produced by common and voltaic electricity, and says that in its magnetic and chemical power it does not seem to be essentially peculiar,—p. 274; but then he says, p. 275, there are other points of difference; and after referring to them, adds, "How are these differences to be explained? Do they admit of explanation similar to that advanced by Mr. Cavendish in his theory of the torpedo; or may we suppose, according to the analogy of the solar ray, that the electrical power, whether excited by the common machine, or by the voltaic battery, or by the torpedo, is not a simple power; but a combination of powers, which may occur variously associated, and produce all the varieties of electricity with which we are acquainted?"

At p. 279 of the same volume of Transactions is Dr. Ritchie's paper, from which the following are extracts: "Common electricity is diffused over the surface of the metal;— voltaic electricity exists within the metal. Free electricity is conducted over the surface of the thinnest gold leaf as effectively as over a mass of metal having the same surface;—voltaic electricity requires thickness of metal for its conduction," p. 280; and again, "The supposed analogy between common and voltaic electricity, which was so eagerly traced after the invention of the pile, completely fails in this case, which was thought to afford the most striking resemblance." p. 291.

---

266. *eliminated*: Here, *released* or *expelled*. In the action of voltaic electricity on water, several researchers, including Davy himself, noticed the appearance of acid products in the water near the positive wire and alkaline products near the negative wire. Were these substances newly *generated* from water? Or were they previously-existing contaminants which had merely been "eliminated" from the materials under action? The chemical action of electricity could not be elucidated so long as doubt remained on this fundamental point. Davy showed that the acid and alkali evolved from existing impurities in the water or from the material of the container. The account is in Davy's Bakerian Lecture, "On Some Chemical Agencies of Electricity" (Phil. Trans., 1807).

the doubtful point, which interfered with the extension of his views…: Like Davy previously, Faraday has reached a stage of investigation where a single

the strictness of his reasoning, has obliged me to ascertain the identity or difference of common and voltaic electricity. I have satisfied myself that they are identical, and I hope the experiments which I have to offer, and the proofs flowing from them, will be found worthy of the attention of the Royal Society.

267. The various phenomena exhibited by electricity may, for the purpose of comparison, be arranged under two heads; namely, those connected with electricity of tension, and those belonging to electricity in motion. This distinction is taken at present not as philosophical, but merely as convenient. The effect of electricity of tension, at rest, is either attraction or repulsion at sensible distances. The effects of electricity in motion or electrical currents may be considered as 1st, Evolution of heat; 2nd, Magnetism; 3rd, Chemical decomposition; 4th, Physiological phenomena; 5th, Spark. It will be my object to compare electricities from different sources, and especially common and voltaic electricities, by their power of producing these effects.

I. *Voltaic Electricity.*

268. Tension.—When a voltaic battery of 100 pairs of plates has its extremities examined by the ordinary electrometer, it is well known that they are found positive and negative, the gold leaves at the same extremity repelling each other, the gold leaves at different extremities attracting each other, even when half an inch or more of air intervenes.

---

undecided question stands as a barrier to further progress; this question is the *identity of electricities*. But why is the matter of identity so important now? One reason might be this: When Faraday first revealed the evolution of electricity from magnetism he made it clear that its chief significance lay in being the *reciprocal* of the production of magnetism from electric currents (paragraph 3)—as we will see again and again in these Researches, the mutual convertibility and fundamental unity of powers is a central article in Faraday's vision of nature. But if the electricity that produces magnetism is different from the electricity that is evolved from it, that reciprocity fails. This question of identity of electricities, then, is more directly connected with the previous researches than might at first have appeared.

267. A "philosophical distinction" is grounded in theory and understanding, in knowledge of a distinction in natures or essences. But the distinction between electricity "at rest" and "in motion" is not philosophical, according to Faraday, but merely *convenient*. We should remember, then, not to take the terms "rest" and "motion" too literally.

269. That ordinary electricity is discharged by points with facility through air; that it is readily transmitted through highly rarefied air; and also through heated air, as for instance a flame; is due to its high tension. I sought, therefore, for similar effects in the discharge of voltaic electricity, using as a test of the passage of the electricity either the galvanometer or chemical action produced by the arrangement hereafter to be described (312. 316.).

270. The voltaic battery I had at my disposal consisted of 140 pairs of plates four inches square, with double coppers. It was insulated throughout, and diverged a gold leaf electrometer about one third of an inch. On endeavouring to discharge this battery by delicate points very nicely arranged and approximated, either in the air or in an exhausted receiver, I could obtain no indications of a current, either by magnetic or chemical action. In this, however, was found no point of discordance between voltaic and common electricity; for when a Leyden battery (291.) was charged so as to deflect the gold leaf electrometer to the same degree, the points were found equally unable to

---

269. The characteristically high tension of "ordinary" (static) electricity permits it to discharge through air, producing a variety of visible and audible effects as described in the introduction to the present Series. Electricity that has been voltaically produced is usually of much lower tension; but he suspects that it too can be made to discharge through air—though without the sounds and visible displays typical of high-tension discharge. In the absence of those sensory indications he will look to magnetic or chemical actions, which have been long associated with voltaic currents, as evidence of the discharges sought.

270. *double coppers* were described in the comment to paragraph 7 (First Series). The *tension* characteristic of his voltaic battery is shown by its ability to separate the leaves of the indicator, and is in this case rather low. He was unable to show discharge of this voltaic battery through air; but *common* (static) electricity, if at the same low tension (as indicated by the gold leaves), was equally incapable of discharging through air. Thus the failure is due to the low tension of the voltaic electricity, and not to its character of being voltaic.

In this paragraph Faraday characterizes the electricity by its "intensity" and its "quantity." The "intensity" of an electrical action appears to be hardly distinguishable from its "tension"; but Faraday usually reserves the term "tension" for references to *attraction and repulsion* (as, for example in 267.). The "quantity" of electricity is readily visualized if a fluid image of electricity is employed, but what meaning can the term have apart from that image? He will offer both *chemical* and *magnetic* expressions of the "quantity" of current later in the present Series (paragraphs 366–367, 377) and again in the Seventh Series.

nicely: that is, *carefully, precisely*. *approximated*: here, made to approach one another.

discharge it with such effect as to produce either magnetic or chemical action. This was not because common electricity could not produce both these effects (307. 310.), but because when of such low intensity the quantity required to make the effects visible (being enormously great (371. 375.),) could not be transmitted in any reasonable time. In conjunction with the other proofs of identity hereafter to be given, these effects of points also prove identity instead of difference between voltaic and common electricity.

271. As heated air discharges common electricity with far greater facility than points, I hoped that voltaic electricity might in this way also be discharged. An apparatus was therefore constructed (Plate III. fig. 46.), in which A B is an insulated glass rod upon which two copper wires, C, D, are fixed firmly; to these wires are soldered two pieces of fine platina wire, the ends of which are brought very close to each other at *e*, but without touching; the copper wire C was connected with the positive pole of a voltaic battery, and the wire D with a decomposing apparatus (312. 316.), from which the communication was completed to the negative pole of the battery. In these experiments only two troughs, or twenty pairs of plates, were used.

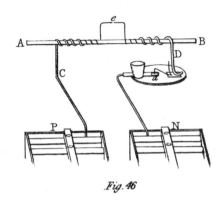

Fig. 46

272. Whilst in the state described, no decomposition took place at the point *a*, but when the side of a spirit-lamp flame was applied to the two platina extremities at *e*, so as to make them bright red-hot, decomposition occurred; iodine soon appeared at the point *a*, and the transference of electricity through the heated air was established. On raising the temperature of the points *e* by a blowpipe, the

---

272. Since heated air facilitates discharge of *common* (static) electricity, perhaps it will do the same for *voltaic* electricity? And indeed, when the gap at *e* (Figure 46) is heated, chemical action at *a* indicates that a current is passing— and thus that there must be discharge at *e*. This discharge through hot air occurred with a tension of only 20 pairs of plates—whereas in the previous paragraph the tension of 140 pairs was insufficient to produce discharge through room-temperature air. The "decomposing apparatus" is discussed in the introduction to the present Series.

An alcohol-burning "spirit-lamp" is illustrated here. With a "blow-pipe," usually made of brass or glass, one can apply a jet of air to the flame, thereby intensifying and directing it.

discharge was rendered still more free, and decomposition took place instantly. On removing the source of heat, the current immediately ceased. On putting the ends of the wires very close by the side of and parallel to each other, but not touching, the effects were perhaps more readily obtained than before. On using a larger voltaic battery (270.), they were also more freely obtained.

273. On removing the decomposing apparatus and interposing a galvanometer instead, heating the points *e* as the needle would swing one way, and removing the heat during the time of its return (302.), feeble deflections were soon obtained: thus also proving the current through heated air; but the instrument used was not so sensible under the circumstances as chemical action.

274. These effects, not hitherto known or expected under this form, are only cases of the discharge which takes place through air between the charcoal terminations of the poles of a powerful battery, when they are gradually separated after contact. Then the passage is through heated air exactly as with common electricity, and Sir H. Davy has recorded that with the original battery of the Royal Institution this discharge passed through a space of at least four inches.[7] In the exhausted receiver the electricity would *strike* through nearly half an inch of space, and the combined effects of rarefaction and heat was such upon the inclosed air as to enable it to conduct the electricity through a space of six or seven inches.

[7] Elements of Chemical Philosophy, p. 153.

---

273. *heating the points … and removing the heat*: The galvanometer is not as "sensible" (sensitive) as the chemical indicator, but by synchronizing the heating and cooling with the swing of the instrument, he is gradually able to build up observable deflections.

274. *These effects … are only cases…*: They are *specific examples* of the more general "discharge … between charcoal terminations," that is, the electric arc discharge (discussed in the editor's introduction).

The original battery of the Royal Institution: This monster battery of 2000 cells had been maintained by the Institution during Humphry Davy's tenure. Davy related that at normal atmospheric pressure it could *strike* through one-thirtieth of an inch; and the resulting arc could then be drawn out to four inches' length. In a rarefied atmosphere these distances increased to one-half inch and six or seven inches, respectively. The powerful, continuing, and elongated arc discharge shows how greatly the conductance of air is enhanced by heating, and also—in the "exhausted receiver"—by rarefaction.

275. The instantaneous charge of a Leyden battery by the poles of a voltaic apparatus is another proof of the tension, and also the quantity of electricity evolved by the latter. Sir H. Davy says,[8] "When the two conductors from the ends of the combination were connected with a Leyden battery, one with the internal, the other with the external coating, the battery instantly became charged; and on removing the wires and making the proper connexions, either a shock or a *spark* could be perceived: and the least possible time of contact was sufficient to renew the charge to its full intensity."

276. *In motion*: i. *Evolution of Heat.*—The evolution of heat in wires and fluids by the voltaic current is matter of general notoriety.

277. ii. *Magnetism.*—No fact is better known to philosophers than the power of the voltaic current to deflect the magnetic needle, and to make magnets according to *certain laws;* and no effect can be more distinctive of an electrical current.

[8] Ibid. p. 154.

---

275. *The instantaneous charge of a Leyden battery by the poles of a voltaic apparatus is another proof of the tension ... of electricity evolved...*: In paragraph 292 below, Faraday implicitly identifies the electric tension as one of two factors enhancing rapidity of electric conduction (the other is the conductivity of the discharge path).

...and also the quantity of electricity evolved...: Better, quantity of electricity evolved *per second*. A quantity of electricity, sufficient to electrify a Leyden jar to a specified degree, might require several turns of a frictional electric machine; but it is developed in an immeasurably brief time by the voltaic apparatus. This instantaneity of electrification also bears directly on the overall question of electric identity; for if voltaic and common electricity were essentially different, conversion of one into the other might be expected to take time.

276. *In motion*: As suggested in the comment to paragraph 267, allusions to *electricity in motion* should not be taken too literally. The phraseology in paragraphs 287, 344, and 353 below is similarly figurative. *notoriety*: Here, simple *repute*; there is no derogatory connotation.

277. *certain laws*: specifically, the direction relation between a current and the magnetism surrounding it. Faraday suggested a mnemonic formulation in his note to paragraph 38; and I described an equivalent alternative, the modern *electromagnetic right-hand rule*, in my comment to Faraday's note.

278. iii. *Chemical decomposition.*—The chemical powers of the voltaic current, and their subjection to *certain laws*, are also perfectly well known.

279. iv. *Physiological effects.*—The power of the voltaic current, when strong, to shock and convulse the whole animal system, and when weak to affect the tongue and the eyes, is very characteristic.

280. v. *Spark*—The brilliant star of light produced by the discharge of a voltaic battery is known to all as the most beautiful light that man can produce by art.

---

281. That these effects may be almost infinitely varied, some being exalted whilst others are diminished, is universally acknowledged; and yet without any doubt of the identity of character of the voltaic currents thus made to differ in their effect. The beautiful explication of these variations afforded by Cavendish's theory of quantity and intensity requires no support at present, as it is not supposed to be doubted.

282. In consequence of the comparisons that will hereafter arise between wires carrying voltaic and ordinary electricities, and also because of certain views of the condition of a wire or any other conducting substance connecting the poles of a voltaic apparatus, it will be necessary to give some definite expression of what is called the voltaic current, in contradistinction to any supposed peculiar state of

---

278. *certain laws*: First, that the products of electro-decomposition are evolved *separately*, at the positive and negative poles of the material, respectively. Faraday tentatively suggests a second in paragraph 329: that the quantity of material decomposed is strictly related to the *quantity* of electricity that passes, regardless of the *tension* of that electricity. It will be impressively confirmed in the Seventh Series.

280. *star of light*: the arc discharge, as in paragraph 274. An illustration is reproduced in the introduction to this Series. Although the arc is not exactly the same as the *spark*, both are instances of discharge in air.

282. *some definite expression of what is called the voltaic current*: He acknowledges a question which up to now has not been explicitly addressed: Does electric current have a *progressive* character (as the name "current" indeed suggests); or is it rather a *condition* which may exhibit different dispositions in different parts of the wire, but not motion or transport? We observe that making or breaking the battery connection in one part of a circuit instantly initiates or halts the magnetic action in another part. This is hard to explain if "current" is a stationary electrical condition, but on the idea of *progress* it is easy to see why a stoppage in one part of the circuit would cause an immediate halt in all parts—or, more generally, how a change at one point can affect conditions throughout the whole.

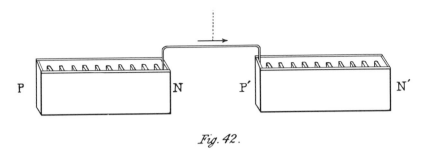

Fig. 42.

arrangement, not progressive, which the wire or the electricity within it may be supposed to assume. If two voltaic troughs PN, P'N', fig. 42, be symmetrically arranged and insulated, and the ends NP' connected by a wire, over which a magnetic needle is suspended, the wire will exert no effect over the needle; but immediately that the ends PN' are connected by another wire, the needle will be deflected, and will remain so as long as the circuit is complete. Now if the troughs merely act by causing a peculiar arrangement in the wire either of its particles or its electricity, that arrangement constituting its electrical and magnetic state, then the wire NP' should be in a similar state of arrangement *before* P and N' were connected, to what it is afterwards, and should have deflected the needle, although less powerfully, perhaps to one half the extent which would result when the communication is complete throughout. But if the magnetic effects depend upon a current, then it is evident why they could not be produced in *any* degree before the circuit was complete; because prior to that no current could exist.

283. By *current*, I mean anything progressive, whether it be a fluid of electricity, or two fluids moving in opposite directions, or merely vibrations, or, speaking still more generally, progressive forces. By *arrangement*, I understand a local adjustment of particles, or fluids, or forces, not progressive. Many other reasons might be urged in support of the view of a *current* rather than an *arrangement*, but I am anxious to avoid stating unnecessarily what will occur to others at the moment.

---

283. *by current, I mean anything progressive*: Note that in appealing to *progression* as distinguished from *flow* he can acknowledge the current's character of having a *direction* without being obliged to invoke fluids, or indeed any *material* imagery. His phrase "progressive forces" may remind us of the *moving lines of force* envisioned in the Second Series (paragraphs 238–239). It may also suggest *wave* motion—an image Faraday follows up with great seriousness in the Twelfth Series.

II. *Ordinary Electricity.*

284. By ordinary electricity I understand that which can be obtained from the common machine, or from the atmosphere, or by pressure, or cleavage of crystals, or by a multitude of other operations; its distinctive character being that of great intensity, and the exertion of attractive and repulsive powers, not merely at sensible but at considerable distances.

285. *Tension.* The attractions and repulsions at sensible distances, caused by ordinary electricity, are well known to be so powerful in certain cases, as to surpass, almost infinitely, the similar phenomena produced by electricity, otherwise excited. But still those attractions and repulsions are exactly of the same nature as those already referred to under the head *Tension, Voltaic electricity* (268.); and the difference in degree between them is not greater than often occurs between cases of ordinary electricity only. I think it will be unnecessary to enter minutely into the proofs of the identity of this character in the two instances. They are abundant; are generally admitted as good; and lie upon the surface of the subject: and whenever in other parts of the comparison I am about to draw, a similar case occurs, I shall content myself with a mere announcement of the similarity, enlarging only upon those parts where the great question of distinction or identity still exists.

286. The discharge of common electricity through heated air is a well-known fact. The parallel case of voltaic electricity has already been described (272, &c.).

287. *In motion.* i. *Evolution of heat.*—The heating power of common electricity, when passed through wires or other substances, is perfectly

---

284. *that which can be obtained from the common machine...*: Thus "ordinary electricity" is what today we commonly call *static electricity*. It is no longer the most "ordinary" kind of electricity in our experience; that epithet would have to be awarded to the electricity we regularly obtain from household receptacles and from chemical batteries. But of course buildings were not wired for power in Faraday's time; and chemical voltaic cells were essentially laboratory devices. In ordinary life, it was *static* electricity that most Victorians would experience routinely.

great intensity: As mentioned previously in the comment to paragraph 270, roughly although not exactly the same as *high tension*.

285. The primary phenomena of "tension" are the same for ordinary as for voltaic electricity, namely, attractions and repulsions. They differ only in degree, the static attractions and repulsions being generally stronger.

upon the surface of the subject: that is, open to view and readily comprehensible.

well known. The accordance between it and voltaic electricity is in this respect complete. Mr. Harris has constructed and described[9] a very beautiful and sensible instrument on this principle, in which the heat produced in a wire by the discharge of a small portion of common electricity is readily shown, and to which I shall have occasion to refer for experimental proof in a future part of this paper (344.).

288. ii. *Magnetism.*—Voltaic electricity has most extraordinary and exalted magnetic powers. If common electricity be identical with it, it ought to have the same powers. In rendering needles or bars magnetic, it is found to agree with voltaic electricity, and the *direction* of the magnetism, in both cases, is the same; but in deflecting the magnetic needle, common electricity has been found deficient, so that sometimes its power has been denied altogether, and at other times distinctions have been hypothetically assumed for the purpose of avoiding the difficulty.[10]

289. M. Colladon, of Geneva, considered that the difference might be due to the use of insufficient quantities of common electricity in all the experiments before made on this head; and in a memoir read to the Academie des Sciences in 1826,[11] describes experiments, in which, by the use of a battery, points, and a delicate galvanometer, he succeeded in obtaining deflections, and thus establishing identity in that respect. MM. Arago, Ampère, and Savary, are mentioned in the paper as having witnessed a successful repetition of the experiments. But as no other one has come forward in confirmation, MM. Arago, Ampère, and Savary, not having themselves published (that I am aware of) their admission of the results, and as some have not been able to obtain

---

[9] Philosophical Transactions, 1827, p. 18. Edinburgh Transactions, 1831 Harris on a New Electrometer; &c. &c.

[10] Demonferrand's Manuel d'Electricité dynamique, p. 121.

[11] Annales de Chimie, xxxiii. p. 62.

---

287. *sensible*: As in paragraph 273 above, *sensitive.* In the article cited by Faraday, Harris describes his instrument as "little more than an air thermometer [having a wire] passed air tight through the bulb. [W]hen an electrical explosion ... is passed through the wire in the bulb, the relative degree of heat it evolves is made evident by the ascent of the fluid along the graduated scale." In a note to paragraph 359 below, Faraday will refer to the device as "Harris's thermo-electrometer."

289. *[Colladon's] experiments ... by the use of a battery*: A battery of Leyden jars, not a voltaic battery, is meant (see the introduction to the First Series).

them, M. Colladon's conclusions have been occasionally doubted or denied; and an important point with me was to establish their accuracy, or remove them entirely from the body of received experimental research. I am happy to say that my results fully confirm those by M. Colladon, and I should have had no occasion to describe them, but that they are essential as proofs of the accuracy of the final and general conclusions I am enabled to draw respecting other magnetic and chemical action of electricity, (360. 366. 367. 377. &c.).

290. The plate electrical machine I have used is fifty inches in diameter; it has two sets of rubbers; its prime conductor consists of two brass cylinders connected by a third, the whole length being twelve feet, and the surface in contact with air about 1422 square inches. When in good excitation, one revolution of the plate will give ten or twelve sparks from the conductors, each an inch in length. Sparks or flashes from ten to fourteen inches in length may easily be drawn from the conductors. Each turn of the machine, when worked moderately, occupies about 4/5ths of a second.

291. The electric battery consisted of fifteen equal jars. They are coated eight inches upwards from the bottom, and are twenty-three inches in circumference, so that each contains one hundred and eighty-four square inches of glass, coated on both sides; this is independent of the bottoms, which are of thicker glass, and contain each about fifty square inches.

292. A good *discharging train* was arranged by connecting metallically a sufficiently thick wire with the metallic gas pipes of the house, with the metallic gas pipes belonging to the public gas works of London; and also with the metallic water pipes of London. It was so effectual in its office as to carry off instantaneously electricity of the feeblest tension, even that of a single voltaic trough, and was essential to many of the experiments.

---

290. *plate electrical machine*: Frictional machines of this type were described in the introduction to the First Series. Faraday's machine is unusually large and powerful, such as to be found in only the best laboratories.

292. *A good discharging train* is a good *ground* connection. Its function here will be to provide rapid discharge of the high-tension electricity produced by the plate machine; but Faraday notes that it will discharge even low tension electrification just as "instantaneously." Thus Faraday implicitly recognizes rapid discharge as the sign either of *high tension* (the usual case), or *high conductivity*, as here. Recall that he asserted the same connection in paragraph 275 above; in the Eleventh and Twelfth Series he will explore the relation between rapidity and tension in much greater depth.

293. The galvanometer was one or the other of those formerly described (87. 205.), but the glass jar covering it and supporting the needle was coated inside and outside with tinfoil, and the upper part (left uncoated, that the motions of the needle might be examined,) was covered with a frame of wire-work, having numerous sharp points projecting from it. When this frame and the two coatings were connected with the discharging train (292.), an insulated point or ball, connected with the machine when most active, might be brought within an inch of any part of the galvanometer, yet without affecting the needle within by ordinary electrical attraction or repulsion.

294. In connexion with these precautions it may be necessary to state that the needle of the galvanometer is very liable to have its magnetic power deranged, diminished, or even inverted by the passage of a shock through the instrument. If the needle be at all oblique, in the wrong direction, to the coils of the galvanometer when the shock passes, effects of this kind are sure to happen.

295. It was to the retarding power of bad conductors, with the intention of diminishing its *intensity* without altering its *quantity*, that I first looked with the hope of being able to make common electricity assume more of the characters and power of voltaic electricity, than it is usually supposed to have.

---

293. *a frame of wire-work, having numerous sharp points projecting...*: In the Eleventh Series Faraday will explain *how* such a conductive network succeeds in shielding the galvanometer from nearby static electrification. The "points," as we have seen, facilitate discharge of any neighboring accumulated electricity in the air (and thence to ground)—they are comparable to miniature lightning rods.

294. Recall that currents were able to magnetize needles (paragraph 13); so it is not surprising that sudden and excessive currents might, while deflecting the needle, also disrupt its magnetic condition.

295. *the retarding power of bad conductors*: Bad conductors certainly interfere with discharge; but does "retarding" denote only general *hindrance*—or the specific sense of *slowing down*? The sense is doubtful here; but in the Twelfth Series—for example, paragraph 1328—Faraday will characterize poor conductors as exhibiting *reduced velocity* in that "progress" which he recognized at paragraphs 282–283 above.

diminishing its intensity without altering its quantity: Since voltaic electricity is normally found at lower tensions than voltaic electricity, he hopes by interposing poor conductors to cause common electricity more nearly to resemble voltaic electricity in its effects. But Faraday does not say, here, *how* poor conductors reduce the intensity (tension) of the electricity they carry, while conserving its overall quantity.

296. The coating and armour of the galvanometer were first connected with the discharging train (292.); the end B (87.) [Pl. I fig. 8]

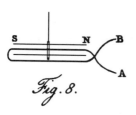

Fig. 8.

of the galvanometer wire was connected with
the outside coating of the battery, and then
both these with the discharging train; the
end A of the galvanometer wire was connected with a discharging rod by a wet
thread four feet long; and, finally, when the
battery (291.) had been positively charged
by about forty turns of the machine, it was discharged by the rod and
the thread through the galvanometer. The needle immediately
moved.

297. During the time that the needle completed its vibration in the
first direction and returned, the machine was worked, and the battery
recharged; and when the needle in vibrating resumed its first direction, the discharge was again made through the galvanometer. By
repeating this action a few times, the vibrations soon extended to
above 40° on each side of the line of rest.

---

296. Refer to the accompanying diagram. B is one end of the galvanometer
coil and C is the outer foil of the Leyden battery (only a single jar is shown).
Both are connected to earth by
means of the "discharging
train." The diagram does not
show the conductive network
which Faraday introduced in
paragraph 293 to shield the galvanometer. Coil end B, foil C,
and the galvanometer shield all
remain grounded throughout
the procedure.

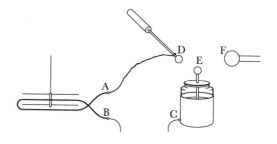

   AD, a length of wet string, joins end A of the galvanometer coil to a discharging rod D. Wet string is a "bad conductor" whose purpose Faraday
described in the previous paragraph.
   Faraday proceeds as follows: He first brings E, the center post of the Leyden
jar, into contact with F, the active terminal of the electrostatic plate machine;
as the machine is operated, the Leyden jar becomes electrified. After
removing the electrified jar from F, he touches rod D to the jar's central post
E. The jar discharges to earth through the string and the galvanometer coil—
and the galvanometer needle deflects.

297. Note here again the alternate charging and discharging, synchronous
with the oscillations of the galvanometer needle. Since the totality of discharges results in a large cumulative deflection, it is certain that *each single
discharge* causes a small, but definite, deflection. Common electricity has thus
exhibited magnetic powers, as he sought to show.

298. This effect could be obtained at pleasure. Nor was it varied, apparently, either in direction or degree, by using a short thick string, or even four short thick strings in place of the long fine thread. With a more delicate galvanometer, an excellent swing of the needle could be obtained by one discharge of the battery.

299. On reversing the galvanometer communications so as to pass the discharge through from B to A, the needle was equally well deflected, but in the opposite direction.

300. The deflections were in the same direction as if a voltaic current had been passed through the galvanometer, *i. e.* the positively charged surface of the electric battery coincided with the positive end of the voltaic apparatus (268.), and the negative surface of the former with the negative end of the latter.

301. The battery was then thrown out of use, and the communications so arranged that the current could be passed from the prime conductor, by the discharging rod held against it, through the wet string, through the galvanometer coil, and into the discharging train (292.), by which it was finally dispersed. This current could be stopped at any moment by removing the discharging rod, and either stopping the machine or connecting the prime conductor by another rod with the discharging train; and could be as instantly renewed. The needle was so adjusted, that whilst vibrating in moderate and small arcs, it required time equal to twenty-five beats of a watch to pass in one direction through the arc, and of coarse an equal time to pass in the other direction.

302. Thus arranged, and the needle being stationary, the current, direct from the machine, was sent through the galvanometer for twenty-five beats, then interrupted for other twenty-five beats, renewed for twenty-five beats more, again interrupted for an equal time, and so on continually. The needle soon began to vibrate visibly, and after several alternations of this kind the vibration increased to 40° or more.

303. On changing the direction of the current through the galvanometer, the direction of the deflection of the needle was also changed. In all cases the motion of the needle was in direction the same as that caused either by the use of the electric battery or a voltaic trough (300.).

---

301. In the previous experiment the machine-evolved electricity was first stored in a Leyden battery, and only then routed to the galvanometer. Now he takes the electric discharge directly from the machine, but reduced (in intensity) by the medium of wet string.

prime conductor: The active terminal of the electrical machine. For the style of machine that was illustrated on page 10, the prime conductor is the brass tube labeled B.

304. I now rejected the wet string, and substituted a copper wire, so that the electricity of the machine passed at once into wires communicating directly with the discharging train, the galvanometer coil being one of the wires used for the discharge. The effects were exactly those obtained above (302.).

305. Instead of passing the electricity through the system, by bringing the discharging rod at the end of it into contact with the conductor, four points were fixed on to the rod; when the current was to pass they were held about twelve inches from the conductor, and when it was not to pass, they were turned away. Then operating as before (302.), except with this variation, the needle was soon powerfully deflected, and in perfect consistency with the former results. Points afforded the means by which Colladon, in all cases, made his discharges.

306. Finally I passed the electricity first through an exhausted receiver, so as to make it there resemble the aurora borealis, and then through the galvanometer to the earth; and it was found still effective in deflecting the needle, and apparently with the same force as before.

307. From all these experiments, it appears that a current of common electricity, whether transmitted through water or metal, or rarefied air, or by means of points in common air, is still able to deflect the needle; the only requisite being, apparently, to allow time for its

---

304. Now the wet string is removed from the circuit. The results are the same as in paragraphs 302–303, *even though the discharge intensity is no longer being reduced.*

305. Finally even the direct copper connection is removed, and *points* discharging through air are substituted. Why does Faraday make this last variation? Of course it is significant as reproducing Colladon's technique (paragraph 289). But notice that it also very beautifully illustrates the dual rhetoric common to all experiments that demonstrate *identity*: On the one hand the electricity discharges through points, hitherto associated with *common* electricity—and on the other hand it exerts magnetic influence, virtually the hallmark of *voltaic* electricity (paragraph 277).

306. *an exhausted receiver*: This is a forerunner of the "electric egg" described in the introduction. Faraday's laboratory Diary for August 30, 1832 reads: "Then put a jar ... onto the air pump, and ... exhausted it until the discharge of electricity through it was ready and in purple streams..."

the aurora borealis: That is, the *northern lights*, also described in the introduction to the present Series. In middle latitudes the aurora usually appears in the form of colored streaks across the sky—appearances which were simulated by the electrical discharges in Faraday's "receiver."

307. *requisite ... to allow time*: that is, it is necessary (because the currents are so small) to use his technique of gradually accumulating resonant vibration

action: that it is, in fact, just as magnetic in every respect as a voltaic current, and that in this character therefore no distinction exists.

308. Imperfect conductors, as water, brine, acids, &c. &c. will be found far more convenient for exhibiting these effects than other modes of discharge, as by points or balls; for the former convert at once the charge of a powerful battery into a feeble spark discharge, or rather continuous current, and involve little or no risk of deranging the magnetism of the needles (294.).

309. iii. *Chemical decomposition.*—The chemical action of voltaic electricity is characteristic of that agent, but not more characteristic than are the *laws* under which the bodies evolved by decomposition arrange themselves at the poles. Dr. Wollaston showed[12] that common electricity resembled it in these effects, and "that they are both essentially the same"; but he mingled with his proofs an experiment having a resemblance, and nothing more, to a case of voltaic decomposition, which however he himself partly distinguished; and this has been more frequently referred to by some, on the one hand, to prove the occurrence of electro-chemical decomposition, like that of the pile, and by others to throw doubt upon the whole paper, than the more numerous and decisive experiments which he has detailed.

310. I take the liberty of describing briefly my results, and of thus adding my testimony to that of Dr. Wollaston on the identity of voltaic and common electricity as to chemical action, not only that I may facilitate the repetition of the experiments, but also lead to some new consequences respecting electrochemical decomposition (376. 377.).

311. I first repeated Wollaston's fourth experiment,[13] in which the ends of coated silver wires are immersed in a drop of sulphate of copper. By passing the electricity of the machine through such an arrangement, that end in the drop which received the electricity became coated with metallic copper. One hundred turns of the

[12] Philosophical Transactions, 1801, p. 427, 434.

[13] Ibid. 1801, p. 429.

---

in the galvanometer needle, as in paragraph 302. Differences in magnitude aside, his main point is established—common and voltaic electricity are shown to be *the same* in their magnetic effects.

311. *that end in the drop which received the electricity became coated with metallic copper:* Faraday refers to the *negative* end of the drop—since the conventional direction of current *in the drop* is from positive to negative. In paragraph 313 below Faraday explicitly identifies the wire attached to the negative conductor of an electrical machine as the site of precipitation of copper.

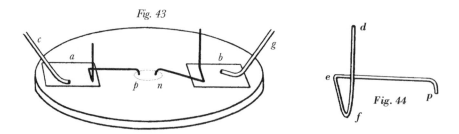

Fig. 43

Fig. 44

machine produced an evident effect; two hundred turns a very sensible one. The decomposing action was however very feeble. Very little copper was precipitated, and no sensible trace of silver from the other pole appeared in the solution.

312. A much more convenient and effectual arrangement for chemical decompositions by common electricity, is the following. Upon a glass plate, fig. 43, placed over, but raised above a piece of white paper, so that shadows may not interfere, put two pieces of tin-foil *a, b*; connect one of these by an insulated wire *c*, or wire and string (301.), with the machine, and the other *g*, with the discharging train (292.) or the negative conductor; provide two pieces of fine platina wire, bent as in fig. 44, so that the part *d, f* shall be nearly upright, whilst the whole is resting on the three bearing points *p, e, f*; place these as in fig. 43; the points *p, n* then become the decomposing poles. In this way surfaces of contact, as minute as possible, can be obtained at pleasure, and the connexion can be broken or renewed in a moment, and the substances acted upon examined with the utmost facility.

---

312. *surfaces of contact*: that is, contact between the metallic conductors and the chemical substance to be decomposed. In Faraday's Figure 43 the surfaces of contact are located where the wire tips *p* and *n* dip into a drop of copper sulfate solution (indicated by the dotted line, as Faraday explains in the next paragraph). It is at these surfaces that the actual decomposition of the solution takes place; thus it is advantageous to secure surfaces "as minute as possible" since small surfaces adjoin small quantities of decomposable material and require correspondingly small quantities of electricity to produce visible effects. Such an apparatus will be very sensitive—in Faraday's words, "much more convenient and effectual."

That electro-chemical decomposition is *confined* to the "surfaces of contact" will lead Faraday to draw far-reaching consequences in the Seventh Series.

the connexion can be broken or renewed in a moment: It is easy to remove and replace either of the bent platinum wires; the upright leg (*df* in Figure 44) serves as a handle.

313. A coarse line was made on the glass with solution of sulphate of copper, and the terminations *p* and *n* put into it; the foil *a* was connected with the positive conductor of the machine by wire and wet string, so that no sparks passed: twenty turns of the machine caused the precipitation of so much copper on the end *n*, that it looked like copper wire; no apparent change took place at *p*.

314. A mixture of equal parts of muriatic acid and water was rendered deep blue by sulphate of indigo, and a large drop put on the glass, fig. 43, so that *p* and *n* were immersed at opposite sides: a single turn of the machine showed bleaching effects round *p*, from evolved chlorine. After twenty revolutions no effect of the kind was visible at *n*, but so much chlorine had been set free at *p*, that when the drop was stirred the whole became colourless.

315. A drop of solution of iodide of potassium mingled with starch was put into the same position at *p* and *n;* on turning the machine, iodine was evolved at *p*, but not at *n*.

316. A still further improvement in this form of apparatus consists in wetting a piece of filtering paper in the solution to be experimented on, and placing that under the points *p* and *n*, on the glass: the paper retains the substance evolved at the point of evolution, by its whiteness renders any change of colour visible, and allows of the point of contact between it and the decomposing wires being contracted to the utmost degree. A piece of paper moistened in the solution of iodide of potassium and starch, or of the iodide alone, with certain precautions (322.), is a most admirable test of electrochemical action; and when thus placed and acted upon by the electric current, will show iodide evolved at *p* by only half a turn of the machine. With these adjustments and the use of iodide of potassium on paper, chemical action is sometimes a more delicate test of electrical currents than the galvanometer (273.). Such cases occur when the bodies traversed by the current are bad conductors, or when the quantity of electricity evolved or transmitted in a given time is very small.

317. A piece of litmus paper moistened in solution of common salt of sulphate of soda, was quickly reddened at *p*. A similar piece moistened in muriatic acid was very soon bleached at *p*. No effects of a similar kind took place at *n*.

---

313. …*wet string, so that no sparks passed…*: Note that interposition of wet string has reduced the tension at the site of the decomposing apparatus. Otherwise, the poles *p*, *n* being so close together, sparks would certainly have passed between them.

318. A piece of turmeric paper moistened in solution of sulphate of soda was reddened at *n* by two or three turns of the machine, and in twenty or thirty turns plenty of alkali was there evolved. On turning the paper round, so that the spot came under *p*, and then working the machine, the alkali soon disappeared, the place became yellow, and a brown alkaline spot appeared in the new part under *n*.

319. On combining a piece of litmus with a piece of turmeric paper, wetting both with solution of sulphate of soda, and putting the paper on the glass, so that *p* was on the litmus and *n* on the turmeric, a very few turns of the machine sufficed to show the evolution of acid at the former and alkali at the latter, exactly in the manner effected by a volta-electric current.

320. All these decompositions took place equally well, whether the electricity passed from the machine to the foil *a*, through water, or through wire only; by *contact* with the conductor, or by *sparks* there; provided the sparks were not so large as to cause the electricity to pass in sparks from *p* to *n*, or towards *n;* and I have seen no reason to believe that in cases of true electro-chemical decomposition by the machine, the electricity passed in sparks from the conductor, or at any part of the current, is able to do more, because of its tension, than that which is made to pass merely as a regular current.

321. Finally, the experiment was extended into the following form, supplying in this case the fullest analogy between common and voltaic electricity. Three compound pieces of litmus and turmeric paper (319.) were moistened in solution of sulphate of soda, and arranged on a plate of glass with platina wires, as in fig. 45. The wire *m* was connected with the prime conductor of the machine, the wire *t* with the discharging train, and the wires *r* and *s* entered into the course of the electrical current by means of the pieces of moistened paper; they were so bent as to rest each on three points, *n, r, p; n, s, p*, the points *r* and *s*

---

319. Thus with discharges of common electricity, just as with continuous voltaic currents, the chemical actions are specific to the positive and negative poles, respectively. Faraday will invoke this specificity as one of the hallmarks of "true electro-chemical decomposition."

320. *the conductor*: that is, the prime conductor or active terminal of the electric machine, as in paragraph 301 above.

321. *wires r and s entered into the course of the electrical current by means of the pieces of moistened paper*: In Figure 45 the current passes from the prime conductor of the machine to wire *m*, thence through the first piece of paper to wire *r*, through the second paper to wire *s*, through the third paper to wire *t*, and thence to earth by means of the discharging train.

being supported by the glass,
and the others by the papers:
the three terminations *p, p, p*
rested on the litmus, and the
other three *n, n, n* on the
turmeric paper. On working
the machine for a short time

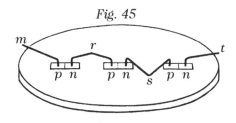

Fig. 45

only, acid was evolved at *all* the poles or terminations *p, p, p,* by which
the electricity entered the solution, and alkali at the other poles, *n, n,
n,* by which the electricity left the solution.

322. In all experiments of electro-chemical decomposition by the
common machine and moistened papers (316.), it is necessary to be
aware of and to avoid the following important source of error. If a
spark passes over moistened litmus and turmeric paper, the litmus
paper (provided it be delicate and not too alkaline,) is reddened by it;
and if several sparks are passed, it becomes powerfully reddened. If the
electricity pass a little way from the wire over the surface of the
moistened paper, before it finds mass and moisture enough to conduct
it, then the reddening extends as far as the ramifications. If similar
ramifications occur at the termination *n,* on the turmeric paper, they
prevent the occurrence of the red spot due to the alkali, which would
otherwise collect there: sparks or ramifications from the points *n* will
also redden litmus paper. If paper moistened by a solution of iodide of
potassium (which is an admirably delicate test of electro-chemical
action,) be exposed to the sparks or ramifications, or even a feeble
stream of electricity through the air from either the point *p* or *n,*
iodine will be immediately evolved.

323. These effects must not be confounded with those due to the
true electro-chemical powers of common electricity, and must be carefully avoided when the latter are to be observed. No sparks should be
passed, therefore, in any part of the current, nor any increase of

---

322. *ramifications*: here, the branching structure usually found in sparks
greater than two or three inches in length. The illustration below and similar
drawings of high-speed phenomena were made possible by a rotating mirror
device invented by Wheatstone. Faraday will refer to Wheatstone's mirror in
the Eleventh Series.

intensity allowed, by which the electricity may be induced to pass between the platina wires and the moistened papers, otherwise than by conduction; for if it burst through the air, the effect referred to above (322.) ensues.

324. The effect itself is due to the formation of nitric acid by the combination of the oxygen and nitrogen of the air, and is, in fact, only a delicate repetition of Cavendish's beautiful experiment. The acid so formed, though small in quantity, is in a high state of concentration as to water, and produces the consequent effects of reddening the litmus paper; or preventing the exhibition of alkali on the turmeric paper; or, by acting on the iodide of potassium, evolving iodine.

325. By moistening a very small slip of litmus paper in solution of caustic potassa, and then passing the electric spark over its length in the air, I gradually neutralized the alkali, and ultimately rendered the paper red; on drying it, I found that nitrate of potassa had resulted from the operation and that the paper had become touch paper.

326. Either litmus paper or white paper moistened in a strong solution of iodide of potassium, offers therefore a very simple, beautiful and ready means of illustrating Cavendish's experiment of the formation of nitric acid from the atmosphere.

---

323. *if [sparks] burst through the air, the effect referred to above ensues*: Since the effects of the spark are the same at both poles, they cannot represent "true electro-chemical powers of common electricity." Faraday will clarify the role of the spark in the next paragraph.

324. The spark triggers combination of nitrogen and oxygen to form nitric oxide—in Faraday's time called "nitric acid" in accordance with the terminology established by Lavoisier. Its solution in atmospheric moisture formed the compound we call "nitric acid" today. Faraday, and chemists generally, viewed that solution not as a new chemical compound but simply as the former acid (that is, the *oxide*) "in a high state of concentration as to water." Cavendish's "beautiful" experiment is identified in paragraph 326 as the formation of nitric acid from the atmosphere.

325. "Touch paper" is paper soaked in potassium nitrate, then dried. Used for firing gunpowder, it easily catches fire from a spark, but it burns very slowly.

Faraday infers that the following has taken place: *Spark* causes formation of nitric acid from the air, as described in the preceding paragraph. The acid reacts with "caustic potassa" (potassium hydroxide—a strong alkali) to produce the neutral salt *potassium nitrate*. The moist paper, having thus been steeped in potassium nitrate solution, will constitute "touch paper" when dried. It is evident that nitric acid was evolved in sufficient quantity to convert *all* the alkali to nitrate, since only thereafter could the water become acidic and so turn the litmus dye red, as observed.

327. I have already had occasion to refer to an experiment (265. 309.) made by Dr. Wollaston, which is insisted upon too much, both by those who oppose and those who agree with the accuracy of his views respecting the identity of voltaic and ordinary electricity. By covering fine wires with glass or other insulating substances, and then removing only so much matter as to expose the point, or a section of the wires, and by passing electricity through two such wires, the guarded points of which were immersed in water, Wollaston found that the water could be decomposed even by the current from the machine, without sparks, and that two streams of gas arose from the points, exactly resembling, in appearance, those produced by voltaic electricity, and, like the latter, giving a mixture of oxygen and hydrogen gases. But Dr. Wollaston himself points out that the effect is different from that of the voltaic pile, inasmuch as both oxygen and hydrogen are evolved from *each* pole; he calls it "a very close *imitation* of the galvanic phenomena," but adds that "in fact the resemblance is not complete," and does not trust to it to establish the principles correctly laid down in his paper.

*   *   *

332. iv. *Physiological effects.*—The power of the common electric current to shock and convulse the animal system, and when weak to affect the tongue and the eyes, may be considered as the same with the

---

327. In contrast to the usual practice of immersing lengths of wire, or even small plates, in water to introduce the current, Wollaston reduced the exposed wires to mere *points*, the rest of the surface being insulated with glass.

Like Cavendish's spark apparatus described in paragraph 322, Wollaston's "guarded points" give identical effects at both poles. Wollaston's results, however, are harder to explain. That an electrical spark could facilitate the *combustion* of a mixture of gases (such as hydrogen and oxygen) had long been known. But how can electrical discharge *produce* such a mixture at a single pole? Besides being completely opposed to the principle of *specific* electrochemical action at the respective poles, any electrical discharge capable of disengaging both gases should suffice to cause their combination again!

Remarkably, Faraday appears untroubled by this apparent departure from standard electro-chemical action and fails to suggest even the outline of an explanation. Evidently his concern is solely to show that the results with "guarded" points cannot represent the "true electro-chemical powers of common electricity" (paragraph 323). I will therefore omit most of Faraday's further references to Wollaston's exceptional experiment; but its explanation is not obvious and still engages the attention of commentators.

Note that *guarded points* are "guarded" by glass or some other insulator. That terminology did not long survive, however; and an electrical "guard" has since come to mean a *conductor* which provides a particular form of shielding in electrometers and other instruments.

similar power of voltaic electricity, account being taken of the intensity of the one electricity and duration of the other. When a wet thread was interposed in the course of the current of common electricity from the battery (291.) charged by eight or ten[15] revolutions of the machine in good action (290.) and the discharge made by platina spatulas through the tongue or the gums, the effect upon the tongue and eyes was exactly that of a momentary feeble voltaic circuit.

333. v. *Spark.*—The beautiful flash of light attending the discharge of common electricity is well known. It rivals in brilliancy, if it does not even very much surpass, the light from the discharge of voltaic electricity; but it endures for an instant only, and is attended by a sharp noise like that of a small explosion. Still no difficulty can arise in recognizing it to be the same spark as that from the voltaic battery, especially under certain circumstances. The eye cannot distinguish the difference between a voltaic and a common electricity spark, if they be taken between amalgamated surfaces of metal at intervals only, and through the same distance of air.

334. When the Leyden battery (291.) was discharged through a wet string placed in some part of the circuit away from the place where the spark was to pass, the spark was yellowish, flamy, having a duration sensibly longer than if the water had not been interposed, was about three-fourths of an inch in length, was accompanied by little or no noise, and whilst losing part of its usual character had approximated in some degree to the voltaic spark. When the electricity retarded by water was discharged between pieces of charcoal, it was exceedingly

[15] Or even from thirty to forty.

---

332. *intensity … duration*: Common electricity is characterized by its typically high intensity (or tension), voltaic electricity by its extended duration. To equalize them, Faraday discharges the Leyden battery through a wet thread to lower the tension, and makes only momentary voltaic connections to shorten the duration.

effect upon the tongue and the eyes: A weak current directed through the tongue can produce a distinctive sensation of taste; directed through the eyeball, the sensation of a bright flash of light may result. I cited these investigations—with a strong cautionary note—in the introduction to the First Series; see the second footnote on page 17.

333. *amalgamated surfaces of metal*: here, metal surfaces coated with mercury.

334. *electricity retarded by water*: See the earlier comment to paragraph 295. As a result of retardation, the discharge is *drawn out in time* (the spark had a "longer duration"), its noise is less sharp, and its flash less brilliant.

luminous and bright upon both surfaces of the charcoal, resembling the brightness of the voltaic discharge on such surfaces. When the discharge of the unretarded electricity was taken upon charcoal, it was bright upon both the surfaces, (in that respect resembling the voltaic spark,) but the noise was loud, sharp, and ringing.

335. I have assumed, in accordance, I believe, with the opinion of every other philosopher, that atmospheric electricity is of the same nature with ordinary electricity (284.), and I might therefore refer to certain published statements of chemical effects produced by the former as proofs that the latter enjoys the power of decomposition in common with voltaic electricity. But the comparison I am drawing is far too rigorous to allow me to use these statements without being fully assured of their accuracy; yet I have no right to suppress them, because, if accurate, they establish what I am labouring to put on an undoubted foundation, and have priority to my results.

336. M. Bonijol of Geneva[16] is said to have constructed very delicate apparatus for the decomposition of water by common electricity. By connecting an insulated lightning rod with his apparatus, the decomposition of the water proceeded in a continuous and rapid manner even when the electricity of the atmosphere was not very powerful. The apparatus is not described; but as the diameter of the wire is mentioned as very small, it appears to have been similar in construction to that of Wollaston (327.); and as that does not furnish a case of true polar electrochemical decomposition (328.), this result of M. Bonijol does not prove the identity in chemical action of common and voltaic electricity.

337. At the same page of the Bibliothèque Universelle, M. Bonijol is said to have decomposed *potash,* and also chloride of silver by putting them into very narrow tubes and passing electric sparks from an ordinary machine over them. It is evident that these offer no analogy to cases of true voltaic decomposition, where the electricity only decomposes when it is *conducted* by the body acted upon, and ceases to decompose, according to its ordinary laws; when it passes in sparks. These effects are probably partly analogous to that which takes place with water in Pearson's or Wollaston's apparatus, and may be due to

---

[16] Bibliothèque Universelle, 1830, tome xlv. p. 213.

---

336. *but as the diameter [is] very small*: Clearly he regards the *smallness* of Wollaston's wire as responsible for the strange results obtained with that apparatus; so that any decomposition device having poles of similar dimensions will be equally suspect. In "true *polar* decomposition" hydrogen and oxygen would be evolved at separate poles—in contrast to Wollaston's results.

very high temperature acting on minute portions of matter; or they may be connected with the results in air (322.). As nitrogen can combine directly with oxygen under the influence of the electric spark (324.), it is not impossible that it should even take it from the potassium of the potash, especially as there would be plenty of potassa in contact with the acting particles to combine with the nitric acid formed. However distinct all these actions may be from true polar electro-chemical decompositions, they are still highly important, and well worthy of investigation.

*   *   *

III. *Magneto-Electricity.*

343. *Tension.*—The attractions and repulsions due to the tension of ordinary electricity have been well observed with that evolved by magneto-electric induction. M. Pixii by using an apparatus, clever in its construction and powerful in its action,[18] was able to obtain great divergence of the gold leaves of an electrometer.[19]

344. *In motion*: i. *Evolution of Heat*—The current produced by magneto-electric induction can heat a wire in the manner of ordinary electricity. At the British Association of Science at Oxford, in June of the present year, I had the pleasure, in conjunction with Mr. Harris, Professor Daniell, Mr. Duncan, and others, of making an experiment, for which the great magnet in the museum, Mr. Harris's new electrometer (287.), and the magneto-electric coil described in my first paper (34.), were put in requisition. The latter had been modified in the manner I have elsewhere described,[20] so as to produce an electric spark when its contact with the magnet was made or broken. The terminations of the spiral, adjusted so as to have their contact with each other broken when the spark was to pass, were connected with the wire in the electrometer, and it was found that each time the magnetic contact was made and broken, expansion of the air within the instrument occurred, indicating an increase, at the moment, of the temperature of the wire.

[18] Annales de Chimie, l. p. 822.

[19] Ibid. li. p. 77.

---

337. "True polar electro-chemical decompositions" are effected by *conduction*, not by influence or proximity. Note that he finally advances a suggestion for explaining the peculiar effect obtained with Wollaston's apparatus.

345. ii. *Magnetism.*—These currents were discovered by their magnetic power.

346. iii. *Chemical decomposition.*—I have made many endeavours to effect chemical decomposition by magneto-electricity, but unavailingly. In July last I received an anonymous letter (which has since been published[21]) describing a magnetoelectric apparatus, by which the decomposition of water was effected. As the term "guarded points" is used, I suppose the apparatus to have been Wollaston's (327. &c.), in which case the results did not indicate polar electro-chemical decomposition. Signor Botto has recently published certain results which he has obtained;[22] but they are, as at present described, inconclusive. The apparatus he used was apparently that of Dr. Wollaston, which gives only fallacious indications (327. &c.). As magneto-electricity can produce sparks, it would be able to show the effects proper to this apparatus. The apparatus of M. Pixii already referred to (343.) has however, in the hands of himself[23] and M. Hachette,[24] given decisive chemical results, so as to complete this link in the chain of evidence. Water was decomposed by it, and the oxygen and hydrogen obtained in separate tubes according to the law governing volta-electric and machine-electric decomposition.

347. iv. *Physiological effects.*—A frog was convulsed in the earliest experiments on these currents (56.). The sensation upon the tongue, and the flash before the eyes, which I at first obtained only in a feeble degree (56.), have been since exalted by more powerful apparatus, so as to become even disagreeable.

348. v. *Spark.*—The feeble spark which I first obtained with these currents (32.), has been varied and strengthened by Signori Nobili and Antinori, and others, so as to leave no doubt as to its identity with the common electric spark.

[20] Phil. Mag. and Annals, 1832, vol. xi. p. 405.

[21] Lond. and Edin. Phil. Mag. and Journ. 1632, vol. i. p. 161.

[22] Ibid. 1832, vol. i. p. 441.

[23] Annales de Chimie, li. p. 77.

[24] Ibid. li. p. 72.

---

345. *discovered*: here, revealed, disclosed.

346. *oxygen and hydrogen obtained in separate tubes*: It is the physical separation of products, not the decomposition itself, that indicates "true polar electro-chemical decomposition" (paragraph 337).

IV. *Thermo-Electricity.*

349. With regard to thermo-electricity, (that beautiful form of electricity discovered by Seebeck,) the very conditions under which it is excited are such as to give no ground for expecting that it can be raised like common electricity to any high degree of tension; the effects, therefore, due to that state are not to be expected. The sum of evidence respecting its analogy to the electricities already described, is, I believe, as follows:—*Tension.* The attractions and repulsions due to a certain degree of tension have not been observed. *In currents*: i. *Evolution of Heat.* I am not aware that its power of raising temperature has been observed. ii. Magnetism. It was discovered, and is best recognised, by its magnetic powers. iii. *Chemical decomposition* has not been effected by it. iv. Physiological effects. Nobili has shown[25] that these currents are able to cause contractions in the limbs of a frog. v. Spark. The spark has not yet been seen.

350. Only those effects are weak or deficient which depend upon a certain high degree of intensity; and if common electricity be reduced in that quality to a similar degree with the thermo-electricity, it can produce no effects beyond the latter.

V. *Animal Electricity.*

351. After an examination of the experiments of Walsh,[26] Ingenhousz,[27] Cavendish,[28] Sir H. Davy,[29] and Dr. Davy,[30] no doubt remains on my mind as to the identity of the electricity of the torpedo with common and voltaic electricity; and I presume that so little will remain on the minds of others as to justify my refraining from entering

[25] Bibliothèque Universelle, xxxvii. 15.

[26] Philosophical Transactions, 1773, p. 461.

[27] Ibid. 1775, p. 1.

[28] Ibid. 1776, p. 196.

[29] Ibid. 1829, p. 15.

[30] Ibid. 1832, p. 259.

---

349. Thermo-electricity is described in the introduction.

351. *the electricity of the torpedo*: See also Faraday's researches on the electric eel ("gymnotus") in the Fifteenth Series.

at length into the philosophical proofs of that identity. The doubts raised by Sir H. Davy have been removed by his brother Dr. Davy; the results of the latter being the reverse of those of the former. At present the sum of evidence is as follows:—

352. *Tension.*—No sensible attractions or repulsions due to tension have been observed.

353. *In motion*: i. *Evolution of Heat;* not yet observed; I have little or no doubt that Harris's electrometer would show it (287. 359.).

354. ii. *Magnetism.*—Perfectly distinct. According to Dr. Davy,[31] the current deflected the needle and made magnets under the same law, as to direction, which governs currents of ordinary and voltaic electricity.

355. iii. *Chemical decomposition.*—Also distinct; and though Dr. Davy used an apparatus of similar construction with that of Dr. Wollaston (327.), still no error in the present case is involved, for the decompositions were polar, and in their nature truly electro-chemical. By the direction of the magnet, it was found that the under surface of the fish was negative, and the upper positive; and in the chemical decompositions, silver and lead were precipitated on the wire connected with the under surface and not on the other; and when these wires were either steel or silver, in solution of common salt, gas (hydrogen?) rose from the negative wire, but none from the positive.

356. Another reason for the decomposition being electrochemical is, that a Wollaston's apparatus constructed with *wires,* coated with

---

[31] Philosophical Transactions, 1832, p. 200.

---

354. *the same law, as to direction*: That is, the electromagnetic right-hand rule or its equivalent.

355. *the decompositions were polar, and in their nature truly electro-chemical*: Thus Davy has obtained electrochemical effects from animal electricity. Note that since he gained these results using Wollaston's apparatus (paragraph 327), it is clear that Wollaston's device is *capable* of causing "polar" decomposition. Why does it exhibit the true "polar" action for electricity obtained from the torpedo, but not for electricity produced by the frictional machine? Faraday suggests a reason in the next paragraph.

By the direction of the magnet...: Faraday refers to the "magnets" mentioned in the previous paragraph. Dr. Davy had magnetized a needle by passing the torpedo's electricity through a coiled wire joining the upper and lower surfaces of the fish. From the direction of magnetization, Davy could deduce which surface was positive and which negative by using the electromagnetic right-hand rule (or its equivalent).

sealing wax, would most probably not have decomposed water, even in its own peculiar way, unless the electricity had risen high enough in intensity to produce sparks in some part of the circuit; whereas the torpedo was not able to produce sensible sparks. A third reason is, that the purer the water in Wollaston's apparatus, the more abundant is the decomposition: and I have found that a machine and wire points which succeeded perfectly well with distilled water, failed altogether when the water was rendered a good conductor by sulphate of soda, common salt, or other saline bodies. But in Dr. Davy's experiments with the torpedo, *strong* solutions of salt nitrate of silver, and superacetate of lead were used successfully, and there is no doubt with more success than weaker ones.

357. iv. *Physiological effects.*—These are so characteristic, that by them the peculiar powers of the torpedo and gymnotus are principally recognised.

358. v. *Spark.*—The electric spark has not yet been obtained, or at least I think not; but perhaps I had better refer to the evidence on this point. Humboldt, speaking of results obtained by M. Fahlberg, of Sweden, says, "This philosopher has seen an electric spark, as Walsh and Ingenhousz had done before him at London, by placing the gymnotus in the air, and interrupting the conducting chain by two gold leaves pasted upon glass, and a line distant from each other."[32] I cannot, however, find any record of such an observation by either Walsh or

---

[32] Edinburgh Phil. Journal, ii. p. 249.

---

356. *would most probably not have decomposed water ... unless the electricity had risen high enough in intensity...*: Faraday implies that the problem with Wollaston's apparatus was probably the combination of its *tiny* junction between conductors and water ("wires coated with sealing wax") and the *high* electrical tension produced by the frictional machine. Since the torpedo's discharge is *low*-tension (it does not spark), the peculiarities of Wollaston's device did not appear.

But why should small dimensions coupled with high tensions deviate from the usual electro-chemical action, as Faraday seems to imply? In the Twelfth Series Faraday will describe how insulating materials tend to *break down* in the vicinity of small or pointed conductors electrified to a high degree. This suggests the idea that Wollaston's nonpolar decomposition involves disruptive mechanical forces, not electro-chemical powers primarily.

357. The "gymnotus" is Faraday's *electric eel*, as he will explain in the next paragraph. See also the editors's introduction to this Series.

Ingenhousz, and do not know where to refer to that by M. Fahlberg. M. Humboldt could not himself perceive any luminous effect.

Again, Sir John Leslie, in his dissertation on the progress of mathematical and physical science, prefixed to the seventh edition of the Encyclopædia Britannica, Edinb. 1830, p. 622, says, "From a healthy specimen" of the *Silurus electricus,* meaning rather the *gymnotus,* "exhibited in London, vivid sparks were drawn in a darkened room"; but he does not say he saw them himself, nor state who did see them; nor can I find any account of such a phenomenon; so that the statement is doubtful.[33]

359. In concluding this summary of the powers of torpedinal electricity, I cannot refrain from pointing out the enormous absolute quantity of electricity which the animal must put in circulation at each effort. It is doubtful whether any common electrical machine has as yet been able to supply electricity sufficient in a reasonable time to cause true electro-chemical decomposition of water (330. 339.), yet the current from the torpedo has done it. The same high proportion is shown by the magnetic effects (296. 371.). These circumstances indicate that the torpedo has power (in the way probably that Cavendish describes) to continue the evolution for a sensible time, so that its successive discharges rather resemble those of a voltaic arrangement, intermitting in its action, than those of a Leyden apparatus, charged and discharged many times in succession. In reality, however, there is no *philosophical difference* between these two cases.

360. The general conclusion which must, I think, be drawn from this collection of facts is that electricity, whatever may be its source,

---

[33] Mr. Brayley, who referred me to these statements, and has extensive knowledge of recorded facts, is unacquainted with any further account relating to them.

---

358. "luminous" refers here to emission of light from attached apparatus, not from the animal itself!

359. If, as Faraday anticipated in paragraph 329 above, the amount of material decomposed depends upon the quantity of electricity passed, the torpedo must be capable of producing electricity in quantities that far exceed what any electric machine can evolve in a comparable time. In that respect, at least, the *voltaic pile* is probably a better image for the animal's electric power than is the *Leyden battery*. Note also that a voltaic cell acts continuously, while a Leyden jar requires successive charging and discharging. Still, he reminds us, the distinction is only figurative—not "philosophical" as defined in the comment to paragraph 267.

is identical in its nature. The phenomena in the five kinds or species quoted, differ, not in their character but only in degree; and in that respect vary in proportion to the variable circumstances of quantity and intensity[34] which can at pleasure be made to change in almost any one of the kinds of electricity, as much as it does between one kind and another.

Table of the experimental Effects common to the Electricities derived from different Sources.[35]

	Physiological Effects.	Magnetic Deflection.	Magnets made.	Spark.	Heating Power.	True chemical Action.	Attraction and Repulsion.	Discharge by Hot Air.
1. Voltaic electricity	×	×	×	×	×	×	×	×
2. Common electricity...	×	×	×	×	×	×	×	×
3. Magneto-Electricity..	×	×	×	×	×	×	×	
4. Thermo-Electricity...	×	×	+	+	+	+		
5. Animal Electricity...	×	×	×	+	+	×		

§ 8. *Relation by Measure of common and voltaic Electricity.*[36]

361. Believing the point of identity to be satisfactorily established, I next endeavoured to obtain a common measure, or a known relation as to quantity, of the electricity excited by a machine, and that from a voltaic pile; for the purpose not only of confirming their

[34] The term *quantity* in electricity is perhaps sufficiently definite as to sense; the term *intensity* is more difficult to define strictly. I am using the terms in their ordinary and accepted meaning.

[35] Many of the spaces in this table originally left blank may now be filled. Thus with *thermo-electricity*, Botto made magnets and obtained polar chemical decomposition: Antinori produced the spark; and if it has not been done before, Mr. Watkins has recently heated a wire in Harris's thermo-electrometer. In respect to *animal electricity*, Matteucci and Linari have obtained the spark from the torpedo, and I have recently procured it from the gymnotus: Dr. Davy has observed the heating power of the current from the torpedo. I have therefore filled up these spaces with crosses, in a different position to the others originally in the table. There remain but five spaces unmarked, two under *attraction* and *repulsion*, and three under *discharge by hot air*; and though these effects have not yet been obtained, it is a necessary conclusion that they must be possible, since the *spark* corresponding to them has been procured. For when a discharge across cold air can occur, that intensity which is the only essential additional requisite for the other effects must be present.—Dec. 18, 1838.

[36] In further illustration of this subject see 855–873 in Series VII.—Dec. 1838.

identity (378.), but also of demonstrating certain general principles (366. 377, &c.), and creating an extension of the means of investigating and applying the chemical powers of this wonderful and subtile agent.

362. The first point to be determined was, whether the same absolute quantity of ordinary electricity, sent through a galvanometer, under different circumstances, would cause the same deflection of the needle. An arbitrary scale was therefore attached to the galvanometer, each division of which was equal to about 4°, and the instrument arranged as in former experiments (296.). The machine (290.), battery (291.), and other parts of the apparatus were brought into good order, and retained for the time as nearly as possible in the same condition. The experiments were alternated so as to indicate any change in the condition of the apparatus and supply the necessary corrections.

363. Seven of the battery jars were removed, and eight retained for present use. It was found that about forty turns would fully charge the eight jars. They were then charged by thirty turns of the machine, and discharged through the galvanometer, a thick wet string, about ten inches long, being included in the circuit. The needle was immediately deflected five divisions and a half, on the one side of the zero, and in vibrating passed as nearly as possible through five divisions and a half on the other side.

---

362. *whether the same absolute quantity of ordinary electricity ... cause the same deflection of the needle*: Faraday does not say why one would even suppose such an equality to hold; but notice that, if established, it will constitute a measure of quantity of electricity based on *magnetic* effects, corresponding to the anticipated measure based on *chemical* effects (paragraph 329).

Since the discharges of ordinary electricity are momentary rather than continuous, the galvanometer "deflection" refers not to a steady but a transitory *peak* displacement. Subsequently, this practice came to be called the *ballistic* mode of the galvanometer, described in the introduction to the First Series. Faraday will recount a ballistic mode measurement in the next paragraph.

363. Forty turns of the machine will "fully charge" eight jars. He therefore charges them with only *thirty* turns, since electrification too near the limiting degree might result in spontaneous or irregular discharge. The "thick wet string" serves, as before (paragraph 313), to lower the tension and avoid spark or other discharge across the galvanometer coil. Discharging the seven jars takes only a fraction of a second, but it sets the galvanometer needle into continued vibration about its central resting position. The amplitude of vibration gradually decays, but it is the initial maximum (5½ divisions) that is significant.

364. The other seven jars were then added to the eight, and the whole fifteen charged by thirty turns of the machine. The Henley's electrometer stood not quite half as high as before; but when the discharge was made through the galvanometer, previously at rest, the needle immediately vibrated, passing *exactly* to the same division as in the former instance. These experiments with eight and with fifteen jars were repeated several times alternately with the same results.

365. Other experiments were than made, in which all the battery was used, and its charge (being fifty turns of the machine,) sent through the galvanometer: but it was modified by being passed sometimes through a mere wet thread, sometimes through thirty-eight inches of thin string wetted by distilled water and sometimes through a string of twelve times the thickness, only twelve inches in length, and soaked in dilute acid (298.). With the thick string the charge passed at once; with the thin string it occupied a sensible time, and with the *thread* it required two or three seconds before the electrometer fell entirely down. The current therefore must have varied extremely in intensity in these different cases, and yet the deflection of the needle was sensibly the same in all of them. If any difference occurred, it was that the thin string and thread caused greatest deflection; and if there is any lateral transmission, as M. Colladon says, through the silk in the galvanometer coil, it ought to have been so, because then the intensity is lower and the lateral transmission less.

---

364. Now 15 jars are charged with what must be the same quantity of electricity as before, being the amount evolved by thirty turns of the machine. A "Henley's electrometer" (illustrated in the editor's introduction to this Series) shows a lowered reading, indicating that when the same quantity of electricity is distributed over a greater number of jars, the tension decreases. This rough *ordinal* relation (expressing only greater, equal, or less) will be superseded in the Twelfth Series when Faraday argues for a relation of inverse *proportion*; see paragraph 1372, for example.

Discharging the 15 jars produces the same throw of the galvanometer needle as before. Thus *equal ballistic deflections of the galvanometer indicate equal quantities of electricity, irrespective of tension*. Faraday will declare this explicitly in paragraph 366.

365. *the battery*: that is, the battery of Leyden jars.

lateral transmission: that is, current passing between adjacent coil windings ("through the silk in the galvanometer coil") and so failing to contribute to the galvanometer deflection. Such "leakage" (as it is sometimes called) would be more likely to develop when the coil experiences higher electrical tensions. In the present case, tension will be highest when the discharge is received through a thick, highly-conductive string. Thus one might expect to obtain

366. Hence it would appear that if the same absolute quantity of electricity pass through the galvanometer, whatever may be its intensity, the deflecting force upon the magnetic needle is the same.

367. The battery of fifteen jars was then charged by sixty revolutions of the machine, and discharged, as before, through the galvanometer. The deflection of the needle was now as nearly as possible to the eleventh division, but the graduation was not accurate enough for me to assert that the arc was exactly double the former arc; to the eye it appeared to be so. The probability is, that *the deflecting force of an electric current is directly proportional to the absolute quantity of electricity passed,* at whatever intensity that electricity may be.[37]

[37] The great and general value of the galvanometer, as an actual measure of the electricity passing through it, either continuously or interruptedly, must be evident from a consideration of these two conclusions. As constructed by Professor Ritchie with glass threads (see Philosophical Transactions, 1830, p. 218, and Quarterly Journal of Science, New Series, vol. i. p. 29), it apparently seems to leave nothing unsupplied in its own department.

---

greater galvanometer deflections when the thin, less conductive string is employed; for then, "the intensity is lower and the lateral transmission less."

In the remark just quoted, the term "intensity" can be taken as equivalent to "tension." But note Faraday's surprising allusion, earlier in the paragraph, to intensity *of current*: "the current therefore must have varied extremely in intensity…" He appears to base this inference on the widely-varying times of discharge, which suggests that he views *intensity of current* as roughly equivalent to the *time rate of discharge.* If so, it would represent a usage quite different from that "intensity" which is nearly synonymous with "tension." On the other hand, it is true that the same measures which varied the *discharge rate* would have altered the *tension* as well; so which quantity does Faraday mean to single out? He is ordinarily careful to avoid such equivocation, at least within a single paragraph; this lapse is unusual.

366. Remember that the rule enunciated here applies only when the galvanometer is used in the *ballistic* mode—that is, when the time of discharge is short compared to the galvanometer's period of oscillation.

deflecting force: here, simply the power to produce a ballistic "throw" of given length. Faraday does not mean "force" in the sense of Newton's "motive force"; but the statement would continue to be true if he did.

Thus the ballistic galvanometer indicates *equality* between quantities of electricity and, by extension, relations of *greater* and *less* as well—sufficient to establish what in measurement theory is called an *ordinal scale.* In the following paragraph, Faraday will show *proportionality* between the galvanometer throw and the quantity of discharge; that in turn will suffice to establish a *ratio scale* for quantity of electricity, although that is of course not his major intention here.

368. Dr. Ritchie has shown that in a case where the intensity of the electricity remained the same, the deflection of the magnetic needle was directly as the quantity of electricity passed through the galvano-meter.[38] Mr. Harris has shown that the *heating* power of common electricity on metallic wires is the same for the same quantity of electricity whatever its intensity might have previously been.[39]

369. The next point was to obtain a *voltaic* arrangement producing an effect equal to that just described (367). A platina and a zinc wire were passed through the same hole of a draw-plate, being then one eighteenth of an inch in diameter; these were fastened to a support, so that their lower ends projected, were parallel, and five sixteenths of an inch apart. The upper ends were well connected with the galvano-meter wires. Some acid was diluted, and, after various preliminary experiments, that adopted as a standard which consisted of one drop strong sulphuric acid in four ounces distilled water. Finally, the time was noted which the needle required in swinging either from right to left or left to right: it was equal to seventeen beats of my watch, the latter giving one hundred and fifty in a minute. The object of these prepara-tions was to arrange a voltaic apparatus, which, by immersion in a given acid for a given time, much less than that required by the needle to swing in one direction, should give equal deflection to the instrument with the discharge of ordinary electricity from the battery (363, 364.); and a new part of the zinc wire having been brought into position with the platina, the comparative experiments were made.

[38] Quarterly Journal of Science, New Series, Vol. i. p. 33.

[39] Plymouth Transactions, page 22.

---

368. *Mr. Harris has shown that the heating power … is the same for the same quantity of electricity whatever its intensity…*: That conclusion, however, is controverted by later investigations which show that the heating power of an electric discharge is proportional to its quantity and intensity (tension) *jointly.*

369. *a voltaic arrangement producing an effect equal to that just described*: He has already shown "common" and "voltaic" electricity to be identical in their natures, since they produce the same general effects. But now that it is clear that the "effect" on a ballistic galvanometer signifies *quantity of electricity*, he will for the first time be able to exhibit "common" and "voltaic" electricity *in equal quantities.*

The galvanometer needle completed one swing in either direction during "seventeen beats of my watch," or $17/150$ minute, which equals 6.8 seconds. He means to apply the voltaic current for a time "much less than" that interval, for galvanometer measurement in the ballistic mode *requires* brevity of current flow, as noted in the editor's introduction to the First Series.

370. On plunging the zinc and platina wires five eighths of an inch deep into the acid, and retaining them there for eight beats of the watch, (after which they were quickly withdrawn,) the needle was deflected, and continued to advance in the same direction some time after the voltaic apparatus had been removed from the acid. It attained the five-and-a-half division, and then returned, swinging an equal distance on the other side. This experiment was repeated many times, and always with the same result.

371. Hence, as an approximation, and judging from *magnetic force* only at present (376.), it would appear that two wires, one of platina and one of zinc, each one eighteenth of an inch in diameter, placed five sixteenths of an inch apart and immersed to the depth of five eighths of an inch in acid, consisting of one drop oil of vitriol and four ounces distilled water, at a temperature about 60°, and connected at the other extremities by a copper wire eighteen feet long and one eighteenth of an inch thick (being the wire of the galvanometer coils), yield as much electricity in eight beats of my watch, or in $8/150$ths of a minute, as the electrical battery charged by thirty turns of the large machine, in excellent order (363. 364.). Notwithstanding this apparently enormous disproportion, the results are perfectly in harmony with

---

370. He finds that if the voltaic current is applied for eight beats of the watch (equal to 3.2 seconds), it will produce the same galvanometer swing as he previously obtained in the discharge of ordinary electricity from the Leyden battery.

The eight-beat duration was found initially by trial and error and is confirmed by multiple repetitions. Moreover it is less than the seventeen beats required for one swing of the needle, and evidently Faraday regards it as sufficiently brief to qualify the galvanometer for ballistic operation. But what are the requirements, exactly? *How* short in duration does the measured current have to be? He will discuss these questions more explicitly in the 28th Series; see paragraphs 3103–3105.

371. The specified voltaic apparatus yields as much electricity in 3.2 seconds as are evolved by thirty turns of his large and powerful electrical machine. But what "enormous disproportion" does Faraday see here? Is it (i) a disproportion in time? Thirty turns of the machine probably required about 24 seconds (paragraph 290), compared with 3.2 seconds for the voltaic current evolving the same quantity of electricity. On the other hand, when the Leyden battery was discharged through strings (paragraph 365), he reported times ranging from 3 seconds down to a fraction of a second ("at once")—but these ratios cannot be styled "enormous." Is it (ii) a disproportion in material? The electrical machine indeed requires vastly more weight and bulk than the little voltaic device to produce comparable quantities of electricity. I suspect that what he has in mind above all, though, is (iii) a disproportion in power. But we will see this more clearly in the next paragraph.

those effects which are known to be produced by variations in the intensity and quantity of the electric fluid.

372. In order to procure a reference to *chemical action*, the wires were now retained immersed in the acid to the depth of five eighths of an inch, and the needle, when stationary, observed; it stood, as nearly as the unassisted eye could decide, at $5\frac{1}{3}$ division. Hence a permanent deflection to that extent might be considered as indicating a constant voltaic current, which in eight beats of my watch (369.), could supply as much electricity as the electrical battery charged by thirty turns of the machine.

373. The following arrangements and results are selected from many that were made and obtained relative to chemical action. A platina wire one twelfth of an inch in diameter, weighing two hundred and sixty grains, had the extremity rendered plain so as to offer a definite surface equal to a circle of the same diameter as the wire; it was then connected in turn with the conductor of the machine, or with the voltaic apparatus (369.), so as always to form the positive pole, and at the same time retain a perpendicular position, that it

---

371, continued. *the electric fluid*: Presumably this is figurative speech. Nowhere else does he appear to accept the fluid vocabulary uncritically.

372. Faraday now supplies to the galvanometer the same voltaic current as before; but instead of disconnecting it after 3.2 seconds, he waits until the galvanometer needle has assumed a *steady* (not a ballistic) deflection. That steady deflection proves to be about $5\frac{1}{3}$ divisions—only slightly less than the previous ballistic maximum (paragraph 370) had been. This in effect calibrates the galvanometer to measure *steady currents*, as he formerly had done for *brief discharges*. We might restate the results as follows: A steady deflection of $5\frac{1}{3}$ divisions represents a continuous current easily obtainable from a tiny voltaic device, but which, to be equaled by Faraday's electrical machine, would require the six-foot glass disk to be turned thirty times in $8/150$ minute, or more than nine times per second! Since Faraday previously characterized $1\frac{1}{4}$ turns per second as a "moderate" rate of working (paragraph 290), this would represent a very strenuous pace. Or, if one man working moderately can turn the machine $1\frac{1}{4}$ times per second, then it would take *seven* men, each working even more vigorously, to operate the machine at the requisite rate! If we consider not only the required speed, but the effort necessary to maintain that speed, perhaps we can better appreciate the "disproportion" he remarked on in the previous paragraph. It will acquire a more fundamental significance in the Seventh Series (paragraph 861).

373. The galvanometer indicated the *magnetic* power of an electric current. Faraday now uses paper moistened with "hydriodate of potassa" (potassium iodide, KI) as an indicator of *chemical* power: when current passes through the KI solution (which is a conductor) iodine is evolved at the positive pole.

might rest, with its whole weight, upon the test paper to be employed. The test paper itself was supported upon a platina spatula, connected either with a discharging train (292.), or with the negative wire of the voltaic apparatus, and it consisted of four thicknesses, moistened at all times to an equal degree in a standard solution of hydriodate of potassa (316.)

374. When the platina wire was connected with the prime conductor of the machine, and the spatula with the discharging train, ten turns of the machine had such decomposing power as to produce a pale round spot of iodine of the diameter of the wire; twenty turns made a much darker mark, and thirty turns made a dark brown spot penetrating to the second thickness of the paper. The difference in effect produced by two or three turns, more or less, could be distinguished with facility.

375. The wire and spatula were then connected with the voltaic apparatus (369.) the galvanometer being also included in the arrangement; and, a stronger acid having been prepared, consisting of nitric acid and water, the voltaic apparatus was immersed so far as to give a permanent deflection of the needle to the 5 1/3 division (372.), the fourfold moistened paper intervening as before.[40] Then by shifting the end of the wire from place to place upon the test paper, the effect of the current for five, six, seven, or any number of the beats of the watch (369.) was observed, and compared with that of the machine. After alternating and repeating the experiments of comparison many times, it was constantly found that this standard current of voltaic electricity, continued for eight beats of the watch, was equal, in chemical effect, to thirty turns of the machine; twenty-eight revolutions of the machine were sensibly too few.

[40] Of course the heightened power of the voltaic battery was necessary to compensate for the bad conductor now interposed.

---

374. The chemical indicator has a precision of 2 or 3 turns; that is, it can distinguish quantities of electricity to within the amount evolved by 2 or 3 turns of the machine.

375. Restating the results: Evolution of iodine at a certain rate requires a current that is easily obtainable from a small voltaic apparatus, but which, in order to be supplied by Faraday's large electrical machine, would require the glass to be turned thirty times in $8/150$ minute—the same outcome as in paragraph 372, judged this time by *chemical* rather than *magnetic* effect. Note that the results are good to better than 2 turns out of 30, just as he established in paragraph 374—a precision better than 7%.

376. Hence it results that both in *magnetic deflection* (371.) and in *chemical force,* the current of electricity of the standard voltaic battery for eight beats of the watch was equal to that of the machine evolved by thirty revolutions.

377. It also follows that for this case of electro-chemical decomposition, and it is probable for all cases, that the *chemical power, like the magnetic force* (366.) *is in direct proportion to the absolute quantity of electricity* which passes.

378. Hence arises still further confirmation, if any were required, of the identity of common and voltaic electricity, and that the differences of intensity and quantity are quite sufficient to account for what were supposed to be their distinctive qualities.

379. The extension which the present investigations have enabled me to make of the facts and views constituting the theory of electro-chemical decomposition; will, with some other points of electrical doctrine, be almost immediately submitted to the Royal Society in another series of these Researches.

Royal Institution,
   15th Dec. 1832.

[A long note which Faraday appended to this Series is here omitted. In paragraph 78 of the First Series he had imputed certain errors to Ampère; he now withdraws those charges, suspecting that he may have misunderstood some of Ampère's descriptions. —Editor]

---

376. The specific equality here established between voltaic and frictional electricity is expressed in highly idiosyncratic units; but what is important is the fact of equality itself. By the means here described, equality can in principle be established among *any* combination of voltaic and frictional electrical sources.

377. We have then an instance confirming the relation he had anticipated in paragraph 329 above—that *chemical power is proportional to quantity of electricity.* Similarly the magnetic power exerted by a current is found to be a measure of the quantity of electricity that passes—or rather the amount that passes *in a given time* (making allowance for the mode of employment of the galvanometer). These propositions represent great advances in the art of measurement of electric quantities. Yet I think we can discern in the foregoing procedure that *measurement,* to Faraday, is not the good-in-itself that modern quantitative science often seems to crave. Rather, when measurements prove to be possible they signify the existence of deeper affinities and equivalencies. It is finally those interdependencies, Faraday considers, which constitute our sound knowledge about the world, and which—as a magnificent and sometimes astonishing blessing—can often be brought to light.

Seventh Series — Editor's Introduction

Electro-chemical decomposition

I mentioned in the Introductions to the First and Second Series that with the voltaic cell came the discovery that electricity was capable of causing chemical decomposition of certain substances; indeed, in the Third Series we saw Faraday make use of chemical effects, both to indicate the existence of electric currents and as a criterion for identity of electricity from different sources.

We must now examine one instance more thoroughly—the electrical decomposition of water. The sketch shows one of Faraday's decomposition vessels. It is filled with water, which may be pure or may contain a small amount of salt, acid, or other substances.* Wires are led into the apparatus through water-tight tubes and soldered to pieces of platinum foil (platinum is desirable because it is not readily attacked by most other substances). When the wires are connected to the positive and negative poles of a voltaic battery, gas bubbles begin to form about both foils and gas gradually collects in the tubes. As the labeling indicates, the gases differ: hydrogen evolves at the negatively-electrified foil, while oxygen is liberated at the positively-electrified foil. As the drawing also shows, the volume of hydrogen gas produced is about double that of oxygen. Evidently water is being decomposed; and this action proceeds for so long—and only so long—as the electric current continues.**

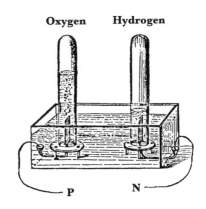

Oxygen   Hydrogen

P   N

---

* Pure water is a rather bad conductor and develops only feeble currents under action of ordinary voltaic batteries, with correspondingly slow rates of decomposition. The addition of even tiny amounts of certain substances greatly enhances both the conduction and the rate of decomposition—usually without consuming the additives themselves. Faraday suggested a tentative explanation for this effect in the Fourth Series; I will quote it on page 188 below.

** Faraday can independently confirm continuance of the current in several ways: bringing a compass near either wire, or interrupting the circuit with pieces of charcoal to strike an electric arc between them, or—if the wires are small enough, by observing the current's heating effect upon them. But of course he had long before learned to recognize the *chemical action itself* as an indication of current flow, as the Third Series makes clear.

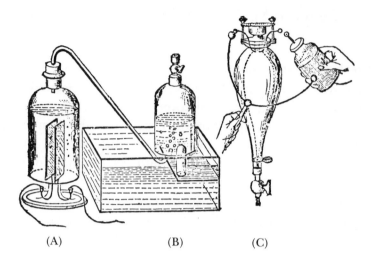

(A)   (B)   (C)

A second variety of decomposition apparatus makes no attempt to collect the hydrogen and oxygen gases separately. The device (A) produces a mixture of hydrogen and oxygen gases from water and collects the mixed gases by means of the *water trough* (B).* Such a mixture is highly explosive—and the apparatus is capable of accumulating it in rather dangerous quantities. For that reason this form of the instrument is no longer in common use, and I strongly caution the reader against assembling one. Faraday, however, seems to have employed it routinely. He evidently used it in the Christmas Lectures of 1860–61** (the drawing shown here is taken from an illustration for one of them).

The danger lies, of course, in the chance of accidental explosion in the collection vessel, which may shatter and cause injury from flying glass. In contrast, the flask (C) is made to withstand the force of an explosion between mixed hydrogen and oxygen gases. It is first evacuated with an air-pump; then the gas mixture is transferred from the collection bottle to the combustion flask by joining their matching connectors. Two wires at the top of the flask are connected to platinum contacts inside; and when a spark from the Leyden jar is passed between them the hydrogen-oxygen mixture explodes with a flash of light, forming water which condenses in tiny droplets on the bottle walls.

It is easy to show that the mixed gases combine completely in the explosion, with no excess of either gas left over: Place the flask's mouth under water and open the stopcock; liquid will pour in to fill the flask completely. Since there is no residue of either gas, it is clear that the

---

* Faraday will employ the water trough at paragraph 866 in the present Series.

** *The Chemical History of a Candle*, London, 1861, 1865, and numerous reprints. Several modern editions feature new illustrations.

proportions in which the gases were evolved by electro-chemical decomposition of water are the same as the proportions in which they combine again by ordinary chemical combustion. Faraday finds this agreement—say, rather, identity—between the laws of ordinary chemical combination and *electro*-chemical decomposition to be profoundly suggestive. To state baldly what he will formulate incrementally and with scrupulous care: If the electrical and chemical powers exhibit identical laws, may they not be in fact *the same power*?

Electro-chemical "poles"

The electrified terminals of the water decomposition apparatus are typically called its *poles*. This terminology is expressive but also somewhat ambiguous. On the one hand, "poles" denotes *a pair of related opposites*, such as the north and south poles of the earth, or the metaphorical "poles" of the political spectrum. But the term sometimes bears an additional, more causative stamp as a *point of guidance* or *center of action*. The "poles" of a magnet, conceived as centers of attraction and repulsion, are examples of this. The electrochemical "poles" were similarly conceived by many investigators as centers of attractive and repulsive force capable of acting "at a distance" to tear water or other substances apart, directing the newly-liberated components towards themselves.

Faraday is highly dissatisfied with such views, fundamentally because he regards such action as incompatible with the unity (and hence the intelligibility) of nature;* but more immediately because he has already shown by experiment that *poles*—in the sense of identifiable centers of force—are wholly unnecessary for electrical decomposition! In 1833, as part of the Fifth Series of Experimental Researches, he set up the apparatus drawn here.

Litmus and turmeric are dyes that turn red in the presence of acids and alkalies, respectively. Faraday

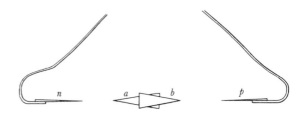

moistened a piece of turmeric paper *a* and a piece of litmus paper *b* with a solution of sodium sulfate. Two needles *p* and *n* were connected to the positive and negative terminals of an electrical

---

* Faraday is generally suspicious of theories that divide phenomena too neatly into the *active* and the *passive*. A potent but indifferent Zeus, hurling down his thunderbolts upon a submissive earth, is not the paradigm for nature! Rather, natural powers are reciprocal and interdependent. He will grow more explicit on this point in later Series.

machine. We saw already that electricity at high tension can discharge from pointed conductors directly into the air; that is what happened here. When the machine was operated, the impregnated papers *a* and *b* turned red at their extremities, showing that the sulfate solution had decomposed into acidic and alkaline products. Faraday writes:*

> These cases of electro-chemical decomposition are in their nature exactly of the same kind as those effected under ordinary circumstances by the voltaic battery, notwithstanding the presence or absence ... of the parts usually called poles... They indicate at once an *internal* action of the parts suffering decomposition, and appear to show that the power which is effectual in separating the elements is exerted *there* [i. e., internally], and not at the poles. ...

> That electro-chemical decomposition does not depend upon any direct attraction and repulsion of the poles ... upon the elements in contact with or near to them, appeared very evident from [these] experiments made in air, when the substances evolved did not collect about any poles, but, in obedience to the direction of the current, were evolved, and I would say *ejected* at the extremities of the decomposing substance. ...

> The poles are merely the surfaces or doors by which the electricity enters into or passes out of the substance suffering decomposition. They limit the extent of that substance in the course of the electric current, being its *terminations* in that direction: hence the elements evolved pass so far and no further.

> Metals make admirable poles, in consequence of their high conducting power, their immiscibility with the substances generally acted upon, their solid form, and the opportunity afforded of selecting such as are not chemically acted upon by ordinary substances.

Faraday is convinced that the decomposing power is exercised throughout the substance by the current as a whole, not by external centers of power acting on passive matter. What the current *is*, is still highly doubtful. I mentioned earlier that Faraday has been struggling to transcend the imagery and phraseology of fluid-motion. He makes another such effort in the Fifth Series at paragraph 517 (Faraday's italics):

> [The current] may perhaps best be conceived of as *an axis of power having contrary forces, exactly equal in amount, in contrary directions.*

---

* Excerpts from paragraphs 471, 493, and 556–557 of the Fifth Series; some italics added for emphasis.

"A new law of conduction"

Even earlier in 1833, Faraday had made a discovery which both surprised and puzzled him. He announced the new phenomenon at the opening of the Fourth Series:

> I was working with ice, and the solids resulting from the freezing of solutions, arranged either as barriers across a substance to be decomposed, or as the actual poles of a voltaic battery, that I might trace and catch certain elements in their transit, when I was suddenly stopped in my progress by finding that *ice was in such circumstances a non-conductor of electricity*; and that as soon as a thin film of it was interposed, in the circuit of a very powerful voltaic battery, the transmission of electricity was prevented, and all decomposition ceased.*

This was a surprise because *pure water* is a conductor, though indeed a very poor one. Why should water lose that small degree of conducting power it possesses, merely as a result of assuming the solid state?

It was a widely-held assumption among electrical researchers of the time that all materials were inherently classifiable as either conductors or insulators. "Inherently" presumably meant "chemically"; and since liquid water and solid ice were chemically identical, it was difficult to see why they should differ in their conducting abilities. Nor was the assumption itself without a rationale: Since electricity was demonstrably capable of causing chemical decomposition, electric forces must be no less powerful than chemical forces. And chemical forces, in turn, are obviously stronger than the forces of solid cohesion, since solids routinely undergo chemical reaction. We can represent the situation this way:

ELECTRIC FORCES ≥ CHEMICAL FORCES > COHESIVE FORCES

According to these relations, then, electric forces must be more powerful than the forces of cohesion. But the insulating behavior of ice would appear to show the contrary—cohesive forces more powerful than electrical forces! Otherwise, how could the mutual cohesion of the particles of a substance prevent the discharge of electric forces through that substance? You can see why the impermanence of conducting power was not only a surprise but an enigma.

Since water loses its conducting power when frozen, would solid insulators *gain* conducting power when melted? Faraday found many substances that did so—including numerous chlorides, iodides, oxides and sulfates. He pointed out that, among substances which do acquire

---

* Fourth Series, paragraph 381, italics added.

conducting power when liquefied, water is the *weakest*. This could help to explain why the low conductivity of pure water is so abundantly increased when even tiny amounts of salt or acid are added:

> With regard to the substances on which conducting power is thus conferred by liquidity, the degree of power so given is generally very great. Water is that body in which this acquired power is feeblest. In the various oxides, chlorides, salts, &c., &c., it is given in a much higher degree. I have not had time to measure the conducting power in these cases, but it is apparently some hundred times that of pure water. The increased conducting power known to be given to water by the addition of salts, would seem to be in a great degree dependent upon the high conducting power of these bodies when in the liquid state, that state being given them for the time, not by heat but solution in the water.*

Faraday proposed a "new law" remarking the "general assumption of conducting power by bodies as soon as they pass from the solid to the liquid state." To be sure, not all substances illustrated this new law; there were many exceptions, among them a variety of organic compounds.** But those substances which *did* insulate when solid and conduct when liquid were also substances which *decomposed when conducting*, their products appearing separately at the positive and negative poles. As Faraday noted,

> When conduction took place, decomposition occurred; when decomposition ceased, conduction ceased also; and it becomes a fair and an important question, Whether the conduction itself may not, wherever the law holds good, be a consequence not merely of the capability, but of the act of decomposition? And that question may be accompanied by another, namely, Whether solidification does not prevent conduction, merely by chaining the particles to their places ... and preventing their final separation in the manner necessary for decomposition?***

More generally, is the decomposition due to conduction; or is the conduction rather due to decomposition? In his Diary on 24 January 1833, Faraday wrote, "If ice will not conduct, is it because it *cannot* decompose?"

* Fourth Series, paragraph 410.

** The liquid non-conductors listed in the Fourth Series include "mixed margaric and oleic acids, artificial camphor; caffeine, sugar, adipocire, stearine of cocoa-nut oil, spermaceti, camphor, naphthaline, resin, gum sandarach, shell-lac."

*** Fourth Series, paragraph 413.

The question contains a strong challenge to fluid theories of electricity. If electric forces are inherently rooted in a specifically electrical fluid, then decomposition under conduction would represent the outcome of a contest between opposing forces—the forces of the electric fluid, in this case, prevailing over the chemical forces of ordinary matter. But if decomposition is *necessary* to conduction, the relation between electrical and chemical forces cannot be simply antagonistic. Instead, mutual symmetry and correlation would characterize them—or, going further, perhaps even outright *identity*.

Gasometry*

In the present Series Faraday will examine an electro-chemical decomposition experiment that requires him to determine the *weight* of the hydrogen gas it produces. Indeed in most experiments involving gases it is usually their weight which we want to know; but accurate methods for weighing gases are generally intricate and time consuming. By contrast, gaseous *volume* measurements are easy and accurate—but the volume of a gas depends not only on its *weight* but also on its *pressure*, *temperature*, and *moisture content*.

Chemists have therefore endeavored to establish the weight of unit volume of each known gas under specified standard conditions. Then whenever a volume measurement is carried out under other than those standard conditions, they mathematically "correct" the measured volume to the volume which that same quantity of the gas would occupy under standard conditions. The investigator can then calculate how much that "corrected" volume would weigh, and the weight of the measured gas is thereby known. This and related techniques for measuring gases are included under the term *gasometry*.

Some years before beginning the Experimental Researches, Faraday had published a detailed manual of laboratory practices.** The techniques of gas measurement are fully delineated in that book; and Faraday refers to those procedures when recounting the decomposition experiment beginning at paragraph 852 in the Seventh Series below.

---

* This section relates to an experiment which Faraday describes in paragraphs 863–866 below. Some readers may wish to defer it until then.

** *Chemical Manipulation*, London, 1827 and subsequent British and American editions. The book is as delightful as it is instructive; and Faraday's Introduction includes a wise discussion of the purposes of experimentation, and of the legitimate role played in it by the *skill* of the experimenter.

Let us then consult his instructions for *correction of the measured volume of gases for temperature, pressure, and moisture*:*

> The [standard conditions] usually adopted in this country, and distinguished as mean temperature and pressure, are for temperature 60° of Fahrenheit's scale, and for pressure 30 inches of mercury. Hence when necessary, gas observed at any other temperature and pressure, has to be reduced to the volume it would occupy at these points. This may be done in the following manner:

> *Correction for temperature.*

> It appears by the experiments of M. M. Gay Lussac and Dalton, that all gases and vapours, of whatever nature, when not in contact with liquids, are affected equally in their volume by changes of temperature, the increase in volume for every additional degree of heat of Fahrenheit's scale being $1/480$ part of the volume at 32° Fahrenheit, and the decrease for every diminution of temperature of one degree being also $1/480$ part of the volume at 32° Fahr. This known, it is easy to calculate how much a volume of gas at a given temperature, 60° Fahr. for instance, would be increased or diminished by a change of one or more degrees. ... For conceive 480 parts of gas at 32°: at 33° they become 481 parts; at 34°, 482 parts; at 60°, 508 parts; the increase at each degree being $1/480$ of the volume at 32°, and consequently such part of the volume at any other temperature, as is indicated by adding the number of degrees above 32° to 480.

> The rule for correction to be applied to an observed volume of gas, is, therefore, to add to 480 the number of degrees above 32°, to divide the observed volume by this sum, which gives the expansion or contraction for each degree at the observed temperature; to multiply this by the number of degrees between the observed temperature and the temperature to which the gas is to be corrected, which will of course indicate the whole expansion or contraction; and then to subtract this, if the observed be above the corrected temperature, or to add it, if the former be below the latter; thus allowing for the contraction or expansion which would actually take place, if the temperature of the gas were really to be brought to the point to which by calculation it may thus be corrected.

Here is an illustration using Faraday's actual measurements in the decomposition experiment at paragraph 866 below. 12.5 cubic inches of gas at 52° F. are to be corrected to mean temperature of 60°. The

---

* From *Chemical Manipulation*, Section XV. I have omitted paragraph numbers from the following passages in order to avoid confusion with similarly-numbered paragraphs in the Experimental Researches.

difference between 52°, the observed temperature, and 32°, is 20, which added to 480 equals 500. The 12.5 cubic inches divided by 500, gives 0.025 cubic inches as the whole expansion for each degree; and this multiplied by 8, the difference between 60° and 52°, gives 0.2 cubic inches as the whole expansion. This, finally, when added to 12.5 cubic inches, gives *12.7 cubic inches* as the volume which would be occupied by the collected hydrogen at 60° F.

Correction for Pressure.

Boyle and Hooke were perhaps the first to observe that the volumes of gases varied inversely in proportion to the pressure exerted upon them, although the law, having been first distinctly announced and enlarged upon by Marriotte, has received his name. ...

A pressure of 30 inches of mercury, as observed by an accurate barometer, has been assumed as the *mean height* or *barometric pressure,* and volumes of gas observed at any other pressure, frequently require to be corrected to what they would be at this point. For this purpose it is only necessary to compare the observed height with the mean height, or 30 inches, and increase or diminish the observed volume inversely in the same proportion. Thus as the mean height of the barometer is to the observed height, so is the observed volume to the volume required.

Continuing the former example, we now have 12.7 cubic inches of gas (corrected for temperature), with the barometer standing at 29.2 inches. Then as 30 inches or mean height is to 29.2 inches or observed height, so is 12.7 cubic inches, the uncorrected volume, to the volume corrected for pressure:

$$\frac{30}{29.2} = \frac{12.7}{\text{volume corrected for pressure}}$$

Since $29.2 \times 12.7 = 370.84$, which divided by 30 equals 12.36 cubic inches, the volume occupied by the gas at 30 inches of barometric pressure would be *12.36 cubic inches.*

[Correction for Moisture.]

Gas when standing over water becomes saturated with aqueous vapour, the quantity being proportional to the temperature. In these cases a part of the volume observed, and also a part of the weight, is due to the vapour, which therefore must be ascertained before the true weight of the gas under examination can be determined. The following table exhibits the proportion by volume of aqueous vapour existing in any gas standing over or in contact with water at the corresponding temperatures, and at mean barometric pressure of 30 inches.

40°	.00933	54°	.01533	68°	.02406
41	.00973	55	.01586	69	.02483
42	.01013	56	.01640	70	.02566
43	.01053	57	.01693	71	.02653
44	.01093	58	.01753	72	.02740
45	.01133	59	.01810	73	.02830
46	.01173	60	.01866	74	.02923
47	.01213	61	.01923	75	.03020
48	.01253	62	.01980	76	.03120
49	.01293	63	.02050	77	.03220
50	.01333	64	.02120	78	.03323
51	.01380	65	.02190	79	.03423
52	.01426	66	.02260	80	.03533
53	.01480	67	.02330		

By reference to this table, which is founded upon the experiments of Mr. Dalton and Dr. Ure, and includes any temperature at which gases are likely to be weighed, the proportions in bulk of vapour present, and consequently of the dry gas, may easily be ascertained. For this purpose the observed temperature of the gas should be looked for, and opposite to it will be found the proportion in bulk of aqueous vapour at a pressure of 30 inches. The volume to which this amounts should be ascertained and corrected to mean temperature. Then the *whole* volume is to be corrected to mean temperature and pressure, and the corrected volume of vapour subtracted from it. This will leave the corrected volume of dry gas.

Completing the example from the present Series, recall that Faraday collected 12.5 cubic inches of hydrogen over water, at temperature 52° F. By reference to his table we find that at 52°, the proportion of water vapor in gas standing over water is .01426, which in the 12.5 cubic inches collected will amount to .17825 cubic inches. This, corrected as above to the temperature of 60°, becomes .18110 cubic inches. But we found that the whole volume, corrected to mean temperature and pressure, was 12.36 cubic inches, from which, if the .18110 cubic inches of water vapor present be subtracted, will leave *12.17890 cubic inches* as the volume that would be occupied by the collected hydrogen gas if it were free of water vapor and measured at mean temperature and pressure.*

---

* In paragraph 866 below, Faraday actually cites 12.15453 cubic inches, rather than our figure of 12.17890. We need not afflict ourselves concerning the discrepancy, which amounts to only about two tenths of a percent.

Chemical equivalents*

One of the theorems of quantitative chemistry is the Law of Definite Proportions, which holds that when elements combine to form a given compound, the ratio of their combining weights is definite and characteristic of that compound. Suppose elements A and B are found to combine in the ratio $m : n$ by weight; and also that elements B and C are found to combine in the ratio $p : q$. Now experiment discloses a remarkable result: If A and C also combine, *their combining ratio is related to the two former ratios*, according to the following steps:

1. In place of m and n, find two other numbers x and y which have the same ratio as m and n. (Symbolized, $m : n :: x : y$.)

2. In place of p and q, find *one* other number z such that the former number y and the new number z will have the same ratio as p and q. (Symbolized, $p : q :: y : z$.)

3. Then A and C will combine in the ratio $x : z$, if they combine at all.**

The ratio $x : z$ formed in this way is what early mathematicians called the *compound* of the two former ratios (Euclid, VI. 23). Thus the result is often stated in the following form: If elements A and B combine in one ratio, and elements B and C combine in a second ratio, then A and C will combine in the compound of the two ratios—if they combine at all.

Notice that we have to find *three* numbers, one of which will be used twice. Passing over for now any practical difficulties in actually finding them, the numbers x, y, and z represent the so-called "equivalent numbers" of the three elements, respectively. They actually signify *combining weights*;*** that is, any two of the elements that combine with one another at all, will combine in the ratio of their corresponding equivalent numbers. And clearly, by bringing additional elements into combination with any of the first three, we can find as many additional equivalent numbers as there are of those elements.

---

* The following account is much oversimplified; I ask forbearance from chemist readers.

** I must emphasize that this remarkable claim is the result of *experiment*; it could not have been predicted based only on the premises laid down so far. In fact it is not unqualifiedly true, because some elements are found to be capable of combining in *multiple* ratios (under different conditions, and forming different compounds). That complicates, but it does not essentially change, the concept of *equivalence* that will emerge from this argument.

*** Then why are they called "equivalent"? Because if two elements combine with some third substance, weights in the ratio of the equivalent numbers can *substitute* for one another, and they are thus "equivalent" in their ability to combine with a fixed quantity of that third element. In other words, they are *equivalent in combining power*.

Here is an example involving four elements (all cited ratios are approximate): Mercury unites with chlorine in the ratio 11.3 : 4 to form what Faraday knew as *chloride of mercury*. Chlorine unites with hydrogen in the ratio 71 : 2 to form *muriatic acid gas*. Finally, hydrogen combines with oxygen in the ratio 1 : 8 to form *water*. We may look for the desired substitutes for these numbers by trial and error. We eventually find

$$1 : 8 \text{ same ratio as } 1 : 8$$
$$71 : 2 \text{ same ratio as } 35.5 : 1$$
$$11.3 : 4 \text{ same ratio as } 100.3 : 35.5$$

In place of the original three ratios we have three new ratios, linked by shared terms. We have thus found four numbers as required; these will be the (provisional*) equivalent weights of the four elements:

Element	Equivalent Number
Hydrogen	1
Oxygen	8
Chlorine	35.5
Mercury	100.3

Then according to the claim I made above, if mercury unites with oxygen it should do so in the ratio of the equivalent numbers found— here, a ratio of 100.3 : 8. Now mercury *does* combine with oxygen; and for the *red oxide of mercury*, experiment discloses a combining ratio of nearly 12.55 : 1 or 100.4 : 8—a close confirmation of the expected ratio!

I limited the foregoing account to equivalent numbers as they apply to elements. Their extension to compounds is just what you would expect: the equivalent number of a compound is the sum of the equivalent numbers of the elements which it contains. For example, since hydrogen and oxygen combine in the ratio of their respective equivalent numbers, then 1 gram, say, of hydrogen will unite with 8 grams of oxygen to form 9 grams of water;** and in paragraph 866 below Faraday will cite 9 as "the equivalent number of water." Other compounds obtain their corresponding equivalents in the same way.

---

* "Provisional" for two reasons. First, as additional elements are brought into the system, the numbers already established are subject to change. Second, *all* the numbers could be increased or decreased in the same proportion and still retain their validity. Chemists therefore choose a number for one element arbitrarily, and the other numbers are expressed relative to that choice.

** Hydrogen will also unite with oxygen in a different ratio, forming the familiar disinfectant *hydrogen peroxide*. Thus oxygen has (as do also many other elements) different equivalent weights under different conditions. As explained in the second footnote on page 193, however, I here ignore that secondary complication.

SEVENTH SERIES.

§ 11. On Electro-chemical Decomposition, continued.[1] ... *§ 13. On the absolute quantity of Electricity associated with the particles or atoms of Matter.*

Received January 9.—Read January 23, February 6 and 13, 1834.

Preliminary.

661. THE theory which I believe to be a true expression of the facts of electro-chemical decomposition, and which I have therefore detailed in a former series of these Researches, is so much at variance with those previously advanced, that I find the greatest difficulty in stating results, as I think, correctly, whilst limited to the use of terms which are current with a certain accepted meaning. Of this kind is the term *pole*, with its prefixes of positive and negative, and the attached ideas of attraction and repulsion. The general phraseology is that the positive pole *attracts* oxygen, acids, &c., or more cautiously, that it *determines* their evolution upon its surface, and that the negative pole acts in an equal manner upon hydrogen, combustibles, metals, and bases. According to my view the determining force is *not* at the poles but *within* the body under decomposition; and the oxygen and acids are rendered at the negative extremity of that body, whilst hydrogen, metals, &c., are evolved at the positive extremity (518. 524.).

[1] Refer to the note after 1047, Series VIII.—Dec. 1838.

---

661. *the determining force is not at the poles but within the body under decomposition*: One ground for this view is his earlier demonstration (cited in the introduction to the present Series) that the presence of "poles" is not at all necessary for electro-chemical decomposition. But it also reflects Faraday's more fundamental conviction that a too-neat division between *active* poles and *passive* bodies is a misleading fiction—that natural powers are instead characterized by *reciprocity* and *interdependence.*

oxygen and acids are rendered at the negative extremity of that body, whilst hydrogen, metals, &c., are evolved at the positive extremity: This may seem a blunder, since in electrochemical decomposition oxygen and acids are found at the *positive* pole and hydrogen, etc. at the *negative* pole. But Faraday is about to distinguish (paragraph 663) between the *extremities* of the body, and the *poles*, which are external to the body and are in effect extremities *of the battery*. That portion of the body which is adjacent to the positive pole must be negative; and that adjacent to the negative pole must be positive—as he will there explain.

662. To avoid, therefore, confusion and circumlocution, and for the sake of greater precision of expression than I can otherwise obtain, I have deliberately considered the subject with two friends, and with their assistance and concurrence in framing them, I purpose henceforward using certain other terms, which I will now define. The *poles*, as they are usually called, are only the doors or ways by which the electric current passes into and out of the decomposing body (556.); and they of course, when in contact with that body, are the limits of its extent in the direction of the current. The term has been generally applied to the metal surfaces in contact with the decomposing substance; but whether philosophers generally would also apply it to the surfaces of air (465. 471.) and water (493.), against which I have effected electro-chemical decomposition, is subject to doubt. In place of the term pole, I propose using that of *Electrode*,[2] and I mean thereby that substance, or rather surface, whether of air, water, metal, or any other body, which bounds the extent of the decomposing matter in the direction of the electric current.

[2] ἤλεκτρον, and ὁδὸς *a way*.

---

662. *I purpose henceforward using certain other terms*: Faraday intends to introduce a new terminology for electro-chemistry, not in order to name *new* phenomena but because the standard terminology of atoms, positive and negative fluids, and poles is so laden with habitual preconceptions as to obscure our thinking about the *old* phenomena. Faraday's concern has been perceptively characterized in a recent book: "Words and images give us the power to describe possibilities beyond our experience. However, familiar, established words can also constrain the way we think. Faraday wanted to say something new about familiar phenomena. ... To do this he had to redescribe the phenomena so as to free them from habitual ways of thinking about them. ... Faraday needed to clear a space for a new theory, even though he could not yet say what [that theory] was." G. Cantor, D. Gooding, and F. A. J. L. James, *Michael Faraday*. Humanities Press, 1996.

the surfaces of air and water, against which I have effected electro-chemical decomposition: In the Fifth Series, Faraday showed that electro-decomposition could proceed even when the *poles*, which are ordinarily solid metal, are replaced by air or water! A few of the relevant passages are given in the introduction to the present Series.

Electrode, [defined as] that substance ... which bounds the extent of the decomposing matter...: Thus electrodes have the place and function of poles but may be of any material. Poles, on the other hand, are *metal* electrodes.

663. The surfaces at which, according to common phraseology, the electric current enters and leaves a decomposing body, are most important places of action, and require to be distinguished apart from the poles, with which they are mostly, and the electrodes, with which they are always, in contact. Wishing for a natural standard of electric direction to which I might refer these, expressive of their difference and at the same time free from all theory, I have thought it might be found in the earth. If the magnetism of the earth be due to electric currents passing round it, the latter must be in a constant direction, which, according to present usage of speech, would be from east to west, or, which will strengthen this help to the memory, that in which the sun appears to move. If in any case of electro-decomposition we consider the decomposing body as placed so that the current passing through it shall be in the same direction, and parallel to that supposed to exist in the earth, then the surfaces at which the electricity is passing into and out of the substance would have an invariable reference, and exhibit constantly the same relations of powers. Upon this notion we purpose calling that towards the east the *anode*,[3] and that towards the west the *cathode*;[4] and whatever changes may take place in our views of the nature of electricity and electrical action, as they must affect the

[3] ἄνω *upwards*, and ὁδòς *a way*; the way which the sun rises.

[4] κατà *downwards*, and ὁδòς *a way*; the way which the sun sets.

---

663. *The surfaces at which ... current enters and leaves a decomposing body ... require to be distinguished apart from the poles ... and the electrodes...:* Faraday wishes to name the surface extremities of the decomposing material, at which the current enters and exits—as distinguished from the *electrodes*, which by definition are their external boundaries and so "always in contact" with them. Since the electrodes are usually solid "poles," the surfaces to be named are "mostly" but not always in contact with poles.

a natural standard of electric direction ... at the same time free from all theory...: The standard he proposes is based on the observable *magnetic* action of the current, not on a hypothetical transport of electric fluid: Imagine a tiny compass needle held above an electro-decomposition cell and deflected by the current according to the straight-wire case of the electromagnetic right-hand rule (see the Second Series' introduction, page 84); the needle's north-seeking pole will be deflected by the current. If the cell is then oriented so that the direction of deflection coincides with *terrestrial north*, then the positive electrode must lie to the east and the negative electrode to the west. Now Faraday proposes to call the easterly extremity of the decomposition material (towards sunrise) the *anode* and the westerly extremity (towards sunset) the *cathode*.

natural standard referred to, in the same direction, and to an equal amount with any decomposing substances to which these terms may at any time be applied, there seems no reason to expect that they will lead to confusion, or tend in any way to support false views. The *anode* is therefore that surface at which the electric current according to our present expression enters: it is the *negative* extremity of the decomposing body; is where oxygen, chlorine, acids, &c., are evolved; and is against or opposite the positive electrode. The *cathode* is that surface at which the current leaves the decomposing body; and is its *positive* extremity, the combustible bodies, metals, alkalies, and bases, are evolved there, and it is in contact with the negative electrode.

664. I shall have occasion in these Researches, also, to class bodies together according to certain relations derived from their electrical actions (822.); and wishing to express those relations without at the same time involving the expression of any hypothetical views I intend using the following names and terms. Many bodies are decomposed directly by the electric current, their elements being set free: these I propose to call *electrolytes*.[5] Water, therefore, is an electrolyte. The bodies which, like nitric or sulphuric acids, are decomposed in a secondary manner

[5] ἤλεκτρον, and λύω, *solvo*. N. Electrolyte, V. Electrolyze.

663, continued. *The anode is … the negative extremity of the decomposing body*: The anode is adjacent to the positive electrode—but it is the *negative* end of the material; similarly the cathode is adjacent to the negative electrode—but the cathode itself is *positive*. As an example, suppose a voltaic battery is connected to a water-decomposition apparatus. The "positive" terminal of the battery is so named because (in the conventional view) *positive electricity exits from it*. That electricity enters the water at the fluid surface adjacent to the positive electrode—this fluid surface is the *anode*. It exits the water through the fluid surface adjacent to the negative electrode—this fluid surface is the *cathode*. The cathode must then be called *positive* by the same criterion as before, inasmuch as it is that surface by which (conventionally positive) electricity is said to exit *from the water*. The relations are shown in the accompanying sketch.

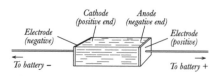

Today Faraday's terms survive but, confusingly, his distinctions do not. Thus in modern nomenclature "anode" is regarded as the *name* of the positive electrode instead of being distinguished from it; and "cathode" is considered to be the name of the negative electrode.

664. *bodies which, like nitric or sulphuric acids, are decomposed in a secondary manner*: In the cited paragraphs (omitted below), Faraday explains that neither nitric acid nor sulfuric acid undergoes electrolysis directly. Instead, when either of those acids is in solution the *water* decomposes, yielding oxygen at

der this term. Then for *electro-chemically* term *electrolyzed*, derived in the same y spoken of is separated into its com- electricity: it is analogous in its sense lerived in a similar manner. The term t once: muriatic acid is electrolytical,

o express those bodies which can pass ually called, the poles. Substances are tro-negative, or *electro-positive*, according influence of a direct attraction to the ese terms are much too significant for o put them; for though the meanings hypothetical, and may be wrong; and ible, but still very dangerous, because reat injury to science, by contracting of those engaged in pursuing it. I propose to distinguish such bodies by calling those *anions*[6] which go to the *anode* of the decomposing body; and those passing to *the cathode*, *cations*;[7] and when I have occasion to speak of these together, I shall call them *ions*. Thus, the chloride of lead is an *electrolyte*, and when *electrolyzed* evolves the two *ions*, chlorine and lead, the former being an *anion*, and the latter a *cation*.

666. These terms being once well defined, will, I hope, in their use enable me to avoid much periphrasis and ambiguity of expression. I do not mean to press them into service more frequently than will be required, for I am fully aware that names are one thing and science another.[8]

[6] ἀνιὼν *that which goes up.* (Neuter participle.)

[7] κατιὼν *that which goes down.*

[8] Since this paper was read, I have changed some of the terms which were first proposed, that I might employ only such as were at the same time simple in their nature, clear in their reference, and [free] from hypothesis.

---

the anode and hydrogen at the cathode; these products subsequently act on the acid in what are called "secondary" reactions—hydrogen reacting with nitric acid, for example, to form nitrous acid or even nitric oxide at the cathode; or with sulfuric acid to deposit sulfur there.

Muriatic acid (hydrochloric acid) conducts a current and is thereby decomposed; *boracic acid* (boric acid) does not conduct and is not electrically decomposed.

665. *too significant*: that is, they carry too many habitual assumptions concerning the supposed attractive power of the "poles."

667. It will be well understood that I am giving no opinion respecting the nature of the electric current now, beyond what I have done on former occasions (283. 517.); and that though I speak of the current as proceeding from the parts which are positive to those which are negative (663.), it is merely in accordance with the conventional, though in some degree tacit, agreement entered into by scientific men, that they may have a constant, certain, and definite means of referring to the direction of the forces of that current.

*   *   *

13. *On the absolute quantity of Electricity associated with the particles or atoms of Matter.*

852. The theory of definite electrolytical or electro-chemical action appears to me to touch immediately upon the *absolute quantity* of electricity or electric power belonging to different bodies. It is impossible, perhaps, to speak on this point without committing oneself beyond what present facts will sustain; and yet it is equally impossible, and perhaps would be impolitic, not to reason upon the subject. Although we know nothing of what an atom is, yet we cannot resist forming some

---

667. *merely in accordance with the conventional ... agreement*: Note his aloofness regarding the conventional ideas of current; he is obliged to use the conventional metaphors of flow and transport but refuses to take them literally. What is really *evident* in current—and what we really want to refer to—is the "direction of its forces." We fall into the imagery of *transport* because it is so easy and familiar, but it obscures the actual phenomena. Forces, or at least their activities, are or can be made visible; but transport of a hypothetical electric fluid cannot. See also his characterization of "current" in paragraph 517; the passage is given on page 186 of the introduction to the present Series.

852. *absolute quantity*: A determination made, not by comparison to an arbitrary sample, but in terms of an implicit natural unit. The experiment described in this section will imply a natural unit of electricity, thereby making an absolute determination, in this sense, possible.

perhaps would be impolitic...: What could be "impolitic" about failing to reason about the subject of electrical ability? If electricity is inherently capable of developing more power than our present techniques permit, it might indeed be impolitic (rash) to ignore an opportunity to wield such increased power. In paragraph 873 below, Faraday envisions new, hugely powerful sources and applications of electricity—perhaps that is what he has in mind here. If so, he has adopted a more engaged attitude than in paragraph 159, where he appeared relatively unconcerned with the prospective development of *magneto-electric* generation.

we know nothing of what an atom is: To be sure, we have the image of a "small

idea of a small particle, which represents it to the mind; and though we are in equal, if not greater, ignorance of electricity, so as to be unable to say whether it is a particular matter or matters, or mere motion of ordinary matter, or some third kind of power or agent, yet there is an immensity of facts which justify us in believing that the atoms of matter are in some way endowed or associated with electrical powers, to which they owe their most striking qualities, and amongst them their mutual chemical affinity. As soon as we perceive, through the teaching of Dalton, that chemical powers are, however varied the circumstances in which they are exerted, definite for each body, we learn to estimate the relative degree of force which resides in such bodies: and when upon that knowledge comes the fact, that the electricity, which we appear to be capable of loosening from its habitation for a while, and conveying from place to place, *whilst it retains its chemical force*, can be measured out, and being so measured is found to be *as definite in its action* as any of *those portions* which, remaining associated with the particles of matter, give them their *chemical relation*; we seem to have found the link which connects the proportion of that we have evolved to the proportion of that belonging to the particles in their natural state.

853. Now it is wonderful to observe how small a quantity of a compound body is decomposed by a certain portion of electricity. Let us, for instance, consider this and a few other points in relation to water. *One grain* of water, acidulated to facilitate conduction, will require an electric current to be continued for three minutes and three quarters

---

particle," but this image grows out of our need for *some* representation, not out of actual knowledge. He will express reservations about atomism again in paragraph 869.

loosening from its habitation for a while: This lovely phrase characterizes the voltaic cell as a device which removes electricity from its proper "habitation" in matter, where it has the characteristics of a *chemical force*. The electricity so disengaged in a voltaic cell is capable of performing *definite chemical action* in a decomposition apparatus. It should then be possible to characterize electricity in terms of the *chemical* power that belongs naturally to matter. All this is anticipatory; Faraday will give a more specific example in paragraph 858 below.

853. *Now it is wonderful to observe...*: The facts cited in this paragraph are experimental results, not theoretical calculations.

The *grain* is a unit of weight shared by the troy, avoirdupois and apothecaries' systems and is equal to .0648 grams in the metric system. It was originally based on an average among grains of wheat or corn.

acidulated: a tiny quantity of acid is added. (By the way, and simply for pleasure's sake, savor the delectable sonorities of the phrase "acidulated to facilitate conduction.")

of time to effect its decomposition, which current must be powerful enough to retain a platina wire $1/104$ of an inch in thickness,[9] red hot, in the air during the whole time; and if interrupted anywhere by charcoal points, will produce a very brilliant and constant star of light. If attention be paid to the instantaneous discharge of electricity of tension, as illustrated in the beautiful experiments of Mr. Wheatstone,[10] and to what I have said elsewhere on the relation of common and voltaic electricity (371. 375.), it will not be too much to say that this necessary quantity of electricity is equal to a very powerful flash of lightning. Yet we have it under perfect command; can evolve, direct, and employ it at pleasure; and when it has performed its full work of electrolyzation, it has only separated the elements of *a single grain of water.*

854. On the other hand, the relation between the conduction of the electricity and the decomposition of the water is so close, that one cannot take place without the other. If the water is altered only in that small degree which consists in its having the solid instead of the fluid state, the conduction is stopped, and the decomposition is stopped with it. Whether the conduction be considered as depending upon the decomposition, or not (413. 703.), still the relation of the two functions is equally intimate and inseparable.

[9] I have not stated the length of wire used, because I find by experiment, as would be expected in theory, that it is indifferent. The same quantity of electricity which, passed in a given time, can heat an inch of platina wire of a certain diameter red hot, can also heat a hundred, a thousand, or any length of the same wire to the same degree, provided the cooling circumstances are the same for every part in all cases. This I have proved by the volta-electrometer. I found that whether half an inch or eight inches were retained at one constant temperature of dull redness, equal quantities of water were decomposed in equal times. When the half-inch was used, only the centre portion of wire was ignited. A fine wire may even be used as a rough but ready regulator of a voltaic current; for if it be made part of the circuit, and the larger wires communicating with it be shifted nearer to or further apart, so as to keep the portion of wire in the circuit sensibly at the same temperature, the current passing through it will be nearly uniform.

[10] Literary Gazette, 1833, March 1 and 8. Philosophical Magazine, 1833, p. 204. L'Institute, 1833, p. 261.

853, continued. *constant star of light*: the current is strong enough to maintain an intense electric arc (see the editor's introduction to the First Series).

854. *having the solid instead of the fluid state*: Previously, Faraday had discovered that electrolytes cease to conduct, and fail to decompose, when frozen; see the introduction to the present Series.

855. Considering this close and twofold relation, namely, that without decomposition transmission of electricity does not occur; and, that for a given definite quantity of electricity passed, an equally definite and constant quantity of water or other matter is decomposed; considering also that the agent, which is electricity, is simply employed in overcoming electrical powers in the body subjected to its action; it seems a probable, and almost a natural consequence, that the quantity which passes is the *equivalent* of, and therefore equal to, that of the particles separated; *i. e.* that if the electrical power which holds the elements of a grain of water in combination, or which makes a grain of oxygen and hydrogen in the right proportions unite into water when they are made to combine, could be thrown into the condition of *a current*, it would exactly equal the current required for the separation of that grain of water into its elements again.

856. This view of the subject gives an almost overwhelming idea of the extraordinary quantity or degree of electric power which naturally belongs to the particles of matter; but it is not inconsistent in the slightest degree with the facts which can be brought to bear on this point. To illustrate this I must say a few words on the voltaic pile.[11]

857. Intending hereafter to apply the results given in this and the preceding series of Researches to a close investigation of the source of electricity in the voltaic instrument, I have refrained from forming any decided opinion on the subject; and without at

---

[11] By the term voltaic pile, I mean such apparatus or arrangement of metals as up to this time have been called so, and which contain water, brine, acids, or other aqueous solutions or decomposable substances (470.), between their plates. Other kinds of electric apparatus may be hereafter invented, and I hope to construct some not belonging to the class of instruments discovered by Volta.

---

855. *equivalent*: Note Faraday's distinctive usage of the word: powers are *equivalent* when they are mutually convertible. This implies an unusual rhetoric of measurement. We are accustomed to pronouncing things equivalent in some respect when we ascertain, often by use of an instrument, that they respond identically to a specified test or analysis. Such is the case with the chemist's *equivalent weights*—a conventional usage which Faraday readily employs (see the introduction). But for him, the real equivalencies are more intimate: instead of being pronounced equal according to some external criterion, powers *display their own equivalence*, so to speak, by transforming themselves into one another. Faraday will emphasize *convertibility* repeatedly in later Series, especially in connection with electricity and magnetism.

all meaning to dismiss metallic contact, or the contact of dissimilar substances, being conductors, but not metallic, as if they had nothing to do with the origin of the current, I still am fully of opinion with Davy, that it is at least continued by chemical action, and that the supply constituting the current is almost entirely from that source.

858. Those bodies which, being interposed between the metals of the voltaic pile, render it active, *are all of them electrolytes* (476.); and it cannot but press upon the attention of every one engaged in considering this subject, that in those bodies (so essential to the pile) decomposition and the transmission of a current are so intimately connected, that one cannot happen without the other. This I have shown abundantly in water, and numerous other cases (402. 476.). If, then, a voltaic trough have its extremities connected by a body capable of being decomposed, as water, we shall have a continuous current through the apparatus; and whilst it remains in this state we may look at the part where the acid is acting upon the plates, and that where the current is acting upon the water, as the reciprocals of each other. In both parts we have the two conditions *inseparable in such bodies as these,* namely, the passing of the current, and decomposition; and this is as true of the cells in the battery as of the water cell; for no voltaic battery has as yet been constructed in which the chemical action is only that of combination: *decomposition is always included,* and is, I believe, an essential chemical part.

---

857. *I still am fully of opinion ... that [the current] is at least continued by chemical action*: In the First Series' introduction I noted that Faraday had misgivings about Volta's contact theory; he expresses one of them here: Even if we accept that mere contact of the metals could explain the *tendency* to develop current, still it must require some *continuing action* if such a current is to be sustained. The obvious candidate for such action is the chemical action in the voltaic cell, which starts and stops consistently with the current.

858. *all of them electrolytes*: A voltaic cell consists of two different metals, separated by a liquid; but the only liquids that enable the cell to develop electric current are liquids which are themselves susceptible to electrolysis. The cited paragraph (476) is not included in this selection; but see the excerpt from paragraph 410 given in the introduction.

reciprocals of each other: In the voltaic cell, the chemical forces of acid and metal are *disengaged* in the form we call electricity (recall his appealing phrase, "loosening from its habitation for a while," in paragraph 852 above). In the decomposition cell, the electricity supplied comprises that same chemical force, here *re-engaged* to decompose the electrolyte.

859. But the difference in the two parts of the connected battery, that is, the decomposition or experimental cell, and the acting cells, is simply this. In the former we urge the current through, but it, apparently of necessity, is accompanied by decomposition: in the latter we cause decompositions by ordinary chemical actions, (which are, however, themselves electrical,) and, as a consequence, have the electrical current; and as the decomposition dependent upon the current is definite in the former case, so is the current associated with the decomposition also definite in the latter (862. &c.).

860. Let us apply this in support of what I have surmised respecting the enormous electric power of each particle or atom of matter (856.). I showed in a former series of these Researches on the relation by measure of common and voltaic electricity, that two wires, one of platina and one of zinc, each one eighteenth of an inch in diameter, placed five sixteenths of an inch apart, and immersed to the depth of five eighths of an inch in acid, consisting of one drop of oil of vitriol and four ounces of distilled water at a temperature of about 60° Fahr., and connected at the other extremities by a copper wire eighteen feet long, and one eighteenth of an inch in thickness, yielded as much electricity in little more than three seconds of time as a Leyden battery charged by thirty turns of a very large and powerful plate electric machine in full action (371.). This quantity, though sufficient if passed at once through the head of a rat or cat to have killed it, as by a flash of lightning, was evolved by the mutual action of so small a portion of the zinc wire and water in contact with it, that the loss of weight sustained by either would be inappreciable by our most delicate instruments; and as to the water which could be decomposed by that current, it must have been insensible in quantity, for no trace of hydrogen appeared upon the surface of the platina during those three seconds.

861. What an enormous quantity of electricity, therefore, is required for the decomposition of a single grain of water! We have

---

859. *the two parts of the connected battery*: Note that the term "battery" here includes the decomposition cell as well as the voltaic cells; they are to be viewed as correlative parts of a whole, not as essentially separate *agent* and *patient*.

definite … definite: A definite quantity of electricity, passing as a current, will decompose a measured weight of electrolyte. Similarly a measured weight of electrolyte decomposed in the voltaic cell will develop a definite quantity of electricity in the form of a current.

860. *platina* is platinum metal; *oil of vitriol* is undiluted sulphuric acid.

already seen that it must be in quantity sufficient to sustain a platina wire $1/104$ of an inch in thickness, red hot, in contact with the air, for three minutes and three quarters (853.), a quantity which is almost infinitely greater than that which could be evolved by the little standard voltaic arrangement to which I have just referred (860. 371.). I have endeavoured to make a comparison by the loss of weight of such a wire in a given time in such an acid, according to a principle and experiment to be almost immediately described (862.); but the proportion is so high that I am almost afraid to mention it. It would appear that 800,000 such charges of the Leyden battery as I have referred to above, would be necessary to supply electricity sufficient to decompose a single grain of water; or, if I am right, to equal the

---

861. *800,000 such charges*: The figure is very roughly estimated. Faraday says he obtains it from an experiment involving "loss of weight of a wire in acid." We will see shortly how the weight loss of a wire bears upon the decomposition of water. But first let us see why Faraday cannot measure the water-electrolysis of the standard "charge" *directly*.

As he recounts in paragraph 860, the "charge" in question was evolved by the "little standard voltaic arrangement" in a little over 3 seconds. 800,000 such charges, therefore, would have been developed in 2,400,000 seconds— about 667 hours. If then Faraday can show that the little voltaic apparatus decomposes water at a rate of 1 grain per 2,400,000 seconds, he will have obtained the figure *800,000*. Of course there is no question of actually running the apparatus for 667 hours! It could not possibly operate for that length of time, certainly not at a steady rate. But in practice the device only has to run long enough to decompose a *measurable* weight of water; this weight, together with the time, suffices to establish the decomposition rate. So how much weight *is* measurable?—that is, how small a weight can Faraday detect? In his book *Chemical Manipulation* he suggests that a decent laboratory balance ought to be capable of detecting weight changes of a few hundredths of a grain; so let us suppose that Faraday can measure a weight loss of as little as .02 grains (a little more than 1 milligram). Still, at the decomposition rate of 1 grain of water per 2,400,000 seconds—the rate that yields the *800,000* figure—it would take .02 × 2,400,000 seconds, or about 13.3 hours, to obtain the result. It is hardly more realistic to expect the voltaic apparatus to perform consistently for 13 hours, than for 667 hours. So a *direct* measurement on water is probably not practical.

I will return to Faraday's figure "800,000 charges" in the comment to paragraph 868 below. Note that he views that proportion—so astonishingly large—as comparing *one and the same power* in its resident (chemical) form and in its mobile (electrical) form. Recall that in the Third Series he noted a "disproportion" between the little voltaic apparatus and the large plate machine (paragraphs 371–372). That appraisal was a forerunner of the present, more fundamental comparison between electrical and chemical measures—the voltaic apparatus providing a chemical standard, the plate machine an electrical one.

quantity of electricity which is naturally associated with the elements of that grain of water, endowing them with their mutual chemical affinity.

862. In further proof of this high electric condition of the particles of matter, and the *identity as to quantity of that belonging to them with that necessary for their separation*, I will describe an experiment of great simplicity but extreme beauty, when viewed in relation to the evolution of an electric current and its decomposing powers.

863. A dilute sulphuric acid, made by adding about one part by measure of oil of vitriol to thirty parts of water, will act energetically upon a piece of zinc plate in its ordinary and simple state: but, as Mr. Sturgeon has shewn,[12] not at all, or scarcely so, if the surface of the metal has in the first instance been amalgamated; yet the amalgamated zinc will act powerfully with platina as an electromotor, hydrogen being evolved on the surface of the latter metal, as the zinc is oxidized and dissolved. The amalgamation is best effected by sprinkling a few drops of mercury upon the surface of the zinc, the latter being moistened with the dilute acid, and rubbing with the fingers or tow so as to extend the liquid metal over the whole of the surface. Any mercury in excess, forming liquid drops upon the zinc, should be wiped off.[13]

[12] Recent Experimental Researches, &c., 1830, p. 74, &c.

[13] The experiment may be made with pure zinc, which, as chemists well know, is but slightly acted upon by dilute sulphuric acid in comparison with ordinary zinc, which during the action is subject to an infinity of voltaic actions. See De la Rive on this subject, Bibliothèque Universelle, 1830, p. 391.

---

endowing them with their mutual chemical affinity: If Faraday is right, the mutual chemical affinity of the elements of water *arises from* the electricity that is naturally associated with them. Equally, the electricity associated with the elements of one grain of water *constitutes* their mutual chemical affinity. Our expressions tend to subordinate either chemical power to electricity, or electricity to chemical affinity; but Faraday's aim is not to rank but to coordinate them.

863. *amalgamated*: here, coated with mercury. *electromotor*: something that moves or tends to move electricity.

rubbing with the fingers or tow: Tow—not *toe!*—is the coarse, broken fibers of flax or hemp. But his directive to use the *fingers* for applying mercury is dangerous— *never handle mercury!* It is a toxic metal with a poisonous vapor that can be absorbed into the body through the skin. Safety precautions regarding the heavy metals were unheard of in Faraday's time; and they have been made standard practice in laboratories and schools only in the past thirty years or so. During his life Faraday suffered recurring attacks of giddiness, memory loss, and exhaustion; some biographers have diagnosed mercury poisoning.

864. Two plates of zinc thus amalgamated were dried and accurately weighed; one, which we will call A, weighed 163.1 grains; the other to be called B, weighed 148.3 grains. They were about five inches long, and 0.4 of an inch wide. An earthenware pneumatic trough was filled with dilute sulphuric acid, of the strength just described (863.), and a gas jar, also filled with the acid, inverted in it.[14] A plate of platina of nearly the same length, but about three times as wide as the zinc plates, was put up into this jar. The zinc plate A was also introduced into the jar, and brought in contact with the platina, and at the same moment the plate B was put into the acid of the trough, but out of contact with other metallic matter.

865. Strong action immediately occurred in the jar upon the contact of the zinc and platina plates. Hydrogen gas rose from the platina, and was collected in the jar, but no hydrogen or other gas rose from *either* zinc plate. In about ten or twelve minutes, sufficient hydrogen having been collected, the experiment was stopped; during its progress a few small bubbles had appeared upon plate B, but none upon plate A. The plates were washed in distilled water, dried, and reweighed. Plate B weighed 148.3 grains, as before, having lost nothing by the direct chemical action of the acid. Plate A weighed 154.65 grains, 8.45 grains of it having been oxidized and dissolved during the experiment.

[14] The acid was left during a night with a small piece of unamalgamated zinc in it, for the purpose of evolving such air as might be inclined to separate, and bringing the whole into a constant state.

---

864. Since 1 grain = .0648 gram, the zinc plate A (163.1 grains) weighs 163.1 × .0648 or 10.56888 gm. Similarly plate B (148.3 grains) weighs 9.60984 gm. "Platina" is platinum metal.

865. Here is a diagram of apparatus similar to Faraday's. Plate B, not in contact with other metal, has undergone no change. Plate A has lost 8.45 grains, equal to 8.45 × .0648 or .54756 gm. Thus .54756 gm of zinc has combined with oxygen from the water, becoming soluble *zinc oxide* and releasing hydrogen. Nevertheless, the hydrogen bubbles do not form at the zinc plate where the oxidation occurs, but rather at the platina plate—that is, at distances of as much as an inch or more away, as Faraday will soon point out (paragraph 867).

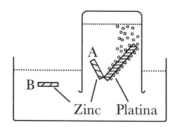

Zinc   Platina

866. The hydrogen gas was next transferred to a water-trough and measured; it amounted to 12.5 cubic inches, the temperature being 52°, and the barometer 29.2 inches. This quantity, corrected for temperature, pressure, and moisture, becomes 12.15453 cubic inches of dry hydrogen at mean temperature and pressure; which, increased by one half for the oxygen that must have gone to the *anode, i. e.* to the zinc, gives 18.232 cubic inches as the quantity of oxygen and hydrogen evolved from the water decomposed by the electric current. According to the estimate of the weight of the mixed gas before adopted (791.), this volume is equal to 2.3535544 grains, which therefore is the weight of water decomposed; and this quantity is to 8.45, the quantity of zinc oxidized, as 9 is to 32.31. Now taking 9 as

---

866. *water trough*: See the illustration on page 184 of the editor's introduction. The device permits collection and measurement of a gas by displacement of water from a graduated cylinder.

corrected for temperature, pressure, and moisture: Hydrogen gas at temperature 52° Fahrenheit is collected over water, where it occupies 12.5 cu. in. under a barometric pressure of 29.2 inches of mercury. Now, weights have been carefully measured for unit volume of *dry* hydrogen and other gases, at "mean" temperature and pressure; but since the volume of 12.5 cu. in. was measured under conditions differing from these—and, moreover, in the presence of water and therefore not *dry*—the measured volume must be "corrected for temperature, pressure, and moisture" as outlined in the introduction. The corrected volume is 12.15453 cu. in., meaning that if the experiment had been performed under mean conditions and the gas somehow protected from all moisture, the quantity of hydrogen yielded would have occupied 12.15453 cu. in.

increased by one half for the oxygen: Gaseous hydrogen and oxygen unite to form water in a ratio of 2:1 by volume. Since decomposition of the water under mean conditions would have yielded 12.15453 cu. in. of (dry) hydrogen, then 6.07727 cu. in. of oxygen would also have been released. In Faraday's apparatus, this oxygen unites with the zinc anode; but had it been liberated as a gas, and the hydrogen and oxygen collected in a single (dry) container, the experiment would have yielded a total volume of 18.2318 cu. in. (Faraday rounds to 18.232) of a *mixture* of hydrogen and oxygen gases, at mean temperature and pressure.

the weight of the mixed gas: In an earlier paragraph here omitted Faraday cited, as a well-established measurement, that under mean conditions 100 cu. in. of a dry mixture of gases, consisting of 66.66 cu. in. hydrogen and 33.34 cu. in. oxygen (the 2:1 proportion suited to form water if ignited) weigh 12.92 grains. Therefore the 18.2318 cu. in. of mixed gas considered above would weigh $18.2318 \div 100 \times 12.92 = 2.35$ grains—which means that the mixture represents, and the experiment yielded, the products of decomposition of 2.35 grains of water.

the equivalent number of water, the number 32.5 is given as the equivalent number of zinc; a coincidence sufficiently near to show, what indeed could not but happen, that for an equivalent of zinc oxidized an equivalent of water must be decomposed.[15]

867. But let us observe *how* the water is decomposed. It is electrolyzed, *i. e.* is decomposed voltaically; and not in the ordinary manner (as to appearance) of chemical decompositions; for the oxygen appears at the *anode* and the hydrogen at the *cathode* of the body under decomposition, and these were in many parts of the experiment above an inch asunder. Again, the ordinary chemical affinity was not enough under the circumstances to effect the decomposition of the water, as was abundantly proved by the inaction on plate B; the voltaic current was essential. And to prevent any idea that the chemical affinity was almost sufficient to decompose the water, and that a smaller current of electricity might under the circumstances, cause the hydrogen to pass to the *cathode*, I need only refer to the results which I have given (807. 813.) to show that the chemical action at the electrodes has not the slightest influence over the *quantities* of water or other substances decomposed between them, but that they are entirely dependent upon the quantity of electricity which passes.

[15] The experiment was repeated several times with the same results.

---

equivalent: The chemically *equivalent numbers* of substances are described in the introduction. They were originally established for ordinary chemical reactions; now they are found to apply even when the driving agency is *electrical*! That would, on its face, be surprising; but Faraday is not surprised ("what indeed could not but happen"), because he already believes that chemical power is electrical in nature.

867. *these were … above an inch asunder*: Since the hydrogen and oxygen make their appearances at two different locations, it is clear they must derive from two different groups of water molecules, respectively. If a water molecule simply breaks up into its constituent parts, we would expect both constituents to appear or act at the same place, namely, the location of the parent molecule. So this is not an ordinary chemical decomposition. Nevertheless Faraday shows that it obeys the same weight relations as ordinary chemical action.

to prevent any idea that the chemical affinity was almost sufficient to decompose the water…: Faraday wishes to claim that the electrical action is *equivalent and essential to* the chemical action. But perhaps the chemical and electrical forces are *two independent agents*, neither one sufficient by itself but both together competent to decompose the water? Against this objection Faraday cites

868. What, then, follows as a necessary consequence of the whole experiment? Why, this: that the chemical action upon 32.31 parts, or one equivalent of zinc, in this simple voltaic circle, was able to evolve such quantity of electricity in the form of a current, as passing through water, should decompose 9 parts, or one equivalent of that substance: and considering the definite relations of electricity as developed in the

---

paragraphs that are omitted in the present selection, but it will suffice to recount the results of paragraph 807: A given quantity of electricity decomposes the same amount of water, liberating the same amount of hydrogen at the cathode, whether the anode is platinum, copper, or zinc—metals which have widely different affinities for oxygen. Thus, higher and lower degrees of chemical affinity neither enhance nor reduce the chemical action of the current.

868. *What, then, follows...? Why, this*: The quantity of electricity associated with decomposition of a given quantity of material is the same, whether the electricity *causes* or *results from* the decomposition; this equality of cause and result reflects the essential identity of chemical and electrical power. Moreover, it gives fresh support to his notion that voltaic electricity must reflect some *process*, that it cannot be sustained by an *unchanging* cause, as the contact theory seems to hold. He will be more explicit about this in paragraph 872 below.

Now we are in a position to appreciate how the experiment just described might bear on Faraday's earlier calculation of "800,000 charges," based, as he stated there (paragraph 861), on *the weight loss of a* [zinc] *wire*. It will be helpful to compare the present discussion with the earlier comments to paragraph 861. There we saw that Faraday's "standard" voltaic device could decompose water at the rate of 1 grain per 2,400,000 seconds. But the experiment just described shows that when zinc and platinum act voltaically upon water, $32.5 \div 9$ or 3.611 grains of zinc dissolve for every grain of water that is decomposed (Faraday cites 32.5 and 9 as the chemical equivalents of zinc and water, respectively). So a *water*-decomposition rate of 1 grain per 2,400,000 seconds will be equivalent, in this experiment, to a *zinc* loss rate of 3.611 grains per 2,400,000 seconds. Then let us calculate how long it would take the standard voltaic arrangement to bring about a weight loss of *.02 grains of zinc*, instead of the .02 grains of water considered in the earlier comment. It is easy to see that, at the rate cited for zinc, the required operating time will be $.02 \div 3.611 \times 2,400,000$ or 13293 seconds—about 3.7 hours, instead of the 13 hours calculated previously. This is more nearly within realistic levels of performance for the voltaic device Faraday describes.

To summarize, then, we might conjecture the following: If Faraday's "standard voltaic arrangement" ran continuously and steadily (as indicated by the galvanometer) for 3.7 hours or 13293 seconds; and if at the end of that time its zinc wire decreased about .02 grains in weight; then .02 grains zinc per 13293 seconds equals $.02 \div 3.611$ grains water per 13293 seconds, which is equivalent to 1 grain water per 2,400,000 seconds. And thus he would have obtained the figure "800,000 charges," cited in paragraph 861. Although these particulars are indeed only conjectural, it is at least clear that Faraday could have obtained that remarkable result in just the way he claims to have done.

preceding parts of the present paper, the results prove that the quantity of electricity which, being naturally associated with the particles of matter, gives them their combining power, is able, when thrown into a current, to separate those particles from their state of combination; or, in other words, that *the electricity which decomposes, and that which is evolved by the decomposition of, a certain quantity of matter, are alike.*

869. The harmony which this theory of the definite evolution and the equivalent definite action of electricity introduces into the associated theories of definite proportions and electro-chemical affinity, is very great. According to it, the equivalent weights of bodies are simply those quantities of them which contain equal quantities of electricity, or have naturally equal electric powers; it being the ELECTRICITY which *determines* the equivalent number, *because* it determines the combining force. Or, if we adopt the atomic theory or phraseology, then the atoms of bodies which are equivalents to each other in their ordinary chemical action, have equal quantities of electricity naturally associated with them. But I must confess I am jealous of the term *atom;* for though it is very easy to talk of atoms, it is very difficult to form a clear idea of their nature, especially when compound bodies are under consideration.

*   *   *

---

869. *harmony:* Faraday views his electro-chemical theory as bringing together two fundamental but hitherto separate theorems of chemistry—the law of definite proportions and the table of chemical affinities. Its bearing on the first of these is clear: *definite chemical combining power* will be the result of a *definite quantity of electricity,* and conversely. The relation of his theory to *affinity,* however, is less apparent. Chemical affinity is a measure of the tendency of two elements to combine, or of the ability of one element to displace another from compounds. Later investigators, however, will argue that this property depends not on the *quantity* of electricity involved but on the electrical *potential.* Although electrical potential (popularly called *voltage*) is similar to Faraday's "tension," a clearly-defined concept of it is not yet available to him, nor does he possess measurement techniques that would disclose it with sufficient accuracy to establish the connection accurately.

atomic theory or phraseology: Conventional atomism may not even merit the status of a *theory*! It is sadly deficient in clarity, and Faraday suspects that it had better be regarded as only a figure of speech, a conventional *phraseology.*

jealous: Here, *suspicious* or *vigilant.* He maintains this watchful attitude because the atomic terminology is in constant danger of becoming empty jargon, lacking clear concepts. It is difficult enough to formulate a clear and coherent idea of a *single* atom with all its powers—no atomic theory has even come close to a satisfactory image of the *compound.*

It may be impossible for a present-day reader to imagine atomistic chemical thinking as being dubious in the slightest degree; but note that as recently as 1908, Wilhelm Ostwald, in the introduction to his *Outline of General*

871. In this exposition of the law of the definite action of electricity, and its corresponding definite proportion in the particles of bodies, I do not pretend to have brought, as yet, every case of chemical or electro-chemical action under its dominion. There are numerous considerations of a theoretical nature, especially respecting the compound particles of matter and the resulting electrical forces which they ought to possess, which I hope will gradually receive their development; and there are numerous experimental cases, as, for instance, those of compounds formed by weak affinities, the simultaneous decomposition of water and salts, &c., which still require investigation. But whatever the results on these and numerous other points may be, I do not believe that the facts which I have advanced, or even the general laws deduced from them, will suffer any serious change; and they are of sufficient importance to justify their publication, though much may yet remain imperfect or undone. Indeed, it is the great beauty of our science, CHEMISTRY, that advancement in it, whether in a degree great or small, instead of exhausting the subjects of research, opens the doors to further and more abundant knowledge, overflowing with beauty and utility, to those who will be at the easy personal pains of undertaking its experimental investigation.

872. The definite production of electricity (868.) in association with its definite action proves, I think, that the current of electricity in the voltaic pile is sustained by chemical decomposition, or rather by chemical action, and not by contact only. But here, as elsewhere (857.), I beg to reserve my opinion as to the real action of contact, not having

---

Chemistry, was congratulating his fellow chemists that *finally* there was beginning to appear some really compelling evidence for the atomic hypothesis. Understand that this was three years *after* Einstein's theory of relativity and eight years after Planck's energy quanta—the two developments most characteristic of "modern" physics!

871. *our science,* CHEMISTRY: Electricity had been viewed as a branch of chemistry ever since the voltaic battery. It would soon acquire a more independent status—largely as a result of Faraday's own work.

872. *the real action of contact*: He continues to voice doubt about the contact theory, as earlier in paragraph 857. In the Eighth Series, *doubt* will turn to *attack*, an attack which becomes crushing in the Sixteenth and Seventeenth Series: "But the contact theory assumes that these particles [of the contacting metals], which have thus by their mutual action acquired opposite electrical states, can discharge those states to one another, and yet remain in the state they were first in, being *in every point* unchanged..." (paragraph 2067). That analysis will reflect a general principle of *reciprocity* that he is already beginning to discern throughout nature: No agent can exercise a power yet be itself *unaltered* in that exercise! See also the comments to paragraphs 661 and 868 above.

yet been able to make up my mind as to whether it is an exciting cause of the current, or merely necessary to allow of the conduction of electricity, otherwise generated, from one metal to the other.

873. But admitting that chemical action is the source of electricity, what an infinitely small fraction of that which is active do we obtain and employ in our voltaic batteries! Zinc and platina wires, one eighteenth of an inch in diameter and about half an inch long, dipped into dilute sulphuric acid, so weak that it is not sensibly sour to the tongue, or scarcely to our most delicate test papers, will evolve more electricity in one twentieth of a minute (860.) than any man would willingly allow to pass through his body at once. The chemical action of a grain of water upon four grains of zinc can evolve electricity equal in quantity to that of a powerful thunder-storm (868. 861.). Nor is it merely true that the quantity is active; it can be directed and made to perform its full equivalent duty (867. &c.). Is there not, then, great reason to hope and believe that, by a closer *experimental* investigation of the principles which govern the development and action of this subtile agent, we shall be able to increase the power of our batteries, or invent new instruments which shall a thousandfold surpass in energy those which we at present possess?

874. Here for a while I must leave the consideration of the *definite chemical action of electricity.* But before I dismiss this series of experimental Researches, I would call to mind that, in a former series, I showed the current of electricity was also *definite in its magnetic action* (216. 366. 367. 376. 377.); and, though this result was not pursued to any extent, I have no doubt that the success which has attended the development of the chemical effects is not more than would accompany an investigation of the magnetic phenomena.

Royal Institution,
December 31*st*, 1833.

---

873. *increase the power of our batteries … thousandfold surpass*: Is this the kind of opportunity it would be "impolitic" to ignore (paragraph 852)? He expressed a decidedly more distant attitude in the Second Series (paragraph 159): "I have rather, however, been desirous of discovering new facts and new relations … than of exalting the force of those already obtained…" But truly, as he makes clear in a number of other writings and speeches, *civic and industrial power*— with their promise of easing the lot of mankind—are very important to Faraday.

Eleventh Series — Editor's Introduction

Electric induction

In the Introduction to the First Series I described some of the pheno-mena associated with *electrostatic induction*. One of them was shown by the electroscope. As the electroscope's top plate is approached by an electrified body the two leaves gradually separate—showing by their mutual repulsion that they have become similarly electrified. Yet since the leaves droop again as soon as the approaching body is withdrawn, their electrified condition cannot have been transferred directly; it is a case of induction, not conduction.

But what is the *nature* of electric induction? One account, widely accepted when Faraday wrote, asserted the mutual attraction and repulsion of independent but imponderable electric fluids. According to that view, all unelectrified bodies were endowed with equal quantities of positive and negative electric fluids. A positively electrified glass rod, then, was thought to carry an excess of positive fluid; and when the rod approached the electroscope that excess would attract some of the instrument's negative fluid towards the upper plate and repel some of its positive fluid towards the leaves (sketch *a*). Both plate and leaves would then be in an electrified state, having excesses of negative and positive fluid, respectively. As the drawing shows, the electrified leaves manifest their condition by mutual repulsion.

(a)

Although in the first stage such "charge by induction" is characteristically *dual*—both positive and negative electrification being exhibited in different regions of the body under induction—the condition can be trans-formed, in conducting bodies at least, to one of homogeneous electrification. With the electrified rod held near the plate, *ground* the plate momentarily—or, as Faraday says, "uninsulate" it. A wire connected to the earth does the job (sketch *b*), or even a touch with one's finger. The leaves fall; but when both the ground con-nection and the charged rod are withdrawn, the leaves separate permanently (sketch *c*).

(b)

The "electric fluid" explanation of this result is ingenious. I described a moment ago how, when the elec-troscope is insulated, the positive rod was said to attract negative fluid to the electroscope plate and repel positive fluid to the leaves. But the *grounded* electroscope

(c)

becomes, electrically, part of the earth. In that context, the positive rod will attract negative fluid *to the electroscope as a whole*, while repelling positive fluid into the depths of the earth (again see sketch *b*). There will then reside an excess of negative fluid in the electroscope as a whole; and when the instrument is again insulated and left to itself that excess becomes distributed about the whole conductor, including the repelling leaves.

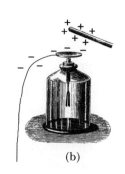

(b)

Notice that by this technique the instrument will be left with a permanent charge having the *opposite sign* to that of the approaching body. The same result must obtain for any body that is electrified by momentary grounding under induction, since a body so electrified must necessarily be small in comparison to the earth. When making measurements with the *balance electrometer* in the present Series, Faraday will employ a small "carrier ball" as the body under induction, and he will obtain charges on it which are opposite in sign to the inducing electrification.

As I have noted many times already, Faraday was suspicious of, if not downright hostile to, explanations in terms of independent electrical fluids of the sort I have just reviewed. These fluids were merely hypothetical and had never been observed—indeed it is not clear what would count as "observation" of a fluid whose sole property was that of attracting and repelling other electric fluids! Moreover, the imagery of electric fluids often proved surprisingly obscure in the elucidation of particulars. How in the previous sketch (b), for example, can the fluid theory explain the disappearance of positive fluid from the electroscope leaves (thus causing them to fall) at the moment when the plate is grounded? In order to pass to the earth, the fluid would first have to flow up the electroscope stalk *towards* the rod, which would be contrary to the direction of repulsion. Or, if negative fluid from the earth neutralizes the positive leaves, it would eventually have to flow down the stalk and away from the positive rod, contrary to the direction of attraction.*

But the most fundamental objection, from Faraday's point of view, was the way the "fluid" theories *exploited* natural phenomena instead of responding to them. As the foregoing account itself illustrates, fluid theories elevated the phenomena of *attraction and repulsion* to paradigm status and subordinated all other electrical actions to attraction and repulsion between electrical charges. This approach was of course modeled on Newton's planetary theory, which found, in the

---

* I do not mean to imply that this difficulty is a fatal one; it is not. It can be answered. But in the effort to do so the electric fluid becomes a hypothesis to be saved, rather than a source of elucidation.

force of gravitational attraction, the key to the motions of heavenly bodies. Now it may very well be that for planets, comets, and moons—whose whole significance, at least for celestial mechanics, lies in their *motions*—the phenomena of attraction *ought* to be regarded as the most fundamental facts about them, from which all other propositions are to be deduced. But do we find the same situation for electricity? Will the true electrical science be a *mechanics* of electricity?

To Faraday, who attends not only to attraction and repulsion but equally to the phenomena of *chemical change, spark and other discharge*, the distinctive *geometrical patterns* of distribution of electrical influence, and—most importantly—the *mutual relations* between electric, magnetic, and other natural powers, it is an inexcusable dogmatism to elevate a single class of powers above all others, simply to emulate Newtonian philosophy. Whether or not the planetary motions reveal themselves as fundamentally attractional, the electrical and magnetic phenomena decidedly *do not*, in Faraday's view; and it is only the comprehensive pattern of these interrelated natural phenomena—not a narrow and dogmatic selection among them—that is capable of disclosing to us their essential "look" as natural powers.

In the present Series, Faraday will survey some familiar inductive phenomena and also bring forth some spectacular new ones. What he finds will impel him to turn the conventional theory on its head: instead of explaining induction and other electrical actions by appealing to electric charge, Faraday will explain *charge* by appealing to *induction*! Several specific findings help to support and shape this novel view. It may be helpful to mention a few of them in advance.

No "absolute" charge

Any one who has worked or played with magnets will have noted the evident *inseparability* of the north-seeking and south-seeking magnetic powers. In his 1859 Christmas Lectures, excerpted in the introduction to the First Series, Faraday emphasized that inseparability to his audience by breaking small bar magnets into pieces—of course each piece, however small, consistently exhibited both powers, one at each end.* The magnetic power appears to be *inherently* dual; no material has ever exhibited either the north-seeking or the south-seeking power exclusively, but rather *both* powers paired and in equal overall amounts.

With electrification, however, the appearances are very different. There it is not only possible but *commonplace* for a body to exhibit a single electrical power. If, for example, we electrify glass by friction with silk, its charge is *positive* exclusively. If the whole surface of the

---

* See his remarks on page 20 pertaining to Figure 40 in that lecture.

glass is rubbed, positive electrification will be found distributed over the whole surface. If the electrified glass touches a small metallic sphere, the sphere will take on positive electrification as a whole. It would seem, therefore, that the positive and negative electrical powers are distinct and separable—a view that obtains perfect expression in the image of independent *electrical fluids* which, however much they may attract or repel one another, are no less separable in principle than are the earth and the moon. And it is true that if you only pay attention to phenomena involving *attraction and repulsion*, there is very little fault to be found with the "fluid" view.

But Faraday has been investigating other phenomena in which electricity shows a different face. To begin with, it appears impossible to electrify a body either positively or negatively without at the same time producing an opposite electrification also—not necessarily on the same body, but *somewhere*. In electrifying glass by friction, for example, the glass becomes *positive*—but the silk or other rubbing material becomes at the same time *negative*.* Moreover, Faraday reports in the present Series, this electrical duality is perfectly general. *There is no known instance* of a body able to be charged with a single electricity "absolutely"—that is, independent of a relation to some other, opposite, electricity. But this universal relation between opposite electricities is the same relation of mutual dependence that we identified, in our example on page 215, between the charged rod and the electroscope plate—the relation called induction. *All electric charge, therefore, participates in induction.* That is what Faraday means when he reports in the present Series that *there is no "absolute" charge of matter.*

The "Faraday Cage"

Faraday built a 12-foot cubical wooden framework and wrapped it about with copper wire and tinfoil so that, except for narrow slits between the foil bands (through which the interior could be viewed), it became a highly conductive hollow cage. This he mounted on glass insulators so that it could be electrified, using the Royal Institution's most powerful plate machine. He then carried out a number of tests; let us consider three of them:

1. He places an electrometer inside the otherwise empty cage; the instrument shows no initial deflection, and it continues to be unaffected *even when he charges the cage walls.* Even though the walls are highly electrified with respect to *exterior* bodies—they spark and hiss!— their intense electrification has no effect on the *interior* of the cage.

---

* Faraday gives an example in the Christmas Lectures; see his remarks pertaining to Figure 34 on page 5, in the editor's introduction to the First Series.

2. If, however, the electrometer had previously been mounted *very close to one of the slits*, it does respond—not only to electrification of the walls but even to movements of bodies *outside* the cage. The electrometer is affected in another circumstance also: The cage walls having been charged and the machine stopped, when an access door is opened—and especially if objects are inserted through the door and held inside the cage—the electrometer responds strongly.*

3. Finally, *Faraday himself enters the cage* with a variety of objects and electrical instruments. The plate machine is put into operation and, with sparks flying out from the cage walls and the smell of ozone pervading the room, Faraday carries out ordinary observations and electrical measurements inside the cage, electrifying some bodies by friction and testing them against one another and with the electrometer—all perform as usual. Thus, not only does charging the cage not electrify its interior, it does not alter the mutual electrical relations of bodies—whether charged or uncharged—within the cage.

The Faraday Cage is surely one of his most spectacular experiments; but Faraday's account in the present Series is surprisingly low-keyed. Perhaps he intentionally slighted its sensational aspects lest they distract readers from the theoretical implications of the experiment. One theoretical consequence, at least, is that electrical "charge" is *essentially relative*; it can neither be produced nor detected except in comparison to something else. In our ordinary electrical measurements it is easy to overlook that a charged sphere, for instance, is not simply electrified "in itself" but sustains inductive relations with the walls of the room and with even more distant objects. The Cage forces him—and us—to become aware of those abiding and neglected background relations, by conspicuously interrupting them.** For example, test (1) shows that a solitary electrometer cannot acquire "charge" in the cage, since it has no access to anything else capable of sustaining the "charge relation" with it. Test (2), on the other hand, reveals that when a flaw or opening in the cage restores those exterior relations, then the electrometer *does* acquire electrification—but of course it is then no longer "solitary"!

---

* This investigation is recorded in the Diary. I see it as integral to the Cage experiments but, curiously, Faraday does not mention it in the published Experimental Researches.

** *How* does the Cage "interrupt" them? The next section explains why an interposed conductor tends to block the inductive relation between electrified bodies; see also Faraday's account at paragraph 1338 in the Twelfth Series.

Finally, by bringing all the usual electrical furnishings into the cage for test (3), Faraday is recreating the "ordinary" electrical world—but within an artificial background that is now too prominent to be ignored. As a result, his view of "ordinary" experience is altered. Perhaps that was the whole point of building an environment large enough that Faraday could insert himself into it; with smaller apparatus he could *imagine* what it would be like to keep the background present in his thinking—but in the Cage he experiences it directly.

Perhaps we are inclined to neglect the distant, permanent relations of electric (and magnetic) bodies because they are so much weaker than the proximate relations among those bodies. For gravity, by contrast, the distant and permanent relations—the attractions of individual ponderable bodies, including ourselves, to the earth—are the *predominating* relations in everyday experience; gravitational attraction *between* everyday bodies is practically undetectable. Perhaps, then, a comparison with gravity can help define what it is that the Cage reveals.

If the Cage were capable of doing for gravity what it does for electricity, then (1) a solitary body within the cage would be weightless; it would not fall if released. But (2) if a door or hatch were opened the body would accelerate immediately!* And (3) if several objects were inside the cage they would exhibit only their gravitation towards *one another*, not towards the earth—until a door was opened! Test (3) in a gravitational "cage" would decidedly *not* resemble "ordinary life"; it would be more like living among the planets. In a gravitational "cage" we would learn, if we did not already know it, that weight is not an inherent property of bodies but a relative one, that bodies have weight only with respect to other bodies. If we actually did live out among the planets, that knowledge would be embedded in our everyday experience; but instead we live on an earth whose ever-present influence can be isolated only by an act of thought. Newton had to *think* himself into the cosmos in order to change his (and everyone's) thinking about gravity; with a gravitational "cage" that insight would have been gained by direct experience.

David Gooding has pointed out** that Faraday is always concerned with the *spatial disposition* of powers and qualities, and that he regularly uses his instruments and apparatus to survey magnetic and electric space in the same way that a terrestrial explorer ranges over geographical territory. Gooding cites Faraday's Cage as the culminating

---

* But in which direction? Towards the floor? The door?

** *Faraday Rediscovered*, ed. D. Gooding and F. A. J. L. James. Macmillan, 1985 and American Institute of Physics, 1989. Chapter 6, "In Nature's School".

instance of occupying the experimental space "with his person as well as his instruments." But why is *personal* occupation so important? I am suggesting that Faraday has to be *in* the experimental arena in order to experience the *usual* in a new and richer perspective—and thereby change the nature of "ordinary" electrical consciousness forever.

The relation between "static" and "current" electricity

According to fluid theories, the relation between "static" and "current" electricity is that between *rest* and *motion*, as indeed their very names signify. But Faraday finds a more fundamental relation, suggested by his study of the transition from one to the other. This transition appears to depend wholly on the material medium changing from an insulator to a conductor. Early in the present Series, Faraday will recall his former electrolysis experiments in which current was prevented by the interposition of a film of ice but instantly commenced when the ice melted. This suggests images of the static or inductive state as one of forcible restraint—a case of *stress* or *tension*, the relief of which constitutes the condition we call *discharge* or *current*. An even more suggestive instance is the abrupt *spark* discharge; for here we can observe that the greater the degree of electrification prior to discharge (an electroscope will show this), the greater the violence and intensity of the spark when it finally passes. Since the intensity of discharge is commensurate with the intensity of electrification, it is easy to see the discharge as manifesting the sudden *collapse* of the electrified or inductive state. Thus instead of respectively identifying "common" and current electricity with *fluid at rest* and *fluid in motion*, Faraday adopts the more general image of *tension in a medium* and its *relaxation*.

From this point of view, an insulator will be a material capable of sustaining high electric tension;* a conductor, on the other hand, will be recognized as a material that *cannot sustain electrical tension*—or one that can bear only minimal degrees of tension before giving way. The condition of induction is therefore a condition of *tension in the insulating medium*. The condition of current is the *continual giving way of that tension* in the *conducting medium*. Thus an interposed conductor tends to block the inductive relation between electrified bodies—since it cannot sustain the tension in which that inductive relation consists. Keep in mind, however, that this sense of "tension" is metaphorical. There may be important differences between electrical "tension" and the mechanical tension which a taut string exhibits.

---

* Even today, electric power lines are still called "high-tension wires." The expression has its origin in the image of electric tension. In Britain the corresponding nomenclature is "electric pressure," expressing a similar idea.

As is often the case with Faraday, and as we saw in the Seventh Series, fresh views of familiar topics demand a new terminology. The customary terms "nonconductor" and "insulator" are no longer satisfactory since they emphasize disconnection and solitude. In Faraday's new vision, electrified surfaces are not *isolated* by these materials; they are, on the contrary, *associated* by induction through them. Faraday will therefore propose the new term *dielectric* (Greek δία, *through*) to express the essential property of the inductive medium—that of sustaining electric tension through itself.

"Charge" is a boundary, not a substance

A third and very important phenomenon again has little to do with attraction and repulsion. Coulomb, using his torsion-balance electrometer, had shown that a conductor can be electrified only on its *outer surface*. This means that when oppositely-electrified conductors are linked by a nonconducting medium the electric power is always disposed at the *boundaries* of the medium—or, we may say, at the boundaries of the *tension* if, as just discussed, we view the medium as being in a state of electrical tension.

Pursuing the imagery of tension, if we compare the dielectric medium to a taut string, then the electric charges will correspond to the forces applied to the string or exercised by the string at each end. Note that such a correspondence immediately disposes of *action-at-a-distance*; for insofar as the applied forces can be said to act, they act *on the string*, which they adjoin—not on each *other*, mutually separated.* But even that formulation is somewhat misleading; surely it is not the *forces* that do the "acting"—they are not independent agents. Of course we routinely speak of the forces as stretching the string, but this is a very abstract way of speaking. It corresponds to the physicists' diagram of a disembodied string with two arrows at the ends to represent the forces applied. In actuality, however, a taut string requires a *bow* or some comparable stretching mechanism *which must itself be similarly strained*. The forces may then be viewed simply as the form in which string tension—or, alternatively, bow tension—manifests itself at the boundaries. In either case, *force is subordinated to stress in a medium.*

Similarly, Faraday will come to regard electric "charge" as merely an *aspect* of induction—the manifestation, at the boundary of a dielectric

---

* The taut string analogy also agrees perfectly with Faraday's declaration that every charge has to be in relation to an *equal and opposite* charge (paragraph 1295); since if the tension along a string is uniform, the forces at the ends will be equal and opposite.

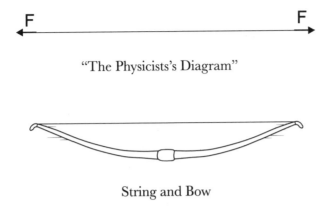

"The Physicists's Diagram"

String and Bow

medium, of electric tension in that medium. Faraday will formulate this view of charge gradually, in stages throughout the Eleventh Series, implicitly at first and later with increasing outspokenness. Be prepared, therefore, for a certain conflation of "charge" and "tension" in Faraday's discussion. Both are *aspects of induction* and are therefore not always easy to distinguish.

Electrical "charge" and "tension" in the electrometer

In connection with the First Series we took note of the electroscope and other electric "indicators." There is a rich variety of such devices, but the principle common to them all is an attraction—or more often, a repulsion—between two electrified parts. Mutual repulsion of a pair of gold leaves, and elevation of a straw pendulum from its vertical support, indicate the degree of electrification in two typical designs.

Efforts were made to standardize and calibrate these appliances, the resulting instruments qualifying as *electrometers* rather than mere indicators; but on the whole they remained rather rough and comparatively insensitive devices. Coulomb constructed a vastly improved electrometer, which he used to measure the force between two electrified spheres.* Faraday in the present Series will put Coulomb's instrument to rather different use. He gives a complete description of its construction and operation beginning at paragraph 1180, but I need to say a few things about it in advance.

---

* Coulomb's measurements disclosed a force that diminished with the square of the distance—thereby heralding, he thought, a mathematical theory of electricity modeled on Newton's gravitational theory. But to Faraday an electrical relation between two *spheres* has no higher status than that between any other pair of bodies; and to elevate it to a paradigm is, from his point of view, a move of surpassing rashness and caprice (paragraph 1303, *comment*).

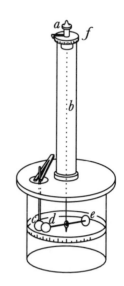

Coulomb's electrometer, shown in the drawing, balances the force (either attractive or repulsive) between two electrified balls *c* and *d* against the torque developed in a twisted thread *b*. With a definite charge given to the pivoting ball *d*, and with a standard distance between the balls, Coulomb had already shown that the force developed—and therefore the degree of twist required to balance it—is proportional to the *charge of the removable ball c*. The electrometer, then, may be said to indicate that charge directly.

On the other hand, ball *c* did not acquire its charge while mounted within the instrument but rather at some external location. If "charge" is indeed the manifestation which electric tension in a dielectric presents at its boundaries, then the charge of ball *c* certainly bears some definite relation to the electric tension in the dielectric surrounding it *at the location where c acquired that charge*. But that relation cannot be assumed to be *fixed* one. Any change—and in particular a change in proximity to other bodies—might alter the degree to which the tension surrounding an electrified body manifests itself as a "charge" belonging to that body. For that reason, Faraday treats the electrometer as an indicator of tension *only* in those cases where the carrier ball *c* acquires all its charges at a single location.

In the present Series, Faraday will describe two electrometer procedures. One is performed repeatedly at a single location and discloses the *tension* sustained in the medium surrounding an electrified body. The second is carried out at multiple locations and focuses on the *charge* which an electrified body is able to induce. I will explain the reasoning behind both procedures; but remember that the distinction between charge and tension is primarily a matter of emphasis; each is only a singular aspect of the fundamental condition of induction.

*Procedure (1): To determine the tension sustained in the medium surrounding an electrified body.** Consider an electrified and insulated metal ball B. By Faraday's account, the electrified body will be related by induction to every other conductor or partial conductor in the universe—but for simplicity let us restrict our consideration to the *earth*. Faraday has determined that induction is a condition of tension; and therefore an overall condition of tension must be sustained through the air or other dielectric materials that occupy the regions between ball B and the earth.

---

* See paragraphs 1187–1214 in the present Series.

Now what will happen if another metal ball *c* is brought into contact with B? I say that electrification will be communicated (that is, tension will be discharged) between B and *c*, until the tension of both balls with respect to the earth is *the same*. Otherwise, a difference between their tensions would have to be sustained by the junction between them; and that is impossible since, as Faraday has seen, a conducting material is practically *incapable of*

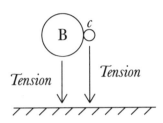

sustaining electric tension. Ball *c*, therefore, will attain some definite degree of electrification while it remains in contact with B; and Faraday assumes it will continue to retain its charge when subsequently removed and insulated again.*

Now as I mentioned earlier, the charge acquired by *c* may be said to indicate the tension (with respect to the earth) which *c* had *when that charge was given to it.* But that tension is the same as the tension of B; so the charge acquired by *c* is similarly related to the tension in the medium between B and the earth. Faraday will therefore make ball *c* the removable ball of the electrometer, carrying it between the experimental apparatus and the instrument. When *c* is mounted in the electrometer, the reading obtained reflects *the tension that subsisted between B and earth* at the last moment *c* was in contact with B—which was the aim of this procedure.**

Note that if the foregoing procedure is valid, electric tension is always exerted and must always be measured *between two points*—there is no electric tension "at" a single point. Faraday is not very explicit about this and sometimes creates a different impression by attributing electric tension to *individual particles* of the medium in question; see, for example, paragraphs 1224 and 1300 in the present Series. But even then, the "tension of a particle" is exerted across the particle's diameter—that is, between two points.

---

* As a practical matter, no one doubts that insulated bodies retain their charges. But this fact, which is obvious on the fluid hypothesis, is not so easy to explain if charge is *not* a fluid. Faraday appears to overlook this theoretical difficulty; see his note to paragraph 1218 below.

** Note that this argument does not rule out additional factors, besides tension, that might affect the electrometer readings. The next procedure will both disclose such an influence and explain why it is not a consideration here.

Procedure (2): To determine the degree of electric induction at various loca-tions about an electrified body.* Consider an insulated and electrified body S, together with a carrier ball *c*. And let *c* be situated anywhere, not necessarily juxtaposed to body S. What will happen if we connect ball *c* to *ground*, instead of touching it to the electrified body as before?

An argument similar to the foregoing will apply. When *c* is grounded, the wire will communicate electrification (that is, will discharge tension) until both *c* and the earth have *equal tension with respect to S*. As before, the charge acquired by *c* will be commensurate with that tension and will be retained even when the wire is removed. When *c* is returned to its proper position in the electrometer, therefore, the reading obtained will indicate the charge it acquired "by induction" in its former location (where it was momentarily grounded)—which was the aim of this procedure.**

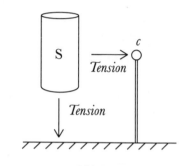

Faraday finds that the carrier ball *c* develops *unequal charges* when the procedure is repeated at different locations. In the first procedure it was assumed that charge and tension increased and decreased together; but that cannot be the case here since, as I explained, the grounded ball *c* sustains *the same tension in all locations*—equal to the (constant) tension between S and earth. Thus in the present pro-cedure we find variability of charge, independent of tension. What does this independence signify?

Inasmuch as charge and tension are both aspects of induction, any variation of one independently of the other must point to a characteristic of the inductive condition itself—what Faraday will call *aptness or capacity* with respect to induction. Roughly speaking, this is the facility with which a particular apparatus manifests its inductive condition under the aspect of *charge*, rather than of *tension*. Faraday becomes gradually more explicit about this in the course of the pre-sent Series. In paragraphs 1214 and 1261 below he recognizes that different charges, sustained, as here, at the same degree of tension, represent proportionally different inductive capacities. Similarly, a constant charge, sustained at different degrees of tension, represents

---

* See paragraphs 1218–1229 in the present Series.

** Notice that the technique of momentary grounding is essentially identical to the electroscope exercise described on page 215. This time, however, we have not framed our account in terms of attractions and repulsions between hypothetical electric fluids.

reciprocally different inductive capacities. Thus the electrometer, employed according to the present procedure, shows that the capacity for induction between two bodies depends on their mutual distance or configuration.*

I said earlier that Faraday's "tension" is metaphorical and is only to be compared to, not identified with, the mechanical tension of, say, a taut string. Later theorists will develop distinctions within the "tension" idea, from which arise more precise concepts such as *field intensity* and *potential difference* (popularly called *voltage*). But it is not clear that Faraday sees any basis—or indeed any need—for such distinctions.

"Sensible" and "latent" electricity

The terms "sensible" and "latent" are borrowed from early doctrines of heat. When heated, a body was thought to accumulate "caloric" (the hypothetical fluid of heat), whose presence would ordinarily be *sensible* as a rise in temperature of the body. But caloric that acted to melt a solid or boil a liquid would not, during the melting or boiling process, change the substance's temperature; it was therefore declared to be hidden, or *latent*, within that substance.

As a boy, Faraday learned about this and other scientific topics from Jane Marcet's *Conversations in Chemistry*. In the 23 dialogues that make up that delightful young people's book, an astonishingly precocious "Caroline" and "Emily" are instructed in scientific methods and reasoning by the genial but mysterious "Mrs. B." The book became an inspiration to the young Faraday, who retained an affection for it throughout his life.** We may enjoy the same apt account:

---

* It was therefore permissible to overlook this spatial dependency in the first procedure, since the carrier ball was electrified exclusively in *one* location—namely, in contact with ball B. Thus there would have been no significant variation of its *inductive capacity* between successive measurements.

** As an old man, Faraday recounted his gratitude in a letter to his close friend Auguste de la Rive, who was planning a short biography of Jane Marcet: "Your subject interested me deeply in every way, for Mrs. Marcet was a good friend to me, as she must have been to many of the human race. I entered the shop of a bookseller and bookbinder at the age of 13, in the year 1804, remained there eight years, and during the chief part of the time bound books. Now it was in those books, in the hours after work, that I found the beginning of my philosophy. There were two that especially helped me, the 'Encyclopaedia Britannica,' from which I gained my first notions of electricity, and Mrs. Marcet's 'Conversations in Chemistry,' which gave me my foundation in that science. ... So when I questioned Mrs. Marcet's book by such little experiments as I could find means to perform, and found it true to the facts as I could understand them, I felt that I had got hold of an anchor in chemical knowledge, and clung fast to it. Thence my deep veneration for Mrs. Marcet: first, as one who had conferred great personal good and pleasure on me..." Bence Jones, *The Life and Letters of Faraday* (1870).

Mrs. B. I shall now show you an experiment which I hope will give you a clear idea of what is understood by latent heat.

The snow which you see in this phial, has been cooled by certain chemical means (which I cannot well explain to you at present), to 5 degrees below the freezing point, as you will find indicated by the thermometer, which is placed in it. We shall expose it to the heat of a lamp, and you will see the thermometer gradually rise, till it reaches the freezing point—

Emily. But there the thermometer stops, Mrs. B., and yet the lamp burns just as well as before. Why is not its heat communicated to the thermometer?

Caroline. And the snow begins to melt, therefore it must be rising above the freezing point?

Mrs. B. The heat no longer affects the thermometer, because it is wholly employed in converting the ice into water. As the ice melts, the caloric becomes latent in the new formed liquid, and therefore cannot raise its temperature; and the thermometer will consequently remain stationary, till the whole of the ice be melted.

Caroline. Now it is all melted, and the thermometer begins to rise again.

Mrs. B. Because the conversion of the ice into water being completed, the caloric no longer becomes latent; and therefore the heat which the water now receives raises its temperature, as you find the thermometer indicates.

In part following a speculative analogy between "electric fluid" and fluid "caloric," some investigators appropriated the thermal vocabulary for electricity. Electrical fluid theorists could understand the term "latent" quite literally, even positing hidden cells and interstices in materials—just as partisans of fluid "caloric" had done in attempting to explain latent heat. But even a disbeliever in electrical fluids, like Faraday, had nevertheless to confront the question. Certainly bodies can acquire electricity in *sensible* form—manifested by its effect on an electrometer or other indicator. Could electrification then also admit a *latent* form, whereby it might be present in a body without affecting an electrometer? It would not be easy to reconcile such an electrical state with Faraday's ideas about the nature and primacy of *induction*.

In the present Series Faraday will give considerable attention to a phenomenon he calls the "return charge." He describes it beginning with paragraph 1233 below and remarks that he *did* at first view the effect as a species of latency. But more thorough study shows it to be not only compatible with his proposed principles of inductive action, but a valuable magnifying lens, as it were, with which to view the process of *conduction*.

ELEVENTH SERIES.

§ 18. *On Induction.* ¶ i. *Induction an action of contiguous particles.* ¶ ii. *Absolute charge of matter.* ¶ iii. *Electrometer and inductive apparatus employed.* ¶ iv. *Induction in curved lines.* ¶ v. *Specific inductive capacity.* ¶ vi. *General results as to induction.*

Received November 30, Read December 21, 1837.

¶ i. *Induction an action of contiguous particles.*

1161. THE science of electricity is in that state in which every part of it requires experimental investigation; not merely for the discovery of new effects, but what is just now of far more importance, the development of the means by which the old effects are produced, and the consequent more accurate determination of the first principles of action of the most extraordinary and universal power in nature: and to those philosophers who pursue the inquiry zealously yet cautiously, combining experiment with analogy, suspicious of their preconceived notions, paying more respect to a fact than a theory, not too hasty to generalize, and above all things, willing at every step to cross-examine their own opinions, both by reasoning and experiment, no branch of knowledge can afford so fine and ready a field for discovery as this. Such is most abundantly shown to be the case by the progress which electricity has made in the last thirty years: Chemistry and Magnetism have successively acknowledged its over-ruling influence; and it is probable that every effect depending upon the powers of inorganic matter, and perhaps most of those related to vegetable and animal life, will ultimately be found subordinate to it.

1162. Amongst the actions of different kinds into which electricity has conventionally been subdivided, there is, I think, none which excels, or even equals in importance that called *Induction*. It is of the most general influence in electrical phenomena, appearing to be concerned in every one of them, and has in reality the character of a first, essential, and fundamental principle. Its comprehension is so important, that I think we cannot proceed much further in the investigation of the laws of electricity without a more thorough understanding of its nature; how otherwise can we hope to comprehend the harmony and even unity of action which doubtless governs electrical excitement by friction, by chemical means, by magnetic influence, by evaporation, and even by the living being?

1163. In the long-continued course of experimental inquiry in which I have been engaged, this general result has pressed upon me constantly, namely, the necessity of admitting two forces, or two forms or directions of a force (516. 517.), combined with the impossibility of separating these two forces (or electricities) from each other, either in the phenomena of statical electricity or those of the current. In association with this, the impossibility under any circumstances, as yet, of absolutely charging matter of any kind with one or the other electricity only, dwelt on my mind, and made me wish and search for a clearer view than any that I was acquainted with, of the way in which electrical powers and the particles of matter are related; especially in inductive actions, upon which almost others appeared to rest.

1164. When I discovered the general fact that electrolytes refused to yield their elements to a current when in the solid state, though they gave them forth freely if in the liquid condition (380. 394. 402.), I thought I saw an opening to the elucidation of inductive action, and the possible subjugation of many dissimilar phenomena to one law. For let the electrolyte be water, a plate of ice being coated with platina foil on its two surfaces, and these coatings connected with any

---

1162. *electrical excitement ... by evaporation*: Faraday has not previously named *evaporation* as a source of electrical excitation, but he may have in mind the reported electrification of high-pressure steam jets; at least one observer thought that the change of state from liquid water to steam was responsible. In the Eighteenth Series, however, Faraday will show that the vapor particles develop electric charge through *friction with the nozzle walls* as they exit at high speed.

1163. *two forms or directions ... two forces (or electricities)*: that is, what customary nomenclature recognizes as *positive and negative electricities*. But Faraday is reluctant to adopt all the connotations of that conventional terminology. He will explain what he means by "absolutely charging matter" beginning at paragraph 1169 below.

1164. *solid state ... liquid condition*: Ice is an insulator and therefore supports induction; but liquid water is a conductor and therefore permits current flow. Since change from the solid to the liquid state involves the *particles* of a material, then it would seem that both conduction and induction are

continued source of the two electrical powers, the ice will charge like a Leyden arrangement, presenting a case of common induction, but no current will pass. If the ice be liquefied, the induction will fall to a certain degree, because a current can now pass; but its passing is dependent upon a *peculiar molecular arrangement* of the particles consistent with the transfer of the elements of the electrolyte in opposite directions, the degree of discharge and the quantity of elements evolved being exactly proportioned to each other (377. 783.). Whether the charging of the metallic coating be effected by a powerful electrical machine, a strong and large voltaic battery, or a single pair of plates, makes no difference in the principle, but only in the degree of action (360.). Common induction takes place in each case if the electrolyte be solid, or if fluid, chemical action and decomposition ensue, provided opposing actions do not interfere; and it is of high importance occasionally thus to compare effects in their extreme degrees, for the purpose of enabling us to comprehend the nature of an action in its weak state, which may be only sufficiently evident to us in its stronger condition (451.). As, therefore, in the electrolytic action, *induction* appeared to be the *first* step, and *decomposition* the *second* (the power of separating these steps from each other by giving the solid or fluid condition to the electrolyte being in our hands);

---

essentially connected with the particles of the material medium. Electrolysis, examined in the Seventh Series, discloses one aspect of *conduction* in relation to its material. In the present Series, Faraday finds "an opening to the elucidation of" *induction* in relation to its material medium.

peculiar molecular arrangement...: that is, passing of the current depends not only upon the particles of the substance having acquired mobility as wholes (which is *liquefaction*), but also upon their *elements* being able to move relatively to one another—in fact in opposite directions. Note that Faraday's "molecule" is a tiny portion of the substance in question, but not necessarily an assemblage of elementary *atoms*, which are still highly questionable in Faraday's eyes (paragraph 869).

induction appeared to be the first step, and decomposition the second...: Induction and decomposition are "first" and "second" not only chronologically but *functionally*. He conceives the ice under induction as being in a state of *readiness* for electrolytic decomposition, which however cannot actually occur until the particles of the electrolyte become mobile—that is, until the ice melts.

in our hands: Because we control the freezing and melting of ice we can artificially separate the stages of *readiness* and *performance* in this instance of decomposition—but he suspects that a similar distinction, between an antecedent condition of induction and a consequent decomposition, applies to electrolytes generally.

as the induction was the same in its nature as that through air, glass, wax, &c. produced by any of the ordinary means; and as the whole effect in the electrolyte appeared to be an action of the particles thrown into a peculiar or polarized state, I was led to suspect that common induction itself was in all cases an *action of contiguous particles*,[1] and that electrical action at a distance (*i. e.* ordinary inductive action) never occurred except through the influence of the intervening matter.

1165. The respect which I entertain towards the names of Epinus, Cavendish, Poisson, and other most eminent men, all of whose theories I believe consider induction as an action at a distance and in straight lines, long indisposed me to the view I have just stated; and though I always watched for opportunities to prove the opposite opinion, and made such experiments occasionally as seemed to bear directly on the point, as, for instance, the examination of electrolytes, solid and fluid,

---

[1] The word contiguous is perhaps not the best that might have been used here and elsewhere; for as particles do not touch each other it is not strictly correct. I was induced to employ it, because in its common acceptation it enabled me to state the theory plainly and with facility. By contiguous particles I mean those which are next. —Dec. 1838

1164, continued. *induction was the same in its nature ... polarized state*: There is no difference between the induction observed in ice and that found in any other substance. The particles of electrolytes evidently become *polarized* with respect to chemical powers (since their constituent elements migrate in opposite directions); he suspects that even in induction that does not involve electrolysis, there must arise some comparable state of *tension* and *separation of opposites* in the particles of the substance under induction. Faraday will say more about this condition—the *polarized* state—in paragraph 1298 below.

through ... intervening matter: The usual conception of electric force supposes a *direct relation* of attraction or repulsion between electrified bodies. According to that doctrine, if air, water, or other media intrude, the intervening particles develop their own relations with the electrified bodies but do not alter the direct relation between the bodies themselves. But Faraday has been led to suspect that, on the contrary, the intervening medium plays an *essential role* in supporting the interaction between electrified bodies. Thus what was conventionally looked upon as "action at a distance" is really, he thinks, a chain of successive proximate relations involving the particles of the medium. But what can Faraday possibly mean by "contiguous particles" *when the intervening medium is a vacuum?* It will be some time before he faces this question.

Comment on Faraday's note 1. Since "particles do not touch"—can in gases be *inches apart* in fact, as one annoyed correspondent pointed out to him—it is clear that Faraday is not denying the "distance" in action-at-a-distance. Rather, he opposes the idea that electric action can leap over intervening matter *if it is there*. Any intervening material must, in his view, play an essential role in the mutual interaction of electricities. Then do the spaces between particles play *no* role? The question will trouble him again at paragraph 1293.

whilst under induction by polarized light (951. 955.), it is only of late, and by degrees, that the extreme generality of the subject has urged me still further to extend my experiments and publish my view. At present I believe ordinary induction in all cases to be an action of contiguous particles consisting in a species of polarity, instead of being an action of either particles or masses at sensible distances; and if this be true, the distinction and establishment of such a truth must be of the greatest consequence to our further progress in the investigation of the nature of electric forces. The linked condition of electrical induction with chemical decomposition; of voltaic excitement with chemical action; the transfer of elements in an electrolyte; the original cause of excitement in all cases; the nature and relation of conduction and insulation; of the direct and lateral or transverse action constituting electricity and magnetism; with many other things more or less incomprehensible at present, would all be affected by it, and perhaps receive a full explication in their reduction under one general law.

1166. I searched for an unexceptionable test of my view, not merely in the accordance of known facts with it, but in the consequences which would flow from it if true; especially in those which would not be consistent with the theory of action at a distance. Such a consequence seemed to me to present itself in the direction in which inductive action could be exerted. If in straight lines only, though not perhaps decisive, it would be against my view; but if in curved lines also, that would be a natural result of the action of contiguous particles, but, as I think, utterly incompatible with action at a distance as assumed by the received theories, which, according to every fact and analogy we are acquainted with, is always in straight lines.

1165. *lateral or transverse action*: Faraday will say more about this later, for example at paragraph 1224, where lines of inductive action appear to be able to nudge one another into new positions.

1166. *straight lines … curved lines*: If the action between two electrified bodies is disposed along the straight line joining them, it fails to show any role for an intervening medium; since two points (the respective locations of the bodies) suffice to determine a straight line. But if the action is disposed in *curves*, it suggests that the medium participates in shaping the action; since the two terminating bodies can no longer determine a specific curve by their positions alone. See also paragraph 1215 below.

the received theories: that is, the theories handed down from earlier investigators and now generally accepted. As mentioned in the introduction to the present Series, these were more or less deliberately modeled on Newtonian gravitation, which acts in straight lines disposed radially about each particle of matter.

1167. Again, if induction be an action of contiguous particles, and also the first step in the process of electrolyzation (1164. 949.), there seemed reason to expect some particular relation of it to the different kinds of matter through which it could be exerted, or something equivalent to a *specific electric induction* for different bodies, which, if it existed, would unequivocally prove the dependence of induction on the particles; and though this, in the theory of Poisson and others, has never been supposed to be the case, I was soon led to doubt the received opinion, and have taken great pains in subjecting this point to close experimental examination.

1168. Another ever-present question on my mind has been, whether electricity has an actual and independent existence as a fluid or fluids, or was a mere power of matter, like what we conceive of the attraction of gravitation. If determined either way it would be an enormous advance in our knowledge; and as having the most direct and influential bearing on my notions, I have always sought for experiments which would in any way tend to elucidate that great inquiry. It was in attempts to prove the existence of electricity separate from matter, by giving an independent charge of either positive or negative power only, to some one substance, and the utter failure of all such attempts, whatever substance was used or whatever means of exciting or *evolving* electricity were employed, that first drove me to look upon

---

1167. *specific electric induction*: If the particles of an intervening medium play an essential role, then the degree of induction between two electrified bodies may be specific to the *kind of material* that is introduced between them.

1168. What is the difference between an *"actual and independent existence as a fluid or fluids"* and a *"mere power of matter"*? Think of a body undergoing electrification to a greater and greater degree, whether by action of a plate machine, or by stroking, or by any other means. If electricity has an independent existence we must conceive the body as acquiring from elsewhere ever-increasing quantities of a specifically electric fluid; if electricity is a power of matter, we view the situation as a continuous *heightening* or *enhancement* of powers which the body already possesses. But Faraday's choice of *gravitation* as an example of a "power of matter" is not, perhaps, the most instructive example; since according to the received theory (Newton's), *all* matter exercises the power of gravitation and its amount *cannot be altered* for a given quantity of matter (mass)—whereas the electrification of bodies is certainly capable of increase and decrease.

an independent charge of either positive or negative power only, to some one substance: He says that attempts to charge a body in this way always fail. But how can that be? Everybody knows how to charge a body positively or negatively! Faraday will acknowledge this in paragraph 1171 below, and will there explain more clearly what he means by "the utter failure of all such attempts".

induction as an action of the particles of matter, each having *both* forces developed in it in exactly equal amount. It is this circumstance, in connection with others, which makes me desirous of placing the remarks on absolute charge first, in the order of proof and argument, which I am about to adduce in favour of my view, that electric induction is an action of the contiguous particles of the insulating medium or *dielectric*.[2]

¶ ii. *On the absolute charge of matter.*

1169. Can matter, either conducting or non-conducting, be charged with one electric force independently of the other, in any degree, either in a sensible or latent state?

1170. The beautiful experiments of Coulomb upon the equality of action of *conductors*, whatever their substance, and the residence of *all* the electricity upon their surfaces,[3] are sufficient, if properly viewed, to prove that *conductors cannot be bodily charged*; and as yet no means of communicating electricity to a conductor so as to place its particles in relation to one electricity, and not at the same time to the other in exactly the same amount, has been discovered.

1171. In regard to electrics or non-conductors, the conclusion does not at first seem so clear. They may easily be electrified bodily, either by communication (1247.) or excitement; but being so charged, every

[2] I use the word *dielectric* to express that substance through or across which the electric forces are acting.—*Dec.* 1838.

[3] Mémoires de l'Académie, 1786, pp. 67. 69. 72; 1787, p. 452.

1169. *either in a sensible or latent state*: that is, either in an observable or a hidden condition. The terminology is borrowed from heat, as discussed in the introduction. Faraday will identify a possibly "latent" state of electricity in paragraph 1235 below.

1170. *conductors cannot be bodily charged*: That is, they can be electrified only on their exterior surfaces—at least, when the electricity is static. But Faraday acknowledges elsewhere that *current* electricity occupies the whole volume of the conducting material (see the Second Series introduction); and in paragraph 1295 below, he will point out that in a mediocre conductor, it may take a significant amount of *time* for all electrification to migrate from the interior to the surface.

1171. *electrics*: This was the antique term for bodies that could be electrified by friction. Such bodies are necessarily nonconductors and therefore, on Faraday's view, they are substances *through which* electrical influence acts. He therefore generally prefers the term *dielectric* (δία, through), as described in his note to paragraph 1168, above.

case in succession, when examined, came out to be a case of induction, and not of absolute charge. Thus, glass within conductors could easily have parts not in contact with the conductor brought into an excited state; but it was always found that a portion of the inner surface of the conductor was in an opposite and equivalent state, or that another part of the glass itself was in an equally opposite state, an *inductive* charge and not an *absolute* charge having been acquired.

1172. Well-purified oil of turpentine, which I find to be an excellent liquid insulator for most purposes, was put into a metallic vessel, and, being insulated, an endeavour was made to charge its particles, sometimes by contact of the metal with the electrical machine, and at others by a wire dipping into the fluid within; but whatever the mode of communication, no electricity of one kind only was retained by the arrangement, except what appeared on the exterior surface of the metal, that portion being present there only by an inductive action through the air to the surrounding conductors. When the oil of turpentine was confined in glass vessels, there were at first some appearances as if the fluid did receive an absolute charge of electricity from the charging wire, but these were quickly reduced to cases of common induction jointly through the fluid, the glass, and the surrounding air.

1171, continued. *a case of induction, and not of absolute charge.* See the examples of "charge by induction" in the introduction. In Faraday's example of placing an electrified glass body within a tin box, it turned out that either (*i*) the glass would display positive and negative electrification in equal degrees at different parts of its surface, or (*ii*) if the glass exhibited positive electrification *only*, then *the inner surface of the box* would show negative electrification to the same degree. Had things been otherwise—if the glass were to exhibit positive electrification only, and the box *no* sign of electrification, neither on its inner or its outer surface, then either this would be a case of "absolute charge" on the glass ball—unrelated to any other charge anywhere else—or it would be genuine *action at a distance* if the ball's charge *were* related to other bodies outside the box, while somehow inexplicably skipping over the box itself. When Faraday said in his note to paragraph 1164 that by contiguous particles he meant "those which are next," he meant to deny that any power could leap over one particle to affect a more distant particle.

1172. *inductive action through the air to the surrounding conductors*: An electrified metal container filled with turpentine exhibited only one kind of charge all over its surface—but in all cases Faraday found *nearby surfaces* that displayed the opposite electrification. When the container was glass, and the turpentine electrified by a wire dipped into it and connected to the electrical machine, the *fluid* exhibited electrification—but always a corresponding opposite electrification was found either elsewhere on the fluid surface, or on the wire, or the glass, or on neighboring objects.

1173. I carried these experiments on with air to a very great extent. I had a chamber built, being a cube of twelve feet. A slight cubical wooden frame was constructed, and copper wire passed along and across it in various directions, so as to make the sides a large network, and then all was covered in with paper, placed in close connexion with the wires, and supplied in every direction with bands of tin foil, that the whole might be brought into good metallic communication, and rendered a free conductor in every part. This chamber was insulated in the lecture-room of the Royal Institution; a glass tube about six feet in length was passed through its side, leaving about four feet within and two feet on the outside, and through this a wire passed from the large electrical machine (290.) to the air within. By working the machine, the air in this chamber could be brought into what is considered a highly electrified state (being, in fact, the same state as that of the air of a room in which a powerful machine is in operation), and at the same time the outside of the insulated cube was everywhere strongly charged. But putting the chamber in communication with the perfect discharging train described in a former series (292.), and working the machine so as to bring the air within to its utmost degree of charge, if I quickly cut off the connexion with the machine, and at the same moment or instantly after insulated the cube, the air within had not the least power to communicate a further charge to it. If any portion of the air was electrified, as glass or other insulators may be charged (1171.), it was accompanied by a corresponding opposite action *within* the cube, the whole effect being merely a case of induction. Every attempt to charge air bodily and independently with the least portion of either electricity failed.

1173. *I had a chamber built…*: This is Faraday's famous Cage, discussed in the introduction. Note its similarity to the turpentine apparatus just discussed. Its insulated mounting corresponds to the glass jar, the air inside to the turpentine, and the wire running into the interior electrifies the air just as, in the previous apparatus, a wire dipped into the turpentine electrified the fluid.

the perfect discharging train: that is, the excellent connection to *earth* effected by means of the city's network of underground water and gas pipes.

communicate a further charge to it: The air inside the Cage is being electrified by the machine, and any electricity that might be transmitted to the Cage from the electrified air will be discharged to earth through the ground connection. He then suddenly disconnects both the *machine* and the *earth* connection. If the electrified air had been capable of accepting a charge, it would have conveyed that charge to the conductive walls; and the cage would show electrification—by raising a straw electrometer, for example. But there was no such sign, even momentarily.

1174. I put a delicate gold-leaf electrometer within the cube, and then charged the whole by an *outside* communication, very strongly, for some time together; but neither during the charge or after the discharge did the electrometer or air within show the least signs of electricity. I charged and discharged the whole arrangement in various ways, but in no case could I obtain the least indication of an absolute charge; or of one by induction in which the electricity of one kind had the smallest superiority in quantity over the other. I went into the cube and lived in it, and using lighted candles, electrometers, and all other tests of electrical states, I could not find the least influence upon them, or indication of anything particular given by them, though all the time the outside of the cube was powerfully charged, and large sparks and brushes were darting off from every part of its outer surface. The conclusion I have come to is, that non-conductors, as well as conductors, have never yet had an absolute and independent charge of one electricity communicated to them, and that to all appearance such a state of matter is impossible.

1175. There is another view of this question which may be taken under the supposition of the existence of an electric fluid or fluids. It may be impossible to have one fluid or state in a free condition without its producing by induction the other, and yet possible to have cases in which an isolated portion of matter in one condition being uncharged, shall, by a change of state, evolve one electricity or the

---

1174. *I went into the cube and lived in it…:* that is, while inside the Cage he carried on normal activities—which for Faraday included making experiments with standard electrical instruments. All of them acted in the usual manner. Why was it important that Faraday actually *enter* the cage? Certainly that was indispensable if experiments were to be performed; but I suspect that Faraday may have had a deeper intention as well, one concerned with shaping the character of his "ordinary" experience. That suggestion is discussed in the introduction.

1175. Could there be a condition *antecedent* to induction, in which a primary charge evolves independently, and only subsequently induces its opposite counterpart? If so, why does Faraday think it would be most likely to occur in conjunction with a *change of state*? For a possible reason, consider electrification by friction. It requires *two bodies*—and in the next paragraph Faraday will point out that friction simultaneously develops one charge on the *rubbed* and an opposite charge on the *rubbing* body. But no external agent is required to produce a change of state—a change merely in internal conditions (temperature, pressure) suffices. Then since only one body need be involved, perhaps only one kind of electricity need develop. However, no such effect is observed, as Faraday will state in the next paragraph.

other: and though such evolved electricity might immediately induce the opposite state in its neighbourhood, yet the mere evolution of one electricity without the other in the *first instance*, would be a very important fact in the theories which assume a fluid or fluids; these theories as I understand them assigning not the slightest reason why such an effect should not occur.

1176. But on searching for such cases I cannot find one. Evolution by friction, as is well known, gives both powers in equal proportion. So does evolution by chemical action, notwithstanding the great diversity of bodies which may be employed, and the enormous quantity of electricity which can in this manner be evolved (371. 376. 861. 868. 961.). The more promising cases of change of state, whether by evaporation, fusion, or the reverse processes, still give both forms of the power in *equal* proportion; and the cases of splitting of mica and other crystals, the breaking of sulphur, &c., are subject to the same law of limitation.

1177. As far as experiment has proceeded, it appears, therefore, impossible either to evolve or make disappear one electric force without equal and corresponding change in the other. It is also equally impossible experimentally to charge a portion of matter with one electric force independently of the other. Charge always implies *induction*, for it can in no instance be effected without; and also the presence of the *two* forms of power, equally at the moment of the development and afterwards. There is no *absolute* charge of matter with one fluid; no latency of a single electricity. This though a negative result is an exceedingly important one, being probably the consequence of a natural impossibility, which will become clear to us when we understand the true condition and theory of the electric power.

1176. *"fusion"* is another name for *melting* (it is what protective electrical fuses do when the current is too great). That and the other processes specified always produce "both powers in equal proportion"—that is, the powers we conventionally call positive and negative "charge." Faraday is beginning to use the term "powers" more frequently, perhaps to de-emphasize the terminology of "charge" with its fluid-derived associations.

1177. There does not exist, nor can there be generated, any quantity of electricity not in relation to another equal and opposite quantity. Unlike the north- and south-seeking poles of a magnet, these corresponding electricities need not reside in the same body. They may be distributed over several bodies; but even so, they will be as intimately related as the two ends of a stick: in order for one of them to be generated or destroyed, the other must be, also.

no absolute charge of matter with one fluid; no latency of a single electricity: Don't overlook the words "one" and "single" here; it is not latency as such, but latency of *a single electricity* that he is rejecting.

1178. The preceding considerations already point to the following conclusions: bodies cannot be charged absolutely, but only relatively, and by a principle which is the same with that of *induction*. All *charge* is sustained by induction. All phenomena of *intensity* include the principle of induction. All *excitation* is dependent on or directly related to induction. All *currents* involve previous intensity and therefore previous induction. INDUCTION appears to be the essential function both in the first development and the consequent phenomena of electricity.

¶ iii. *Electrometer and inductive apparatus employed.*

1179. Leaving for a time the further consideration of the preceding facts until they can be collated with other results bearing directly on the great question of the nature of induction, I will now describe the apparatus I have had occasion to use; and in proportion to the importance of the principles sought to be established is the necessity of doing this so clearly, as to leave no doubt of the results behind.

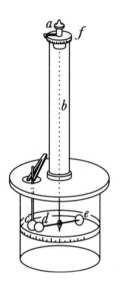

1180. *Electrometer.* The measuring instrument I have employed has been the torsion balance electrometer of Coulomb, constructed, generally, according to his directions,[4] but with certain variations and additions, which I will briefly describe. The lower part was a glass cylinder eight inches in height and eight

[4] Mémoires de l'Académie, 1785, p. 570.

1178. *All charge is sustained by induction*: Induction is then prior to charge (in explanatory order), not charge to induction. Electricity is not an independent substance; rather it appears only insofar as a corresponding *medium* is under inductive action. The view is discussed in the introduction.

intensity: It is perhaps not very clear at this point what the connection is between intensity and induction, but Faraday will cite an example in paragraph 1230 and explain very effectively in paragraph 1299 below. *Current* is simply the relaxation or giving way of that state of tension which constitutes induction.

1180. Faraday does not furnish a drawing of his apparatus. I have supplied a sketch of Coulomb's balance, upon which Faraday's was closely patterned. Of course it does not illustrate the modifications Faraday describes later in the paragraph.

inches in diameter; the tube for the torsion thread was seventeen inches in length. The torsion thread itself [b] was not of metal, but glass, according to the excellent suggestion of the late Dr. Ritchie.[5] It was twenty inches in length, and of such tenuity that when the shell-lac lever and attached ball, &c. were connected with it, they made about ten vibrations in a minute. It would bear torsion through four revolutions or 1440°, and yet, when released, return accurately to its position; probably it would have borne considerably more than this without injury. The repelled ball [d] was of pith, gilt, and was 0.3 of an inch in diameter. The horizontal stem or lever supporting it was of shell-lac, according to Coulomb's direction, the arm carrying the ball being 2.4 inches long, and the other only 1.2 inches: to this was attached the vane [e], also described by Coulomb, which I found to answer admirably its purpose of quickly destroying vibrations. That the inductive action within the electrometer might be uniform in all positions of the repelled ball and in all states of the apparatus, two bands of tin foil, about an inch wide each, were attached to the inner surface of the glass cylinder, going entirely round it, at the distance of 0.4 of an inch from each other, and at such a height that the intermediate clear surface was in the same horizontal plane with the lever and ball. These bands were connected with each other and with the earth, and, being perfect conductors, always exerted a uniform influence on the electrified balls within, which the glass surface, from its irregularity of condition at different times, I found, did not. For the purpose of keeping the air within the electrometer in a constant state as to dryness, a glass dish, of such size as to enter easily within the cylinder,

[5] Philosophical Transactions, 1830.

The torsion thread itself was not of metal, but glass: Although we commonly think of glass as a rigid material, a thin glass thread is flexible and highly elastic, and it can bear twisting through many revolutions. The thread is clamped at the top to the knob *a*, and the lever assembly is firmly affixed to its lower end.

tenuity: slenderness, here implying lack of resistance to twisting, as we see from the slow rate at which the lever oscillates ("ten vibrations in a minute") when set in motion from side to side. In practice, such oscillations are undesirable; it is the function of vane *e* to provide air resistance to damp the vibrations and permit the lever arm to take its equilibrium position quickly.

two bands of tin foil, not shown in the drawing, encircle the inner surface of the glass cylinder and are connected to ground. As Faraday explains, an uncoated glass surface readily becomes electrified and might exert an irregular and unpredictable influence on the balls.

had a layer of fused potash placed within it, and this being covered with a disc of fine wire-gauze to render its inductive action uniform at all parts, was placed within the instrument at the bottom and left there.

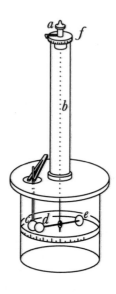

1181. The moveable ball used to take and measure the portion of electricity under examination, and which may be called the *repelling*, or the *carrier*, ball [*c*], was of soft alder wood, well and smoothly gilt. It was attached to a fine shell-lac stem, and introduced through a hole into the electrometer according to Coulomb's method: the stem was fixed at its upper end in a block or vice, supported on three short feet; and on the surface of the glass cover above was a plate of lead with stops on it, so that when the carrier ball was adjusted in its right position, with the vice above bearing at the same time against these stops, it was perfectly easy to bring away the carrier-ball and restore it to its place again very accurately, without any loss of time.

1182. It is quite necessary to attend to certain precautions respecting these balls. If of pith alone they are bad; for when very dry, that substance is so imperfect a conductor that it neither receives nor gives a charge freely, and so, after contact with a charged conductor, it is liable to be in an uncertain condition. Again, it is difficult to turn pith so smooth as to leave the ball, even when gilt, as free from irregularities of form, as to retain its charge undiminished for a considerable length of time. When, therefore, the balls are finally prepared and gilt they should be examined; and being electrified, unless they can hold their charge with very little diminution for a considerable time, and yet be discharged instantly and perfectly by the touch of an uninsulated conductor, they should be dismissed.

1180, continued. *fused potash*: the hydrated protoxide of potassium, purified by alcohol, melted and cast into bars or wafers. "It quickly absorbs moisture from the air," according to an 1829 chemistry manual.

1181. *bearing … against these stops*: The drawing does not show these stops, which insure that the carrier-ball *c* always takes the same position within the instrument when it is mounted. *gilt*: gilded, covered with gold leaf.

1182. *free from irregularities of form*: Points or irregularities would facilitate discharge into the air, as we saw in the Third Series.

1183. It is, perhaps, unnecessary to refer to the graduation of the instrument, further than to explain how the observations were made. On a circle or ring of paper on the outside of the glass cylinder, fixed so as to cover the internal lower ring of tinfoil, were marked four points corresponding to angles of 90°; four other points exactly corresponding to these points being marked on the upper ring of tinfoil within. By these and the adjusting screws on which the whole instrument stands, the glass torsion thread could be brought accurately into the centre of the instrument and of the graduations on it. From one of the four points on the exterior of the cylinder a graduation of 90° was set off, and a corresponding graduation was placed upon the upper tinfoil on the opposite side of the cylinder within; and a dot being marked on that point of the surface of the repelled ball nearest to the side of the electrometer, it was easy, by observing the line which this dot made with the lines of the two graduations just referred to, to ascertain accurately the position of the ball. The upper end of the glass thread was attached, as in Coulomb's original electrometer, to an index [a], which had its appropriate graduated circle [f], upon which the degree of torsion was ultimately to be read off.

1184. After the levelling of the instrument and adjustment of the glass thread, the blocks which determine the place of the *carrier ball* are to be regulated (1181.) so that when the carrier arrangement is placed against them, the centre of the ball may be in the radius of the instrument corresponding to 0° on the lower graduation or that on the

---

1183. *four points corresponding to angles of 90°*:   The first of these points coincides with the position subsequently to be occupied by the carrier ball *c*. Each is marked on both the exterior and the interior of the cylinder, to define perpendicular sight lines; and Faraday levels the instrument so that the torsion thread passes through the center of the cylinder, where the sight lines intersect. Similarly, degree scales marked opposite one another on the inner and outer faces of the cylinder permit Faraday to sight the repelled ball *d* against both scales simultaneously, and so measure the angular position of the repelled ball with respect to the carrier ball.

1184. To prepare the electrometer for use, Faraday positions the carrier ball *c* at the same height and radial distance from the central axis as the suspended ball *d*, with *c* at the 0° mark on the degree scale that runs around the large glass cylinder. He then fixes the "stops" described in paragraph 1181 so that ball *c* may be quickly and reliably returned to that same position henceforth. Both balls are of course unelectrified during this initial adjustment; the lever assembly hangs freely at an arbitrary angular position, and the thread does not sustain any twist.

side of the electrometer, and at the same level and distance from the centre as the *repelled ball* on the suspended torsion lever. Then the torsion index is to be turned until the ball connected with it (the repelled ball) is accurately at 30°, and finally the graduated arc [*f*] belonging to the torsion index is to be adjusted so as to bring 0° upon it to the index. This state of the instrument was adopted as that which gave the most direct expression of the experimental results, and in the form having fewest variable errors; the angular distance of 30° being always retained as the standard distance to which the balls were in every case to be brought, and the whole of the torsion being read off at once on the graduated circle above.

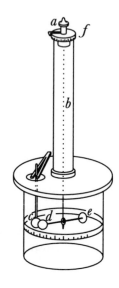

Under these circumstances the distance of the balls from each other was not merely the same in degree, but their position in the instrument, and in relation to every part of it, was actually the same every time that a measurement was made; so that all irregularities arising from slight difference of form and action in the instrument and the bodies around were avoided. The only difference which could occur in the position of anything within, consisted in the deflection of the torsion thread from a vertical position, more or less, according to the force of repulsion of the balls; but this was so slight as to cause no interfering difference in the symmetry of form within the instrument, and gave no error in the amount of torsion force indicated on the graduation above.

1184, continued. Faraday's "torsion index" is the pointer of suspension *a*. It rides on the "graduated circle," that is, on the adjustable scale *f*. Since he has selected an angular distance of 30° as the separation at which measurements are to be made, he turns the torsion index until lever ball *d* occupies the 30° mark on the cylinder's degree scale. Note that this adjustment merely establishes the initial position of lever *d e*. During its execution, the suspension, thread, and lever turn together as a unit; the thread does not suffer any twist and it continues to hang freely afterwards.

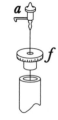

Now, holding the suspension and index *a* in place, he adjusts the graduated circle *f* so that the index points to 0°. Thus, in the initial state of the instrument, the graduated circle *f* reads 0° when the lever ball occupies the 30° position and the thread sustains no twist.

1185. Although the constant angular distance of 30° between the centres of the balls was adopted, and found abundantly sensible, for all ordinary purposes, yet the facility of rendering the instrument far more sensible by diminishing this distance was at perfect command; the results at different distances being very easily compared with each other either by experiment, or, as they are inversely as the squares of the distances, by calculation.

1186. The Coulomb balance electrometer requires experience to be understood; but I think it a very valuable instrument in the hands of those who will take pains by practice and attention to learn the precautions needful in its use. Its insulating condition varies with circumstances, and should be examined before it is employed in experiments. In an ordinary and fair condition, when the balls were so electrified as to give a repulsive torsion force of 400° at the standard distance of 30°, it took nearly four hours to sink to 50° at the same distance; the average loss from 400° to 300° being at the rate of 2.7° per minute, from 300° to 200° of 1.7° per minute, from 200° to 100° of 1.3° per minute, and from 100° to 50° of 0.87° per minute. As a complete measurement by the instrument may be made in much less than a minute, the amount of loss in that time is but small, and can easily be taken into account.

1187. *The inductive apparatus.* My object was to examine inductive action carefully when taking place through different media, for which purpose it was necessary to subject these media to it in exactly similar circumstances, and in such quantities as should suffice to eliminate any

---

1186. *a repulsive torsion force of 400°*...: In order to test the instrument, Faraday electrifies both balls (he will describe the technique later in paragraph 1197). When carrier-ball *c* is placed in position, ball *d* is repelled to some position beyond the original 30° mark, thereby twisting the thread. He then rotates the suspension *a* in the *opposite* direction, twisting the glass thread even more and building up sufficient torsion to return the repelled ball *d* back to the 30° position. The pointer on *a* records the total angle of rotation of the suspension. If, for example, *a* has to be rotated 400° (one complete turn plus 40°), then, overall, the upper end of the thread will have been turned through 400° while the lower end will not have been turned at all (since it occupies the same original 30° position it did at the outset). Thus the overall twist sustained by the thread will be 400°. Since the torsion built up in the thread as a result of this twist is just sufficient to balance the repulsive force between the balls, Faraday will express that force as "a repulsive torsion force of 400° at the standard distance of 30°."

variations they might present. The requisites of the apparatus to be constructed were, therefore, that the inducing surface of the conductors should have a constant form and state, and be at a constant distance from each other; and that either solids, fluids, or gases might be placed and retained between these surfaces with readiness and certainty, and for any length of time.

1188. The apparatus used may be described in general terms as consisting of two metallic spheres of unequal diameter, placed, the smaller within the larger, and concentric with it; the interval between the two being the space through which the induction was to take place. A section of it is given (Plate VII. fig. 104.) on a scale of one half: a, a are the two halves of a brass sphere, with an air-tight joint at b, like that of the Magdeburg hemispheres, made perfectly flush and smooth inside so as to present no irregularity; c is a connecting piece by which the apparatus is joined to a good stop-cock d, which is itself attached either to the metallic foot e, or to an air pump. The aperture within the hemisphere at f is very small: g is a brass collar fitted to the upper hemisphere, through which the shell-lac support of the inner ball and its stem passes; h is the inner ball, also of brass; it screws on to a brass stem i, terminated above by a brass ball B; l, l is a mass of shell-lac, moulded carefully on to i, and serving both to support and insulate it and its balls h, B. The shell-lac stem l is fitted into the socket g, by a little ordinary resinous cement, more fusible than shell-lac, applied at $m\ m$ in such a way as to give sufficient strength and render the apparatus air-tight there, yet leave as much as possible of the lower part of the shell-lac stem untouched, as an insulation between the ball h and the surrounding sphere a, a. The ball h has a small aperture at n, so that when the apparatus is exhausted of one gas and filled with another, the ball h may itself also be exhausted and filled, that no variation of the gas in the interval o may occur during the course of an experiment.

1189. It will be unnecessary to give the dimensions of all the parts, since the drawing is to a scale of one half: the inner ball has a diameter of 2.33 inches, and the surrounding sphere an internal diameter of 3.57 inches. Hence the width of the intervening space, through which the induction is to take place, is 0.62 of an inch; and the extent of this

---

1188. *on a scale of one half*: The present reproduction is not exactly half scale. One must look very carefully to discern Faraday's label marking the brass rod i which runs through the shell-lac stem and which is threaded at both ends to mate with balls h and B. The drawing fails to show the aperture at n, though it does indicate the location of n correctly.

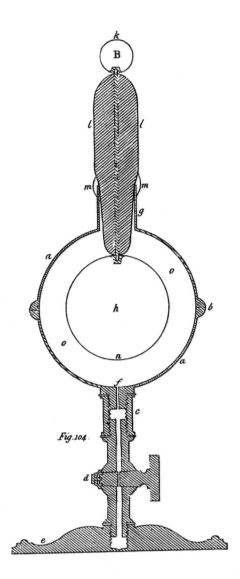

Fig. 104.

place or plate, *i. e.* the surface of a medium sphere, may be taken as twenty-seven square inches, a quantity considered as sufficiently large for the comparison of different substances. Great care was taken in finishing well the inducing surfaces of the ball *h* and sphere *a, a*; and no varnish or lacquer was applied to them, or to any part of the metal of the apparatus.

1190. The attachment and adjustment of the shell-lac stem was a matter requiring considerable care, especially as, in consequence of its cracking, it had frequently to be renewed. The best lac was chosen and applied to the wire *i*, so as to be in good contact with it everywhere, and in perfect continuity throughout its own mass. It was not smaller than is given by scale in the drawing, for when less it frequently cracked within a few hours after it was cold. I think that very slow cooling or annealing improved its quality in this respect. The collar *g* was made as thin as could be, that the lac might be as wide there as possible. In order that at every re-attachment of the stem to the upper hemisphere the ball *h* might have the same relative position, a gauge *p* (fig. 105.) was made of wood, and this being applied to the ball and hemisphere whilst the cement at *m* was still soft, the bearings of the ball at *qq*, and the hemisphere at *rr*, were forced home, and the whole left until cold. Thus all difficulty in the adjustment of the ball in the sphere was avoided.

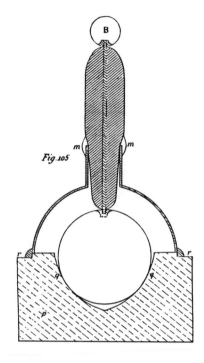

Fig. 105

1191. I had occasion at first to attach the stem to the socket by other means, as a band of paper or a plugging of white silk thread; but these were very inferior to the cement, interfering much with the insulating power of the apparatus.

1192. The retentive power of this apparatus was, when in good condition, better than that of the electrometer (1186.), *i. e.* the proportion of loss of power was less. Thus when the apparatus was electrified, and also the balls in the electrometer, to such a degree, that after the inner ball had been in contact with the top *k* of the ball of the apparatus, it caused a repulsion indicated by 600° of torsion force, then in falling from 600° to 400° the average loss was 8.6° per minute; from 400° to 300° the average loss was 2.6° per minute; from 300° to 200° it was 1.7° per minute; from 200° to 170° it was 1° per minute. This was after the apparatus had been charged for a short time; at the first instant of charging there is an apparent loss of electricity, which can only be comprehended hereafter (1207. 1250.).

1193. When the apparatus loses its insulating power suddenly, it is almost always from a crack near to or within the brass socket. These cracks are usually transverse to the stem. If they occur at the part attached by common cement to the socket, the air cannot enter, and

---

1190. "Annealing" is a process of heating followed by slow cooling. It is effective in reducing brittleness in metals and other materials.

thus constituting vacua, they conduct away the electricity and lower the charge, as fast almost as if a piece of metal had been introduced there. Occasionally stems in this state, being taken out and cleared from the common cement, may, by the careful application of the heat of a spirit-lamp, be so far softened and melted as to restore the perfect continuity of the parts; but if that does not succeed in replacing things in a good condition, the remedy is a new shell-lac stem.

1194. The apparatus when in order could easily be exhausted of air and filled with any given gas; but when that gas was acid or alkaline, it could not properly be removed by the air-pump, and yet required to be perfectly cleared away. In such cases the apparatus was opened and emptied of gas; and with respect to the inner ball h, it was washed out two or three times with distilled water introduced at the screw-hole, and then being heated above 212°, air was blown through to render the interior perfectly dry.

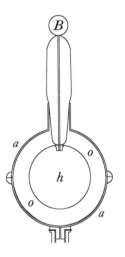

1195. The inductive apparatus described is evidently a Leyden phial, with the advantage, however, of having the *dielectric* or insulating medium changed at pleasure. The balls h and B, with the connecting-wire i, constitute the charged conductor, upon the surface of which all the electric force is resident by virtue of induction (1178.). Now though the largest portion of this induction is between the ball h and the surrounding sphere $a\,a$, yet the wire i and the ball B determine a part of the induction from their surfaces towards the external surrounding conductors. Still, as all things in that respect remain the same, whilst the medium within at $o\,o$, may be varied, any changes exhibited by the whole apparatus will in such cases depend upon the variations made in the interior; and these were the changes I was in search of, the negation or establishment of such differences being the great object of my inquiry. I considered that these differences, if they existed, would be most distinctly set forth by having two apparatus of the kind described, precisely similar in every respect; and then, *different insulating media* being within, to charge one and measure it, and after dividing the charge with the other, to observe what the ultimate conditions of both were. If insulating media really had any specific

<hr />

1194. Acid or alkaline gas *"could not properly be removed by the air-pump"*—probably because of chemical residues, as his practice of *washing* the surfaces suggests. The "screw hole" he mentions is located at the top of ball h where the rod i enters. Refer to the sketch, which reproduces part of Faraday's Figure 104.

differences in favouring or opposing inductive action through them, such differences, I conceived, could not fail of being developed by such a process.

1196. I will wind up this description of the apparatus, and explain the precautions necessary to their use, by describing the form and order of the experiments made to prove their equality when both contained common air. In order to facilitate reference I will distinguish the two by the terms App. i. and App. ii.

1197. The electrometer is first to be adjusted and examined (1184.), and the app. i. and ii. are to be perfectly discharged. A Leyden phial is to be charged to such a degree that it would give a spark of about one-sixteenth or one-twentieth of an inch in length between two balls of half an inch diameter; and the carrier ball of the electrometer being charged by this phial, is to be introduced into the electrometer, and the lever ball brought by the motion of the torsion index against it; the charge is thus divided between the balls, and repulsion ensues. It is useful then to bring the repelled ball to the standard distance of 30° by the motion of the torsion index, and observe the force in degrees required for this purpose; this force will in future experiments be called *repulsion of the balls.*

1196. *prove their equality when both contained common air*: Since the two devices are to reveal differences associated with the insulating materials contained, he must verify that their properties are identical when those materials are identical. In addition, this preliminary procedure will establish the general comparative method to be followed subsequently.

1197. In order to make a measurement, the electrometer's lever ball must be given an initial charge. Faraday gives a more detailed account here than he did in paragraph 1186 above. First the electrometer's removable ball (*c* in the diagram) is electrified by contact with a charged Leyden jar. It is then introduced into the electrometer and the lever ball *d* forced up against it, so as to acquire a share of its electrification; this constitutes the initial charge of ball *d.*

Since balls *c* and *d* are now similarly charged, they will repel one another. Faraday says it will be "useful" to measure this initial repulsion. This he does by the technique already outlined in paragraph 1186: when the suspension is rotated so as to return ball *d* to the standard 30° mark, the total angle of rotation is a measure of the repulsive force. He calls it the (initial) *repulsion of the balls.*

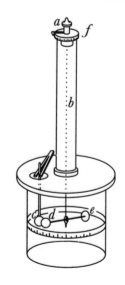

1198. One of the inductive apparatus, as, for instance, app. i., is now to be charged from the Leyden phial, the latter being in the state it was in when used to charge the balls; the carrier ball is to be brought into contact with the top of its upper ball [B], then introduced into the electrometer, and the repulsive force (at the distance of 30°) measured. Again, the carrier should be applied to the app. i. and the measurement repeated; the apparatus i. and ii. are then to be joined, so as to *divide* the charge, and afterwards the force of each measured by the carrier ball, applied as before, and the results carefully noted. After this

---

1198. The inductive devices are now compared. As stated in the previous paragraph, both must be discharged initially; then he carries out the following steps:

(*i*) One apparatus is charged by bringing the Leyden jar in contact with its upper ball B. Faraday notes that the jar is "in the same state as when it was used to charge the [electrometer] balls"; by this he means that the jar is still electrified, and that the *quality* of its electrification (positive or negative) has not changed. Therefore the charges which the Leyden jar gives to the inner sphere of the inductive apparatus and to the lever ball *d* of the electrometer are similar in kind: they are either both positive or both negative, and portions of each will repel the other. This is important since a balance such as Faraday here describes can measure *repulsion* far more easily than it can *attraction.*

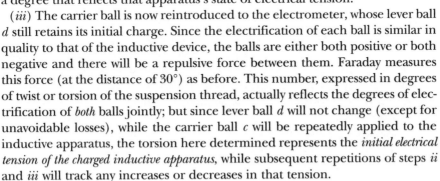

(*ii*) The electrometer carrier ball *c* is removed, then touched momentarily to B. As described under Procedure (1) in the editor's introduction, pages 224–225, the carrier ball thereby becomes electrified similarly to the inductive apparatus, and to a degree that reflects that apparatus's state of electrical tension.

(*iii*) The carrier ball is now reintroduced to the electrometer, whose lever ball *d* still retains its initial charge. Since the electrification of each ball is similar in quality to that of the inductive device, the balls are either both positive or both negative and there will be a repulsive force between them. Faraday measures this force (at the distance of 30°) as before. This number, expressed in degrees of twist or torsion of the suspension thread, actually reflects the degrees of electrification of *both* balls jointly; but since lever ball *d* will not change (except for unavoidable losses), while the carrier ball *c* will be repeatedly applied to the inductive apparatus, the torsion here determined represents the *initial electrical tension of the charged inductive apparatus,* while subsequent repetitions of steps *ii* and *iii* will track any increases or decreases in that tension.

(*iv*) He repeats steps *ii* and *iii* again, without offering any reasons—but see the comment to paragraph 1208 below.

(*v*) *apparatus i. and ii. are then to be joined…*: that is, their upper balls B are connected momentarily. If then the devices are *identical*, Faraday reasonably expects the electricity resident in one to *divide equally* between both and bring them to new, but identical, degrees of electrification.

(*vi*) *afterwards the force of each measured*: He brings the carrier ball *c* to each inductive apparatus in turn and measures each device's degree of electrification.

both i. and ii. are to be discharged; then app. ii. charged, measured, divided with app. i., and the force of each again measured and noted. If in each case the half charges of app. i. and ii. are equal, and are together equal to the whole charge before division, then it may be considered as proved that the two apparatus are precisely equal in power, and fit to be used in cases of comparison between different insulating media or *dielectrics*.

1199. But the *precautions* necessary to obtain accurate results are numerous. The apparatus i. and ii. must always be placed on a thoroughly uninsulating medium. A mahogany table, for instance, is far from satisfactory in this respect, and therefore a sheet of tinfoil, connected with an extensive discharging train (292.), is what I have used. They must be so placed also as not to be too near each other, and yet equally exposed to the inductive influence of surrounding objects; and these objects, again, should not be disturbed in their position during an experiment, or else variations of induction upon the external ball B of the apparatus may occur, and so errors be introduced into the results. The carrier ball, when receiving its portion of electricity from the apparatus, should always be applied at the same part of the ball, as, for instance, the summit *k*, and always in the same way; variable induction from the vicinity of the head, hands, &c. being avoided, and the ball after contact being withdrawn upwards in a regular and constant manner.

1200. As the stem had occasionally to be changed (1190.), and the change might occasion slight variations in the position of the ball within,

---

1198, continued. *the force of each again measured and noted*: As an additional precaution, the whole operation is repeated with the inductive devices exchanged. This will reveal whether any inequalities found are specifically associated with one apparatus or the other. But if the devices are identical, the results of steps *v* should also be identical, and each equal to one-half the result obtained in step *iv*. (Faraday will consistently choose step *iv* and not step *iii* as the standard of comparison. Even though step *iv* is but a repetition of step *iii* it typically yields a different result, as we shall see in paragraphs 1208 and 1210. The comment to paragraph 1208 suggests a reason for the difference.)

1199. *The apparatus … must always be placed on a thoroughly uninsulating medium*: That is, placed on a *grounded* work surface such as the tinfoil sheet here specified. The sheet functions here just as the foil bands do in the electrometer (paragraph 1180). Note also that when the apparatus rests on the sheet, its metal base, stopcock, and collar will be grounded, and thereby also its outer sphere *a a*—as is required.

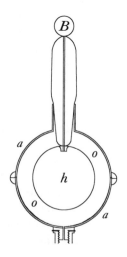

[Historical reproductions of Faraday's inductive apparatus, constructed at Oldenburg University. Photo by kind permission of Dietmar Höttecke.]

I made such a variation purposely, to the amount of an eighth of an inch (which is far more than ever could occur in practice), but did not find that it sensibly altered the relation of the apparatus, or its inductive condition *as a whole.* Another trial of the apparatus was made as to the effect of dampness in the air, one being filled with very dry air, and the other with air from over water. Though this produced no change in the result, except an occasional tendency to more rapid dissipation, yet the precaution was always taken when working with gases (1290.) to dry them perfectly.

1201. It is essential that the interior of the apparatus should be perfectly free from *dust or small loose particles,* for these very rapidly lower the charge and interfere on occasions when their presence and action would hardly be expected. To breathe on the interior of the apparatus and wipe it out quietly with a clean silk handkerchief, is an effectual way of removing them; but then the intrusion of other particles should be carefully guarded against, and a dusty atmosphere should for this and several other reasons be avoided.

1202. The shell-lac stem requires occasionally to be well wiped, to remove in the first instance, the film of wax and adhering matter which is upon it; and afterwards to displace dirt and dust which will gradually attach to it in the course of experiments. I have found much to depend upon this precaution, and a silk handkerchief is the best wiper.

1203. But wiping and some other circumstances tend to give a charge to the surface of the shellac stem. This should be removed, for, if allowed to remain, it very seriously affects the degree of charge given to the carrier ball by the apparatus (1232.). This condition of the stem is best observed by discharging the apparatus, applying the carrier ball to the stem, touching it with the finger, insulating and removing it, and examining whether it has received any charge (by induction) from the stem; if it has, the stem itself is in a charged state. The best method of removing the charge I have found to be, to cover the finger with a single fold of a silk handkerchief, and breathing on the stem, to wipe it immediately after with the finger; the ball B and its connected wire, &c. being at the same time *uninsulated:* the wiping place of the silk must not be changed; it then becomes sufficiently damp not to excite the stem, and is yet dry enough to leave it in a clean and excellent insulating condition. If the air be dusty, it will be found that a single charge of the apparatus will bring on an electric state of the outside of the stem, in consequence of the carrying power of the particles of dust; whereas in the morning, and in a room which has been left quiet, several experiments can be made in succession without the stem assuming the least degree of charge.

1204. Experiments should not be made by candle or lamp light except with much care, for flames have great and yet unsteady powers of affecting and dissipating electrical charges.

1205. As a final observation on the state of the apparatus, they should retain their charges well and uniformly, and alike for both, and at the same time allow of a perfect and instantaneous discharge, giving afterwards no charge to the carrier ball, whatever part of the ball B it may be applied to (1218.).

1203. *touching [the carrier ball] with the finger, insulating and removing it*: By touching it momentarily with the finger, Faraday *grounds* and then *ungrounds* (insulates) the carrier ball, after which it is removed to the electrometer and examined for electrification. This is the technique described as Procedure (2) on page 226 of the editor's introduction. If the carrier ball exhibits charge, Faraday will know that the stem is electrified and must be discharged before the experiment can proceed.

The best method of removing the charge I have found to be…: Faraday describes an elaborate procedure of breathing and wiping. Some of his steps have an evident rationale. For example, if the moisture which condenses from breath is conductive, one can see how it might help protect the stem from excitation by the silk. But his phrase "I have found" suggests that his procedure was developed by trial and error, and that he does not necessarily have a theoretical reason in mind for each stage of the process.

1205. *perfect and instantaneous discharge, giving afterwards no charge to the carrier…*: Each apparatus should be capable of being discharged completely, leaving no residual electrification on any portion of the upper ball. However, such "perfect" discharge cannot always be expected, as he acknowledges in paragraph 1207 below. The most intractable problem is electrification of the

1206. With respect to the balance electrometer, all the precautions that need be mentioned, are, that the carrier ball is to be preserved during the first part of an experiment in its electrified state, the loss of electricity which would follow upon its discharge being avoided; and that in introducing it into the electrometer through the hole in the glass plate above, care should be taken that it do not touch, or even come near to, the edge of the glass.

1207. When the whole charge in one apparatus is divided between the two, the gradual fall, apparently from dissipation, in the apparatus which has *received* the half charge is greater than in the one *originally* charged. This is due to a peculiar effect to be described hereafter (1250. 1251.), the interfering influence of which may be avoided to a great extent by going through the steps of the process regularly and quickly; therefore, after the original charge has been measured, in app. i. for instance, i. and ii. are to be symmetrically joined by their balls B, the carrier touching one of these balls at the same time; it is first to be removed, and then the apparatus separated from each other; app. ii. is next quickly to be measured by the carrier, then app. i.; lastly, ii. is to be discharged and the discharged carrier applied to it to ascertain whether any residual effect is present (1205.), and app. i. being discharged is also to be examined in the same manner and for the same purpose.

1208. The following is an example of the division of a charge by the two apparatus, air being the dielectric in both of them. The observations are set down one under the other in the order in which they were taken, the left hand numbers representing the observations made on app. i., and the right hand numbers those on app. ii. App. i. is that which was originally charged, and after two measurements, the charge was divided with app. ii.

---

shell-lac stem, as described in paragraph 1203 above. Another interfering effect is the phenomenon called "return charge," wherein a Leyden jar or similar device, having been charged and then discharged, appears spontaneously to regain a portion of its former electrified state! Faraday will begin an investigation of that effect at paragraph 1234.

1207. *dissipation* (sometimes called *leakage*) is the gradual discharge of electrified bodies into the air or through imperfect insulators to ground. But its effects may be disguised by stem electrification as well by as the "return charge" mentioned in my previous comment. This prompts Faraday to add additional steps to the method outlined in paragraph 1198: After completing steps *i* through *vi* as there reviewed, Faraday directs us (*vii*) to *discharge* both devices and then measure each of them with the electrometer, determining thereby how much residual charge or other disturbing influence may remain.

	App. i.		App. ii.
	Balls 160°		
		. . . .	0°
254°		. . . .	
250°		. . . .	
divided and instantly taken			
		. . . .	122
124		. . . .	
1		. . . .	
		. . . .	2

1	. . . .			after being discharged
	. . . .	2		after being discharged

1209. Without endeavouring to allow for the loss which must have been gradually going on during the time of the experiment, let us observe the results of the numbers as they stand. As 1° remained in app. i. in an undischargeable state, 249° may be taken as the utmost amount of the transferable or divisible charge, the half of which is 124.5°. As app. ii. was free of charge in the first instance, and immediately after the division was found with 122°, this amount *at least* may be taken as what it had received. On the

---

1208 (table). Faraday will offer an analytical account of this table in the next paragraph, but it might be helpful to give a more chronological review first. "Balls 160°" is ellipsis for what Faraday called "*repulsion of the balls*" in paragraph 1197 above; it indicates that the initial electrification of the electrometer, described in the comment to that paragraph, was such as to sustain a twist of 160° in the suspension thread. The figure 254° in column App. i represents step *ii* (in the comment to paragraph 1198, page 251), the first measurement of the electrified apparatus. Below it, the entry 250° represents step *iv*, the same measurement repeated a short time later. Evidently the original charge of App. i suffers gradual dissipation, and the difference between 254° and 250° gives a rough idea of the rate of loss; we will see him make use of such information in paragraph 1263 below. By "division" of the charge between the two devices—step *v*— App. ii acquires 122° while App. i retains 124°; these figures represent step *vi*. Finally, both devices are discharged, App. i retaining a charge of 1°, App. ii of 2°; these values represent the additional step *vii* introduced in paragraph 1207.

1209. *As 1° remained ... 249° may be taken as the utmost amount of the transferable or divisible charge*: Faraday has subtracted 1° from 250° to obtain this figure, indicating that he regards 250°—that is, the result of step *iv*—as the best measure of the apparatus's initial charge.

this amount at least may be taken as what it had received...: Measurement of the receiving apparatus showed a gain of 122° of electrification. But if the

other hand 124° minus 1°, or 123°, may be taken as the half of the transferable charge retained by app. i. Now these do not differ much from each other, or from 124.5°, the half of the full amount of transferable charge; and when the gradual loss of charge evident in the difference between 254° and 250° of app. i. is also taken into account, there is every reason to admit the result as showing an equal division of charge, *unattended by any disappearance of power* except that due to dissipation.

*   *   *

1212. The experiments were repeated with charges of negative electricity with the same general results.

1213. That I might be sure of the sensibility and action of the apparatus, I made such a change in one as ought upon principle to increase its inductive force, *i. e.* I put a metallic lining into the lower hemisphere of app. i., so as to diminish the thickness of the intervening air in that part, from 0.62 to 0.435 of an inch: this lining was carefully shaped and rounded so that it should not present a sudden projection within at its edge, but a gradual transition from the reduced interval in the lower part of the sphere to the larger one in the upper.

---

apparatus is subject to leakage or dissipation, it may have actually received greater electrification than the measurement discloses. Ignoring such small inequalities as appear to be attributable to slow leakage, and correcting for the inherently undischargeable residues, the results are symmetrical and show that the two inductive devices are electrically identical.

unattended by any disappearance of power...: Note that the electrification lost by one apparatus is equal to that gained by the other. That is just what we would expect if the electrometer reading were an indication of the *quantity of electricity* held in each device—but see the comment to paragraph 1214 below.

1213. *I put a metallic lining into the lower hemisphere ... so as to diminish the thickness of the intervening air*: Although Faraday characterizes this change as one which "ought upon principle to increase its inductive force," he has not identified such a principle explicitly. Does he think it is *self-evident* that diminishing the distance should increase the induction between two surfaces?

Faraday will struggle to formulate a principle of induction throughout the Twelfth Series; see especially paragraphs 1369–1374. In the next paragraph, though, he brings to light a related aspect of induction, that of "capacity for induction"—a disclosure that is perhaps richer in consequences than the mere formulation of a principle might be.

1214. This change immediately caused app. i. to produce effects indicating that it had a greater aptness or capacity for induction than app. ii. Thus, when a transferable charge in app. ii. of 469° was divided with app. i., the former retained a charge of 225°, whilst the latter allowed one of 227°, *i. e.* the former had lost 244° in communicating 227° to the latter: On the other hand, when app. i. had a transferable charge in it of 381° divided by contact with app. ii., it lost 181° only, whilst it gave to app. ii. as many as 194: the sum of the divided forces being in the first instance *less*, and in the second instance *greater* than the original undivided charge. These results are the more striking, as only one half of the interior of app. i. was modified, and they allow that the instruments are capable of bringing out differences in inductive force from amongst

---

1214. *greater aptness or capacity for induction*: In the previous paragraph Faraday described how he partially lined one of the devices with metal, in order to create a *nonidentical* pair. Here he reports that the lined apparatus exhibits "greater aptness or capacity for induction" than the unaltered device. What are the signs of "greater capacity"? They are, at first, somewhat startling: When electrification is exchanged between the devices, the lined apparatus either loses less electrification than the unlined device gains, or gains less than the other loses. Unlike the earlier division between identical devices, the electrification lost by one apparatus *does not equal* the electrification gained by the other! That should remind us, if we had forgotten, that when the electrometer is used in the way here described, it does not measure the *quantity of electricity* residing in the induction devices—for if electricity really has *quantity*, that quantity presumably cannot spontaneously increase or decrease as it passes from one apparatus to the other. Instead, Faraday understands the present procedure to measure what is ordinarily called the *intensity* or *tension* of the electrification each apparatus sustains, as he states explicitly in paragraphs 1250, 1258, 1260, 1284 and elsewhere. The introduction to the present Series reviews some of the grounds for that understanding. We must therefore interpret the experiments differently.

In Faraday's first example, some quantity of electricity passed from the unlined apparatus to the lined one. In leaving the unlined device, it reduced that apparatus's tension by 244 degrees; but in entering the lined device it raised the tension by only 227 degrees. One and the same quantity of electricity, therefore, must be associated with a *lower tension* when it resides in lined App. i than when it resides in unlined App. ii. Thus "greater capacity for induction" means the ability to sustain *the same induction* (retain the same quantity of electricity) *at lower tension* than another device.

Equivalently, greater capacity for induction means the ability to sustain *greater induction at the same tension*. For consider Faraday's lined and unlined devices immediately after their momentary joining. They exhibit essentially equal tensions of 225 and 227 degrees respectively; but their quantities of electricity must be substantially unequal, since App. ii lost 244 ÷ 469 or 52% of its charge to App. i and retains only 225 ÷ 469 or 48%. Thus App. i sustains a greater quantity of electricity than App. ii at the same tension.

the errors of experiment, when these differences are much less than that produced by the alteration made in the present instance.

iv. *Induction in curved lines.*

1215. Amongst those results deduced from the molecular view of induction (1166.), which, being of a peculiar nature, are the best tests of the truth or error of the theory, the expected action in curved lines is, I think, the most important at present; for if shown to take place in an unexceptionable manner, I do not see how the old theory of action at a distance and in straight lines can stand, or how the conclusion that ordinary induction is an action of contiguous particles can be resisted.

1216. There are many forms of old experiments which might be quoted as favourable to, and consistent with the view I have adopted. Such are most cases of electro-chemical decomposition, electrical brushes, auras, sparks, &c.; but as these might be considered equivocal evidence, inasmuch as they include a current and discharge, (though they have long been to me indications of prior molecular action (1230.)) I endeavoured to devise such experiments for first proofs as should not include transfer, but relate altogether to the pure simple inductive action of statical electricity.

1217. It was also of importance to make these experiments in the simplest possible manner, using not more than one insulating medium or dielectric at a time, lest differences of slow conduction should produce effects which might erroneously be supposed to result from induction in curved lines. It will be unnecessary to describe the steps of the investigation minutely; I will at once proceed to the simplest mode of proving the facts, first in air and then in other insulating media.

1218. A cylinder of solid shell-lac, 0.9 of an inch in diameter and seven inches in length, was fixed upright in a wooden foot (fig. 106.): it was made concave or cupped at its upper extremity so that a brass ball or other small arrangement could stand upon it. The upper half of the stem having been excited *negatively* by friction with warm flannel, a brass ball, B, 1 inch in diameter,

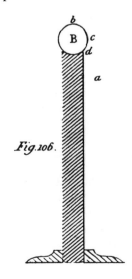

Fig. 106.

1215. *I do not see how the old theory of action at a distance and in straight lines can stand*: The comment to paragraph 1166 suggested a reason why *action in curved lines* would be difficult to reconcile with conventional "action at a distance" thinking.

was placed on the top, and then the whole arrangement examined by the carrier ball and Coulomb's electrometer (1180. &c.). For this purpose the balls of the electrometer were charged *positively* to about 360°, and then the carrier being applied to various parts of the ball B, the two were uninsulated whilst in contact or in position, then insulated,[6] separated, and the charge of the carrier examined as to its nature and force. Its electricity was always positive, and its force at the different positions *a*, *b*, c, *d*, &c. (figs. 106. and 107.) observed in succession, was as follows:

at *a*	. .	above 1000°
b	it was . . .	149
c	. . . . . .	270
d	. . . . .	512
b	. . . . . .	130

[6] It can hardly be necessary for me to say here, that whatever general state the carrier ball acquired in any place where it was uninsulated and then insulated, it retained on removal from that place, notwithstanding that it might pass through other places that would have given to it, if uninsulated, a different condition.

1218. *the carrier being applied to various parts of the ball B, the two were uninsulated* [that is, grounded] *whilst in contact … then insulated…*: This is different from his earlier practice (paragraphs 1207–1214), where the carrier and the test apparatus were brought into mutual contact, *but not grounded.* The technique of momentary grounding is highly significant, as Faraday will explain in the next paragraph. The introduction discusses it as well as the earlier method.

Its [the carrier's] *electricity was always positive*: That is, *opposite* to that of the negatively-electrified shell-lac stem. This is an emblem of induction, as Faraday will note in the next paragraph. The introduction discusses a similar example.

its force at the different positions a, b, c, d, &c. … observed in succession, was as follows: The table shows two values recorded for position *b* while *e* is absent entirely. One might suspect a typographical error; but an entry in Faraday's *Diary* confirms that the final entry really does apply to position *b*, not *e*. Faraday does not comment on the discrepancy between the readings 149° and 130°.

Comment on Faraday's note 6: Faraday regards it as evident that the carrier ball, once insulated, will *retain* its state of electrification so long as it continues to be insulated. From one point of view, such constancy is a mere consequence of the *definition* of an insulator as a material that cannot convey the electrified condition. But we have also become aware that "electrification" is not a simple state—it can be characterized by the two aspects of "quantity" and "tension" and perhaps others as well. *Which* of these aspects, specifically, is conserved in an insulated electrified body? On the "electric fluid" hypothesis it would be obvious that *quantity* is conserved, since "fluid" can neither pour in or leak out. But once fluid ideas are abandoned it is not so easy to specify which aspect or aspects of the electrified condition we would expect to be maintained by insulation.

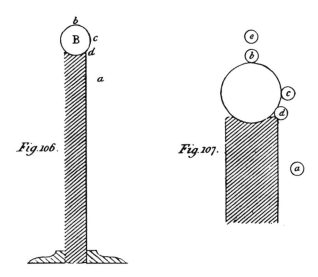

Fig. 106.

Fig. 107.

1219. To comprehend the full force of these results, it must first be understood, that all the charges of the ball B and the carrier are charges by induction, from the action of the excited surface of the shell-lac cylinder; for whatever electricity the ball B received by *communication* from the shell-lac, either in the first instance or afterwards, was removed by the uninsulating contacts, only that due to induction remaining; and this is shown by the charges taken from the ball in this its uninsulated state being always positive, or of the contrary character to the electricity of the shell-lac. In the next place, the charges at *a, c,* and *d* were of such a nature as might be expected from an inductive action in straight lines, but that obtained at *b* is *not so*: it is clearly a charge by induction, but *induction* in *a curved line*, for the carrier ball whilst applied to *b,* and after its removal to a distance of six inches or more from B, could not, in consequence of the size of B, be connected by a straight line with any part of the excited and inducing shell-lac.

1219. *all the charges of the ball B and the carrier are charges by induction...*: That is, their electrification was developed only by the influence of the shell-lac cylinder. The technique of momentary grounding, described in paragraph 1218, insures that any charge remaining on the ball and carrier must be *sustained by induction with the shell-lac cylinder*—for had the excited cylinder not been present, a momentary grounding of the balls would have resulted in their complete discharge. Moreover the *opposite signs* of electrification are consistent with induction. This being so, the inductive action between the cylinder and point *b,* at least, must propagate in curved lines for there is no straight line path between any point on the cylinder and point *b.*

1220. To suppose that the upper part of the *uninsulated* ball B, should in some way be retained in an electrified state by that portion of the surface of the ball which is in sight of the shell-lac, would be in opposition to what we know already of the subject. Electricity is retained upon the surface of conductors only by induction (1178.); and though some persons may not be prepared as yet to admit this with respect to insulated conductors, all will as regards uninsulated conductors like the ball B; and to decide the matter we have only to place the carrier ball at *e* (fig. 107.), so that it shall not come in contact with B, uninsulate it by a metallic rod descending perpendicularly, insulate it, remove it, and examine its state; it will be found charged with the same kind of electricity as, and even to a *higher degree* (1224.) than, if it had been in contact with the summit of B.

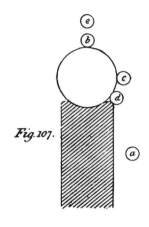

Fig. 107.

1221. To suppose, again, that induction acts in some way *through or across* the metal of the ball, is negatived by the simplest considerations; but a fact in proof will be better. If instead of the ball B a small disc of metal be used, the carrier may be charged at, or above the middle of

---

1220. *To suppose that the upper part of the uninsulated ball B...*: Someone might object that the electrification detected at *b* need not be sustained by curvilinear induction through the air, but might instead be conveyed to *b* along the conducting surface of the ball, from regions such as *c* or *d* which are in direct straight-line relations to the cylinder. Against this, Faraday repeats the declaration of paragraph 1178; that electricity does not reside upon a conductor independently, but is retained "only by induction." He is confident that "all will [admit this] as regards uninsulated conductors"; for even those who believe in an independent electric fluid acknowledge that such a fluid cannot maintain itself on a *grounded* conductor by its own efforts—mutual repulsion among all portions of the fluid would cause it to disperse through the grounding path, into the earth.

But the hypothetical objection is not about a grounded ball, but about the ungrounded ball B. Faraday therefore offers an experimental rejoinder: Simply place the carrier ball at *e, not* in contact with the ball B; ground it momentarily with a separate wire, and test with the electrometer. The electrification obtained at *e* has the same sign as at *b*, indicating that both *e* and *b* are electrified by induction from the same source, namely, the shell-lac cylinder—that is, through the air; and therefore by *curved* lines. Moreover, Faraday reports, the electrification obtained at *e* is *greater* than at *b*. He will explain the significance of this "higher degree" of electrification in paragraph 1224 below.

its upper surface: but if the plate be enlarged to about 1½ or 2 inches in diameter, C (fig. 108.), then no charge will be given to the carrier at *f*, though when applied nearer to the edge at *g*, or even *above the middle* at *h*, a charge will be obtained; and this is true though the plate may be a mere thin film of gold-leaf. Hence it is clear that the induction is not *through* the metal, but through the surrounding air or *dielectric*, and that in curved lines.

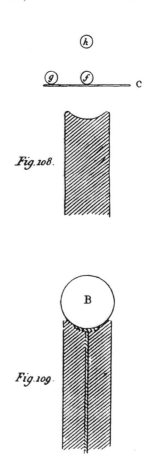

Fig. 108.

Fig. 109.

1222. I had another arrangement, in which a wire passing downwards through the middle of the shell-lac cylinder to the earth, was connected with the ball B (fig. 109.) so as to keep it in a constantly uninsulated state. This was a very convenient form of apparatus, and the results with it were the same as those just described.

1223. In another case the ball B was supported by a shell-lac stem, independently of the excited cylinder of shell-lac, and at half an inch distance from it; but the effects

---

1221. *Hence it is clear that the induction is not through the metal...*: Faraday is answering another objection—that perhaps the induction acts *through* the metal ball, and therefore in straight lines after all. Against this Faraday alleges the "simplest considerations"—I will suggest shortly what these may be. But once again, the best rebuttal is an *experiment* ("a fact in proof"). He mounts a large metal disk C as shown in Figure 108. No electrification is detected at location *f*, showing that *induction is not acting through the conducting material.* But electrification *is* found at *g* and *h*—locations which can be related to the cylinder only along *curved* paths through the air. The disk's failure to transmit inductive action through itself illustrates the general idea, discussed in the introduction, that *conductors cannot sustain tension.* Perhaps when Faraday speaks of "the simplest considerations" he has that idea in mind.

1222. In the arrangement illustrated by Figure 109, a permanently grounded ball exhibits charge by induction from the shell-lac column, producing *the same results* as did the ball depicted in Figure 107, which had been only momentarily grounded. Recall that in paragraph 1220 Faraday noted that even objectors admit that the electricity on *grounded* conducting surfaces is retained by induction; they will therefore have to acknowledge the present exercise as an example of induction in curved lines.

were the same. Then the brass ball of a charged Leyden jar was used in place of the excited shell-lac to produce induction; but this caused no alteration of the phenomena. Both positive and negative inducing charges were tried with the same general results. Finally, the arrangement was inverted in the air for the purpose of removing every possible objection to the conclusions, but they came out exactly the same.

1224. Some results obtained with a brass hemisphere instead of the ball B were exceedingly interesting. It was 1.36 of an inch in diameter, (fig. 110.), and being placed on the top of the excited shell-lac cylinder, the carrier ball was applied, as in the former experiments (1218.), at the respective positions delineated in the figure. At i the force was 112°, at k 108°, at l 65°, at m 35°; the inductive force gradually diminishing, as might have been expected, to this point. But on raising the carrier to the position n, the charge increased to 87°;

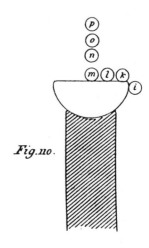

Fig. 110.

and on raising it still higher to o, the charge still further increased to 105°: at a higher point still, p, the charge taken was smaller in amount, being 98°, and continued to diminish for more elevated positions. Here the induction fairly turned a corner. Nothing, in fact, can better show both the curved lines or courses of the inductive

---

1224. The results with the brass hemisphere are a blend, so to speak, of those obtained with the ball (Figure 107) and the disk (Figure 108).

the induction fairly turned a corner: That is, the path of induction changed its direction by 90° or more. There are two separate examples of this in figure 110. First, detection of *any* induction at points k, l, and m means that some lines of induction must run from the cylinder surface, around corner i, to each of those points; and even the gentlest of such paths involves at least a 90° bend. But these observations essentially repeat what Faraday has reported for the preceding figures. When he now speaks of "turning a corner" it is not the measurements at k, l, and m but rather the conditions at points n, o, and p that suggest that phrase. What happens at those places to suggest a "corner"?

At point o, Faraday measures 105° of induction. Not only is that almost as much as the 112° measured at i—and i has direct straight-line exposure to much of the cylinder's surface!—but o is an induction *maximum* compared to n and p, points vertically adjacent to it. This is significant because it shows that the induction here measured cannot be explained away as some kind of leakage or other accidental penetration of inductive action into regions normally inaccessible by straight-line paths; for if such were the case we should expect

action, disturbed as they are from their rectilineal form by the shape, position, and condition of the metallic hemisphere; and also a *lateral tension,* so to speak, of these lines on one another: all depending, as I conceive, on induction being an action of the contiguous particles of the dielectric, which being thrown into a state of polarity and tension, are in mutual relation by their forces in all directions.

1225. As another proof that the whole of these actions were inductive I may state a result which was exactly what might be expected, namely, that if uninsulated conducting matter was brought round and near to the excited shell-lac stem, then the inductive force was directed towards it, and could not be found on the top of the hemisphere. Removing this matter the lines of force resumed their former direction. The experiment affords proofs of the lateral tension of these lines, and supplies a warning to remove such matter in repeating the above investigation.

1226. After these results on curved inductive action in air I extended the experiments to other gases, using first carbonic acid and then hydrogen: the phenomena were precisely those already described. In these experiments I found that if the gases were confined in vessels they required to be very large, for whether of glass or earthenware, the conducting power of such materials is so great that the induction of

---

the induction to *decrease continually* from *m*, to *n*, to *o*, to *p*. Instead, the lines of induction appear to be *concentrated* near *o*. For a visual representation of this concentration, as well as the "corner" which the lines thereby turn, see the comment to paragraph 1231 below.

Note that Faraday characterizes the lines of action not merely as *curved*, but as having been "disturbed ... from their rectilineal form". This highly evocative language represents the lines of action as *beings in their own right*, capable of suffering alteration and disturbance by the brass hemisphere, as well as by other lines. He extends the same imagery with the phrase, "*lateral tension,* so to speak..." In this image, lines of action squeeze up against one another; and if one line is displaced, it in turn pushes its neighbors aside. Does Faraday's phraseology represent only a whimsical indulgence in figurative language? Or are there sound reasons for ascribing *substantiality* to the lines of action?

1225. Another powerful image of *substantiality* in the lines of induction! Faraday places a grounded conductor next to, but not touching, the shell-lac cylinder of Figure 110. He finds inductive charge on its surface facing the shell-lac; but the electrification formerly observed *above* the hemisphere diminishes or even vanishes. Faraday's language leaves no doubt that he sees the newly-introduced body as having drawn some lines of action towards itself, away from their former places.

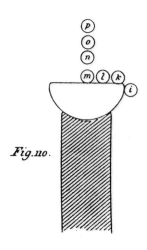

Fig. 110.

the excited shell-lac cylinder towards them is as much as if they were metal; and if the vessels be small, so great a portion of the inductive force is determined towards them that the lateral tension or mutual repulsion of the lines of force before spoken of, (1224.) by which their inflexion is caused, is so much relieved in other directions, that no inductive charge will be given to the carrier ball in the positions *k, l, m, n, o, p,* (fig. 110.). A very good mode of making the experiment is to let large currents of the gases ascend or descend through the air, and carry on the experiments in these currents.

* * *

1228. Lastly, I used a few solid dielectrics for the same purpose, and with the same results. These were shell-lac, sulphur, fused and cast borate of lead, flint glass well covered with a film of lac, and spermaceti. The following was the form of experiment with sulphur, and all were of the same kind. A square plate of the substance, two inches in extent and 0.6 of an inch in thickness, was cast with a small hole or depression in the middle of one surface to receive the carrier ball. This was placed upon the surface of the metal hemisphere (fig. 112.) arranged on the excited lac as in former cases, and observations were made at *n, o, p,* and *q.* Great care was required in these experiments to free the sulphur or other solid substance from any charge it might previously have received. This was done by breathing and wiping (1203.), and the substance being found free from all electrical excitement, was then used in the experiment; after which it was removed and again examined, to ascertain that it had received no charge, but had acted really as a dielectric. With all these precautions the results were the same: and it is thus very satisfactory to obtain the curved inductive action through *solid bodies,* as any possible effect from the translation of charged particles in fluids or gases, which some persons might imagine to be the case, is here entirely negatived.

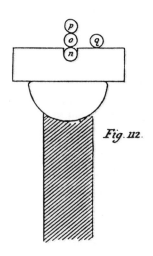

Fig. 112.

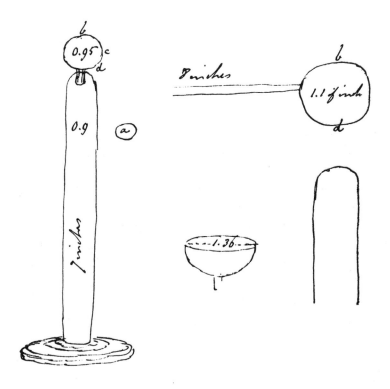

[Sketches from Faraday's laboratory *Diary*, 14 October 1837, showing some of the inductive configurations described in the text. They were not published with the *Experimental Researches*.]

1229. In these experiments with solid dielectrics, the degree of charge assumed by the carrier ball at the situations *n, o, p* (fig. 112.), was decidedly greater than that given to the ball at the same places when air only intervened between it and the metal hemisphere. This effect is consistent with what will hereafter be found to be the respective relations of these bodies, as to their power of facilitating induction through them (1269. 1273. 1277.).

1229. It is significant that the solid dielectrics *increase* the degree of electrification found at the locations specified. It indicates, first, that these materials carry on induction more effectively than air does—an idea that is confirmed in the later paragraphs cited. But in paragraph 1231 below, Faraday will offer a more visual interpretation when he suggests that lines of force are *gathered into the dielectric* when it is present, and therefore exhibit greater concentration at locations *n, o,* and *p* than before. *Concentration of lines of force* will become a principal image for the action of dielectric materials in "facilitating induction" through themselves.

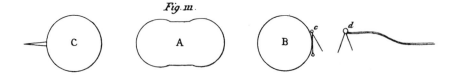

Fig. m.

1230. I might quote *many* other forms of experiment, some old and some new, in which induction in curved or contorted lines takes place, but think it unnecessary after the preceding results; I shall therefore mention but two. If a conductor A, (fig. 111.) be electrified, and an uninsulated metallic ball B, or even a plate, provided the edges be not too thin, be held before it, a small electrometer at *c* or at *d*, uninsulated, will give signs of electricity, opposite in its nature to that of A, and therefore caused by induction, although the influencing and influenced bodies cannot be joined by a right line passing through the air. Or if, the electrometers being removed, a point be fixed at the back of the ball in its uninsulated state as at C, this point will become luminous and discharge the conductor A. The latter experiment is described by Nicholson,[7] who, however, reasons erroneously upon it. As to its introduction here, though it is a case of discharge, the discharge is preceded by induction, and that induction must be in curved lines.

1231. As argument against the received theory of induction and in favour of that which I have ventured to put forth, I cannot see how the preceding results can be avoided. The effects are clearly inductive effects produced by electricity, not in currents but in its statical state,

[7] Encyclopædia Britannica, vol. vi. p. 504.

1230. Refer to Faraday's figure 111. The inductive action between the electrified conductor A and the electrometer at *c* or *d* must be through the air, as he has shown already that induction cannot act through a conductor. But if the paths of action are through the air, they must be *curved* paths.

On the other hand, the electrified conductor A and the grounded conductor C are related by induction, and certainly there are straight line paths between them. But he shows that the straight paths do not sustain *all* the induction, since A and C can be mutually discharged through the air by mounting a pointed electrode on C, facing *away* from A. It is obvious that the *discharge* had to have taken place along curved paths between the point and A; therefore there must have been a previous condition of *induction* along those same curved paths *even though straight paths were also available*! The pointed electrode becomes "luminous" during discharge: this refers to the electric *glow* mentioned in the Third Series (paragraph 306). (A fuller investigation in the Twelfth Series is not included in the present selection.)

and this induction is exerted in lines of force which, though in many experiments they may be straight, are here curved more or less according to circumstances. I use the term *line of inductive force* merely as a temporary conventional mode of expressing the direction of the power in cases of induction; and in the experiments with the hemisphere (1224.), it is curious to see how, when certain lines have terminated on the under surface and edge of the metal, those which were before lateral to them *expand and open out from each other,* some bending round and terminating their action on the upper surface of the hemisphere, and others meeting, as it were, above in their progress outwards, uniting their forces to give an increased charge to the carrier ball, at an *increased distance* from the source of power, and influencing each

---

1231. *"lines of force"*: This will become a central term in Faraday's visual imagery for both electric and magnetic action. Despite his cautious disclaimer—"merely ... a temporary conventional mode of expressing the direction of the power"—which would suggest a retreat from the exuberance of earlier passages, Faraday continues to cultivate in the latter parts of this very paragraph a highly concrete and substantial language for the lines of induction.

lines ... expand and open out from each other...: Much is left to the imagination here. To be sure, we have seen such patterns of expansion among *magnetic lines,* which are readily visualized by means of iron filings; but Faraday has specified no comparable means for making *lines of electric induction* visible. In the following Series, however, he will show how bits of silk thread suspended in oil can make manifest the lines of electric action, as iron filings do for lines of magnetic action. He will also note expansion and dilatation in *electrical discharge patterns,* which are perceptible to the eye directly. None of these techniques is capable of raising electric action to the high level of literal visibility that magnetic lines enjoy; nevertheless, Faraday now gives an unhesitatingly visual interpretation to the observations he reported previously in paragraph 1224.

some bending round and terminating their action on the upper surface of the hemisphere, and others meeting, as it were, above in their progress outwards, uniting their forces...: Faraday here names two sets of lines of induction that may be said to "turn a corner": The first set terminates on the hemisphere. Additionally, from the inductive maximum observed at *o* Faraday infers another set of lines, sketched here as *concentrated* (not literally "meeting") there. Note his phrase "uniting their forces," which suggests that each line represents a determinate share of the total inductive action.

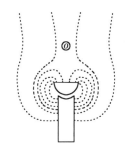

at an increased distance from the source of power: It would be difficult for action-at-a-distance partisans to explain how an electrified body could exert stronger action at a more distant point than it does at a nearer one. But Faraday sees that since the lines of force *curve* they can be *focused,* as it were, at any distance—and so exert increased influence at that distance.

other so as to cause a second flexure in the contrary direction from the first one. All this appears to me to prove that the whole action is one of contiguous particles, related to each other, not merely in the lines which they may be conceived to form through the dielectric, between the *inductric* and the *inducteous* surfaces (1483.), but in other lateral directions also. It is this which gives an effect equivalent to a lateral repulsion or expansion in the lines of force I have spoken of, and enables induction to turn a corner (1304.). The power, instead of being like that of gravity, which causes particles to act on each other through straight lines, whatever other particles may be between them, is more analogous to that of a series of magnetic needles, or to the condition of the particles considered as forming the whole of a straight or a curved magnet. So that in whatever way I view it, and with great suspicion of the influence of favourite notions over myself, I cannot perceive how the ordinary theory applied to explain induction can be a correct representation of that great natural principle of electrical action.

1232. I have had occasion in describing the precautions necessary in the use of the inductive apparatus, to refer to one founded on induction in curved lines (1203.); and after the experiments already described, it will easily be seen how great an influence the shell-lac stem may exert upon the charge of the carrier ball when applied to the apparatus (1218.), unless that precaution be attended to.

1233. I think it expedient, next in the course of these experimental researches, to describe some effects due to *conduction,* obtained with such bodies as glass, lac, sulphur, &c., which had not been anticipated. Being

---

1231, continued. *"inductric" and "inducteous" surfaces*: As will be defined in paragraph 1483, the "inductric surface" *originates* the inductive relation and thereby brings the "inducteous surface" into an oppositely-charged condition.

analogous to that of a series of magnetic needles…: Notice that in this comparison, the magnetic needles (or iron filings) are not merely passive indicators of the magnetic lines of force. They indicate the forces *because they carry them onward.* In a comparable way, the particles of any dielectric medium transmit the electrical forces from each to the next. (But here again we wonder how to fit the *vacuum*—which has no particles—into this scheme.)

suspicion of … favourite notions: He acknowledges a natural tendency to favor one's own ideas and is doing his best to guard against it.

1232. *how great an influence the shell-lac stem may exert…*: One example of this influence arises in an experiment described in paragraph 1258 below.

1233. *some effects due to conduction…*: Faraday is about to study the so-called "return charge." He will give reasons for regarding it as a species of *conduction* in paragraph 1240 below.

understood, they will make us acquainted with certain precautions necessary in investigating the great question of specific inductive capacity.

1234. One of the inductive apparatus already described (1187, &c.) had a hemispherical cup of shell-lac introduced, which being in the interval between the inner ball and the lower hemisphere, nearly occupied the space there; consequently when the apparatus was charged, the lac was the dielectric or insulating medium through which the induction took place in that part. When this apparatus was first charged with electricity (1198.) up to a certain intensity, as 400°, measured by the COULOMB's electrometer (1180.), it sank much faster from that degree than if it had been previously charged to a higher point, and had gradually fallen to 400°; or than it would do if the charge were, by a second application, raised up again to 400°; all other things remaining the same. Again, if after having been charged for some time, as fifteen or twenty minutes, it was suddenly and perfectly discharged, even the stem having all electricity removed from it (1203.), then the apparatus being left to itself, would gradually recover a charge, which in nine or ten minutes would rise up to 50° or 60°, and in one instance to 80°.

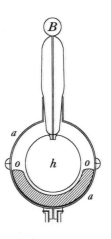

1234. *a hemispherical cup of shell-lac introduced*: The shell-lac cup occupies the lower hemispere of the inductive apparatus, as illustrated in the sketch.

intensity … measured by the … electrometer: The electrometer is being employed in the mode of direct contact and therefore measures what is ordinarily called *intensity* or *tension* (see the introduction). In paragraph 1299, however, Faraday will suggest that "intensity" and "tension" are not quite the same thing.

it sank much faster from that degree than if it had been previously charged…: An apparatus that has been newly electrified gradually loses some of its charge. We might suppose that the loss merely indicates dissipation or imperfect insulation, except that when the charge is restored to its former degree the subsequent losses are greatly diminished. Since the rates of ordinary leakage and dissipation would not be affected by *how long* it has been since the apparatus was last charged, some other mechanism must be at work. Recall Faraday's earlier observation (paragraph 1207) that when electricity was divided between two devices, the apparatus *newly receiving* the charge appeared to suffer greater dissipation than did the previously-electrified, but otherwise identical, apparatus *from which* the charge was obtained. Faraday will explicitly link the two cases in paragraph 1250 below.

the apparatus … would gradually recover a charge: In a reverse manifestation of the previous effect, a recently-*discharged* apparatus will gradually regain electrification—in one case as much as 20% of its previous charge! This recovery of electrification is the phenomenon he will call "return charge" or, less often, "residual charge."

1235. The electricity, which in these cases returned from an apparently latent to a sensible state, was always of the same kind as that which had been given by the charge. The return took place at both the inducing surfaces; for if after the perfect discharge of the apparatus the whole was insulated, as the inner ball resumed a positive state the outer sphere acquired a negative condition.

1236. This effect was at once distinguished from that produced by the excited stem acting in curved lines of induction (1203. 1232.), by the circumstance that all the returned electricity could be perfectly and instantly discharged. It appeared to depend upon the shell-lac within, and to be, in some way, due to electricity evolved from it in consequence of a previous condition into which it had been brought by the charge of the metallic coatings or balls.

1237. To examine this state more accurately, the apparatus, with the hemispherical cup of shell-lac in it, was charged for about forty-five minutes to above 600° with positive electricity at the balls *h* and B (fig. 104.) above and within. It was then discharged, opened, the shell-lac taken out, and its state examined; this was done by bringing the carrier ball near the shell-lac, uninsulating it, insulating it, and then observing what charge it had acquired. As it would be a charge by induction, the state of the ball would indicate the opposite state of electricity in that surface of the shell-lac which had produced it. At first the lac appeared

---

1235. Since the charge regained is "always of the same kind as that which had been given," Faraday speaks of the electricity as having "returned" from a "latent" or hidden state to a "sensible" or observable one. This does not contradict his declaration in paragraph 1177 that there is "no latency of a single electricity," for the present example is not latency of a *single* electricity—the active surfaces regain *opposite* charges; furthermore, they do so *simultaneously*. No *independent* return of either positive or negative electrification is observed.

1236. *perfectly and instantly discharged*: But not *permanently* discharged; a residual charge continues to reappear even after several discharges. And since each reappearance can be discharged (by grounding), the "returned" charge cannot be due to induction from other bodies—for we saw already that charge by induction can be sustained even on a *permanently grounded* body (paragraph 1222).

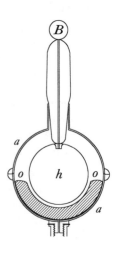

1237. *the balls h and B. (fig. 104.)*: Faraday's Figure 104 appears on page 247. The sketch depicts the same apparatus with shell-lac hemisphere installed.

As it would be a charge by induction, the state of the ball would indicate the opposite state of electricity...: The reason why momentary grounding under induction gives an opposite electrification is explained in the introduction.

quite free from any charge; but gradually its two surfaces assumed opposite states of electricity, the concave surface, which had been next the inner and positive ball, assuming a positive state, and the convex surface, which had been in contact with the negative coating, acquiring a negative state; these states gradually increased in intensity for some time.

1238. As the return action was evidently greatest instantly after the discharge, I again put the apparatus together, and charged it for fifteen minutes as before, the inner ball positively. I then discharged it, instantly removing the upper hemisphere with the interior ball, and, leaving the shell-lac cup in the lower uninsulated hemisphere, examined its inner surface by the carrier ball as before (1237.). In this way I found the surface of the shell-lac actually *negative,* or in the reverse state to the ball which had been in it; this state quickly disappeared, and was succeeded by a positive condition, gradually increasing in intensity for some time, in the same manner as before. The first negative condition of the surface opposite the positive charging ball is a natural consequence of the state of things, the charging ball being in contact with the shell-lac only in a few points. It does not interfere with the general result and peculiar state now under consideration, except that it assists in illustrating in a very marked manner the ultimate assumption by the surfaces of the shell-lac of an electrified condition, similar to that of the metallic surfaces opposed to or against them.

1239. *Glass* was then examined with respect to its power of assuming this peculiar state. I had a thick flint-glass hemispherical cup formed, which would fit easily into the space *o* of the lower hemisphere (1188. 1189.); it had been heated and varnished with a solution of shell-lac in alcohol, for the purpose of destroying the conducting power of the vitreous surface (1254.). Being then well warmed and experimented with, I found it could also assume the *same state,* but not apparently to the same degree, the return action amounting in different cases to quantities from 6° to 18°.

1238. *the return action was evidently greatest instantly after the discharge*: He means the *rate* of return, which is greatest immediately after discharge. A return charge continues to accumulate for some time after discharge ceases, but more and more slowly.

the first negative condition of the surface … is a natural consequence: Immediately after discharge, the shell-lac surface which had faced the positively-charged ball is found to be *negative*. The fact of opposite electrification points to induction across the tiny *air gap* between the ball and the shell-lac's inner surface. But Faraday offers this explanation only for the "*first*" negative condition of the surface; induction cannot explain its *retention* of a negative condition after the positive ball is removed. Ultimately, though, the (positive) return charge predominates.

1240. *Spermaceti* experimented with in the same manner gave striking results. When the original charge had been sustained for fifteen or twenty minutes at about 500°, the return charge was equal to 95° or 100°, and was about fourteen minutes arriving at the maximum effect. A charge continued for not more than two or three seconds was here succeeded by a return charge of 50° or 60°. The observations formerly made (1234.) held good with this substance. Spermaceti, though it will insulate a low charge for some time, is a better conductor than shell-lac, glass, and sulphur; and this conducting power is connected with the readiness with which it exhibits the particular effect under consideration.

1241. *Sulphur.* I was anxious to obtain the amount of effect with this substance, first, because it is an excellent insulator, and in that respect would illustrate the relation of the effect to the degree of conducting power possessed by the dielectric (1247.); and in the next place, that I might obtain that body giving the smallest degree of the effect now under consideration, for the investigation of the question of specific inductive capacity (1277.).

1242. With a good hemispherical cup of sulphur cast solid and sound, I obtained the return charge, but only to an amount of 17° or 18°. Thus glass and sulphur, which are bodily very bad conductors of electricity, and indeed almost perfect insulators, gave very little of this return charge.

1240. *Spermaceti,* a white, waxy substance obtained from the head of the sperm whale or other cetaceans, was used in making candles, ointments, and cosmetics.

this conducting power is connected with the readiness with which it exhibits the particular effect: Though a poor conductor in relation to materials generally, spermaceti is the best conductor among the materials tested, and so too does it exhibit the greatest return charge—100° out of an original 500°. With shell-lac, the return was at most equal (80° out of 400°) and typically less (paragraph 1234); while glass showed the least effect of all. Faraday already signaled such a correlation between the *degree of returned charge* and the *conducting ability of the substance*, when he first introduced the topic of return charge under the heading "effects due to conduction" in paragraph 1233 above. He will explicitly accept return charge as a *conduction* phenomenon in paragraph 1245 below.

1241. *Sulphur*: If the return charge is indeed a phenomenon of *conduction*, then sulphur, one of the best insulators, should develop only a minimal return charge. On the level of theory, if such a result ensues it will support the conductive interpretation. Practically, it will identify sulphur as a suitable material for use in subsequent experiments; since a sizable return charge would seriously complicate the investigation of specific inductive capacity which Faraday will undertake in the next section.

1242. Faraday confirms that sulphur and glass, the two best solid insulators, also exhibit about the same low degree of return charge. In the next

1243. I tried the same experiment having *air* only in the inductive apparatus. After a continued high charge for some time I could obtain a little effect of return action, but it was ultimately traced to the shell-lac of the stem.

1244. I sought to produce something like this state with one electric power and without induction; for upon the theory of an electric fluid or fluids, that did not seem impossible, and then I should have obtained an absolute charge (1169. 1177.), or something equivalent to it. In this I could not succeed. I excited the outside of a cylinder of shell-lac very highly for some time, and then quickly discharging it (1203.), waited and watched whether any return charge would appear, but such was not the case. This is another fact in favour of the inseparability of the two electric forces (1177.), and another argument for the view that induction and its concomitant phenomena depend upon a polarity of the particles of matter.

1245. Although inclined at first to refer these effects to a peculiar masked condition of a certain portion of the forces, I think I have since correctly traced them to known principles of electrical action. The effects appear to be due to an actual penetration of the charge to some distance within the electric, at each of its two surfaces, by what we call *conduction*; so that, to use the ordinary phrase, the electric forces sustaining the induction are not upon the metallic surfaces only, but upon and within the dielectric also, extending to a greater or smaller

---

paragraph he will point out that *air*—which except in cases involving pointed electrodes or very high tensions is an excellent insulator—shows no detectible return charge whatever! But inasmuch as air is a gas and not a solid, there may be more fundamental reasons than poor conductivity for its failure to develop a return charge.

1244. Evidently the return charge results only when a susceptible dielectric has served as the medium *between* inductively-related charges. Direct excitation of the shell-lac surface establishes induction not within the shell-lac itself but through the surrounding *air*, towards neighboring bodies—and he pointed out in the previous paragraph that air does not exhibit the return effect.

1245. *the electric*: the insulating material, as in paragraph 1171 above. For some reason Faraday in this one paragraph alternates between "electric," the antique term, and "dielectric", the term he proposed in his note to paragraph 1168. Though it is commonly classified as an insulator, Faraday supposes this dielectric to have *some* conducting ability. In the Twelfth Series he will represent insulation and conduction as just extreme degrees of the ability to sustain tension, which is *induction*. So *every* dielectric is a conductor to some degree.

depth from the metal linings. Let *c*
(fig. 113.) be the section of a plate
of any dielectric, *a* and *b* being the
metallic coatings; let *b* be uninsulated, and *a* be charged
positively; after ten or fifteen minutes,
if *a* and *b* be discharged, insulated,
and immediately examined, no
electricity will appear in them; but
in a short time, upon a second
examination, they will appear
charged in the same way, though
not to the same degree, as they

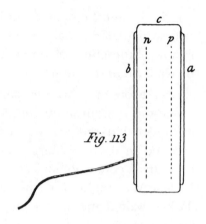

Fig. 113

were at first. Now suppose that a portion of the positive force has,
under the coercing influence of all the forces concerned, penetrated
the dielectric and taken up its place at the line *p*, a corresponding
portion of the negative force having also assumed its position at the
line *n*; that in fact the electric at these two parts has become charged
positive and negative; then it is clear that the induction of these two
forces will be much greater one towards the other, and less in an
external direction, now that they are at the small distance *n p* from

---

1245, continued. *suppose that a portion of [each] force has … penetrated the
dielectric…*: Faraday conjectures that electrification may spread from the metal
foils *a* and *b* into the adjacent shell-lac or other dielectric, "under the coercing
influence of all the forces concerned" (this phrase indicates that he believes
the condition to be a forced one that will give way again when the coercion is
removed). Note that Faraday is not trying to explain *why* such penetration
should occur; this is indeed *conjecture*—albeit conjecture guided by *experiment*,
as he will declare at the opening of paragraph 1246.

   taken up its place at the line p … assumed its position at the line n: Does Faraday
envision two isolated concentrations of positive and negative charge, situated
like puddles at *p* and *n* respectively? Probably not. By specifying that only "a
portion" of each force has migrated to the depths specified, he seems rather
to imply that other portions pass to other depths—for a *continuous distribution*
of positive charge between *a* and *p* (and negative charge between *b* and *n*), or
beyond.

   *the induction of these two forces will be much greater one towards
the other, and less in an external direction*: The two electrified
regions will be inductively related to each other through
short, direct internal paths, as well as though longer, round-
about "external" paths. Faraday supposes that the internal
relations are "greater"—perhaps in the sense that the lines
of induction are more highly concentrated there. Sketch *i*
illustrates this interpretation.

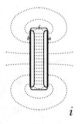

i

each other, than when they were at the larger interval *a b*. Then let *a* and *b* be discharged; the discharge destroys or neutralizes all external induction, and the coatings are therefore found by the carrier ball unelectrified; but it also removes almost the whole of the forces by which the electric charge was driven into the dielectric, and though probably a part of that charge goes forward in its passage and terminates in what we call discharge, the greater portion returns on its course to the surfaces of *c*, and consequently to the conductors *a* and *b*, and constitutes the recharge observed.

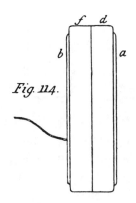

Fig. 114.

1246. The following is the experiment on which I rest for the truth of this view. Two plates of spermaceti, *d* and *f* (fig. 114.), were put together to form the dielectric, *a* and *b* being the metallic coatings of this compound

---

the discharge desctroys or neutralizes all external induction…: The highly-conductive discharge wire cannot sustain tension, and thus there can be no lines of induction running between its terminii, that is, between plates *a* and *b*. But such are the "external" lines of induction, which are therefore represented in sketch *ii* as having been discharged. If plates *a* and *b* were measured under these conditions (the wire having been removed), they would be found to be *uncharged*.

ii

…but it also removes almost the whole of the forces by which the electric charge was driven into the dielectric: The distribution of lines of induction between "internal" and "external" paths is a matter of equilibrium between their mutual tendencies to nudge and displace one another (recall the "lateral action" of paragraph 1224). Therefore at least some of the internal lines take the paths they do because they have been pushed into place by external lines. But now that the external lines are gone, some at least of the internal lines can be expected to migrate back to external courses, establishing a new equilibrium (sketch *iii*). This is the "return charge."

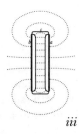

iii

probably a part of that charge … terminates in what we call discharge: Why doesn't *all* the electrification discharge? After all, the dielectric has been assumed to be conductive; and a conductor cannot support electric tension (see the editor's introduction, page 221); so not just some, but all the "internal" lines of force illustrated in sketch *iii* should disappear. Faraday will outline the basis for at least a partial answer in the Twelfth Series: Conduction takes *time*; and in a poor conductor such as is supposed here, the time required for a significant internal discharge could be very long indeed.

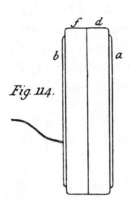

Fig. 114.

plate, as before. The system was charged, then discharged, insulated, examined, and found to give no indications of electricity to the carrier ball. The plates *d* and *f* were then separated from each other, and instantly *a* with *d* was found in a positive state, and *b* with *f* in a negative state, nearly all the electricity being in the linings *a* and *b*. Hence it is clear that, of the forces sought for, the positive was in one half of the compound plate and the negative in the other half; for when removed bodily with the plates from each other's inductive influence, they appeared in separate places, and resumed of necessity their power of acting by induction on the electricity of surrounding bodies. Had the effect depended upon a peculiar relation of the contiguous particles of matter only, then each half plate, *d* and *f*, should have shown positive force on one surface and negative on the other.

1247. Thus it would appear that the best solid insulators, such as shell-lac, glass, and sulphur, have conductive properties to such an extent, that electricity can penetrate them bodily, though always subject to the over-ruling condition of induction (1178.). As to the depth to which the forces penetrate in this form of charge of the particles, theoretically, it should be throughout the mass, for what the charge of the metal does for the portion of dielectric next to it, should be done by the charged dielectric for the portion next beyond it again; but probably in the best insulators the sensible charge is to a very small depth only in the dielectric, for

---

1246. *the system was charged*: That is, plate *a* was charged positively and plate *b* was grounded, just as in the previous paragraph.

the positive was in one half … and the negative in the other: There must have been "actual penetration" (the wording of paragraph 1245) of electricity into the dielectric since when the dielectric is charged, then discharged, and then divided in two, the separate halves bear positive and negative charges, respectively.

a peculiar relation of the contiguous particles…: Faraday is referring to one preva-lent explanation of the return charge: a supposed *polar condition* of the particles, formally comparable to the dual condition of magnetic needles in his example of paragraph 1231 above—but also *partially inelastic*; so that they could never lose their polarity completely, and therefore discharge could never fully restore the dielectric to a wholly unpolarized condition. But Faraday points out that, were that view correct, we should expect to find evidence of *permanent polarization* in each half of the dielectric. As pieces of a broken bar magnet exhibit north and south magnetic powers at contrary extremities, so each piece of dielectric should exhibit positive and negative electrification on opposite surfaces. Yet no such pattern of electrification is observed.

otherwise more would disappear in the first instance whilst the original charge is sustained, less time would be required for the assumption of the particular state, and more electricity would re-appear as return charge.

1248. The condition of *time* required for this penetration of the charge is important, both as respects the general relation of the cases to conduction, and also the removal of an objection that might otherwise properly be raised to certain results respecting specific inductive capacities, hereafter to be given (1269. 1277.).

1249. It is the assumption for a time of this charged state of the glass between the coatings in the Leyden jar, which gives origin to a well-known phenomenon, usually referred to the diffusion of electricity over the uncoated portion of the glass, namely, the *residual charge*. The extent of charge which can spontaneously be recovered by a large battery, after perfect uninsulation of both surfaces, is very considerable, and by far the largest portion of this is due to the return of electricity in the manner described. A plate of shell-lac six inches square, and half an inch thick, or a similar plate of spermaceti an inch thick, being coated on the sides with tinfoil as a Leyden arrangement, will show this effect exceedingly well.

1250. The peculiar condition of dielectrics which has now been described, is evidently capable of producing an effect interfering with the results and conclusions drawn from the use of the two inductive apparatus, when shell-lac, glass, &c. is used in one or both of them (1192. 1207.); for upon dividing the charge in such cases according to the method described (1198. 1207.), it is evident that the apparatus just receiving its half charge must fall faster in its tension than the other. For suppose app. i. first charged, and app. ii. used to divide with it; though both may actually lose alike, yet app. i., which has been diminished one half, will be sustained by a certain degree of return action or charge (1234.), whilst app. ii. will sink the more rapidly from the coming on of the particular state. I have endeavoured to avoid this interference by

---

1247. *the particular state*: that is, the state *of the particles*, whatever that state may prove to be.

1249. *large battery*: that is, a battery of Leyden jars.

1250. *the apparatus just receiving its half charge*: that is, the initially unelectrified apparatus, which now receives a charge for the first time. Compare his distinction between the induction apparatus's first charge and its subsequent charges in paragraph 1234 above.

performing the whole process of comparison as quickly as possible, and taking the force of app. ii. immediately after the division, before any sensible diminution of the tension arising from the assumption of the peculiar state could be produced; and I have assumed that as about three minutes pass between the first charge of app. i. and the division, and three minutes between the division and discharge, when the force of the non-transferable electricity is measured, the contrary tendencies for those periods would keep that apparatus in a moderately steady and uniform condition for the latter portion of time.

1251. The particular action described occurs in the shell-lac of the stems, as well as in the *dielectric* used within the apparatus. It therefore constitutes a cause by which the outside of the stems may in some operations become charged with electricity, independent of the action of dust or carrying particles (1203.).

¶ v. *On specific induction, or specific inductive capacity.*

1252. I now proceed to examine the great question of specific inductive capacity, i. e. whether different dielectric bodies actually do possess any influence over the degree of induction which takes place through them. If any such difference should exist, it appeared to me not only of high importance in the further comprehension of the laws and results of induction, but an additional and very powerful argument for the theory I have ventured to put forth, that the whole depends upon a molecular action, in contradistinction to one at sensible distances.

The question may be stated thus: suppose A an electrified plate of metal suspended in the air, and B and C two exactly similar plates, placed parallel to and on each side of A at equal distances and uninsulated; A will then induce equally toward B and C. If in this position of the plates some other dielectric than air, as shell-lac, be introduced between A and C, will the induction between them remain the same? Will the relation of C and B to A be unaltered, notwithstanding the difference of the dielectrics interposed between them?[8]

1253. As far as I recollect, it is assumed that no change will occur under such variation of circumstances, and that the relations of B and C to A

---

[8] Refer for the practical illustration of this statement to the supplementary note commencing 1307, &c. —Dec. 1838.

1253. *it is assumed*: That is, according to the conventional theory of action at a distance, it is assumed that the charge induced by an electrified body upon a neighboring surface depends only on the distance between them, and not on the intervening material.

depend entirely upon their distance. I only remember one experimental illustration of the question, and that is by Coulomb,[9] in which he shows that a wire surrounded by shell-lac took exactly the same quantity of electricity from a charged body as the same wire in air. The experiment offered to me no proof of the truth of the supposition: for it is not the mere films of dielectric substances surrounding the charged body which have to be examined and compared, but the *whole mass* between that body and the surrounding conductors at which the induction terminates. Charge depends upon induction (1171. 1178.); and if induction is related to the particles of the surrounding dielectric, then it is related to *all* the particles of that dielectric inclosed by the surrounding conductors, and not merely to the few situated next to the charged body. Whether the difference I sought for existed or not, I soon found reason to doubt the conclusion that might be drawn from Coulomb's result; and therefore had the apparatus made, which, with its use, has been already described (1187, &c.), and which appears to me well suited for the investigation of the question.

1254. Glass, and many bodies which might at first be considered as very fit to test the principle, proved exceedingly unfit for that purpose. Glass, principally in consequence of the alkali it contains, however well warmed and dried it may be, has a certain degree of conducting power upon its surface, dependent upon the moisture of the atmosphere, which renders it unfit for a test experiment. Resin, wax, naphtha, oil of turpentine, and many other substances were in turn rejected, because of a slight degree of conducting power possessed by them; and ultimately shell-lac and sulphur were chosen, after many experiments, as the dielectrics best fitted for the investigation. No difficulty can arise in perceiving how the possession of a feeble degree of conducting power tends to make a body produce effects, which would seem to indicate that it had a greater capability of allowing induction through it than another

---

[9] Mémoires de l'Académie, 1787, pp. 452, 453.

1254. *Glass:* Since glass is regarded as a good insulator, one would expect it to be a suitable material in which to study induction. But glass turns out not to be a good choice after all, since its *surface* tends to be significantly conductive.

No difficulty can arise...: It seems clear that in any study of induction that assumes perfect insulation, the presence of significant conductivity would cause *some* erroneous results. But why would it present the appearance of *enhanced* induction, specifically? In paragraph 1214 above, Faraday identified the "capacity for induction" of a body as being *inversely related* to the tension that sustains a given quantity of electricity in that body. Since conductors cannot sustain electric tension, any degree of conductivity in a dielectric will result in reduced tension and so present an appearance of increased induction.

body perfect in its insulation. This source of error has been that which I have found most difficult to obviate in the proving experiments.

*   *   *

1256. *Shell-lac and air* were compared in the first place. For this purpose a thick hemispherical cup of shell-lac was introduced into the lower hemisphere of one of the inductive apparatus (1187, &c.), so as nearly to fill the lower half of the space *o, o* (fig. 104.) between it and the inner ball; and then charges were divided in the manner already described (1198. 1207.), each apparatus being used in turn to receive the first charge before its division by the other. As the apparatus were known to have equal inductive power when air was in both (1209. 1211.), any differences resulting from the introduction of the shell-lac would show a peculiar action in it, and if unequivocally referable to a specific inductive influence, would establish the point sought to be sustained. I have already referred to the precautions necessary in making the experiments (1199, &c.); and with respect to the error which might be introduced by the assumption of the peculiar state, it was guarded against, as far as possible, in the first place, by operating quickly (1248.); and, afterwards, by using that dielectric as glass or sulphur, which assumed the peculiar state most slowly, and in the least degree (1239. 1241.).

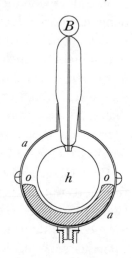

1257. The shell-lac hemisphere was put into app. i., and app. ii. left filled with air. The results of an experiment in which the charge through air was divided and reduced by the shell-lac app. were as follows:

App. i. Lac		App. ii. Air	
	Balls 255°		
0°	. . . .		
	. . . .	304°	
	. . . .	297	
	Charge divided.		
113	. . . .		
	. . . .	121	
0	. . . .		after being discharged
	. . . .	7	after being discharged

1256. The sketch depicts the apparatus with shell-lac hemisphere installed.

the peculiar state: here, the state resulting in a *return charge*.

1258. Here 297°, minus 7°, or 290°, may be taken as the divisible charge of app. ii. (the 7° being fixed stem action (1203. 1232.)), of which 145° is the half. The lac app. i. gave 113° as the power or tension it had acquired after division; and the air app. ii. gave 121°, minus 7°, or 114°, as the force it possessed from what it retained of the divisible charge of 290°. These two numbers should evidently be alike, and they are very nearly so, indeed far within the errors of experiment and observation. But these numbers differ very much from 145°, or the force which the half charge would have had if app. i. had contained air instead of shell-lac; and it appears that whilst in the division the induction through the air has lost 176° of force, that through the lac has only gained 113°.

1259. If this difference be assumed as depending entirely on the greater facility possessed by shell-lac of allowing or causing inductive action through its substance than that possessed by air, then this capacity for electric induction would be inversely as the respective loss and gain indicated above; and assuming the capacity of the air apparatus as 1, that of the shell-lac apparatus would be $176/113$ or 1.55.

1260. This extraordinary difference was so unexpected in its amount, as to excite the greatest suspicion of the general accuracy of the experiment, though the perfect discharge of app. i. after the division, showed that the 113° had been taken and given up readily. It was evident that, if it really existed, it ought to produce corresponding effects in the reverse order; and that when induction through shell-lac was converted into induction through air, the force or tension of the whole ought to be *increased*. The

---

1258. *fixed stem action*: That is, electrification of the shell-lac stem, as noted earlier in the comments to paragraphs 1205 and 1255 above. The 7° figure does not represent return charge in *air*, the principal dielectric in this apparatus, since air has been shown to be free of such residual effects.

These two numbers: 113° and 114° represent the tensions of the respective devices (corrected as necessary for stem electrification) after momentary connection. They "should evidently be alike" because when the two devices were joined, they must necessarily have been brought to the same degree of tension.

induction through the air has lost 176° of force, that through the lac has only gained 113°: When the devices were connected, some of the electricity residing in app. ii (air) passed to app. i (lac), thereby lowering the tension of app. ii from 290° to 114° (corrected values) and raising the tension of app. i from 0° to 113°. Now 290° minus 114° equals 176°; thus a quantity of electricity that was sustained in air by a tension of 176°, is sustained in shell-lac by a tension of only 113°.

1260. *induction through shell-lac … converted into induction through air…*: Faraday here undertakes a reversal of the previous experiment. Since electricity when transferred from an air environment to a shell-lac one occasioned a *reduction* in tension, the transfer from shell-lac to air should necessitate an *increase* in tension.

app. i. was therefore charged in the first place, and its force divided
with app. ii. The following were the results:

App. i. Lac		App. ii. Air
	. . . .	0°
215°	. . . .	
204	. . . .	
	Charge divided.	
	. . . .	118
118	. . . .	
	. . . .	0 after being discharged
0	. . . .	after being discharged

1261. Here 204° must be the utmost of the divisible charge. The
app. i. and app. ii. present 118° as their respective forces; both now
much *above* the half of the first force, or 102°, whereas in the former
case they were below it. The lac app. i. has lost only 86°, yet it has given
to the air app. ii. 118°, so that the lac still appears much to surpass the
air, the capacity of the lac app. i. to the air app. ii. being as 1.37 to 1.

1262. The difference of 1.55 and 1.37 as the expression of the
capacity for the induction of shell-lac seems considerable, but is in
reality very admissible under the circumstances, for both are in error
in *contrary directions*. Thus in the last experiment the charge fell from
215° to 204° by the joint effects of dissipation and absorption (1192.
1250.), during the time which elapsed in the electrometer operations,
between the applications of the carrier ball required to give those two
results. Nearly an equal time must have elapsed between the application of the carrier which gave the 204° result, and the division of the
charge between the two apparatus; and as the fall in force progressively
decreases in amount (1192.), if in this case it be taken at 6° only, it will
reduce the whole transferable charge at the time of division to 198°
instead of 204°; this diminishes the loss of the shell-lac charge to 80°
instead of 86°; and then the expression of specific capacity for it is
increased, and, instead of 1.37, is 1.47 times that of air.

1263. Applying the same correction to the former experiment in
which air was *first* charged, the result is of the *contrary* kind. No shell-lac
hemisphere was then in the apparatus, and therefore the loss would be

---

1261. *the capacity of the lac ... being as 1.37 to 1:* that is, the ratio of 118 (tension
in air) to 86 (tension in lac). Thus the "capacity for induction" of a specific
material is *inversely proportional* to the tension that sustains a given quantity of
electricity in that material. Compare paragraph 1214 above.

principally from dissipation, and not from absorption: hence it would be nearer to the degree of loss shown by the numbers 304° and 297°, and being assumed as 6° would reduce the divisible charge to 284°. In that case the air would have lost 170°, and communicated only 118° to the shell-lac; and the relative specific capacity of the latter would appear to be 1.50, which is very little indeed removed from 1.47, the expression given by the second experiment when corrected in the same way.

* * *

1266. These four expressions of 1.47, 1.50, 1.55, and 1.49 for the power of the shell-lac apparatus, through the different variations of the experiment, are very near to each other; the average is close upon 1.5, which may hereafter be used as the expression of the result. It is a very important result; and, showing for this particular piece of shell-lac a decided superiority over air in allowing or causing the act of induction, it proved the growing necessity of a more close and rigid examination of the whole question.

1267. The shell-lac was of the best quality, and had been carefully selected and cleaned; but as the action of any conducting particles in it would tend, virtually, to diminish the quantity or thickness of the dielectric used, and produce effects as if the two inducing surfaces of the conductors in that apparatus were nearer together than in the one with air only, I prepared another shell-lac hemisphere, of which the material had been dissolved in strong spirit of wine, the solution filtered, and then carefully evaporated. This is not an easy operation, for it is difficult to drive off the last portions of alcohol without injuring the lac by the heat applied; and unless they be dissipated, the substance left conducts too well to be used in these experiments. I prepared two hemispheres this way, one of them unexceptionable; and with it I repeated the former experiments with all precautions.

1266. *These four expressions*: In paragraphs 1264–1265, here omitted, he interchanges the devices and carries out the same measurements, which (corrected as above) yield two additional values. The four numbers obtained in all are the "four expressions" cited.

the average is close upon 1.5, … showing for this particular piece of shell-lac a decided superiority over air: An apparatus only partially filled with shell-lac developed more induction at a given tension than an identical apparatus containing air alone. Hence shell-lac *as such* must be superior to air *as such*, with respect to capacity for induction. But the ratio "1.5" represents unequal volumes of shell-lac and air; it would be different if the shell-lac volume were different, and in that sense it applies only to "this particular piece." In paragraph 1270 Faraday will estimate the ratio of capacities for *equal* volumes of shell-lac and air.

The results were exactly of the same kind; the following expressions for the capacity of the shell-lac apparatus, whether it were app. i. or ii., being given directly by the experiments, 1.46, 1.50, 1.52, 1.51; the average of these and several others being very nearly 1.5.

1268. As a final check upon the general conclusion, I then actually brought the surfaces of the air apparatus, corresponding to the place of the shell-lac in its apparatus, nearer together, by putting a metallic lining into the lower hemisphere of the one not containing the lac (1213.). The distance of the metal surface from the carrier ball was in this way diminished from 0.62 of an inch to 0.435 of an inch, whilst the interval occupied by the lac in the other apparatus remained 0.62 of an inch as before. Notwithstanding this change, the lac apparatus showed its former superiority; and whether it or the air apparatus was charged first, the capacity of the lac apparatus to the air apparatus was by the experimental results as 1.45 to 1.

1269. From all the experiments I have made, and their constant results, I cannot resist the conclusion that shell-lac does exhibit a case of *specific inductive capacity*. I have tried to check the trials in every way, and if not remove, at least estimate, every source of error. That the final result is not due to common conduction is shown by the capability of the apparatus to retain the communicated charge; that it is not due to the conductive power of inclosed small particles, by which they could acquire a polarized condition as conductors, is shown by the effects of the shell-lac purified by alcohol; and, that it is not due to any influence of the charged state, formerly described (1250.), first absorbing and then evolving electricity, is indicated by the *instantaneous* assumption and discharge of those portions of the

---

1268. After reducing the air space, Faraday finds that the capacity ratio of lac apparatus to air apparatus *decreases* from about 1.5 to a consistent 1.45. Decreasing the air distance has thus *increased* the capacity of the air apparatus. Faraday drew a similar inference in paragraphs 1213-1214; and in general, decreasing the distance between inducing surfaces (the dielectric material remaining unchanged) will increase their mutual capacity for induction.

1269. *it is not due to the conductive power of inclosed small particles...*: Faraday excluded metallic contamination of the shell-lac by the means described in paragraph 1267 above.

it is not due to any influence of the charged state...: Nor are the results simply a side-effect of the return charge, for induction measurements can be made quickly, whereas the return charge phenomena develop only slowly. To be sure, the phenomenon of return charge affects the *accuracy* of the induction measurements—but it cannot be considered responsible for *producing* the whole measured effect.

power which are concerned in the phenomena, that instantaneous effect occurring in these cases, as in all others of ordinary induction by charged conductors. The latter argument is the more striking in the case where the air apparatus is employed to divide the charge with the lac apparatus, for it obtains its portion of electricity in an *instant,* and yet is charged far above the *mean.*

1270. Admitting for the present the general fact sought to be proved; then 1.5, though it expresses the capacity of the apparatus containing the hemisphere of shell-lac, by no means expresses the relation of lac to air. The lac only occupies one half of the space *o, o,* of the apparatus containing it, through which the induction is sustained; the rest is filled with air, as in the other apparatus; and if the effect of the two upper halves of the globes be abstracted, then the comparison of the shell-lac powers in the lower half of the one, with the power of the air in the lower half of the other, will be as 2:1; and even this must be less than the truth, for the

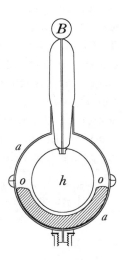

induction of the upper part of the apparatus, i. e. of the wire and ball B. (fig. 104.) to external objects, must be the same in both, and considerably diminish the difference dependent upon, and really producible by, the influence of the shell-lac within.

*　　*　　*

1283. A most interesting class of substances, in relation to specific inductive capacity, now came under review, namely, the gases or aeriform bodies. These are so peculiarly constituted, and are bound together by so many striking physical and chemical relations, that I expected some remarkable results from them: air in various states was selected for the first experiments.

1284. *Air, rare and dense.* Some experiments of division (1208.) seemed to show that dense and rare air were alike in the property

---

the case where the air apparatus is employed to divide the charge with the lac apparatus: That is, the case where the uncharged air apparatus receives electrification from the charged shell-lac apparatus. Faraday described such an instance in paragraphs 1260-1261 above.

1270. The sketch depicts the apparatus with shell-lac hemisphere installed.

2:1; and even this must be less than the truth…: Later investigators will put the specific inductive capacity of shell-lac at about 2.74.

under examination. A simple and better process was to attach one of the apparatus to an air pump, to charge it, and then examine the tension of the charge when the air within was more or less rarefied. Under these circumstances it was found, that commencing with a certain charge, that charge did not change in its tension or force as the air was rarefied, until the rarefaction was such that *discharge* across the space *o, o* (fig. 104.) occurred. This discharge was proportionate to the rarefaction; but having taken place, and lowered the tension to a certain degree, that degree was not at all affected by restoring the pressure and density of the air to their first quantities.

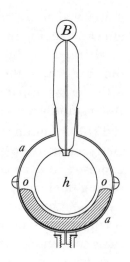

inches of mercury.

Thus at a pressure of	30	the charge was	88°
Again	30	the charge was	88
Again	30	the charge was	87
Reduced to	14	the charge was	87
Raised again to . . .	30	the charge was	86
Being now reduced to	3.4	the charge fell to . . .	81
Raised again to . . .	30	the charge was still . . .	81

1285. The charges were low in these experiments, first that they might not pass off at low pressure, and next that little loss by dissipation might occur. I now reduced them still lower, that I might rarefy further, and for this purpose in the following experiment used a measuring interval in the electrometer of only 15° (1185.). The pressure of air within the apparatus being reduced to 1.9 inches of mercury, the charge was found to be 29°; then letting in air till the pressure was 30 inches, the charge was still 29°.

1285. *pass off at low pressure*: As described in the previous paragraph, the degree of electrical tension that air can sustain is reduced when the pressure is reduced. In order to experiment at lower pressures ("rarefy still further"), therefore, Faraday must use smaller charges. This in turn demands a more sensitive electrometer, which he achieves by reducing the separation between the electrometer spheres (to 15° instead of 30°). When the air pressure is again increased, there is no change in the electrometer reading. Thus air's capacity for induction is not affected by pressure, even though its maximum sustainable electrical *tension* depends strongly on pressure.

1286. These experiments were repeated with pure oxygen with the same consequences.

1287. This result of no *variation* in the electric tension being produced by variation in the density or pressure of the air, agrees perfectly with those obtained by Mr. Harris,[10] and described in his beautiful and important investigations contained in the Philosophical Transactions; namely that induction is the same in rare and dense air, and that the divergence of an electrometer under such variations of the air continues the same, provided no electricity pass away from it. The effect is one entirely independent of that power which dense air has of causing a higher charge to be *retained* upon the surface of conductors in it than can be retained by the same conductors in rare air; a point I propose considering hereafter.

1288. I then compared *hot and cold air* together, by raising the temperature of one of the inductive apparatus as high as it could be without injury, and then dividing charges between it and the other apparatus containing cold air. The temperatures were about 50° and 200°. Still the power or capacity appeared to be unchanged; and when I endeavoured to vary the experiment, by charging a cold apparatus and then warming it by a spirit lamp, I could obtain no proof that the inductive capacity underwent any alteration.

1289. I compared *damp and dry air* together, but could find no difference in the results.

1290. *Gases.* A very long series of experiments was then undertaken for the purpose of comparing *different gases* one with another. They were all found to insulate well, except such as acted on the shell-lac of the supporting stem; these were chlorine, ammonia, and muriatic acid. They were all dried by appropriate means before being

[10] Philosophical Transactions, 1834, pp. 223, 224, 237, 244.

1287 *The effect is one entirely independent of that power which dense air has...*: We must distinguish between air's *capacity for induction*—which is not measurably affected by atmospheric pressure—and its dielectric *strength*, which generally increases with increased pressure (and also with greatly reduced pressure: vacuum is the strongest dielectric of all!). Inductive capacity is determined by the quantity of electrification a given conductor exhibits at a given degree of tension. Dielectric strength is indicated by how much tension the air can sustain before giving way to spark or other disruptive discharge.

1290. *these were chlorine, ammonia, and muriatic acid*: That is, these are the gases which "acted on the shell-lac of the supporting stem."

introduced into the apparatus. It would have been sufficient to have compared each with air; but, in consequence of the striking result which came out, namely, that *all had the same power of* or *capacity for,* sustaining induction through them, (which perhaps might have been expected after it was found that no variation of density or pressure produced any effect,) I was induced to compare them, experimentally, two and two in various ways, that no difference might escape me, and that the sameness of result might stand in full opposition to the contrast of property, composition, and condition which the gases themselves presented.

1291. The experiments were made upon the following pairs of gases.

    1.   Nitrogen and . . . . . Oxygen.
    2.   Oxygen . . . . Air.
    3.   Hydrogen . . . . Air.
    4.   Muriatic acid gas . . Air.

1290, contined. *all [gases] had the same power of … sustaining induction*: By contrast, solids exhibit characteristic *differences* in their inductive capacities. Shell-lac had about twice that of air (paragraph 1270); glass and sulphur, in measurements omitted from this selection, exhibited capacities of about $7/4$ and $5/4$ that of air, respectively. Note that for Faraday, such inequality is to be expected; for if the particles of the intervening medium play an essential role, then the capacity of two electrified surfaces for induction should specifically depend upon the material that is introduced between them. Yet gases apparently show no such specificity!

no variation of density … produced any effect: As Faraday reported in paragraphs 1284–1286, a charged apparatus containing air or oxygen did not change its tension, even when the apparatus was exhausted to $1/16$ of normal atmospheric pressure and the number of gas particles was thereby reduced proportionately.

But how can removing $15/16$ of the gas particles fail to affect a property which, on Faraday's view, depends *intimately* on the particles? Here is a possible explanation. Suppose the tension in a gaseous medium is distributed over its *entire volume,* particles and intervening spaces alike. If so the gas particles, representing such a small fraction of the total gas volume, would bear a correspondingly small share—undetectable perhaps—of the electric tension between mutually inducing surfaces; by far the major part of the electric tension in a gas would be sustained across the *spaces* between its particles. Any subsequent decrease in the number of gas particles, therefore, would only reduce their contribution further and so cause no measurable change in the inductive capacity of the gas.

might have been expected: If, as suggested, the particles of all gases bear an unmeasurably small share of electrical tension, then we would *expect* gases to be indistinguishable with respect to inductive capacity—not only from one another, but from *vacuum* as well. In paragraph 1293 below, Faraday will acknowledge the possibility that the differences are simply too small to be detected with his present apparatus.

5.	Oxygen	Hydrogen.
6.	Oxygen	Carbonic acid.
7.	Oxygen	Olefiant gas.
8.	Oxygen	Nitrous gas.
9.	Oxygen	Sulphurous acid.
10.	Oxygen	Ammonia.
11.	Hydrogen	Carbonic acid.
12.	Hydrogen	Olefiant gas.
13.	Hydrogen	Sulphurous acid.
14.	Hydrogen	Fluo-silicic acid.
15.	Hydrogen	Ammonia.
16.	Hydrogen	Arseniuretted hydrogen.
17.	Hydrogen	Sulphuretted hydrogen.
18.	Nitrogen	Olefiant gas.
19.	Nitrogen	Nitrous gas.
20.	Nitrogen	Nitrous oxide.
21.	Nitrogen	Ammonia.
22.	Carbonic oxide ...	Carbonic acid.
23.	Carbonic oxide ...	Olefiant gas.
24.	Nitrous oxide	Nitrous gas.
25.	Ammonia	Sulphurous acid.

1292. Notwithstanding the striking contrasts of all kinds which these gases present of property, of density, whether simple or compound, anions or cathions (665.), of high or low pressure (1284. 1286.), hot or cold (1288.), not the least difference in their capacity to favour or admit electrical induction through them could be perceived. Considering the point established, that in all these gases induction takes place by an action of contiguous particles, this is the more important, and adds one to the many striking relations which hold between bodies having the gaseous condition and form. Another equally important electrical relation, which will be examined in the next paper,[11] is that which the different gases have to each other at the *same pressure* of causing the retention of the *same or different degrees of charge* upon conductors in them. These two results appear to bear importantly upon the subject of electrochemical excitation and decomposition; for as *all* these phenomena, different as they seem to be, must depend upon the electrical forces of the particles of matter, the very distance at which they seem to stand from each other will do much, if properly considered, to illustrate the principle by which they are held in one common bond, and subject, as they must be, to one common law.

[11] See in relation to this point 1382. &c. —*Dec.* 1838.

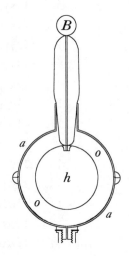

1293. It is just possible that the gases may differ from each other in their specific inductive capacity, and yet by quantities so small as not to be distinguished in the apparatus I have. It must be remembered, however, that in the gaseous experiments the gases occupy all the space *o, o,* (fig. 104.) between the inner and the outer ball, except the small portion filled by the stem; and the results, therefore, are twice as delicate as those with solid dielectrics.

1294. The insulation was good in all the experiments recorded, except Nos. 10, 15, 21, and 25, being those in which ammonia was compared with other gases. When shell-lac is put into ammoniacal gas its surface gradually acquires conducting power, and in this way the lac part of the stem within was so altered, that the ammonia apparatus could not retain a charge with sufficient steadiness to allow of division. In these experiments, therefore, the other apparatus was charged; its charge measured and divided with the ammonia apparatus by a quick contact, and what remained untaken away by the division again measured (1281.). It was so nearly one half of the original charge, as to

---

1293. *It is just possible that the gases may differ ... by quantities so small as not to be distinguished...*: Why "just possible"? It would seem to be not only a *strong* possibility but the very basis of Faraday's remark that the observed equality among gases might have been "expected," as I suggested in a comment to paragraph 1290. But now Faraday seems conspicuously lukewarm to the idea. Perhaps that is because the suggested explanation commits him to ascribing *power* to the spaces which supposedly surround gaseous particles—specifically, the *power to sustain tension.* In his note to paragraph 1164, Faraday acknowledged that "the particles do not touch"; but whether it is *space* that surrounds them, and whether that space can *act* in any sense, are questions that will trouble him throughout these Researches. Many phenomena, both electric and magnetic, invite him to attribute agency to space. But as often as he responds, he draws back. As late as the Twenty-fifth Series (paragraph 2787) Faraday will declare: "Mere space cannot act as matter acts."

the gases occupy all the space o, o, (fig. 104.): the sketch reproduces part of Faraday's Figure 104.

1294. *the other apparatus was charged*: Ammonia gas makes the surface of shell-lac conductive and the apparatus too leaky to retain a charge for very long. Since in Faraday's procedure the apparatus receiving the *initial* charge suffers the greatest time lapse between measurements, he gives the initial charge to "the other apparatus," which does not contain ammonia.

authorize, with this reservation, the insertion of ammoniacal gas amongst the other gases, as having equal power with them.

¶ vi. *General results as to induction.*

1295. Thus *induction* appears to be essentially an action of contiguous particles, through the intermediation of which the electric force, originating or appearing at a certain place, is propagated to or sustained at a distance, appearing there as a force of the same kind exactly equal in amount, but opposite in its direction and tendencies (1164.). Induction requires no sensible thickness in the conductors which may be used to limit its extent; an uninsulated leaf of gold may be made very highly positive on one surface, and as highly negative on the other, without the least interference of the two states whilst the inductions continue. Nor is it affected by the nature of the limiting conductors, provided time be allowed, in the case of those which conduct slowly, for them to assume their final state (1170.).

1296. But with regard to the *dielectrics* or insulating media, matters are very different (1167.). Their thickness has an immediate and important influence on the degree of induction. As to their quality, though all gases and vapours are alike, whatever their state; yet amongst solid bodies, and between them and gases, there are differences which prove the existence of *specific inductive capacities,* these differences being in some cases very great.

1297. The direct inductive force, which may be conceived to be exerted in lines between the two limiting and charged conducting surfaces, is accompanied by a lateral or transverse force equivalent to

---

1295. *time … to assume their final state*: Conductors can be stably electrified only on their surfaces (paragraph 1170); but a mediocre conductor requires *time* for all its electrification to move to the surface.

1296. *gases and vapours … whatever their state*: that is, their state as to *temperature and pressure.*

1297. *lateral or transverse force*: In paragraph 1224 Faraday interpreted the roughly parallel paths of adjacent lines of force—even as they negotiated curves and "turned a corner"—as expressing *mutual accomodation,* an equilibrium among individual lines' squeezing and pushing one another aside. This view has far-reaching consequences; for if lines of force are capable of nudging one another and shouldering one another aside, they are to be accounted as physical beings, not just symbolic or mnemonic devices. Moreover, if the lines resist mutual approach, then any region where they are forced to accumulate (such as a *corner* or *point*) will be a region of stress—or what Faraday will call a "high condition" in paragraph 1302 below.

a dilatation or repulsion of these representative lines (1224.); or the attractive force which exists amongst the particles of the dielectric in the direction of the induction is accompanied by a repulsive or a diverging force in the transverse direction (1304.).

1298. Induction appears to consist in a certain polarized state of the particles, into which they are thrown by the electrified body sustaining the action, the particles assuming positive and negative points or parts, which are symmetrically arranged with respect to each other and the inducing surfaces or particles.[12] The state must be a forced one, for it is originated and sustained only by *force*, and sinks to the normal or quiescent state when that force is removed. It can be *continued* only in insulators by the same portion of electricity, because they only can retain this state of the particles (1304).

[12] The theory of induction which I am stating does not pretend to decide whether electricity be a fluid or fluids, or a mere power or condition of recognized matter. That is a question which I may be induced to consider in the next or following series of these researches.

1298. *Induction appears to consist in a certain polarized state of the particles…*: Faraday ascribed a "polarized" condition to the particles of a material under induction in paragraph 1164 above. Now he specifies that condition more fully: the particles develop positive and negative extremities, which stretch out along the axis of negative and positive inducing powers. But what of the *spaces* surrounding the individual particles of a dielectric? Do they indeed sustain tension—the troublesome hypothesis (paragraph 1293, *comment*)? And if they do, must he attribute a "polarized condition" to those *spaces* as well?

the particles assuming positive and negative points or parts: The word *assuming* is essential. Faraday does not view the particles as possessing *permanent* positive and negative regions, which are subsequently displaced by attractions towards the negative and positive inducing surfaces. That would amount to a return, on the molecular level, to the idea of *attraction between electric substances (fluids)* as the fundamental electrical action—the very picture Faraday has been seeking to transcend. In paragraph 1304 below, Faraday will emphasize again that the polarized molecule "acquires opposite powers on different parts"—that is, as a manifestation, not a cause, of induction.

Comment on Faraday's note 12: It is a strangely cautious statement from one who seems already to have overturned the "fluid" theory completely! Can he *really* be so open-minded towards the fluid ideas, at this late stage? But see his opening sentences in the very next paragraph (1299). Readers may also pause to enjoy Faraday's mild pun ("The theory of induction … a question I may be induced to consider")—a light-hearted note he seldom permits himself, if indeed it is intentional here.

1299. The principle of induction is of the utmost generality in electric action. It constitutes charge in every ordinary case, and probably in every case; it appears to be the cause of all excitement, and to precede every current. The degree to which the particles are affected in this their forced state, before discharge of one kind or another supervenes, appears to constitute what we call *intensity*.

1300. When a Leyden jar is *charged,* the particles of the glass are forced into this polarized and constrained condition by the electricity of the charging apparatus. *Discharge* is the return of these particles to their natural state from their state of tension, whenever the two electric forces are allowed to be disposed of in some other direction.

1301. All charge of conductors is on their surface, because being essentially inductive, it is there only that the medium capable of sustaining the necessary inductive state begins. If the conductors are hollow and contain air or any other dielectric, still no *charge* can appear upon that internal surface, because the dielectric there cannot assume the polarized state throughout, in consequence of the opposing actions in different directions.

1299. *induction ... constitutes charge...*: Surely this means that "charge" is only an aspect, not an independent cause, of induction. Notwithstanding his earlier note to paragraph 1298, Faraday appears firmly to repudiate the fluid ideas.

The degree to which the particles are affected ... appears to constitute what we call intensity: Note that "intensity" so identified pertains to *individual particles*, suggesting a distinction from "tension," which pertains to the dielectric as a whole. The words "are affected" suggest that "intensity" does not refer to a force which merely tends to elongate or polarize the particles; rather it designates their *actual degree of polarization*.

1300. The terms "charge" and "discharge" are now fully freed from the fluid language. Note his lovely phrase "forces ... disposed of in some other direction," in place of the more customary phraseology, "charges are permitted to flow."

1301. *All charge of conductors is on their surface, because ... it is there only that the medium ... begins*: Since "charge" is the terminus of induction, it cannot reside within a solid conductor, for conductors cannot sustain induction. Nor can charge reside in the *interior surface* of a hollow conductor, because the enclosed dielectric "cannot assume the polarized state throughout." For example, if A were a positive region of the interior surface it would have to send lines of force into the enclosed space, and those lines would have to terminate at some negative region B of the same surface. But the lines of force are associated with electric *tension*; and hence the conductor joining A and B would have to sustain equal tension—like the string and bow discussed in the introduction. But that is impossible, since conductors cannot sustain tension. Note that explanations in terms of induction and lines of force make no use of "attraction" or "repulsion" as explanatory factors.

1302. The known influence of *form* is perfectly consistent with the corpuscular view of induction set forth. An electrified cylinder is more affected by the influence of the surrounding conductors (which complete the condition of charge) at the ends than at the middle, because the ends are exposed to a greater sum of inductive forces than the middle; and a point is brought to a higher condition than a ball, because by relation to the conductors around, more inductive force terminates on its surface than on an equal surface of the ball with which it is compared. Here too, especially, can be perceived the influence of the lateral or transverse force (1297.), which, being a

---

1302. *an electrified cylinder is more affected ... at the ends than at the middle*: The drawing shows a popular 19th-century demonstration of the phenomenon Faraday describes. The horizontal plate is strongly charged. A brass cylinder, supported above it by an insulating handle and equipped with miniature electric indicators, exhibits an electrified condition at its ends, but not in the middle. (The charge manifested at the lower end is *opposite* to the charge of the plate while that shown at the upper end is the *same* as that of the plate, thus indicating a case of charge by induction.) Evidently this favoring of the ends cannot be explained by mere proximity, since the upper end of the cylinder is actually further from the inducing plate and yet it develops  greater electrification than the middle regions. Rather, the *geometrical form itself* appears to play a role in determining the electrical state of each part of the body.

surrounding conductors (which complete the condition of charge):   Since a "charged" surface is but one terminus for lines of force, a second surface is also required to "complete" the condition by terminating the lines at their other ends.

a point is brought to a higher condition than a ball: Faraday gave an example of this phenomenon in paragraph 1230 and Figure 11, a portion of which is reproduced here. *C* consists of a pointed conductor mounted upon a smooth, grounded metal ball, and *A* represents the knob of a charged Leyden jar. When the two approach in the positions shown, the jar discharges to the *point*, not 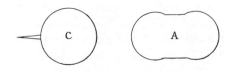 to the ball—as evidenced by a glow which surrounds the point. Thus the path of discharge favored the point over the ball, even though the point was further away, and masked by the ball! Here, too, form appears to play a determining role.

power of the nature of or equivalent to repulsion, causes such a disposition of the lines of inductive force in their course across the dielectric, that they must accumulate upon the point, the end of the cylinder, or any projecting part.

1303. The influence of *distance* is also in harmony with the same view. There is perhaps no distance so great that induction cannot take place through it;[13] but with the same constraining force (1298.) it takes place the more easily, according as the extent of dielectric through which it is exerted is lessened. And as it is assumed by the theory that the particles of the dielectric, though tending to remain in a normal state, are thrown into a forced condition during the induction; so it would seem to follow

---

[13] I have traced it experimentally from a ball placed in the middle of the large cube formerly described (1173.) to the side of the cube six feet distant, and also from the same ball placed in the middle of our large lecture-room to the walls of the room at twenty-six feet distance, the charge sustained upon the ball in these cases being solely due to induction through these distances.

---

the lines … must accumulate upon the point, the end of the cylinder, or any projecting part: Faraday implies that relatively sharp forms, like the ends of a rod or the point of a pin, naturally experience more inductive force than blunter surfaces. But why should that be? Faraday has not yet formulated principles sufficient to deduce the result strictly, but we can see this much: The lines of force manifest transverse action (paragraph 1224), nudging and pushing one another laterally. If, in the course of such mutual action, lines become displaced from the sides of a cylinder to its ends, or from a ball to a pinpoint, they must necessarily become more concentrated, as they are moving from a more spacious surface to a more confined one. If such a concentration of *lines* represents a concentration of *force*, then it will be clear why "more inductive force terminates" on a sharply curved surface than on an equal but more gently curved area.

1303. *influence of distance*: Note that on his view there is nothing especially significant about an inverse-square relation. It is merely a consequence of spherical symmetry, when such symmetry happens to obtain in a particular experimental setup; it is *not* fundamental to or inherent in electrification itself.

with the same constraining force … it takes place the more easily, according as the extent of dielectric through which it is exerted is lessened: Faraday's reference to paragraph 1298 does not wholly clarify his phrase "same constraining force." If (as is likely here) he refers to the *overall* tension sustained by a dielectric, then he means that *when the tension between electrified surfaces is held constant, the induction between them increases* ("takes place more easily") *as their distance is diminished*. Faraday showed a substantially identical result experimentally in paragraphs 1213–1214; with the next sentence he will offer a theoretical explanation.

that the fewer there are of these intervening particles opposing their tendency to the assumption of the new state, the greater degree of change will they suffer, i. e. the higher will be the condition they assume, and the larger the amount of inductive action exerted through them.

1304. I have used the phrases *lines of inductive force* and *curved lines* of force (1231. 1297. 1298. 1302.) in a general sense only, just as we speak of the lines of magnetic force. The lines are imaginary, and the force in any part of them is of course the resultant of compound forces, every molecule being related to every other molecule in *all* directions by the tension and reaction of those which are contiguous. The transverse force is merely this relation considered in a direction oblique to the lines of inductive force, and at present I mean no more than that by the phrase. With respect to the term *polarity* also, I mean at present only a disposition of force by which the same

---

the fewer there are of these intervening particles ... the higher will be the condition they assume: Faraday is considering the mutual approach of bodies so electrified as to maintain *constant tension between themselves.* Two metal spheres A and B, connected respectively to the inner conductor and the grounded outer foil of a charged Leyden jar, would present an example of such bodies, since the Leyden jar maintains an essentially constant tension between its inner and outer conductors so long as it does not experience any discharge. As the spheres approach one another, the tension between them is necessarily distributed over a decreasing volume of dielectric and consequently a decreasing number of particles. The tension sustained by *each particle* (or each equal portion of dielectric) increases; as Faraday expresses it, each will assume a "higher condition."

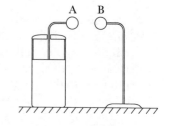

Note that the present case differs from that discussed in paragraph 1290. There Faraday strove to understand why the tension in a gas remained unchanged when its density, and hence the number of its particles, was diminished. In those experiments (paragraphs 1284–1286), the gas was confined between surfaces that were electrified *and insulated*; since the tension was subject to no constraints it was incumbent upon Faraday to try to explain its apparent constancy. But in the present case the electrified surfaces *are* constrained. The tension between them is imposed and maintained independently. Instead of being a *problem* as in paragraph 1290, the constancy of tension is here a *given*; for that reason Faraday's inference about the particles in the present case is not relevant to the earlier discussion; it neither contradicts nor substitutes for the suggestion that the particles of gases bear an inconsiderably small fraction of the overall dielectric tension.

1304. *The lines are imaginary*: This is cautious language, but elsewhere Faraday has spoken of the lines of force as though they were physical and causative, as well as explanatory, agents. Compare paragraphs 1224 and 1225, for example.

molecule acquires opposite powers on different parts. The particular way in which this disposition is made will come into consideration hereafter, and probably varies in different bodies, and so produces variety of electrical relations.[14] All I am anxious about at present is, that a more particular meaning should not be attached to the expressions used than I contemplate. Further inquiry, I trust, will enable us by degrees to restrict the sense more and more, and so render the explanation of electrical phenomena day by day more and more definite.

1305. As a test of the probable accuracy of my views, I have throughout this experimental examination compared them with the conclusions drawn by M. Poisson from his beautiful mathematical inquiries.[15] I am quite unfit to form a judgment of these admirable papers; but as far as I can perceive, the theory I have set forth and the results I have obtained are not in opposition to such of those conclusions as represent the final disposition and state of the forces in the limited number of cases he has considered. His theory assumes a very different mode of action in induction to that which I have ventured to support, and would probably find its mathematical test in the endeavour to apply it to cases of induction in curved lines. To my feeling it is insufficient in accounting for the retention of electricity upon the surface of conductors by the pressure of the air, an effect

[14] See now 1685. &c. —*Dec.* 1838.

[15] Mémoires de l'Institut, 1811, tom. xii. the first page 1, and the second paging 163.

1305. *His theory assumes a very different mode of action...*: It assumes inverse-square forces propagated in straight lines, as in Newton's gravity theory. But as I mentioned in a comment to paragraph 1303, Faraday finds nothing especially significant, and certainly nothing *fundamental,* in the inverse-square relation.

would probably find its mathematical test in the endeavour to apply it to cases of induction in curved lines: As Faraday stated in the previous paragraph, the lines are "of course the resultant of compound forces." Poisson's or any similar theory must be tested against experiment to see whether the forces *it* posits can account for all cases of curved lines. But so far Poisson has treated only a "limited number of cases."

insufficient in accounting for the retention of electricity...: Poisson and others supposed that electricity is subject to atmospheric pressure and for that reason dissipates more rapidly in a partial vacuum than under normal pressures (paragraph 1284); a liquid, similarly, evaporates more rapidly under low atmospheric pressure. But Faraday will offer an account that is not only simpler but also avoids the dubious "liquid" metaphor. He cites the relevant passages in his note 16 later in this paragraph.

which I hope to show is simple and consistent according to the present view;[16] and it does not touch voltaic electricity, or in any way associate it and what is called ordinary electricity under one common principle.

I have also looked with some anxiety to the results which that indefatigable philosopher Harris has obtained in his investigation of the laws of induction,[17] knowing that they were experimental, and having a full conviction of their exactness; but I am happy in perceiving no collision at present between them and the views I have taken.

1306. Finally, I beg to say that I put forth my particular view with doubt and fear, lest it should not bear the test of general examination, for unless true it will only embarrass the progress of electrical science. It has long been on my mind, but I hesitated to publish it until the increasing persuasion of its accordance with all known facts, and the manner in which it linked together effects apparently very different in kind, urged me to write the present paper. I as yet see no inconsistency between it and nature, but, on the contrary, think I perceive much new light thrown by it on her operations; and my next papers will be devoted to a review of the phenomena of conduction, electrolyzation, current, magnetism, retention, discharge, and some other points, with an application of the theory to these effects, and an examination of it by them.

Royal Institution,
November 16, 1837.

[Editor's note: A lengthy *Supplementary Note*, which Faraday later appended to this Series, is here omitted.]

[16] Refer to 1377, 1378, 1379, 1398. —*Dec.* 1838.

[17] Philosophical Transactions, 1834, p. 213.

1305, continued. *does not touch voltaic electricity*:   Faraday also criticizes Poisson's theory for failing to explain the identity of voltaic and ordinary electricities. This criticism is not justified; Poisson's theory, like all fluid theories, associates them as being *at rest* and *in motion,* respectively. While that view may or may not be a satisfactory one, it is inaccurate to say that the theory takes *no* view of the subject.

Twelfth Series — Editor's Introduction

The electric brush and glow

Spark, *brush*, and *glow* are three species of what Faraday calls "disruptive discharge." We are all familiar with the spark; and if you have access to a Wimshurst or other electric machine, you can probably produce the brush as well. It is best to try this in a darkened room, and on a dry day.

Separate the spherical discharge electrodes by 4 or 5 inches and operate the machine vigorously, but not enough to draw a spark. Look for a diverging radiance of pale blue light from one or both electrodes, probably accompanied by a hissing or crackling sound. This is the *brush*.

Under examination by Wheatstone's rotating mirror (described below), the brush proves to consist of innumerable spark-like branches, as shown in the drawing.* Its brightness and size depend to some extent on the size and shape of the discharging electrode. The brush from a large sphere tends to be larger and more diffuse, while that from a blunt wire is generally smaller but more distinct. At a positive electrode, the brush tends to be larger and more highly ramified—but it forms more readily at the negative electrode. Metallic dust is always torn away from an electrode that sustains the brush.

If the machine is fitted with pointed electrodes instead of blunt or rounded ones the brush does not form. Instead, a silent and continuous *glow* forms about the point. It is often possible to detect an accompanying "wind" as electrified air streams away from the electrode tip—it is these electrified air particles that actually emit the light. At a negative electrode the glow is sometimes separated from the electrode by a dark space, which Faraday called the "dark discharge" to emphasize that it was just as much a part of the discharge path as the more spectacular luminous region. The glow may transform into a brush if the stream of electrified air is interfered with; similarly the spark may degenerate into a brush if the electrode surface becomes soiled or pitted.

---

* After S. P. Thompson. Brush at a positive electrode is shown.

The foregoing phenomena characterize brush and glow in air at normal atmospheric pressure. The appearances change, and additional phenomena appear, in other gases and at lower pressures. Faraday studied such effects extensively, but I have not included those investigations in the present selections.

Induction and conduction

With the preceding Series Faraday began to outline a view of induction as a state of *tension* in a medium. No doubt the term is highly metaphorical. Nevertheless the *spark* and other phenomena of sudden discharge acquire a natural interpretation, on this idea, as the collapse or release of an antecedent condition of inductive tension. Partly based on this understanding of induction, and partly stemming from the earlier electrolysis experiments, Faraday now proposes to extend his treatment of *sudden discharge* to include *continuous conduction*. You will recall that the conventional language of current flow, with its undisguised metaphors of fluid transport, has never been satisfactory in Faraday's view; but there was no other comparably-developed terminology available. Now, though, he is in a position to replace the fluid image with a paradigm that reflects a much wider range of phenomena. If sudden discharge represents the precipitous collapse or release of electric tension, then may not a continuous current represent the *continual and successive collapse* of ever-renewed increments of electric tension?

Consider, for example, a single discharge through air between an electric machine's prime conductor and ground. Faraday reasons that when the conductor is electrified, the air will be thrown into a state of tension. Each particle of the air must then sustain a share of the overall tension which will not, in general, be equally distributed among them. As electrification increases, and tension in the individual particles builds up, eventually one particle will reach a critical "turning point" (paragraph 1370) or state of collapse, whereupon it becomes unable to sustain tension in any degree whatsoever. Its share will then be thrown upon an adjacent particle, which may as a result reach its own critical point, collapse, and transfer its share of the tension on to the next particle—and so on. Thus a wave of discharge will be propagated along the lines of tension in the air. This process will of course happen very quickly, being to all appearances sudden and even instantaneous; but according to the view here outlined the process is actually successive and temporal.

"Continuous" conduction would, on this treatment, merely amount to a *succession* of such discharges. Thus Faraday will argue for a kind of propagation of discharge, quite appropriately compared to a wave, along every "current-carrying" conductor.

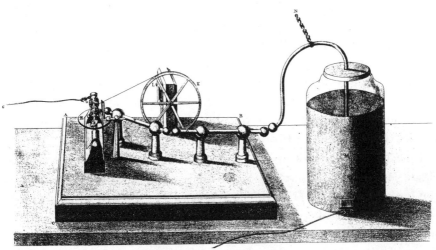

Wheatstone's revolving mirror

The velocity of discharge

Faraday's colleague and friend, the prolific inventor, engineer, and professor Charles Wheatstone,* had employed a high-speed rotating mirror to study the spark and other transient phenomena. In one experiment, the image of a spark was reflected from a rapidly rotating mirror to the eye of a viewer. If the flash were instantaneous its reflected image would be essentially identical to the spark seen directly; but if the discharge occupied sensible time its reflected image would appear elongated because of *persistence of vision*—that peculiarity of the eye which causes the path of a rapidly moving luminous point to appear as a continuous line. The reflected image was in fact elongated, showing that the spark occupied time—as much as $1/24{,}000$ second in one trial.**

* Wheatstone's hand and name appear in an amazingly rich variety of endeavors. He carried out important research in acoustics, musical instruments, and synthetic speech; he pursued studies of vision which led to his introduction of the stereoscope. Wheatstone designed and built new electrical generating machines and contributed greatly to the improvement of the telegraph. He developed precise methods in electrical measurement—the Wheatstone Bridge circuit is ubiquitous in its many variations (although Wheatstone did not actually originate the circuit). Together with all this he carried on the family musical instrument business for many years and during that time invented the Wheatstone concertina, pictured here, whose distinctive fingering pattern achieved such acceptance in England that it effectively lost the Wheatstone name to become universally known as the English concertina.

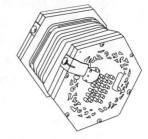

** By estimating the degree of elongation Wheatstone could reckon the angle through which the mirror must have rotated during reflection; he could then calculate the spark duration from the known rate of mirror rotation.

In a related experiment, Wheatstone interrupted a half-mile-long coil of copper wire with spark gaps at three points, two near each end and one in the middle. The gaps were aligned parallel to the axis of mirror rotation, so that if all sparks passed simultaneously their reflected images would match one another. But when discharges from a Leyden jar were synchronized with the rotations of the mirror, Wheatstone found that the middle spark took place some $1/1,152,000$ of a second later than the two end sparks.*

In Wheatstone's results Faraday found confirmation of his own views. He reasoned that when the charged Leyden jar is applied to the ends of the wire, (i) a tension develops first at the ends and subsequently propagates towards the middle, (ii) that shortly after the tension develops at a point it begins to collapse there, and (iii) that this collapse also propagates from both ends towards the middle of the wire. If the wire is interrupted by air gaps, the same propagation of tension and discharge occurs in the air as in the wire—though at a much slower rate. Thus in the experiment cited, the Leyden jar discharge would have traversed each copper wire in about $1/1,152,000$ second and each air gap in about $1/24,000$ second. Note that in Faraday's view, these are not times of *flow of electrical substance* through the copper and air, as fluid theories would assume, but times of *propagation of tension*.

Faraday makes the interesting argument that if the wire (together with its spark gaps) were a perfect insulator it would *never* discharge the Leyden jar, while if it were a poor insulator it would indeed discharge the jar, but slowly.** Therefore, he infers, it is reasonable that copper wire (a good conductor) interrupted by *small* spark gaps, should prove to discharge the jar as rapidly as it does—and a *perfect* conductor ought to discharge electrification *instantaneously*.

What property of a conductor determines the rapidity of discharge, that is, the speed of propagation of electric tension through it? At paragraph 1328 below Faraday will suggest that the speed depends on the degree of tension adjacent particles are capable of withstanding before they transfer their forces to one another. In good conductors this critical degree is presumably small, since if the tension does not build up much before transfer occurs, then each individual transfer will take place comparatively quickly—and the overall succession of transfers

---

* The reflected image of the middle spark was displaced by about half a degree from the end sparks, which would correspond to one-quarter degree of mirror rotation. At the impressive speed of 800 rotations or 288,000 degrees per second, the mirror would turn through a one-quarter-degree angle in $1/1,152,000$ second.

** As Faraday neatly puts it in paragraph 1328 below, retardation *is* insulation!

for a discharge through the whole wire will occupy a comparatively short time. Poorer conductors (better insulators), on the other hand, are presumably capable of sustaining greater inter-particle tensions; so that it takes longer for each succeeding particle to reach the critical point, and longer time overall for a discharge through the whole body. Similarly, increased lengths of any material would naturally result in slower discharge rates, since proportionally more numerous acts of transfer would have to take place.

Discharge through multiple paths

But what if multiple discharge paths are available, as in sheets or bulks of conductive materials? Will *all* of the discharge be directed along the quickest path, or will it be distributed among the various alternate routes? An exercise conducted in the present Series might seem to imply a single path of discharge. In paragraph 1350 Faraday writes:*

> Put into a glass vessel some clear rectified oil of turpentine, and introduce two wires passing through glass tubes where they coincide with the surface of the fluid, and terminating either in balls or points. Cut some very clean dry white silk into small particles, and put these also into the liquid: Then electrify one of the wires by an ordinary machine and discharge [to the earth] by the other. The silk will immediately gather from all parts of  the liquid, and form a band of particles reaching from wire to wire, and if touched by a glass rod will show considerable tenacity; yet the moment the supply of electricity ceases the band will fall away and disappear...

The tiny pieces of silk may be compared to iron filings, delineating "electrical curves" as the iron particles have been found to map magnetic curves. But instead of describing multiple paths, as iron filings do when scattered about a magnet, the silk threads form a single chain. Does this mean there exists only a *solitary line of force* in the oil? If so, one might conjecture that if the oil were a conductor it would sustain only a *single path of discharge.*

But from earlier experiments of Cavendish and others, Faraday already knows that conduction through surfaces—or volumes—is never confined to the shortest path but spreads out to energize the

---

* Faraday did not supply a drawing of his apparatus. The accompanying sketch is adapted from S. P. Thompson.

whole conductive space. Thus when a
voltaic battery is connected to points A
and B at the edges of a flat metal sheet,
discharge takes place along all the co-
terminous paths shown. If the current
flow is not too great, we can assume that
the paths of discharge are essentially
identical to the lines of induction—or
static electrical lines of force—that
would have existed between A and B if

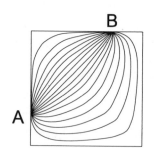

the metal sheet had been an insulator instead. Faraday has elsewhere
called this spreading tendency the "lateral action" of the lines of
force.*

What, then, is the true significance of the *single line* which the silk
threads form in Faraday's experiment? Faraday will take up that
question in the Twenty-first Series, and there offer a new interpretation
of the threads' pattern of migration.

Why does Faraday so seldom illustrate lines of force?

I have called attention to Faraday's characteristically visual idiom
several times already, and Faraday himself has made frequent and
effective use of diagrams in recording and interpreting a variety of
phenomena. But you may have noticed a certain frugality in his visual
representation of *lines of force*. At several places, and especially in the
present Series, Faraday describes the lines verbally, but conspicuously
refrains from illustrating them pictorially. I have supplied sketches in
some of those instances; but always with the nagging question, why
should Faraday have exhibited reluctance in depicting these, of all his
subjects perhaps the most visually inviting and suggestive?

Yet on reflection, it is not the lines of force that are actually made
visible. To be sure, their presence and disposition can often be made
evident by material indicators such as iron filings and compass needles
for magnetic lines, or bits of silk for electric lines; but the lines themselves remain, to an extent, objects of inference and imagination. That
is especially true of electrical lines of force, for which the delineation
techniques are sharply limited. Their courses must usually be deduced
piecemeal from surveys with the electrometer, as in the preceding
Series.

It is this quasi-inferential character, I think, that curbs Faraday in his
use of sketches, especially of electric lines. Though he is a powerful
proponent of the imagination, I sense in Faraday a persistent reluctance

---

* For example, at paragraph 1224 in the preceding Series.

to picture its contents.* Pictures, it almost seems, are for him *sacred to Fact.* When imaginative constructs are to be conveyed, Faraday employs his incomparable gift for verbal narrative instead.

How many lines?

The lines of force are *everywhere*—and this creates a more practical problem for both electric and magnetic diagrams. No matter which indicators Faraday uses—magnetic compass or iron filings, electrometer or silk threads—wherever one line of force is detected an indefinite number of neighboring lines will also be detected—in general, a line of force will be found wherever an indicator is placed! Clearly, then, there can be no such thing as "counting" the lines of force by these methods; and in any diagrams that are based on them, the number of lines drawn can claim no experimental significance.

Nevertheless, it is almost impossible to view certain line-of-force diagrams without feeling invited to *count.* The sketch shows two representations of an electrified sphere placed in the center of a cubical room—a configuration mentioned in paragraph 1378 below. Panel (a) is adapted from a respected turn-of-the-century textbook; its symmetrical and evenly-distributed lines strongly imply that both the electrification on the sphere, and the induction through the air, are symmetrically distributed.

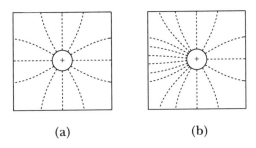

(a)   (b)

But the beautiful symmetry of that diagram depends on the investigator's having chosen angularly-symmetrical locations at which to begin the line-tracing process. Had the inquirer merely drawn a few additional lines, favoring positions on one side of the sphere, the result would be as depicted in panel (b). There the line *shapes* are consistent with those in (a) but their *number*—or rather, their numerical distribution—would seem to indicate irregular electrification and induction, contrary to (a). But it is the line shapes, not their numbers, that are significant in these diagrams. Indeed, an experienced viewer

---

* At least this seems to be his attitude towards *published* pictures. His laboratory Diary, by contrast, contains many imaginative sketches that never made their way to publication.

can frequently deduce the relative concentration of the lines *from* their shapes, at least roughly. In the present sketches, for example, it is not too hard to infer symmetry from the shapes of a few typical lines, whether they are drawn at equal intervals or not.

In the 28th Series Faraday will develop a device—or rather a class of devices he refers to collectively as "the Moving Wire"—that will permit magnetic lines of force, at least, to be *counted* almost literally. But until then, whenever I supply diagrams representing lines of force I will try not to assert any conclusions based on their numerical properties alone.

TWELFTH SERIES.

§ 18. *On Induction (continued).* ¶ vii. *Conduction, or conductive discharge.* ¶ ix. *Disruptive discharge—Insulation—Spark*

Received January 11,—Read February 8, 1838.

1318. I PROCEED now, according to my promise, to examine, by the great facts of electrical science, that theory of induction which I have ventured to put forth (1165. 1295. &c.). The principle of induction is so universal that it pervades all electrical phenomena; but the general case which I purpose at present to go into consists of insulation traced into and terminating with discharge, with the accompanying effects. This case includes the various *modes* of discharge, and also the condition and characters of a current; the elements of magnetic action being amongst the latter. I shall necessarily have occasion to speak theoretically, and even hypothetically; and though these papers profess to be experimental researches, I hope that, considering the facts and investigations contained in the last series in support of the particular view advanced, I shall not be considered as taking too much liberty on the present occasion, or as departing too far from the character which they ought to have, especially as I shall use every opportunity which presents itself of returning to that strong test of truth, experiment.

1319. Induction has as yet been considered in these papers only in cases of insulation; opposed to insulation is *discharge*. The action or effect which may be expressed by the general term *discharge*, may take place, as far as we are aware at present, in several modes. Thus, that which is called simply *conduction* involves no chemical action, and apparently no displacement of the particles concerned. A second mode may be called *electrolytic discharge;* in it chemical action does occur, and particles must, to a certain degree, be displaced. A third mode, namely, that by sparks or brushes, may, because of its violent displacement of the particles of the *dielectric* in its course, be called

---

1319. *sparks or brushes*: Both of these are instances of what Faraday calls "disruptive discharge." The electric *brush*, along with its close relative the *glow*, is described in the introduction to the present Series.

the *disruptive discharge*; and a fourth may, perhaps, be conveniently distinguished for a time by the words *convection*, or *carrying discharge*, being that in which discharge is effected either by the carrying power of solid particles, or those of gases and liquids. Hereafter, perhaps, all these modes may appear as the result of one common principle, but at present they require to be considered apart; and I will now speak of the *first* mode, for amongst all the forms of discharge that which we express by the term conduction appears the most simple and the most directly in contrast with insulation.

¶ vii. *Conduction, or conductive discharge.*

1320. Though assumed to be essentially different, yet neither Cavendish nor Poisson attempt to explain by, or even state in, their theories, what the essential difference between insulation and conduction is. Nor have I anything, perhaps, to offer in this respect, *except* that, according to my view of induction, insulation and conduction depend upon the same molecular action of the dielectrics concerned; are only extreme degrees of *one common condition* or effect; and in any sufficient mathematical theory of electricity must be taken as cases of the same kind. Hence the importance of the endeavour to show the connection between them under my theory of the electrical relations of contiguous particles.

1321. Though the action of the insulating dielectric in the charged Leyden jar, and that of the wire in discharging it, may seem very different, they may be associated by numerous intermediate links, which carry us on from one to the other, leaving, I think, no necessary connection unsupplied. We may observe some of these in succession for information respecting the whole case.

1322. Spermaceti has been examined and found to be a dielectric, through which induction can take place (1240. 1246.), its specific inductive capacity being about or above 1.8 (1279.), and the inductive action has been considered in it, as in all other substances, an action of contiguous particles.

1321. *associated by numerous intermediate links*: Faraday argues for the essential relatedness of two characteristics by exhibiting a succession of gradations between them. In paragraphs 1322–1325 he cites: (i) ordinary water, (ii) spermaceti, (iii) glass and shell-lac, (iv) distilled water and ice—which delineates a gradual transition from *moderate conductivity* to *excellent insulation*. Therefore, he will infer, "ordinary insulation and conduction are closely associated together or rather are extreme cases of one common condition" (paragraph 1324).

1323. But spermaceti is also a *conductor*, though in so low a degree that we can trace the process of conduction, as it were, step by step through the mass (1247.); and even when the electric force has travelled through it to a certain distance, we can, by removing the coercitive (which is at the same time the inductive) force, cause it to return upon its path and reappear in its first place (1245. 1246.). Here induction appears to be a necessary preliminary to conduction. It of itself brings the contiguous particles of the dielectric into a certain condition, which, if retained by them, constitutes *insulation*, but if lowered by the communication of power from one particle to another, constitutes *conduction*.

1324. If *glass* or *shell-lac* be the substances under consideration, the same capabilities of suffering either induction or conduction through

---

1323. *we can trace the process of conduction ... step by step*: Faraday is referring to the "return charge" reported in the preceding Series. In that phenomenon a slow and incomplete conduction actually reversed itself after an applied electric force was removed. He now sees that reversal as revealing something very important about the conduction process at its very beginning—midway between the *initial inductive condition*, on the one hand, and a *completed and irreversible discharge*, on the other: evidently the applied force has to overcome an opposing tendency even in the earliest moments of conduction. Extending that argument back to the beginning, then, the applied force must be a cause *prior* to the conduction itself.

when the electric force has traveled ... to a certain distance: Here is an advance over the seemingly fluid-tainted language with which Faraday described the phenomenon in paragraph 1245 above. Now he makes it clear that what moves through the conductor is not a puddle of electric fluid; rather, *force* (the condition of tension) is what moves. The mechanism of electric force transfer in a conductor is discussed in the introduction; Faraday will outline its particulars in paragraph 1338 below.

induction ... a necessary preliminary to conduction: In the preceding Series Faraday made an analogous point. In electrolysis, he reported, "*induction* appeared to be the *first* step, and *decomposition* the *second*..." (paragraph 1164). But the present locution of "a necessary preliminary" goes further. The connection between induction and conduction is now more intimate than chronological order or even ordinary cause and effect. Faraday sees the inductive condition as a state of *readiness* of which conduction is the *fulfillment*. In the next paragraph he will begin to approach this relation through a metaphor of tension and its discharge.

coercitive force: here, the applied electrical force. The term may be misleading because "coercitive force" was also the name for a hypothetical power some theorists invoked as opposing increase or decrease of magnetization in materials like hard steel. Faraday is not referring to that conjectural force here. Rather, he speaks of "coercion" whenever a change is effected against a continuing *restorative* tendency. A good example of this appears at paragraph 3222 in the 29th Series.

them appear (1233. 1239. 1247.), but not in the same degree. The conduction almost disappears (1239. 1242.); the induction therefore is sustained, *i. e.* the polarized state into which the inductive force has brought the contiguous particles is retained, there being little discharge action between them, and therefore the *insulation* continues. But, what discharge there is, appears to be consequent upon that condition of the particles into which the induction throws them; and thus it is that ordinary insulation and conduction are closely associated together or rather are extreme cases of one common condition.

1325. In ice or water we have a better conductor than spermaceti, and the phenomena of induction and insulation therefore rapidly disappear, because conduction quickly follows upon the assumption of the inductive state. But let a plate of cold ice have metallic coatings on its sides, and connect one of these with a good electrical machine in work, and the other with the ground, and it then becomes easy to observe the phenomena of induction through the ice, by the electrical tension which can be obtained and continued on both the coatings (419. 426.). For although that portion of power which at one moment gave the inductive condition to the particles is at the next lowered by the consequent discharge due to the conductive act, it is succeeded by another portion of force from the machine to restore the inductive state. If the ice be converted into water the same succession of actions can be just as easily proved, provided the water be distilled, and (if the machine be not powerful enough) a voltaic battery be employed.

1324. *discharge ... appears to be consequent upon that condition of the particles into which the induction throws them*:   By means of what phenomena does it so "appear?" The indications will be summarized in the next paragraph, which also describes more fully the presumed relation between induction and consequent discharge.

1325. *in work*:   that is, in action.

the same succession of actions: that is, *initial induction, subsequent discharge*, and *renewed induction*. In a cell with plate electrodes and filled with distilled water, connect the electrodes, momentarily only, to a voltaic battery. By the *initial induction* (through the water) the electrodes develop a charge, which can be demonstrated by connecting them to an electrometer. Their *discharge* (through the water) is shown in that the initial charge immediately begins to decline— indeed more rapidly than if the cell had been filled with air alone. Finally *renewed induction* is confirmed thus: Reattach the battery to the plates; the electrometer will maintain a steady reading for so long as the battery remains connected. But since we have just confirmed that discharge is certainly taking place during that time, a constant reading can only be the result of a continual renewal of electrification, at a rate equal to the rate of discharge.

1326. All these considerations impress my mind strongly with the conviction, that insulation and ordinary conduction cannot be properly separated when we are examining into their nature; that is, into the general law or laws under which their phenomena are produced. They appear to me to consist in an action of contiguous particles dependent on the forces developed in electrical excitement; these forces bring the particles into a state of tension or polarity, which constitutes both *induction* and *insulation*; and being in this state, the continuous particles have a power or capability of communicating their forces one to the other, by which they are lowered, and discharge occurs. Every body appears to discharge (444. 987.); but the possession of this capability in a *greater or smaller degree* in different bodies, makes them better or worse conductors, worse or better insulators; and both *induction* and *conduction* appear to be the same in their principle and action (1320.), except that in the latter an effect common to both is raised to the highest degree, whereas in the former it occurs in the best cases, in only an almost insensible quantity.

1327. That in our attempts to penetrate into the nature of electrical action, and to deduce laws more general than those we are at present acquainted with, we should endeavour to bring apparently opposite effects to stand side by side in harmonious arrangement, is an opinion of long standing, and sanctioned by the ablest philosophers. I hope, therefore, I may be excused the attempt to look at the highest cases of conduction as analogous to, or even the same in kind with, those of induction and insulation.

1328. If we consider the slight penetration of sulphur (1241. 1242.) or shell-lac (1234.) by electricity, or the feebler insulation sustained by spermaceti (1279. 1240.), as essential consequences and indications of their *conducting* power, then may we look on the resistance of metallic wires to the passage of electricity through them as *insulating* power. Of the numerous well-known cases fitted to show this resistance in what are called the perfect conductors, the experiments of Professor

---

provided the water be distilled: Undistilled water is bound to contain impurities that would increase *conduction*, and so make *induction* difficult to distinguish.

1326. *the continuous particles have a power…*: The word "continuous" appears to be a misprint for "contiguous," used earlier in the paragraph; but one commentator accepts it, suggesting a possible reference by Faraday to Boscovich's theory of *point atoms* surrounded by continuous distributions of force.

1328. *penetration*: The term refers to the "return charge," which Faraday discussed in the Eleventh Series.

Wheatstone best serve my present purpose, since they were carried to such an extent as to show that *time* entered as an element into the conditions of conduction[1] even in metals. When discharge was made through a copper wire 2640 feet in length, and $^1/_{15}$th of an inch in diameter, so that the luminous sparks at each end of the wire, and at the middle, could be observed in the same place, the latter was found to be sensibly behind the two former in time, they being by the conditions of the experiment simultaneous. Hence a proof of retardation; and what reason can be given why this retardation should not be of the same kind as that in spermaceti, or in lac, or sulphur? But as, in them, retardation is insulation, and insulation is induction, why should we refuse the same relation to the same exhibitions of force in the metals?

1329. We learn from the experiment, that if *time* be allowed the retardation is gradually overcome; and the same thing obtains for the spermaceti, the lac, and glass (1248.); give but time in proportion to the retardation and the latter is at last vanquished. But if that be the case and all the results are alike in kind, the only difference being in the length of time, why should we refuse to metals the previous inductive action, which is admitted to occur in the other bodies? The diminution of *time* is no negation of the action; nor is the lower degree of tension requisite to cause the forces to traverse the metal, as compared to that necessary in the cases of water, spermaceti, or lac. These differences would only point to the conclusion, that in metals the particles under induction can transfer their forces when at a lower degree of tension or polarity, and with greater facility than in the instances of the other bodies.

[1] Philosophical Transactions, 1834, p. 583.

1328 (continued). *perfect conductors*: A "perfect" conductor would be an instantaneous one, as explained in the introduction. Hence it is noteworthy that Wheatstone has been able to show that conduction occupies time "even in metals." Thus even the very best conductors are, in some small degree, insulators—as even the very best insulators are, to a slight degree, conductive.

retardation is insulation: For both are instances of *inhibited* discharge. True, a leaky insulator is usually associated with mild and steady discharge, rather than the delayed but intense spark of Wheatstone's apparatus; nevertheless Faraday invites us to see the two as essentially the same.

1329. *if time be allowed…*: The *nature* of electrical discharge through slow conductors is the same as through rapid ones. Time is the only difference between them.

in metals the particles … transfer their forces when at a lower degree of tension or polarity, and with greater facility…: As discussed in the introduction, if the tension between adjacent particles of a metal has only to build up to a small

1330. Let us look at Mr. Wheatstone's beautiful experiment in another point of view. If, leaving the arrangement at the middle and two ends of the long copper wire unaltered, we remove the two intervening portions and replace them by wires of iron or platina, we shall have a much greater retardation of the middle spark than before. If, removing the iron, we were to substitute for it only five or six feet of water in a cylinder of the same diameter as the metal, we should have still greater retardation. If from water we passed to spermaceti, either directly or by gradual steps through other bodies, (even though we might vastly enlarge the bulk, for the purpose of evading the occurrence of a spark elsewhere (1331.) than at the three proper intervals,) we should have still greater retardation, until at last we might arrive, by degrees so small as to be inseparable from each other, at actual and permanent insulation. What, then, is to separate the principle of these two extremes, perfect conduction and perfect insulation, from each other; since the moment we leave in the smallest degree perfection at either extremity, we involve the element of perfection at the opposite end? Especially too, as we have not in nature the case of perfection either at one extremity or the other, either of insulation or conduction.

---

degree before those particles transfer their forces to one another, then the cycle of buildup and transfer will occupy a correspondingly short time. Thus the succession of transfers through the material—which is what constitutes *conduction*—will take place more rapidly in metals than in other materials.

1330. *Mr. Wheatstone's beautiful experiment*: The sketch gives a highly schematized representation. Each coil contains a quarter mile of wire; thus spark gaps BC are at the ends and gap DD at the middle of a half-mile length of wire. When the gap at A is momentarily closed by an arm of the revolving mirror (not shown), the Leyden jar discharges through the whole wire. In actuality, the gaps BC and DD are small enough to appear

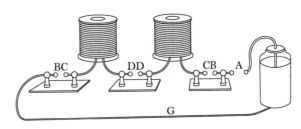

simultaneously in the mirror at one angle of its rotation. The "two intervening portions" are the coiled lengths of wire between C and D.

evading the occurrence of a spark elsewhere: The inserted pieces of spermaceti should be large; otherwise sparks might simply leap across them instead of being confined to the three proper spark gaps.

we have not in nature the case of perfection … either of insulation or conduction: The absence of any demonstrable instance of *absolute* conduction or insulation supports the idea that both are merely correlative degrees of the same action.

1331. Again, to return to this beautiful experiment in the various forms which may be given to it: the forces are not all in the wire (after they have left the Leyden jar) during the whole time (1328.) occupied by the discharge; they are disposed in part through the surrounding dielectric under the well-known form of induction; and if that dielectric be air, induction takes place from the wire through the air to surrounding conductors, until the ends of the wire are electrically related through its length, and discharge has occurred, *i. e.* for the *time* during which the middle spark is retarded beyond the others. This is well shown by the old experiment, in which a long wire is so bent that two parts (fig. 115.) *a. b.* near its extremities shall approach within a short distance, as a quarter of an inch, of each other in the air. If the discharge of a Leyden jar, charged to a sufficient degree, be sent through such a wire, by far the largest portion of the electricity will pass as a spark across the air at the interval, and not by the metal. Does not the middle part of the wire, therefore, act here as an insulating medium, though it be of metal? And is not the spark through the air an indication of the tension (simultaneous with *induction*) of the electricity in the ends of this single wire? Why should not the wire and the air both be regarded as dielectrics; and the action at its commencement and whilst there is tension, as an inductive action? If it acts through the contorted lines of the wire, so it also does in curved and contorted lines through air (1219. 1224. 1231.), and other insulating dielectrics (1228.); and we can apparently go so far in the analogy, whilst limiting the case to the inductive action only, as to show that amongst insulating dielectrics some lead away

---

1331. *they are disposed in part through the surrounding dielectric*: Air or other insulating material surrounding the wire constitutes an alternate path parallel to the wire and will therefore bear its proportionate share of induction and discharge—see paragraph 1334 below. Moreover, even as it discharges electrical force from the battery, the wire is in an electrified condition (though not a *static* one), so it must simultaneously sustain an inductive relation, through the surrounding air, with the walls of the room or with other juxtaposed conductors.

If the discharge of a Leyden jar ... be sent through such a wire...: Figure 115 fails to make clear that the discharge is applied at the *ends* of the wire, near *a* and *b*. It is remarkable that the long wire is unable to "short-circuit" the spark that develops at *ab*.

is not the spark ... an indication of the tension ... in the ends of this single wire? Throughout the duration of the spark, the wire sustains tension at its extremities without itself discharging that tension. But that is how an insulator acts! Hence, as Faraday goes on to state, *"the retardation is for the time insulation."*

the lines of force from others (1229.), as the wire will do from worse conductors, though in it the principal effect is no doubt due to the ready discharge between the particles whilst in a low state of tension. The retardation is for the time insulation; and it seems to me we may just as fairly compare the air at the interval *a, b.* (fig. 115.) and the wire in the circuit, as two bodies of the same kind and acting upon the same principles, as far as the first inductive phenomena are concerned, notwithstanding the different forms of discharge which ultimately follow,[2] as we may compare, according to Coulomb's investigations,[3] *different lengths* of different insulating bodies required to produce the same amount of insulating effect.

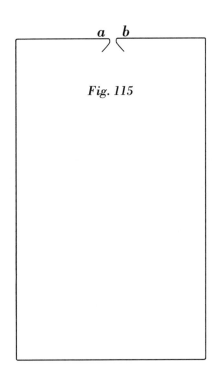

Fig. 115

1332. This comparison is still more striking when we take into consideration the experiment of Mr. Harris, in which he stretched a fine wire across a glass globe, the air within being rarefied.[4] On sending a charge through the joint arrangement of metal and rare air, as much, if not more, electricity passed by the latter as by the former. In the air, rarefied as it was, there can be no doubt the discharge was preceded by induction (1284.); and to my mind all the circumstances indicate that the same was the case with the metal; that, in fact, both substances are dielectrics, exhibiting the same effects in consequence of the action of the same causes, the only variation being one of degree in the different substances employed.

1333. Judging on these principles, velocity of discharge through the *same wire* may be varied greatly by attending to the circumstances which cause variations of discharge through spermaceti or sulphur. Thus, for

---

[2] These will be examined hereafter (1348, &c.).

[3] Mémoires de l'Académie, 1785, p. 612. or Ency. Britann. First. Supp. vol. i. p. 611.

[4] Philosophical Transactions, 1834, p. 242.

instance, it must vary with the tension or intensity of the first urging force (1234. 1240.), which tension is charge and induction. So if the two ends of the wire, in Professor Wheatstone's experiment, were immediately connected with two large insulated metallic surfaces exposed to the air, so that the primary act of induction, after making the contact for discharge, might be in part removed from the internal portion of the wire at the first instant, and disposed for the moment on its surface jointly with the air and surrounding conductors, then I venture to anticipate that the middle spark would be more retarded than before; and if these two plates were the inner and outer coating of a large jar or a Leyden battery, then the retardation of that spark would be still greater.

1333. *it must vary with the tension ... of the first urging force*: Faraday showed (in the cited paragraphs) that the *rate* of discharge through poor conductors like spermaceti rises and falls with the tension of the applied electric force; and he anticipates that the same will hold for good conductors also. That electric tension and rate of discharge are *proportional* is very nearly a statement of *Ohm's Law* (see the Second Series introduction); but it seems clear that Faraday regards the relation only as an interesting characteristic of certain conductors—not a fundamental law of electric action.

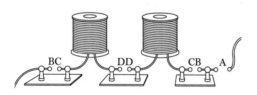

two large insulated metallic surfaces...: Two large facing surfaces will sustain considerable induction through the air between them. If the surfaces are connected to the respective wire ends C C, the combined induction over both wire and air will be much greater than formerly, through the wire material alone.

removed from the internal portion of the wire: During the first moments after the wire is energized, induction will be sustained through both the *copper* (the "internal portion" of the wire) and through the *air*, between the external surfaces of the two coils. But if large inducing surfaces could then be instantly connected, much of the induction that had been sustained through the wire will now be sustained through the air, between the two added surfaces. Induction along the new path must be at the expense of the old, since Wheatstone's machine makes only momentary connection at A; thus the Leyden jar is no longer connected and cannot supply additional electrification until the next rotation of the mirror axle.

Faraday implies that this addition would *immediately lessen the tension of the whole system*. In this he is somewhat anticipatory; but in paragraph 1372 below he will argue that when the total surface area under induction increases, the *tension* falls proportionately. Thus he infers that the wire would discharge this lower tension *at a slower rate*—equivalent, Faraday thinks, to a *greater delay in the propagation* of the discharge. And if a more highly inductive device, such as a Leyden jar, were connected instead, he predicts that the delay would be even greater.

1334. Cavendish was perhaps the first to show distinctly that discharge was not always by one channel,[5] but, if several are present, by many at once. We may make these different channels of different bodies, and by proportioning their thicknesses and lengths, may include such substances as air, lac, spermaceti, water, protoxide of iron, iron and silver, and by *one* discharge make each convey its proportion of the electric force. Perhaps the air ought to be excepted, as its discharge by conduction is questionable at present (1336.); but the others may all be limited in their mode of discharge to pure conduction. Yet several of them suffer previous induction, precisely like the induction through the air, it being a necessary preliminary to their discharging action. How can we therefore separate any one of these bodies from the others, as to the *principles and mode* of insulating and conducting, except by mere degree? All seem to me to be dielectrics acting alike, and under the same common laws.

1335. I might draw another argument in favour of the general sameness, in nature and action, of good and bad conductors (and all the bodies I refer to are conductors more or less), from the perfect equipoise in action of very different bodies when opposed to each other in magneto-electric inductive action, as formerly described (213.), but am anxious to be as brief as is consistent with the clear examination of the probable truth of my views.

1336. With regard to the possession by the gases of any conducting power of the simple kind now under consideration, the question is a very difficult one to determine at present. Experiments seem to indicate that they do insulate certain low degrees of tension perfectly, and that the effects which may have appeared to be occasioned by

[5] Philosophical Transactions, 1776; p. 197.

1334. *discharge … not always by one channel*: As described in the introduction, electrical discharge does not take the path of least resistance exclusively but fans out into multiple paths. Even paths through traditionally "insulating" materials, therefore, will sustain a proportionate amount of the discharge. Surely the principles and general manner of action must be the same for *all* the paths along which discharge actually occurs.

1335. *perfect equipoise in action of very different bodies*: In the experiments cited (paragraph 213), wires of very different conductivities developed the *same current* under electromagnetic action. Perhaps Faraday would argue that this equality shows the conduction process to be at least comparable in both wires; for if not, how could two inherently unequal currents have been reduced to equality by the same cause? But the 28th Series will controvert the supposed inequality of these currents.

conduction have been the result of the carrying power of the charged particles, either of the air or of dust, in it. It is equally certain, however, that with higher degrees of tension or charge the particles discharge to one another, and that is conduction. If the gases possess the power of insulating a certain low degree of tension continuously and perfectly, such a result may be due to their peculiar physical state, and the condition of separation under which their particles are placed. But in that, or in any case, we must not forget the fine experiments of Cagniard de la Tour,[6] in which he has shown that liquids and their vapours can be made to pass gradually into each other, to the entire removal of any marked distinction of the two states. Thus, hot dry steam and cold water pass by insensible gradations into each other; yet the one is amongst the gases as an insulator, and the other a comparatively good conductor. As to conducting power, therefore, the transition from metals even up to gases is gradual; substances make but one series in this respect, and the various cases must come under one condition and law (444.). The specific differences of bodies as to conducting power only serves to strengthen the general argument, that conduction, like insulation, is a result of induction, and is an action of contiguous particles.

1337. I might go on now to consider induction and its concomitant, *conduction,* through mixed dielectrics, as, for instance, when a charged body, instead of acting across air to a distant uninsulated conductor, acts jointly through it and an interposed insulated conductor. In such a case, the air and the conducting body are the mixed dielectrics; and the latter assumes a polarized condition as a mass, like that which my theory assumes *each particle* of the air to possess at the same time (1679.). But I fear to be tedious in the present condition of the subject, and hasten to the consideration of other matter.

[6] Annales de Chimie, xxi. pp. 127, 178; or Quarterly Journal of Science, xv. 145.

1336. *carrying power of the charged particles*: If particles of a gas are electrified, they may move in streams (the process called *convection*), resulting in a translation of charge. But such is not true conduction because it is essentially the particles, and only incidentally the electrification, that moves. In true conduction the tension is propagated from particle to particle—as Faraday puts it, "the particles discharge to one another"; no transfer of ponderable mass is necessary or relevant. The stream of electrified particles associated with brush discharge (described in the introduction) is an example of *carrying* or *convective discharge.*

The specific differences of bodies as to conducting power...: Since each material has a degree of conducting power specific to it, conduction must involve the particles of the body—for it is in the particles individually that the (chemical) identity of each material resides.

1337. *the latter*: that is, the interposed conducting body.

1338. To sum up, in some degree, what has been said, I look upon the first effect of an excited body upon neighbouring matters to be the production of a polarized state of their particles, which constitutes *induction*; and this arises from its action upon the particles in immediate contact with it, which again act upon those contiguous to them, and thus the forces are transferred to a distance. If the induction remain undiminished, then perfect insulation is the consequence; and the higher the polarized condition which the particles can acquire or maintain, the higher is the intensity which may be given to the acting forces. If, on the contrary, the contiguous particles, upon acquiring the polarized state, have the power to communicate their forces, then conduction occurs, and the tension is lowered; conduction being a distinct act of discharge between neighbouring particles. The lower the state of tension at which this discharge between the particles of a body takes place, the better

---

1338. *the higher the polarized condition that the particles can acquire or maintain, the higher is the intensity which may be given to the acting forces*: There is much to appreciate in this superlatively clear expression. Electric force applied to a body is, to the extent possible, balanced by the collective internal tensions of the individual particles of that body—which are then said to be in a "polarized" condition. If a body's particles are capable of sustaining high levels of tension, then correspondingly high electric forces may be applied without significant discharge: the body is an *insulator*. If the particles have low sustaining ability they will transfer their forces to one another—a wave process in which tension and discharge propagate with a characteristic velocity along the lines of tension and which brings about the continuous and gradual discharge of the applied electric force. That propagation of discharge is what was conventionally called a current. The body in which it occurs is a *conductor* and therefore, as discussed previously in the Eleventh Series introduction, a conductor is necessarily incapable of sustaining any significant degree of electric tension. Note that Faraday has at last succeeded in giving an account of current, conductor, and insulator that is completely free of fluid-flow language.

Do not be misled by Faraday's term "polarized." Readers sometimes assume that the term invokes electrical "poles" in the sense of *permanent regions of positive and negative charge*—a retreat into the imagery of mutually attracting and repelling electric fluids if true. But that is not Faraday's meaning. Remember that a "particle" is merely a tiny portion of the inductive medium; so that induction presents the same character in a particle as in the medium as a whole. In the Eleventh Series' introduction I suggested that "charge" was but the manifestation, at the boundary of a medium, of electric tension in that medium. Similarly, the tension sustained by an individual particle will be manifest as "charge" at the opposite boundaries *of the particle*. Such are the "poles" of a polarized particle. They do not represent permanent stores or puddles of electric fluid. They are not the cause but the manifestation of tension in the particle.

conductor is that body. In this view, insulators may be said to be bodies whose particles can retain the polarized state; whilst conductors are those whose particles cannot be permanently polarized. If I be right in my view of induction, then I consider the reduction of these two effects (which have been so long held distinct) to an action of contiguous particles obedient to one common law, as a very important result; and, on the other hand, the identity of character which the two acquire when viewed by the theory (1326.), is additional presumptive proof in favour of the correctness of the latter.

*   *   *

¶ ix. *Disruptive discharge and insulation.*

1359. The next form of discharge has been distinguished by the adjective *disruptive* (1319.), as it in every case displaces more or less the particles amongst and across which it suddenly breaks. I include under it discharge in the form of sparks, brushes, and glow (1405.), but exclude the cases of currents of air, fluids, &c., which, though frequently accompanying the former, are essentially distinct in their nature.

1360. The conditions requisite for the production of an electric spark in its simplest form are well known. An insulating dielectric must be interposed between two conducting surfaces in opposite states of electricity, and then if the actions be continually increased in strength, or otherwise favoured, either by exalting the electric state of the two conductors, or bringing them nearer to each other, or diminishing the density of the dielectric, a *spark* at last appears, and the two forces are for the time annihilated, for *discharge* has occurred.

1361. The conductors (which may be considered as the termini of the inductive action) are in ordinary cases most generally metals, whilst the dielectrics usually employed are common air and glass. In my view of induction, however, every dielectric becomes of importance, for as the results are considered essentially dependent on these bodies, it was to be expected that differences of action never before suspected would be evident upon close examination, and so at once give fresh confirmation of the theory, and open new doors of discovery into the extensive and varied fields of our science. This hope was especially entertained with respect to the gases, because of their high degree of insulation, their uniformity in physical condition, and great difference in chemical properties.

1359. *sparks, brushes, and glow* are described in the introduction. Note that *disruptive* discharge takes place only in material media—not in vacuum, which cannot be "disrupted" (broken apart).

1362. All the effects prior to the discharge are inductive; and the degree of tension which it is necessary to attain before the spark passes is therefore, in the examination I am now making of the new view of induction, a very important point. It is the limit of the influence which the dielectric exerts in resisting discharge; it is a measure, consequently, of the conservative power of the dielectric, which in its turn may be considered as becoming a measure, and therefore a representative of the intensity of the electric forces in activity.

1363. Many philosophers have examined the circumstances of this limiting action in air, but, as far as I know, none have come near Mr. Harris as to the accuracy with, and the extent to, which he has carried on his investigations.[7] Some of his results I must very briefly notice, premising that they are all obtained with the use of air as the *dielectric* between the conducting surfaces.

[7] Philosophical Transactions, 1834, p. 225.

1362. *the degree of tension which it is necessary to attain before the spark passes*: That is, the overall electric tension between the opposed surfaces, as measured by the electrometer.

the limit of the influence which the dielectric exerts in resisting discharge: That is, the maximum stress that the dielectric can bear. Both of these are aspects of the same inductive condition but reflect different perspectives. The overall electric tension between opposed surfaces represents the sum of tensions borne by *all* the particles that lie along a line of force joining the surfaces. The stress limit of a given dielectric, on the other hand, represents the maximum tension that a *single* particle along that line can sustain. This provides grounds for distinguishing between electric *tension* and *intensity*, as suggested in the comment to paragraph 1299 in the previous Series. "Tension" ordinarily refers to the *overall* tension between electrified surfaces, as measured by the electrometer. "Intensity," when distinguished from "tension," refers to that share of the overall tension which must be borne by a *single particle* of the medium. When a single particle gives way under stress (thereby triggering discharge through the whole dielectric), we can be sure that the *intensity* of electric action has reached the characteristic limit of the individual particles of the dielectric. But Faraday will not hold to the distinction invariably; he occasionally reverts to his earlier synonymous usage of "intensity" and "tension." Nor does he explain whether he intends the distinction to hold in vacuum, which has no "particles."

a measure, and therefore a representative...: Notice how Faraday subordinates measurement to representation! Another writer might have reversed the phrase—"a representative, and therefore a measure"—under the impression that a *measure* is what we want primarily. But in Faraday's visual paradigm for science (see the Nineteenth Series introduction), it is the interpretative and representational activities that rank highest. Measurements are valuable *because* they provide imagery (though often rudimentary, even crude) by which to advance our effort to see nature and nature's powers in their essential terms.

1364. First as to the *distance* between the two balls used, or in other words, the *thickness* of the dielectric across which the induction was sustained. The quantity of electricity, measured by a unit jar, or otherwise on the same principle with the unit jar, in the charged or inductive ball, necessary to produce spark discharge, was found to vary exactly with the distance between the balls, or between the discharging points, and that under very varied and exact forms of experiment.[8]

1365. Then with respect to variation in the *pressure* or *density* of the air. The quantities of electricity required to produce discharge across a *constant* interval varied exactly with variations of the density; the quantity of electricity and density of the air being in the same simple ratio. Or, if the quantity was retained the same, whilst the interval and density of the air were varied, then these were found in the inverse simple ratio of each other, the same quantity passing across twice the distance with air rarefied to one half.[9]

1366. It must be remembered that these effects take place without any variation of the *inductive* force by condensation or rarefaction of

[8] Philosophical Transactions, 1834, p. 225.

[9] Ibid. p. 229.

1364. *the two balls used*: that is, the spherical electrodes between which sparks were produced. In the sketch, ball A atop the Leyden jar is the "charged or inductive ball," ball B is grounded. In the "distance" experiments, the Leyden jar is repeatedly charged, then the distance AB diminished until a spark passes. Harris found that the Leyden jar's charges and the corresponding spark distances were proportional. The same results occurred when the balls were replaced by points.

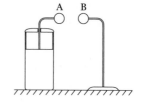

unit jar: Typically a small Leyden jar which, being charged to a specified degree (according to the electrometer) and discharged, would thereby measure out a fixed quantity of electricity, by comparison to which the charge on the large Leyden jar could be determined. Harris was among the first investigators to employ such an electrical standard.

[the quantity of electricity] necessary to produce spark discharge: Such was the customary phraseology; but on Faraday's principles it is not quantity of electricity *per se* that triggers discharge, but rather the *tension* which that quantity develops in the inductive medium and its individual particles. Nevertheless Faraday has reason to believe that the quantity and the tension are proportional; see paragraphs 1372–1373 below.

1366. *without any variation of the inductive force by condensation or rarefaction*: Recall that in the Eleventh Series Faraday had found the specific inductive capacity of air and all other gases to be the same, and not measurably affected

the air. That force remains the same in air,[10] and in all gases (1284. 1292.), whatever their rarefaction may be.

1367. Variations of the *temperature* of the air produced no variation of the quantity of electricity required to cause discharge across a given interval.[11]

Such are the general results, which I have occasion for at present, obtained by Mr. Harris, and they appear to me to be unexceptionable.

1368. In the theory of induction founded upon a molecular action of the dielectric, we have to look to the state of that body principally for the cause and determination of the above effects. Whilst the induction continues, it is assumed that the particles of the dielectric are in a certain polarized state, the tension of this state rising higher in each particle as the induction is raised to a higher degree, either by approximation of the inducing surfaces, variation of form, increase of the original force, or other means; until at last, the tension of the particles

[10] Ibid. pp. 237, 244.

[11] Ibid. p. 230.

---

by varying their densities. So the tension associated with a fixed quantity of electricity *does not change* when the air density varies; consequently such a change cannot be alleged to explain why the quantity of charge required to obtain spark depends on air pressure (or density).

1368. *molecular action*: Recall that by *molecule* Faraday means only a *tiny bit* of material, not a unit of chemical identity or a fixed aggregate of Daltonian atoms.

the tension of this state rising higher in each particle...: Since the dielectric as a whole sustains a condition of electric tension, each of its particles will bear a share of that tension.

the induction is raised to a higher degree ... by approximation of the inducing surfaces, variation of form, increase of the original force...: Refer to the sketch accompanying the comment to paragraph 1364. It is clear that prior to discharge, "increase of the original force"—that is, increased electrification of the Leyden jar—is tantamount to increased induction between the spherical electrodes. How do we know that "approximation of the inducing surfaces"—that is, decreasing distance AB—will similarly increase the induction? We saw in paragraphs 1213–1214 and again in paragraph 1268 that decreasing the air space between surfaces (by lining one of them with metal) raised their "aptness or capacity for induction." In particular, they sustained the same induction at a lower electric tension. But in the discharge apparatus, ball A is maintained at a constant tension by the charged Leyden jar (an electrometer will show this). Thus when the distance AB is diminished and the ratio of induction to tension thereby increased, it follows that with constant tension the induction must *increase*. But note that while these experiments establish the fact, they do not explain it. Faraday will attempt to do that in paragraph 1371 below.

having reached the utmost degree which they can sustain without subversion of the whole arrangement, discharge immediately after takes place.

1369. The theory does not assume, however, that *all* the particles of the dielectric subject to the inductive action are affected to the same amount or acquire the same tension. What has been called the lateral action of the lines of inductive force (1231. 1297.), and the diverging and occasionally curved form of these lines is against such a notion. The idea is, that any section taken through the dielectric across the lines of inductive force, and including *all of them,* would be equal, in the sum of the forces, to the sum of the forces in any other section; and that, therefore, the whole amount of tension for each such section would be the same.

1369. Faraday's argument in this paragraph takes as a premise that *each line of force represents an equal increment of force.* We have seen intimations of this idea before—for example at paragraph 1231 where he spoke of the lines as "meeting ... [and] uniting their forces," and there is similarly indirect language at paragraph 1302. But the present paragraph marks Faraday's most explicit appeal to the principle so far.

The theory does not assume ... that all the particles of the dielectric ... acquire the same tension: Since lines of force are generally curved, their concentration usually varies from place to place. A particle located at a place of high concentration will therefore intersect more lines—and be subject to proportionally greater tension—than an identical particle lying in a region of low concentration.

any section ... would be equal ... to the sum of the forces in any other section: Within any dielectric body, imagine a thin slice, as A in the sketch. Since it is penetrated by electric lines of force, the section will necessarily be in a state of electric tension. Since each line contributes equally to the total action, if the same set of lines also penetrates any other section of the dielectric, as B, the two sections will experience equal overall electrical effect ("sum of the forces"). This will be so whether the surfaces A, B are equal or, as in the sketch, highly unequal in area. In the 26th Series Faraday will make a similar argument for magnetic lines as well.

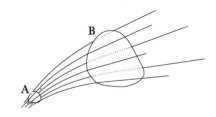

therefore, the whole amount of tension for each such section would be the same: That is, the tension collectively sustained by the particles that make up each slice of dielectric will be the same. The stress borne by each particle depends on the number of lines it intersects; but the total stress borne by *all* the particles in each section is equal, because the number of lines intersecting each section is equal.

1370. Discharge probably occurs, not when all the particles have attained to a certain degree of tension, but when that particle which is most affected has been exalted to the subverting or turning point (1410.). For though *all* the particles in the line of induction resist charge, and are associated in their actions so as to give a sum of resisting force, yet when any one is brought up to the overturning point, *all* must give way in the case of a spark between ball and ball. The breaking down of that one must of necessity cause the whole barrier to be overturned, for it was at its utmost degree of resistance when it possessed the aiding power of that one particle, in addition to the power of the rest, and the power of that one is now lost. Hence *tension* or *intensity*[12] may, according to the theory, be considered as represented by the particular condition of the particles, or the amount in them of forced variation from their normal state (1298. 1308.).

1371. The whole effect produced by a charged conductor on a distant conductor, insulated or not, is by my theory assumed to be due

[12] See Harris on proposed particular meaning of these terms, Philosophical Transactions, 1834, p. 222.

1370. *The breaking down of that one must of necessity cause the whole barrier to be overturned...*: With each particle in the line strained nearly to the limit, if even one were to give way and shift its burden to a neighbor, that neighbor would then be stressed above its limit and would give way in turn, and so on—the collapse of tension (discharge) thus progressing along the whole line. Similarly when fabric is under tension and one tiny rent gives way, a tear spreads progressively and rapidly through the whole. (Note, though, that in fabric the propagation direction is *perpendicular* to the lines of tension; whereas Faraday conceives the collapse of electric tension to propagate *parallel* to the lines of force.)

tension or intensity may be ... represented by the [forced] condition of the particles: As he has already urged (paragraphs 1362, 1368), "tension or intensity" in its original sense as the overall electric tension between inducing surfaces (measured by the electrometer) may be viewed from another perspective as the stress borne by each particle in the dielectric. That does not, however, imply simple proportionality between them. The electric tension between two surfaces relates not only to the stress per particle of the dielectric, but also to the shapes, areas, and separation of the inducing surfaces, as well as to the intervening dielectric itself. But Faraday regards the connection between *overall electric tension* and *stress in the medium* as the *essential* connection—in comparison to which the specifics of configuration and geometry are only incidental.

1371. *The whole effect produced by a charged conductor on a distant conductor...*: As before (paragraph 1364), in the discharge apparatus "a charged conductor" is not an *insulated* conductor but one that is connected to an electrified Leyden jar and therefore maintained at a constant tension with respect to earth—at least, until discharge takes place.

to an action propagated from particle to particle of the intervening and insulating dielectric, all the particles being considered as thrown for the time into a forced condition, from which they endeavour to return to their normal or natural state. The theory, therefore, seems to supply an easy explanation of the influence of *distance* in affecting induction (1303. 1364.). As the distance is diminished induction increases; for there are then fewer particles in the line of inductive force to oppose their united resistance to the assumption of the forced or polarized state, and *vice versâ*. Again, as the distance diminishes, discharge across happens with a lower charge of electricity; for if as in Harris's experiments (1364.), the interval be diminished to one half, then half the electricity required to discharge across the first interval is

---

1371, continued. Faraday's arguments in this and subsequent paragraphs draw upon two principles which he never explicitly identifies. It will therefore be helpful to state them in advance. Principle (*i*): *The overall electric tension between opposed surfaces* (in other words, what the electrometer measures) represents the sum of tensions borne by *all the particles that lie along a single line of force* joining the surfaces. Faraday employed this principle earlier at paragraph 1362. Principle (*ii*): *The "charge" of opposed electrified surfaces* manifests the collective tension in *those particles of the dielectric that are adjacent to the surfaces.* As suggested in the Eleventh Series' introduction, "charge" is the manifestation of electric tension at the boundaries of a dielectric. But the boundaries of the *dielectric* are identical with the boundaries of those *particles* which lie along its surfaces. Thus the tension in the medium, of which the "charge" of conductors is the manifestation, appears specifically as the collective tension of the particles that lie adjacent to the conductors. Again, however, it is unclear how, or even whether, these principles can apply to vacuum. The question will prove troublesome at paragraph 1375 below.

As the distance is diminished induction increases...:   The *fact* was established already—see the comment to paragraph 1368; now Faraday offers a theoretical reason. If the distance between conductors decreases, the lines of force become shorter and fewer particles will lie along them. But the overall tension between the surfaces is held constant, because of the Leyden jar. Then by principle (*i*), since a smaller number of particles must make up the same total tension as before, clearly the stress *of every particle* must increase. Among them, the particles *adjacent to the conductors* must sustain an increase in tension. But by principle (*ii*), it is these very particles whose tension is manifested as the "charge" of the conductors. Thus as distance decreases, the total charge of the conductors increases—even as the tension between them remains the same. Q.E.D.

as the distance diminishes, discharge ... happens with a lower charge...:   Again by principle (*i*), when there are fewer particles lying along each line of force, each particle must bear a larger share of stress, thereby approaching nearer to or possibly exceeding its limit (overcoming its "resistance"). Thus a charge— better, a tension—that failed to spark between surfaces separated by a given distance may eventually succeed in sparking when the distance decreases.

sufficient to strike across the second; and it is evident, also, that at that time there are only half the number of interposed molecules uniting their forces to resist the discharge.

1372. The effect of enlarging the conducting surfaces which are opposed to each other in the act of induction, is, if the electricity be limited in its supply, to lower the intensity of action; and this follows as a very natural consequence from the increased area of the dielectric across which the induction is effected. For by diffusing the inductive action, which at first was exerted through one square inch of sectional area of the dielectric, over two or three square inches of such area, twice or three times the number of molecules of the dielectric are brought into the polarized condition, and employed in sustaining the inductive action, and consequently the tension belonging to the smaller number on which the limited force was originally accumulated, must fall in a proportionate degree.

1373. For the same reason diminishing these opposing surfaces must increase the intensity, and the effect will increase until the surfaces become points. But in this case, the tension of the particles of

---

1372. *electricity ... limited in its supply*: In contrast to the discharge apparatus (paragraph 1364), whose electrodes are permanently connected to a Leyden jar, Faraday now considers a pair of conductors that have been charged *and then insulated.* Their electricity is "limited"—it can neither increase nor decrease.

The effect of enlarging the conducting surfaces [is] to lower the intensity of action: If the surfaces increase in area, a greater number of polarized particles will be situated adjacent to the surfaces. By principle (*ii*) (paragraph 1371, *comment*) it is these particles whose tensions are collectively manifested as charge on the conducting surfaces; and by hypothesis that charge is constant. Thus a greater number of particles contributes to a constant charge; therefore the contribution of each particle must be less, and the tension per particle smaller, than before. But if the tension of particles adjacent to the surfaces has decreased, so must *all* the particles similarly suffer a decrease in tension; and a reduction in tension per particle constitutes a reduction in *intensity of action*, according to the distinction introduced in paragraph 1362. Q.E.D.

consequently the tension ... must fall in a proportionate degree...: Furthermore, if *all* the particles suffer a decrease in tension, that will of course include the particles that coincide with a single line of force. By principle (*i*), it is *these very particles* whose tensions sum to the overall tension between conductors. Thus as the area of the conductors increases, the overall tension between them decreases.

Finally it will be clear—though again Faraday does not say so explicitly—that if plate area, separation, and dielectric material all remain unaltered, then the *overall tension*, as well as *intensity of action*, will be proportional to the *quantity of electrification*, as suggested at paragraph 1364 above.

the dielectric next the points is higher than that of particles midway, because of the lateral action and consequent bulging, as it were, of the lines of inductive force at the middle distance (1369.).

1374. The more exalted effects of induction on a point p, or any small surface, as the rounded end of a rod, when it is opposed to a large surface, as that of a ball or plate, rather than to another point or end, the distance being in both cases the same, fall into harmonious relation with my theory (1302.). For in the latter case, the small surface p is affected only by those particles which are brought into the inductive condition by the equally small surface of the opposed conductor, whereas when that is a ball or plate the lines of inductive force from the latter, are concentrated, as it were, upon the end p. Now though the molecules of the dielectric against the large surface may have a much lower state of

---

1373, continued. *because of the lateral action and consequent bulging, as it were, of the lines...:* In addition to the places Faraday cites, paragraphs 1165 and 1224 illustrated a tendency of adjacent lines of force to push each other away laterally. There is not much scope for this lateral action where lines are constrained to terminate on small conductors; but in more distant regions they have room to "bulge." As the sketch shows, the lines of force will thus be more highly concentrated at and near the electrified conductors than elsewhere; and these regions will then be subject to the greatest stress, as Faraday argues in paragraph 1369.

Thus dielectric stress will be intense near a pointed electrode but will diminish rapidly with distance. It will be less intense, and will diminish more gradually, near blunted or rounded shapes. This may help to explain the observed dependence of brush and glow on electrode shape, as well as their restriction to regions *near* the electrodes (see the introduction to the present Series).

1374. *The more exalted effects of induction on a point ... when it is opposed to a large surface...:* Faraday gives the following example of "exalted effects" in an omitted section of this Series: Let A be a small spherical electrode connected to a plate machine while B, an equal sphere, is grounded. This, Faraday says, will generally produce the *brush* at A. But "if the machine be not in good action," substitute a large conducting surface C in place of sphere B, and the brush will often form. How so? On Faraday's principles, lines of force will be about equally concentrated near equal spherical electrodes but very unequally distributed when one electrode is a large sheet. In the latter case, although the overall tension developed by the machine is the same as before, particles of the dielectric near A will have to bear a greater share of that tension. The dielectric is thus more likely to reach its limit, break down, and form the brush.

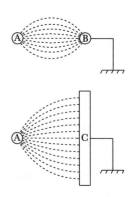

tension than those against the corresponding smaller surface, yet they are also far more numerous, and, as the lines of inductive force converge towards a point, are able to communicate to the particles contained in any cross section (1369.) nearer the small surface an amount of tension equal to their own, and consequently much higher for each individual particle; so that at the surface of the smaller conductor, the tension of a particle rises much, and if that conductor were to terminate in a point, the tension would rise to an infinite degree, except that it is limited, as before (1368.), by discharge. The nature of the discharge from small surfaces and points under induction will be resumed hereafter (1425. &c.).

1375. *Rarefaction* of the air does not alter the *intensity* of inductive action (1284. 1287.); nor is there any reason, as far as I can perceive, why it should. If the quantity of electricity and the distance remain the same, and the air be rarefied one half, then, though one half of the particles of the dielectric are removed, the other half assume a double degree of tension in their polarity, and therefore the inductive forces are balanced, and the result remains unaltered as long as the induction and insulation are sustained. But the case of *discharge* is very different; for as there are only half the number of dielectric particles in the rarefied atmosphere, so these are brought up to the discharging intensity by half the former quantity of electricity; discharge, therefore, ensues, and such a consequence of the theory is in perfect accordance with Mr. Harris's results (1365.).

1375. *Rarefaction of the air does not alter the intensity of inductive action*: Here, "intensity" has its former sense, referring to the overall tension between the charged surfaces. Later in the paragraph, Faraday shifts to the more recent sense of *tension in a single particle*.

nor is there any reason ... why it should...: Faraday's account is questionable. If the number of particles occupying a cubical volume between charged surfaces, is halved, then the number residing *along each line of force* will be reduced to the cube root of one-half (about 79%) and the number *adjacent to each surface* reduced to the square of the cube root (about 63%). According to principle (*ii*) (paragraph 1371, *comment*), the tension sustained by each particle must therefore rise to $^{100}/_{63}$ times its former value; and by principle (*i*) the total tension between the surfaces will be 79% of that amount, or $^{79}/_{63}$ times its former value, *not* equal to it. The difficulty in applying these principles consistently to media of varying density is a manifestation of the far deeper problem of understanding how *vacuum* can sustain a condition of electrical tension.

but the case of discharge is very different: Rarefaction does alter the tension at which *disruptive discharge* through the medium takes place. Discharge occurs at a lower tension through rarefied air than through air at ordinary pressure. At extremely low pressures, however, the effect reverses again, so that *vacuum* appears to be invulnerable to disruptive discharge!

1376. The *increase* of electricity required to cause discharge over the same distance, when the pressure of the air or its density is increased, flows in a similar manner, and on the same principle (1375.), from the molecular theory.

1377. Here I think my view of induction has a decided advantage over others, especially over that which refers the retention of electricity on the surface of conductors in air to the *pressure of the atmosphere* (1305.). The latter is the view which, being adopted by Poisson and Biot,[13] is also, I believe, that generally received; and it associates two such dissimilar things, as the ponderous air and the subtile and even hypothetical fluid or fluids of electricity, by gross mechanical relations; by the bonds of mere static pressure. My theory, on the contrary, sets out at once by connecting the electric forces with the particles of matter; it derives all its proofs, and even its origin in the first instance, from experiment; and then, without any further assumption, seems to offer at once a full explanation of these and many other singular, peculiar, and, I think, heretofore unconnected effects.

1378. An important assisting experimental argument may here be adduced, derived from the difference of specific inductive capacity of different dielectrics (1269. 1274. 1278.). Consider an insulated sphere electrified positively and placed in the centre of another and larger sphere uninsulated, a uniform dielectric, as air, intervening. The case is really that of my apparatus (1187.), and also, in effect, that of any ball electrified in a room and removed to some distance from irregularly-formed conductors. Whilst things remain in this state the electricity is distributed (so to speak) uniformly over the surface of the electrified sphere. But introduce such a dielectric as sulphur or lac, into the space

[13] Encyclopædia Britannica, Supplement, vol. iv. Article Electricity, pp. 76, 81, &c.

1377. *associates two such dissimilar things…*: Some theorists had supposed the pressure of the atmosphere to hinder electric discharge by mechanically *pressing upon electricity*, just as it weighs upon the surfaces of liquids. But even apart from that theory's inability to explain such phenomena as the *insulating power of vacuum* (paragraph 1375, comment), its dismal failure to appreciate the essential incommensurability between "ponderous air" and "subtile … electricity" is highly exasperating. We seldom find Faraday so openly indignant as when he here dismisses "gross mechanical relations"!

1378. *so to speak*: Note his explicit reservation when describing electricity as being "distributed," an epithet which might otherwise be thought to convey fluid imagery.

between the two conductors on one side only, or opposite one part of the inner sphere, and immediately the electricity on the latter is diffused unequally (1229. 1270. 1309.), although the form of the conducting surfaces, their distances, and the *pressure* of the atmosphere remain perfectly unchanged.

*   *   *

Royal Institution,
December 23rd, 1837

But introduce such a dielectric … on one side only, … and immediately the electricity on the latter is diffused unequally: The left-hand sketch depicts an electrified sphere placed at the center of a cubical room. Lines of force have been traced from the sphere by means of an electrometer, as in Faraday's note to paragraph 1304, and are shaped very symmetrically. In the right-hand sketch, a ball of sulfur has been brought

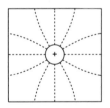 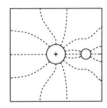

into the room. As in Faraday's earlier experiment, where a shell-lac block bent and concentrated lines of electric force towards itself (paragraphs 1224, 1231), so here will the lines bend and be displaced towards and even onto the sulfur. These altered line shapes suggest a general shift of force in the air, towards the sulfur ball, as well as a displacement of electrification on the electrified sphere—as Faraday states, "the electricity on the latter is diffused unequally." It is important to realize that such suggestions arise solely from the line *shapes* and are therefore limited to qualitative content—matters of *more and less*. The drawings inevitably suggest a possibility of quantitative inferences based on *counting* the lines—but that is highly misleading, as the introduction explains. Except in special cases Faraday has no way, at present, to determine the *number* of electric lines of force in a region.

Fifteenth Series — Editor's Introduction

This wonderful animal

The present Series, containing Faraday's report on the Gymnotus (electric eel), begins the second and middle volume of the *Experimental Researches*. Being moreover the fifteenth of twenty-nine numbered Series, it can be said in two senses to occupy the midpoint of that comprehensive work. The Gymnotus investigation also marks another kind of dividing point in Faraday's researches: *electricity* was the predominating topic of the first fourteen Series; but for the greater part of the last fourteen the focus will be *magnetism*.

To Faraday, the Gymnotus is "wonderful" (paragraph 1769)—an epithet by which he expresses not only marvel but also promise. In two areas, at least, animal electricity promises to be especially illuminating. First, in animal electricity we have an instance of *one identical power* exercised both by living and nonliving agents. The relation between an agent and the power it exercises may be more discernible when it is viewed in the comparison between a living and a nonliving system; and if so, knowledge of the animal may contribute as much to our knowledge of the inorganic system as the other way around.

Second, a living creature's ability to respond to and alter its environment *by intention or habit* adds a new interpretive dimension to the animal's electrical relations with its surroundings. The general relation between an agent and its surrounding medium may therefore stand forth more prominently when exemplified by a *living* agent. In fact Faraday reports a striking example of this when he describes the "coiling incident" at paragraph 1785 in the present Series.

Agent and power

We regularly attribute electric power to charged bodies, magnetic power to magnets, and chemical as well as gravitational powers to matter generally. But how does power "belong" to a body? What relation holds between an agent and the power it exercises? Concerning this question, we seem to possess neither an adequate imagery nor a commensurate terminology. There appears to be an

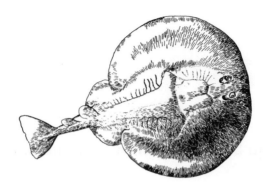

Torpedo-fish (drawing by J. L. Howard)

impenetrable intimacy between the powers conventionally ascribed to matter, and the matter itself. Gravitation, for example, is thought to be wholly inseparable from material bodies. And though electricity and magnetism may seem to be less firmly bound to their hosts—substances can be electrified and unelectrified, and iron can be magnetized and demagnetized almost at will—nevertheless, so long as an electrified body *is* electrified, it exerts its electrical power continuously; and so long as a piece of magnetized iron *is* magnetized, it exercises magnetic influence unceasingly. In such cases it seems impossible to distinguish between having a power and exercising it.

In contrast, the distinction is conspicuous when powers are exercised voluntarily by living agents.* The exercise of animal electricity may therefore prove to be clearer, and potentially more intelligible, than inorganic electric activity in which that distinction is absent or blurred. I believe it is these and similar considerations which move Faraday to cultivate a fresh vision of the Gymnotus; hopefully, such an image will come to our aid as we strive to characterize voltaic batteries, magnets, and other inorganic sources of power. Much of Faraday's activity in the present Series is devoted to this quest. His experiments with Gymnotus are as much concerned with eliciting *images* of the animal as with establishing factual information about him.

The body electric

The present investigation is of course not Faraday's first encounter with animal electricity. He had surveyed animal electricity in the Third Series, when establishing the probable identicality of all electricities;

---

* Aristotle gives a notable example in the *Physics* (Book II, Chapter 3), when he distinguishes between a "house-builder"—one who has acquired the ability to build—and a "house-builder *building*"—one in whom that ability is both present and *active*, not only acquired but *at work*.

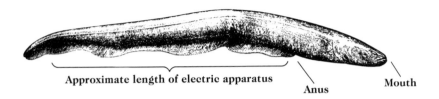

Gymnotus (drawing by J. L. Howard)

and as a young man he had assisted Humphrey Davy in tests, at that time inconclusive, to see whether the shock of the torpedo-fish could decompose water.*

The torpedo's natural element is the salty water of the Mediterranean. In each of his semicircular "wings" there is a kidney-shaped electric organ which renders an area on the upper wing surface momentarily positive, the lower surface negative; thus each discharge produces a current having conventional direction *from* the upper surface, through the surrounding water, *to* the lower surface.

Gymnotus, by contrast, is a freshwater animal. Its electric apparatus runs longitudinally for nearly 80% of his length. Except for electrification and locomotion, all of Gymnotus's life functions are carried on in the roughly 20% of his body length between mouth and anus.

Gymnotus's anal fin, which also runs some 4/5 of the length of the body, is that animal's principal locomotory organ. The fish propels itself forward or backward by sending a sinusoidal wave in the appropriate direction along the fin. But obviously the fin achieves nothing except when the fish is surrounded by its watery medium. Likewise for land animals; hands and feet achieve nothing in the way of locomotion except in reaction to a resisting medium or surface. Bearing that in mind, I hope you will not think it too fanciful of me to suggest that, from a locomotory point of view, the *medium* ought to be counted as *part of the body*. Faraday, I hasten to say, makes no such interpretation of the mechanics of animal locomotion. But electrically, at least, his researches with Gymnotus will contribute to a new image of body—unified with the medium, extended continuously throughout it, and

---

* The torpedo-fish, whose shock plunges his prey into a stupid paralysis, is without doubt the most literary of the electric animals, having been immortalized in Plato's dialogue *Meno*. There the fish provides a simile for the energetically philosophical Socrates; and torpidity in the fish's victim represents the perplexity and ineptitude displayed by one who has been forced under Socratic questioning to acknowledge his own ignorance.

contiguous with all other bodies through its own activity. The Body Electric will possess a distinctive shape and will call for new principles of anatomy.

EXPERIMENTAL RESEARCHES
IN
ELECTRICITY.

FIFTEENTH SERIES.

§ 23. *Notice of the character and direction of the electric force
of the Gymnotus.*

Received November 15,—Read December 6, 1838.

1749. **WONDERFUL** as are the laws and phenomena of electricity when
made evident to us in inorganic or dead matter, their interest can bear
scarcely any comparison with that which attaches to the same force
when connected with the nervous system and with life; and though the
obscurity which for the present surrounds the subject may for the time
also veil its importance, every advance in our knowledge of this mighty
power in relation to inert things, helps to dissipate that obscurity, and

---

1749. **WONDERFUL**: The exuberant typographical treatment this word here
receives is that of the three-volume edition of *Experimental Researches*. As the
very first word of the central Series (see the introduction to the present
Series), the word "wonderful" is in a certain sense the *central word* in the whole
Researches. But what kind of "wonder" does the animal provoke? Does it
suggest the arcane and supernatural, as though an "electric animal" might
somehow go beyond ordinary nature?

*every advance in our knowledge of this mighty power in relation to inert things, helps
to dissipate that obscurity...*: No, what Faraday finds compelling is precisely the
conformity between Gymnotus's living power and the more prosaic electrical
phenomena associated with inorganic bodies. Electrical powers formerly
thought to be confined to "inert" matter are here seen to be exercised by
living beings also; such a communion of powers holds promise for the expansion of our existing knowledge. This is a statement about the order of
discovery in nature. Faraday here notes that advances in our understanding of
inorganic powers will shed light in turn upon *living* processes. Just as inorganic
forces lie well within the domain of standard science, so an understanding of
living forces stands as a merely more distant, but nonetheless assured, prize.

to set forth more prominently the surpassing interest of this very high branch of Physical Philosophy. We are indeed but upon the threshold of what we may, without presumption, believe man is permitted to know of this matter; and the many eminent philosophers who have assisted in making this subject known have, as is very evident in their writings, felt up to the latest moment that such is the case.

1750. The existence of animals able to give the same concussion to the living system as the electrical machine, the voltaic battery, and the thunder storm, being with their habits made known to us by Richer, S'Gravesende, Firmin, Walsh, Humboldt, &c. &c., it became of growing importance to identify the living power which they possess, with that which man can call into action from inert matter, and by him named electricity (265. 351.). With the *Torpedo* this has been done to perfection, and the direction of the current of force determined by the united and successive labours of Walsh,[1] Cavendish,[2] Galvani,[3] Gardini,[4] Humboldt and Gay-Lussac,[5] Todd,[6] Sir Humphry Davy,[7] Dr. Davy,[8] Becquerel,[9] and Matteucci.[10]

[1] Philosophical Transactions, 1773, p. 461.

[2] Ibid. 1776, p. 196.

[3] Aldini's Essai sur la Galvanism, ii. 61.

[4] De Electrici ignis Natura, § 71. Mantua, 1792.

[5] Annales de Chimie, xiv. 15.

[6] Philosophical Transactions, 1816, p. 120.

[7] Ibid. 1829, p. 15.

[8] Ibid. 1832, p. 259; and 1834, p. 531.

[9] Traité de l'Electricité, iv. 264.

[10] Bibliothèque Universelle, 1837, tom. xii. 163.

1749, continued. *to set forth more prominently the surpassing interest…*: But the epithet "surpassing" presents animal processes as more than mere extensions of inorganic ones. "Surpassing" interest suggests almost a reverse order of discovery—that the exercise of a power by a *living being* may prove to be visible and intelligible in ways that power exercised by inert matter alone is not. Faraday will discover that the electric eel richly fulfills such a promise.

1750. *With the Torpedo this has been done to perfection, and the direction of the current of force determined…*: Each of Torpedo's wings generates a current from the positive upper surface, through the water, to its negative lower surface.

1751. The Gymnotus has also been experimented with for the same purpose, and the investigations of Williamson,[11] Garden,[12] Humboldt,[13] Fahlberg,[14] and Guisan,[15] have gone very far in showing the identity of the electric force in this animal with the electricity excited by ordinary means; and the two latter philosophers have even obtained the spark.

1752. As an animal fitted for the further investigation of this refined branch of science, the Gymnotus seems, in certain respects, better adapted than the Torpedo, especially (as Humboldt has remarked) in its power of bearing confinement, and capability of being preserved alive and in health for a long period. A Gymnotus has been kept for several months in activity, whereas Dr. Davy could not preserve Torpedos above twelve or fifteen days; and Matteucci was not able out of 116 such fish to keep one living above three days, though every circumstance favourable to their preservation was attended to.[16] To obtain Gymnoti has therefore been a matter of consequence; and being stimulated, as much as I was honoured, by very kind communications from Baron Humboldt, I in the year 1835 applied to the Colonial Office, where I was promised every assistance in procuring some of these fishes, and continually expect to receive either news of them or the animals themselves.

1753. Since that time Sir Everard Home has also moved a friend to send some Gymnoti over, which are to be consigned to His Royal Highness our late President; and other gentlemen are also engaged in the same work. This spirit induces me to insert in the present communication that part of the letter from Baron Humboldt which I received as an answer to my inquiry of how they were best to be conveyed across the Atlantic. He says, "The Gymnotus, which is common in the

[11] Philosophical Transactions, 1775, p. 94.

[12] Ibid. 1775, p. 102.

[13] Personal Narrative, chap. xvii.

[14] Swedish Transactions, 1801, pp. 122. 156.

[15] De Gymnoto Electrico. Tubingen, 1819.

[16] Bibliothèque Universelle, 1837, xii. p. 174.

1751. *Gymnotus*: from *gymno-* + *notos*: "naked back"—it has no dorsal or ventral fins. The Linnaean classification is *Gymnotus electricus*; the term *electric eel* is a misnomer as the animal is not, taxonomically, an eel (*Anguilla*). In this century, and especially in America, Faraday's *Gymnotus* is customarily called *Electrophorus* in order to avoid confusion with *Gymnotus carapo*, a weakly electric member of the gymnotoidae.

Llanos de Caracas (near Calabozo), in all the small rivers which flow into the Orinoco, in English, French or Dutch Guiana, is not of difficult transportation. We lost them so soon at Paris because they were too much fatigued (by experiments) immediately after their arrival. MM. Norderling and Fahlberg retained them alive at Paris above four months. I would advise that they be transported from Surinam (from Essequibo, Demerara, Cayenne) in summer, for the Gymnotus in its native country lives in water of 25° centigrade (or 77° Fahr.). Some are five feet in height, but I would advise that such as are about twenty-seven or twenty-eight inches in length be chosen. Their power varies with their food, and their state of rest. Having but a small stomach they eat little and often, their food being cooked meat, *not salted,* small fish, or even bread. Trial should be made of their strength and the fit kind of nourishment before they are shipped, and those fish only selected already accustomed to their prison. I retained them in a box or trough about four feet long, and sixteen inches wide and deep. The water must be *fresh,* and be changed every three or four days: the fish must not be prevented from coming to the surface, for they like to swallow air. A net should be put over and round the trough, for the Gymnotus often springs out of the water. These are all the directions that I can give you. It is, however, *important* that the animal should not be tormented or fatigued, for it becomes exhausted by frequent electric explosions. Several Gymnoti may be retained in the same trough."

1754. A Gymnotus has lately been brought to this country by Mr. Porter, and purchased by the proprietors of the Gallery in Adelaide Street: they immediately most liberally offered me the liberty of experimenting with the fish for scientific purposes; they placed it for the time exclusively at my disposal, that (in accordance with Humboldt's directions (1753.)) its powers might not be impaired; only desiring me to have a regard for its life and health. I was not slow to take advantage of their wish to forward the interests of science, and with many thanks accepted their offer. With this Gymnotus, having the kind assistance of Mr. Bradley of the Gallery, Mr. Gassiot, and occasionally other gentlemen, as Professors Daniell, Owen and Wheatstone, I have obtained every proof of the identity of its power with common electricity (265. 351, &c.). All of these had been obtained before with the Torpedo (1750.), and some, as the shock, circuit, and spark (1751.), with the

---

1754. *shock, circuit, and spark*: In addition to achieving *shock* and *spark* with the Gymnotus, earlier investigators obtained *"circuit"* effects, that is, results customarily ascribed to circulating electric currents—for example, deflection of a galvanometer or magnetization of a needle. Faraday will describe similar effects in paragraphs 1761–1765.

Gymnotus; but still I think a brief account of the results will be acceptable to the Royal Society, and I give them as necessary preliminary experiments to the investigations which we may hope to institute when the expected supply of animals arrives (1752.).

1755. The fish is forty inches long. It was caught about March 1838; was brought to the Gallery on the 15th of August, but did not feed from the time of its capture up to the 19th of October. From the 24th of August Mr. Bradley nightly put some blood into the water, which was changed for fresh water next morning, and in this way the animal perhaps obtained some nourishment. On the 19th of October it killed and eat four small fish; since then the blood has been discontinued, and the animal has been improving ever since, consuming upon an average one fish daily.[17]

1756. I first experimented with it on the 3rd of September, when it was apparently languid, but gave strong shocks when the hands were favourably disposed on the body (1760. 1773, &c.). The experiments were made on four different days, allowing periods of rest from a month to a week between each. His health seemed to improve continually, and it was during this period, between the third and fourth days of experiment, that he began to eat.

1757. Beside the hands two kinds of collectors were used. The one sort consisted each of a copper rod fifteen inches long, having a copper disc one inch and a half in diameter brazed to one extremity, and a copper cylinder to serve as a handle, with large contact to the hand, fixed to the other, the rod from the disc upwards being well covered with a thick caoutchouc tube to insulate that part from the water. By these the states of particular parts of the fish whilst in the water could be ascertained.

1758. The other kind of collectors were intended to meet the difficulty presented by the complete immersion of the fish in water;

---

[17] The fish eaten were gudgeons, carp, and perch.

1755. *it killed and eat four small fish*: In nineteenth-century British writing, *eat* is an alternate spelling for *ate*, the past tense of the verb; ordinarily pronounced "et."

1757. *collectors*: These are devices for sampling the electric condition of the fish's body or the surrounding water. Notice that Faraday includes the experimenters' own *hands* as a species of "collectors"!

caoutchouc: The term refers to *natural rubber*—somewhat sticky and not very durable compared to modern vulcanized rubber.

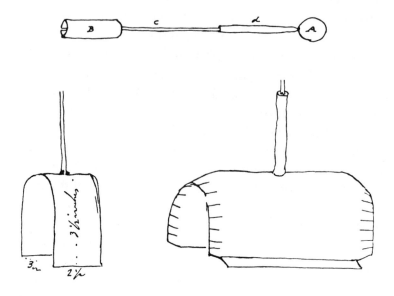

[Top: the "disk" collector (1757.). Bottom: the "saddle" collector and insulating cover (1758.). The sketches are from Faraday's *Diary*; they were not published with the *Experimental Researches*.]

for even when obtaining the spark itself I did not think myself justified in asking for the removal of the animal into air. A plate of copper eight inches long by two inches and a half wide, was bent into a saddle shape, that it might pass over the fish, and inclose a certain extent of the back and sides, and a thick copper wire was brazed to it, to convey the electric force to the experimental apparatus; a jacket of sheet caoutchouc was put over the saddle, the edges projecting at the bottom and the ends; the ends were made to converge so as to fit in some degree the body of the fish, and the bottom edges were made to

---

1758. *I did not think myself justified in asking for the removal of the animal into air*: This is a departure from what had been standard practice in torpedofish researches, which frequently emphasized the strength and quality of shocks delivered to a handler by a fish held in the air. Certainly Faraday's refusal to do likewise was in part a reflection of concern for the welfare of the animal (paragraph 1754); but it may also indicate that his view of the fish is already striving for unity in the treatment of *agent* and *medium* (see the introduction). If so, it would follow that a study of the animal *in its accustomed element* would better reveal the nature of its characteristic action. While this principle is not exactly the same as that of the animal ethologist, nevertheless we shall find in paragraph 1785 below that the fish's habitual behavior can provide rich guidance to Faraday in the interpretation of its electrical activity.

spring against any horizontal surface on which the saddles were placed. The part of the wire liable to be in the water was covered with caoutchouc.

1759. These conductors being put over the fish, collected power sufficient to produce many electric effects; but when, as in obtaining the spark, every possible advantage was needful, then glass plates were placed at the bottom of the water, and the fish being over them, the conductors were put over it until the lower caoutchouc edges rested on the glass, so that the part of the animal within the caoutchouc was thus almost as well insulated as if the Gymnotus had been in the air.

1760. *Shock.*—The shock of this animal was very powerful when the hands were placed in a favourable position, i. e. one on the body near the head, and the other near the tail; the nearer the hands were together within certain limits the less powerful was the shock. The disc conductors (1757.) conveyed the shock very well when the hands were wetted and applied in close contact with the cylindrical handles; but scarcely at all if the handles were held in the dry hands in an ordinary way.

1761. *Galvanometer.*—Using the saddle conductors (1758.) applied to the anterior and posterior parts of the Gymnotus, a galvanometer was readily affected. It was not particularly delicate; for zinc and platina plates on the upper and lower surface of the tongue did not cause a permanent deflection of more than 25°; yet when the fish gave a powerful discharge the deflection was as much as 30°, and in one case even 40°. The deflection was constantly in a given direction, the electric current being always from the anterior parts of the animal through the galvanometer wire to the posterior parts. The former were therefore for the time externally positive, and the latter negative.

1760–1767. *Shock,* magnetic action, chemical action, heat, and *spark* were some of the signs of the *identity of electricities* investigated in the Third Series. The experiments narrated in this and the following paragraphs manage to coax each of these effects from the Gymnotus and are therefore *experiments of identity,* as Faraday pointed out in paragraph 1754.

1761. *It* [the galvanometer] *was not particularly delicate* [sensitive], inasmuch as a voltaic cell formed by zinc and platinum plates (using Faraday's tongue as the electrolyte!) produced a "permanent deflection"—that is, a *steady* deflection—of only 25°. Even so, the fish was able to produce ballistic deflections exceeding that amount.

anterior … posterior: The terms denote *fore* and *hind* ends, respectively. The direction of galvanometer deflection (paragraph 19, *comment*) discloses that the current is directed from the head to the tail. According to the conventional current direction, then, the head is *positive,* the tail *negative.*

1762. *Making a magnet.*—When a large helix containing twenty-two feet of silked wire wound on a quill was put into the circuit, and an annealed steel needle placed in the helix, the needle became a magnet, and the direction of its polarity in every case indicated a current from the anterior to the posterior parts of the Gymnotus through the conductors used.

1763. *Chemical decomposition.*—Polar decomposition of a solution of iodide of potassium was easily obtained. Three or four folds of paper moistened in the solution (322.) were placed between a platina plate and the end of a wire also of platina, these being respectively connected with the two saddle conductors (1768.). Whenever the wire was in conjunction with the conductor at the fore part of the Gymnotus, iodine appeared at its extremity; but when connected with the other conductor, none was evolved at the place on the paper where it before appeared. So that here again the direction of the current proved to be the same as that given by the former tests.

1764. By this test I compared the middle part of the fish with other portions before and behind it, and found that the conductor A, which being applied to the middle was negative to the conductor B applied to the anterior parts, was, on the contrary, positive to it when B was applied to places near the tail. So that within certain limits the condition of the fish externally at the time of the shock appears to be such, that any given part is negative to other parts anterior to it, and positive to such as are behind it.

1765. *Evolution of heat.*—Using a Harris's thermo-electrometer belonging to Mr. Gassiot, we thought we were able in one case, namely, that when the deflection of the galvanometer was 40° (1761.), to observe a feeble elevation of temperature. I was not observing the instrument myself, and one of those who at first believed they saw the effect now doubts the result.[18]

[18] In more recent experiments of the same kind we could not obtain the effect.

1762. *silked*: overwound with silk thread as insulation.

1763. *Whenever the wire was in conjunction with the conductor at the fore part of the Gymnotus, iodine appeared at its extremity*: In the electro-decomposition of potassium iodide, iodine is evolved at the *positive* electrode.

1765. Harris's thermo-electrometer was described in the comment to paragraph 287 in the Third Series.

1766. *Spark.*—The electric spark was obtained thus. A good magneto-electric coil, with a core of soft iron wire, had one extremity made fast to the end of one of the saddle collectors (1758.), and the other fixed to a new steel file; another file was made fast to the end of the other collector. One person then rubbed the point of one of these files over the face of the other, whilst another person put the collectors over the fish, and endeavoured to excite it to action. By the action of the files contact was made and broken very frequently; and the object was to catch the moment of the current through the wire and helix, and by breaking contact *during the current* to make the electricity sensible as a spark.

1767. The spark was obtained four times, and nearly all who were present saw it. That it was not due to the mere attrition of the two files was shown by its not occurring when the files were rubbed together, independently of the animal. Since then I have substituted for the

---

1766. *Spark*: Faraday here completes a demonstration which had been left incomplete in the Third Series; for at paragraph 358 he reported that "the electric spark has not yet been obtained" from the electric eel.

A good magneto-electric coil... In the First Series (paragraph 32), Faraday produced a spark between the ends of one coil by making or breaking current in an adjacent coil. As he subsequently explained in the Second Series, when current commences or ceases, magnetic lines of force expand outward from the current-carrying coil or contract back into it; when these shifting lines are cut by the stationary windings of the adjacent coil, that coil will develop an "induced" current, which may be sufficiently powerful to spark across an air gap. In the Ninth Series (not included in the present selection), he confirmed that a substantially similar process takes place even in a *single* coil—especially if it contains many turns of wire wound on an iron core: When the current in such a coil is suddenly interrupted, a spark develops across the interrupting gap. Thus in the present arrangement, the fish provides the initial current, the intermittent contact between the files repeatedly interrupts it, and the spark appears in the tiny gap between the contacting surfaces of the two files.

One might object to crediting such a spark to Gymnotus himself, since it is not clear how much of the effect is due to the *fish* and how much to the induction *coil*—which enhances a current's sparking ability. But in a note to the next paragraph, Faraday reports that he was subsequently able to obtain the spark even without employing the coil.

1767. *not due to the mere attrition of the two files*: An example of "attrition" (scraping or wearing away) is the spark produced by flint and steel, which arises through abrasion of the flint. But the present spark cannot be due to attrition. If it were, merely scaping the files together would be sufficient to produce it.

lower file a revolving steel plate, cut file fashion on its face, and for the upper file wires of iron, copper and silver, with all of which the spark was obtained.[19]

1768. Such were the general electric phenomena obtained from this Gymnotus whilst living and active in his native element. On several occasions many of them were obtained together; thus a magnet was made, the galvanometer deflected, and perhaps a wire heated, by one single discharge of the electric force of the animal.

1769. I think a few further but brief details of experiments relating to the quantity and disposition of the electricity in and about this wonderful animal will not be out of place in this short account of its powers.

1770. When the shock is strong, it is like that of a large Leyden battery charged to a low degree, or that of a good voltaic battery of perhaps one hundred or more pair of plates, of which the circuit is completed for a moment only. I endeavoured to form some idea of the *quantity* of electricity by connecting a large Leyden battery (291.) with two brass balls, above three inches in diameter, placed seven inches apart in a tub of water, so that they might represent the parts of the Gymnotus to which the collectors had been applied; but to lower the intensity of the discharge, eight inches in length of six-fold thick wetted string were interposed elsewhere in the circuit, this being

[19] At a later meeting, at which attempts were made to cause the attraction of gold leaves, the spark was obtained directly between fixed surfaces, the inductive coil (1766.) being removed, and only short wires (by comparison) employed.

1768. Note that the preceding *identity* experiments propound a rhetoric of *mobility*. In them the electric power is conveyed away from the fish and its habitat; it is transferred through conductors to other venues, where it proceeds to display the same phenomena of magnetic action, chemical action, shock, spark, and so on, as do conventional electricities. Gymnotus's power is thereby treated as separable, as having a nature of its own that can be studied independently of the fish and in comparison to other "electricities," similarly abstracted from their respective sources.

1769. Having demonstrated in the previous experiments the electrical identity of Gymnotus' power, Faraday will now investigate the "quantity and disposition" of that power. Paragraph 1770 describes the *quantity* experiment; paragraphs 1773–1783 the *disposition* experiments.

1770. *that they might represent the parts of the Gymnotus:* He is constructing an imitation Gymnotus, to be energized by a battery of Leyden jars!

to lower the intensity of the discharge, ... wetted string [was] interposed: This was to prevent sparking, since he has already determined that the Gymnotus's

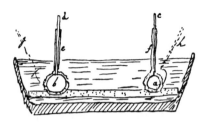

[Sketch of Faraday's discharge arrangement (1770.), from his *Diary*. It was not published with the *Experimental Researches*.]

found necessary to prevent the easy occurrence of the spark at the ends of the collectors (1758.), when they were applied in the water near to the balls, as they had been before to the fish. Being thus arranged, when the battery was strongly charged and discharged, and the hands put into the water near the balls, a shock was felt, much resembling that from the fish; and though the experiments have no pretension to accuracy, yet as the tension could be in some degree imitated by reference to the more or less ready production of a spark, and after that the shock be used to indicate whether the quantity was about the same, I think we may conclude that a single medium discharge of the fish is at least equal to the electricity of a Leyden battery of fifteen jars, containing 3500 square inches of glass coated on both sides, charged to its highest degree (291.). This conclusion respecting the great quantity of electricity in a single Gymnotus shock, is in perfect accordance with the degree of deflection which it can produce in a galvanometer needle (367. 860. 1761.), and also with the amount of chemical decomposition produced (374. 860. 1763.) in the electrolyzing experiments.

---

electrical intensity is too low for a spark to appear except under the most favorable conditions or with the spark coil (paragraphs 1766–1767). Faraday used the same technique in the Third Series, for a similar reason (paragraphs 296, 298).

shock ... much resembling that from the fish: The sensation of shock reflects both quantity and intensity of electricity. The *intensity* has been reduced below the sparking point, so it is at least roughly equal to the intensity of Gymnotus's discharge. Then since the shock from the discharge apparatus resembles that from Gymnotus, Faraday infers that both discharges involve roughly equal *quantities* of electricity.

a single medium discharge of the fish is at least equal to the electricity of a Leyden battery of fifteen jars...: This describes the battery and state of charge that was in fact used to energize the discharge apparatus in this experiment.

1771. Great as is the force in a single discharge, the Gymnotus, as Humboldt describes, and as I have frequently experienced, gives a double and even a triple shock; and this capability of immediately repeating the effect with scarcely a sensible interval of time, is very important in the considerations which must arise hereafter respecting the origin and excitement of the power in the animal. Walsh, Humboldt, Gay-Lussac, and Matteucci have remarked the same thing of the Torpedo, but in a far more striking degree.

1772. As, at the moment when the fish wills the shock, the anterior parts are positive and the posterior parts negative, it may be concluded that there is a current from the former to the latter through every part of the water which surrounds the animal, to a considerable distance from its body. The shock which is felt, therefore, when the hands are in the most favourable position, is the effect of a very small portion only of the electricity which the animal discharges at the moment, by far the largest portion passing through the surrounding water. This enormous external current must be accompanied by some effect within the fish *equivalent* to a current, the direction of which is from the tail towards the head, and equal to the sum of *all these external forces*. Whether the process of evolving or exciting the electricity within the fish includes the production of this internal current (which need not of necessity be as quick and momentary as the external one), we cannot at present say; but at the time of the shock the animal does not apparently feel the electric sensation which he causes in those around him.

1773. By the help of the accompanying diagram I will state a few experimental results which illustrate the current around the fish, and show the cause of the difference in character of the shock occasioned

---

1772. *it may be concluded that there is a current ... through every part of the water...*: It is interesting that Faraday derives this conclusion from the fact that electrification is *distributed* over a large area of the fish's body. In the Twelfth Series he gave a different reason for a very similar inference (paragraph 1334); see also the comment to paragraph 1786 below.

the effect of a very small portion only...: Since the fish's discharge must energize the water *as a whole*, the quantity of electricity sampled by the hands can only be a fraction of the total effect Gymnotus produces.

This enormous external current must be accompanied by some effect within the fish equivalent to a current ... from the tail towards the head: The fish's external discharge must ultimately be the result of some internal physiological action which makes the head positive and the tail negative. Such electrification of head and tail is just what would result if there existed a current, directed from tail to head, within the fish. Whatever the real nature of the fish's internal action may be, therefore, it may be counted as *equivalent* to a current having the direction specified.

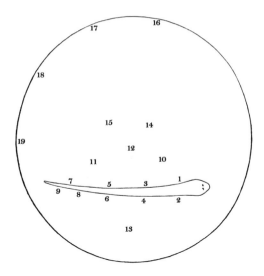

by the various ways in which the person is connected with the animal, or his position altered with respect to it. The large circle represents the tub in which the animal is confined; its diameter is forty-six inches, and the depth of water in it three inches and a half; it is supported on dry wooden legs. The figures represent the places where the hands or the disc conductors (1757.) were applied, and where they are close to the figure of the animal, it implies that contact with the fish was made. I will designate different persons by A, B, C, &c., A being the person who excited the fish to action.

1774. When one hand was in the water the shock was felt in that hand only, whatever part of the fish it was applied to; it was not very strong, and was only in the part immersed in the water. When the hand and part of the arm was in, the shock was felt in all the parts immersed.

1775. When *both* hands were in the water at the *same* part of the fish, still the shock was comparatively weak, and only in the parts immersed. If the hands were on opposite sides, as at 1, 2, or at 3, 4, or 5, 6, or if one was above and the other below at the same part, the effect was the same. When the disc collectors were used in these positions no effect

---

1773. *the places where the hands or the disc conductors … were applied*: Faraday here begins the "disposition" experiments. Notice that they are *mapping* exercises; they employ a rhetoric of *residence*. Unlike the identity experiments, the fish's electric power is not here conveyed to a remote observer; rather the observers make full ingression into the scene of action and quite literally immerse themselves in the place of habitation of the power. In this respect the disposition experiments might be compared to the great Cage of the Eleventh Series, in which Faraday occupied the experimental arena with his own body, in an even more spectacular way.

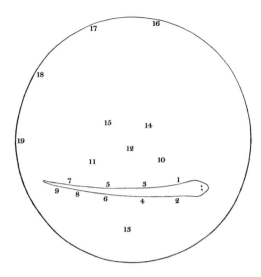

was felt by the person holding them (and this corresponds with the observation of Gay-Lussac on Torpedos[20]), whilst other persons, with both hands in at a distance from the fish, felt considerable shocks.

1776. When both hands or the disc collectors were applied at places separated by a part of the length of the animal, as at 1, 3, or 4, 6, or 3, 6, then strong shocks extending up the arms, and even to the breast of the experimenter, occurred, though another person with a single hand in at any of these places, felt comparatively little. The shock could be obtained at parts very near the tail, as at 8, 9. I think it was strongest at about 1 and 8. As the hands were brought nearer together the effect diminished, until being in the same cross plane, it was, as before described, only sensible in the parts immersed (1775.).

1777. B placed his hands at 10, 11, at least four inches from the fish, whilst A touched the animal with a glass rod to excite it to action; B quickly received a powerful shock. In another experiment of a similar kind, as respects the non-necessity of touching the fish, several persons received shocks independently of each other; thus A was at 4, 6; B at 10, 11; C at 16, 17; and D at 18, 19; all were shocked at once, A and B very strongly, C and D feebly. It is very useful, whilst experimenting with the galvanometer or other instrumental arrangements, for one person to keep his hands in the water at a moderate distance from the animal, that he may know and give information when a discharge has taken place.

1778. When B had both hands at 10, 11, or at 14, 15, whilst A had but one hand at 1, or 3, or 6, the former felt a strong shock, whilst the latter had but a weak one, though in contact with the fish. Or if A had both hands in at 1, 2, or 3, 4, or 5, 6, the effect was the same.

---

[20] Annales de Chimie, xiv. p. 18.

1779. If A had the hands at 3, 5, B at 14, 15, and C at 16, 17, A received the most powerful shock, B the next powerful, and C the feeblest.

1780. When A excited the Gymnotus by his hands at 8, 9, whilst B was at 10, 11, the latter had a much stronger shock than the former, though the former touched and excited the animal.

1781. A excited the fish by one hand at 3, whilst B had both hands at 10, 11 (or along), and C had the hands at 12, 13 (or across); A had the pricking shock in the immersed hand only (1774.); B had a strong shock up the arms; C felt but a slight effect in the immersed parts.

1782. The experiments I have just described are of such a nature as to require many repetitions before the general results drawn from them can be considered as established; nor do I pretend to say that they are anything more than indications of the direction of the force. It is not at all impossible that the fish may have the power of throwing each of its four electric organs separately into action, and so to a certain degree direct the shock, i. e. he may have the capability of causing the electric current to emanate from one side, and at the same time bring the other side of his body into such a condition, that it shall be as a non-conductor in that direction. But I think the appearances and results are such as to forbid the supposition, that he has any control over the direction of the currents after they have entered the fluid and substances around him.

1783. The statements also have reference to the fish when in a straight form; if it assume a bent shape, then the lines of force around it vary in their intensity in a manner that may be anticipated theoretically. Thus if the hands were applied at 1, 7, a feebler shock in the arms would be expected if the animal were curved with that side inwards, than if it were straight, because the distance between the parts would be diminished, and the intervening water therefore conduct more of the force. But with respect to the parts *immersed,* or to animals, as fish *in the water* between 1 and 7, they would be more powerfully, instead of less powerfully, shocked.

1781. *the pricking shock*: Some writers compare this sensation to the prickling one feels in an extremity that has "fallen asleep."

1783. *a feebler shock in the arms would be expected*: With hands applied at 1 and 7, the experimenter's body would constitute an alternate path of discharge—the other path being through the water. If the fish bends as described, the distance of the water path will diminish, while the path through hands, arms, and chest remains unchanged. *Less* current, therefore, will pass through the arms, and *more* through the water or anything immersed in the water.

1784. It is evident from all the experiments, as well as from simple considerations, that all the water and all the conducting matter around the fish through which a discharge circuit can in any way be completed, is filled at the moment with circulating electric power; and this state might be easily represented generally in a diagram by drawing the lines of inductive action (1231. 1304. 1338.) upon it: in the case of a Gymnotus, surrounded equally in all directions by water, these would resemble generally, in disposition, the magnetic curves of a magnet, having the same straight or curved shape as the animal, i. e. provided he, in such cases, employed, as may be expected, his four electric organs at once.

1785. This Gymnotus can stun and kill fish which are in very various positions to its own body; but on one day when I saw it eat, its action seemed to me to be peculiar. A live fish about five inches in length, caught not half a minute before, was dropped into the tub. The Gymnotus instantly turned round in such a manner as to form a coil inclosing the fish, the latter representing a diameter across it; a shock passed, and there in an instant was the fish struck motionless, as if by lightning, in the midst of the waters, its side floating to the light. The Gymnotus made a turn or two to look for its prey, which

---

1784. *all the water and all the conducting matter around the fish … is filled at the moment with circulating electric power…*:   The fish is presented as the bearer of an activity that fills space, an agent that *occupies space through his peculiar action.* Where, then, are the boundaries of his body? What is his real shape?

these would resemble generally, in disposition, the magnetic curves…:   The Gymnotus in action is surrounded by lines of *electric* induction—but Faraday invokes the *magnet*, with its beautiful system of curves, to expound the disposition of electrical power about the fish. The two systems impose comparable geometries upon their surroundings, but the magnetic shapes are more easily made visible. Therefore Faraday portrays Gymnotus with an image of the *fish as magnet.* (In a beautiful later essay he will reverse that order, portraying the magnet as comparable to the Gymnotus or Torpedo—an image of the *magnet as fish!*—see paragraph 3276 in "On the Physical Character of the Lines of Magnetic Force," which follows the 29th Series.)

1785. *The Gymnotus instantly turned round in such a manner as to form a coil inclosing the fish*:   This is the remarkable "coiling incident." Notice Faraday's efforts to convey what is evidently for him the preeminent *readability* of Gymnotus's behavior. The theme of *concentration of ambient power* is evidenced by the unusually sudden and intense convulsion delivered to the prey— emphatically conveyed in Faraday's phraseology: "in an instant … struck motionless, as if by lightning…" *Electrical* readability in this episode derives also from the *volitional* readability of the coiling gesture. Since Gymnotus's shock is generally for the sake of killing his prey, a gesture that enhances

[The coiling incident. The sketch is from Faraday's Diary; it was not published with the *Experimental Researches.*]

having found he bolted, and then went searching about for more. A second smaller fish was given him, which being hurt in the conveyance, showed but little signs of life, and this he swallowed at once, apparently without shocking it. The coiling of the Gymnotus round its prey had, in this case, every appearance of being intentional on its part, to increase the force of the shock, and the action is evidently exceedingly well suited for that purpose (1783.), being in full accordance with the well-known laws of the discharge of currents in masses of conducting matter; and though the fish may not always put this artifice in practice, it is very probable he is aware of its advantage, and may resort to it in cases of need.

---

his habitual hunting behavior implies also an *enhancement of lethal power*—hence a *concentration* of force onto the prey. That the animal must *bend its own body* in order to effect an apparent focusing of its external power suggests, if it does not actually imply, a definite though flexible *structure* in the external action, itself a kind of body or extension of body; a body, moreover, whose substance is not matter but *force*. Gymnotus's shock is not to be viewed as a separable armament, but as a functional extension of the body. It is not a weapon wielded, but a limb employed.

The twin anatomical principles of this new body are *contiguity* and *coherence.* In contrast to the specialized organs, ligaments, and conduits of a physiological body, in this new Body Electric action is *everywhere.* It is voluminous and fills space, yet is not contained either by a membrane or a vessel. It is shaped, but not by a container—rather by its own relations of equilibrium. It is, at the present stage of Faraday's researches, an admittedly enthusiastic and somewhat fantastic metaphor; yet less than 12 years hence Faraday will be speaking essentially the same language—honed, disciplined, and enriched by a series of brilliant magnetic researches—about the *lines of magnetic force,* that most profound, pervasive, and fertile of all his images. See his paper "On the Physical Character of the Lines of Magnetic Force."

bolted: that is, swallowed.

1786. Living as this animal does in the midst of such a good conductor as water, the first thoughts are thoughts of surprise that it can sensibly electrify anything, but a little consideration soon makes one conscious of many points of great beauty, illustrating the wisdom of the whole arrangement. Thus the very conducting power which the water has; that which it gives to the moistened skin of the fish or animal to be struck; the extent of surface by which the fish and the water conducting the charge to it are in contact; all conduce to favour and increase the shock upon the doomed animal, and are in the most perfect contrast with the inefficient state of things which would exist if the Gymnotus and the fish were surrounded by air; and at the same time that the power is one of low intensity, so that a dry skin wards it off, though a moist one conducts it (1760.); so is it one of great quantity (1770.), that though the surrounding water does conduct away much, enough to produce a full effect may take its course through the body of the fish that is to be caught for food, or the enemy that is to be conquered.

1787. Another remarkable result of the relation of the Gymnotus and its prey to the medium around them is, that the larger the fish to be killed or stunned, the greater will be the shock to which it is subject, though the Gymnotus may exert only an equal power; for the large fish has passing through its body those currents of electricity, which, in the case of a smaller one, would have been conveyed harmless by the water at its sides.

1788. The Gymnotus appears to be sensible when he has shocked an animal, being made conscious of it, probably, by the *mechanical impulse* he receives, caused by the spasms into which it is thrown.

1786. *Living ... in the midst of such a good conductor as water*: Gymnotus's freshwater medium is a "good conductor" only in comparison to the air environment of land animals; it would not be considered "good" in other contexts. It is a poorer conductor than saltwater, though better than pure distilled water.

surprise that it can sensibly electrify anything: Such an attitude reflects the supposition that a "good" conductor would discharge the animal's electric organ *right at the body surface*—in what we now would call a "short circuit"— and thus forestall any transmission of the discharge to more distant bodies. But in the Twelfth Series (paragraphs 1331, 1334) Faraday recognized that conduction and induction occur in parallel paths *simultaneously*—which means that in any homogeneous material, electrification can never be confined to a single avenue but must spread throughout the conducting substance. Thus water surrounding the animal will become electrified even at considerable distances.

When I touched him with my hands, he gave me shock after shock; but when I touched him with glass rods, or the insulated conductors, he gave one or two shocks, felt by others having their hands in at a distance, but then ceased to exert the influence, as if made aware it had not the desired effect. Again, when he has been touched with the conductors several times, for experiments on the galvanometer or other apparatus, and appears to be languid or indifferent, and not willing to give shocks, yet being touched by the hands, they, by convulsive motion, have informed him that a sensitive thing was present, and he has quickly shown his power and his willingness to astonish the experimenter.

1789. It has been remarked by Geoffroy St. Hilaire, that the electric organs of the Torpedo, Gymnotus, and similar fishes, cannot be considered as essentially connected with those which are of high and direct importance to the life of the animal, but to belong rather to the common teguments; and it has also been found that such Torpedos as have been deprived of the use of their peculiar organs, have continued the functions of life quite as well as those in which they were allowed to remain. These, with other considerations, lead me to look at these parts with a hope that they may upon close investigation prove to be a species of natural apparatus, by means of which we may apply the principles of *action and reaction* in the investigation of the nature of the *nervous influence*.

1790. The anatomical relation of the nervous system to the electric organ; the evident exhaustion of the nervous energy during the production of electricity in that organ; the apparently equivalent production of electricity in proportion to the quantity of nervous force consumed; the constant direction of the current produced, with its relation to what we may believe to be an equally constant direction of the nervous energy thrown into action at the same time; all induce me to believe, that it is not

---

1788. *astonish*: Here it carries the archaic meaning *to benumb or paralyze*—to "turn to stone," as it were (compare the antique but related participle "astonied"). But the word's modern meaning—*to strike with great wonder or surprise*—is also highly appropriate!

1789. The electric organs are of muscular derivation and are therefore not classified with the higher life functions but among "the common teguments"—that is, the ordinary tissues (*tegument*, or *integument*, is a covering, sheath, hide, or husk). What this means is that the fish's electric apparatus is comparable in its office to any of the ordinary muscular organs, for example to the locomotory structures, the *fins*.

impossible but that, on passing electricity per force through the organ, a reaction back upon the nervous system belonging to it might take place, and that a restoration, to a greater or smaller degree, of that which the animal expends in the act of exciting a current, might perhaps be effected. We have the analogy in relation to heat and magnetism. Seebeck taught us how to commute heat into electricity; and Peltier has more lately given us the strict converse of this, and shown us how to convert the electricity into heat, including both its relation of hot and cold. Oersted showed how we were to convert electric into magnetic forces, and I had the delight of adding the other member of the full relation, by reacting back again and converting magnetic into electric forces. So perhaps in these organs, where nature has provided the apparatus by means of which the animal can exert and convert nervous into electric force, we may be able, possessing in that point of view a power far beyond that of the fish itself, to reconvert the electric into the nervous force.

1791. This may seem to some a very wild notion, as assuming that the nervous power is in some degree analogous to such powers as heat, electricity, and magnetism. I am only assuming it, however, as a reason for making certain experiments, which, according as they give positive or negative results, will regulate further expectation. And with respect to the nature of nervous power, that exertion of it which is conveyed along the nerves to the various organs which they excite into action, is not the direct principle of *life*; and therefore I see no natural reason why we should not be allowed in certain cases to *determine* as well as observe its course. Many philosophers think the power is electricity. Priestley put forth this view in 1774 in a very striking and distinct form, both as

---

1790. *passing electricity per force through the organ*: Faraday proposes nothing less than to recharge the fish! But which direction *through the organ* is "per force"— that of the animal's natural internal process, from tail to head (paragraph 1772), or the reverse direction from head to tail? Writers have too hastily assumed that Faraday means a *reverse current*—which is indeed the way we recharge a modern storage battery. But Faraday nowhere actually specifies a current in the reverse direction; see paragraph 1792 below. Nor is restoration always associated with reverse action: When we wish to restore a degraded bar-magnet to its former power we must apply magnetic power in the *same* direction as the magnet's own internal action, not the reverse. Restoration by reversal is appropriate for things that "run down," like a voltaic battery or a spring motor. But it is far from certain that such is Faraday's view of the living creature.

1791. *not the direct principle of life...*: If the fish's electrical faculty is a natural power like other powers, it should be open to artful experimentation. Even if *life* involves a transcendent principle not subject to experimental control and direction, it would not militate against the proposed experiments.

to determine as well as observe...: Here *determine* has the sense *to govern*.

regards ordinary animals and those which are electric, like the Torpedo.[21] Dr. Wilson Philip considers that the agent in certain nerves is electricity modified by vital action.[22] Matteucci thinks that the nervous fluid or energy, in the nerves belonging to the electric organ at least, is electricity.[23] MM. Prevost and Dumas are of opinion that electricity moves in the nerves belonging to the muscles; and M. Prevost adduces a beautiful experiment, in which steel was magnetized, in proof of this view; which, if it should be confirmed by further observation and by other philosophers, is of the utmost consequence to the progress of this high branch of knowledge.[24] Now though I am not as yet convinced by the facts that the nervous fluid is only electricity, still I think that the agent in the nervous system may be an inorganic force; and if there be reasons for supposing that magnetism is a higher relation of force than electricity (1664. 1731. 1734.), so it may well be imagined that the nervous power may be of a still more exalted character, and yet within the reach of experiment.

1792. The kind of experiment I am bold enough to suggest is as follows. If a Gymnotus or Torpedo has been fatigued by frequent exertion of the electric organs, would the sending of currents of similar force to those he emits, or of other degrees of force, either continuously or intermittingly in the same direction as those he sends forth, restore him his powers and strength more rapidly than if he were left to his natural repose?

---

[21] Priestley on Air, vol. i. p. 277, Edition of 1774.

[22] Dr. Wilson Philip is of opinion, that the nerves which excite the muscles and effect the chemical changes of the vital functions, operate by the electric power supplied by the brain and spinal marrow, in its effects, modified by the vital powers of the living animal; because he found, as he informs me, as early as 1815, that while the vital powers remain, all these functions can be as well performed by voltaic electricity after the removal of the nervous influence, as by that influence itself; and in the end of that year he presented a paper to the Royal Society, which was read at one of their meetings, giving an account of the experiments on which this position was founded.

[23] Bibliothèque Universelle, 1837, tom. xii. 192.

[24] Ibid., 1837, xii. 202; xiv. 200.

1792. *currents … in the same direction as those he sends forth…*: Faraday conjectures that such currents might prove restorative. But which direction is "the same"? The animal "sends forth" currents from the positive head, through the water, to the negative tail (paragraph 1761). Does "the same direction" then mean a *continuation* of this external direction into the interior of the animal—that is, through the organ from *tail* to *head*? Or does "the same direction" mean from *head* to *tail*—but through the organ rather than through the water?

1793. Would sending currents through in the contrary direction exhaust the animal rapidly? There is, I think, reason to believe the Torpedo (and perhaps the Gymnotus) is not much disturbed or excited by electric currents sent only through the electric organ; so that these experiments do not appear very difficult to make.

1794. The disposition of the organs in the Torpedo suggest still further experiments on the same principle. Thus when a current is sent in the natural direction, i. e. from below upwards through the organ on one side of the fish, will it excite the organ on the other side into action? or if sent through in the contrary direction, will it produce the same or any effect on that organ? Will it do so if the nerves proceeding to the organ or organs be tied? and will it do so after the animal has been so far exhausted by previous shocks as to be unable to throw the organ into action in any, or in a similar, degree of his own will?

1795. Such are some of the experiments which the conformation and relation of the electric organs of these fishes suggest, as being rational in their performance, and promising in anticipation. Others may not think of them as I do; but I can only say for myself, that were the means in my power, they are the very first that I would make.

Royal Institution,
November 9th, 1838.

1794. Here Faraday explicitly considers "the natural direction" of the current within Torpedo to be "from below upwards." Since Torpedo's external current is from his upper wing surface, through the water, to the lower wing surface (paragraph 1750), this direction is in accordance with the animal's normal discharge. Perhaps this suggests that for Gymnotus, too, Faraday's proposed restorative currents "in the same direction as those he sends forth" (paragraph 1792) are currents having the "natural" direction—that is, the direction that *agrees* with the animal's normal activity.

1795. *the very first [experiments] that I would make*: Coming from the most celebrated experimentalist of the day, this is extraordinarily urgent language! What specific "promise" does he anticipate? What pressing questions might he hope by such means to address?

Nineteenth Series — Editor's Introduction

"The illumination of magnetic lines of force"

The Nineteenth Series, which opens Volume III of the Experimental Researches, begins with an introductory note clarifying Faraday's title, *On the magnetization of light and the illumination of magnetic lines of force.* The phrase "illumination of magnetic lines" had created an impression, prior to publication, that Faraday claimed to have rendered the lines of force luminous. "This was not within my thought," he explains.

> I intended to express that the line of magnetic force is illuminated as the earth is illuminated by the sun, or the spider's web illuminated by the astronomer's lamp. Employing a ray of light we can tell, *by the eye*, the direction of the magnetic lines through a body; and by the alteration of the ray and its optical effect on the eye, can see the course of the lines just as we can see the course of a thread of glass, or any other transparent substance, rendered visible by light; and this is what I meant by illumination...

Thomas Simpson called appreciative attention to this remark in 1968,[*] pointing out that what Faraday expresses here—in a footnote!—is nothing less than his paradigm of experimental science: *to make visible to the eye the powers of nature.* Indeed, Faraday's images of "the spider's web" and "a thread of glass" seem to attribute *literal visibility* to the magnetic lines—even if not a luminance of their own. But whatever the extent of Faraday's claim, he certainly regards *light* as a powerful investigative instrument; and by the end of the present Series, light may perhaps be credited with having made a *new magnetic condition of matter* more visible, at least to the mind's eye. Both the *light* and the *matter* used in the present experiments are rather special, so I will say something about each in turn.

---

[*] *A Critical Study of Maxwell's Dynamic Theory of the Electromagnetic Field in the Treatise on Electricity and Magnetism.* A new edition of this work is forthcoming from Green Lion Press under the title *Figures of Thought: A Study of Maxwell's* Treatise.

"Polarized" light

You have probably viewed polarized light if you have ever seen day-light reflected from the surface of a lake, or sun glare from a plate glass window. In fact all transparent materials tend to polarize any light that is reflected from their surfaces. On the other hand a few minerals like tourmaline, as well as the synthetic Polaroid material, polarize the light that passes through them. We can show this by passing light through a pair of such materials in succession.

  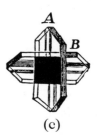

(a)   (b)   (c)

The sketch illustrates a pair of identical tourmaline crystals A and B, illuminated from behind. When held together with their axes parallel (a), light passes through both crystals. But as they are turned more and more out of coincidence (b), less and less light passes through the pair—although it passes readily through either crystal alone, in those areas where they do not overlap. Finally, with the crystals aligned perpendicularly (c), *no* light passes through the overlapping area. It seems that as a result of passing through the first crystal, A, the light acquires a kind of directionality related to the crystal axis. Its subsequent transmission through a second crystal, B, depends on how nearly B is aligned with the preferred direction.

Light possessing such directionality was called "polarized" by Newton.* Thus we say that crystal A *polarizes* the light, while crystal B *detects* the direction of polarization. Newton made the further point that since the polarizer and the detector are identical, they must both perform the same function. Now the detector, B, seems to act as a *filter*, inasmuch as it completely blocks light having a certain direction of polarization, while permitting other directions to pass more or less

* From experiments with Iceland spar (calcite), Newton first diagnosed such light as having "sides," like a flat ribbon. Finding an analogy between the two *sides* of the light beam and the two ends of a magnet that constitute its *poles*, Newton called the light exhibiting this property *polarized* light. The analogy seems farfetched today, but the term "polarized" has stuck, nevertheless. That is unfortunate, since there is almost no similarity between "polarization" as applied to light and "polarization" in the electric or magnetic sense, as used by Faraday in earlier Series.

freely. Therefore the polarizer, A, must similarly act as a filter. But a filter does not *modify* the entities that pass through it; it does not bestow upon them properties they did not have before—rather, it *separates* entities which already possess certain properties from those which do not. From such considerations arises the following understanding: An ordinary light beam is thought to possess independent "directionalities" along every radius about its axis. To be "polarized" therefore means to have been subjected to a selection process that eliminates or attenuates all but one of these directions. Both the fact and the direction of polarization may then be "detected" by causing the light to undergo a second selection.

Thus when polarized light is viewed through a detector (which is just a second polarizer), rotating the detector about the line of sight as axis discloses two positions, 180° apart, for which the transmitted light is a maximum. If the detector is then turned clockwise or counterclockwise away from either position, the transmitted light gradually diminishes, and is extinguished altogether when the material is turned 90° from either maximum. Ordinary light exhibits no such selectivity of transmission.

For his work in the present Series, Faraday will obtain polarized light by reflection from a sheet of glass.* Instead of a tourmaline crystal he will employ the *Nicol prism*, a more sophisticated detector made from two crystals of Iceland spar cemented together. In the proper orientation, it transmits polarized light with much greater intensity than tourmaline does.

Faraday's "heavy glass"

Optical lenses are possible because glass and other transparent materials have the power to refract—that is, to *bend*—the rays of light that enter and leave them. The degree of bending depends on several factors. One of them is the *color* of the light (that is, the *wavelength*, according the wave theory of light). This dependence, called *dispersion*, has some unhappy optical consequences that I will describe in a moment. But when all the pertinent factors are equal, materials that bend light through a greater angle are said to have greater *refractivity* than other materials.

---

* When light is polarized by reflection from a surface, the thoroughness of its polarization depends on the angle of reflection. Polarization becomes *total* for an angle that is specific to the materials involved. For reflections from glass in air, the angle for total polarization is about 58° with respect to the normal line; that is, 32° with respect to the reflecting plane.

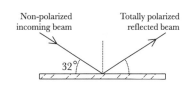

The art of grinding lenses consists in shaping the surfaces of the lens material so that light rays radiating from one point of a source shall be brought together, or *focused*, at a corresponding point somewhere else, thereby contributing to the formation of an *image* of the source. But the fact of *dispersion* makes it impossible to have a single focus for rays of all colors. Different colors in the source will be imaged at different distances; and white light will be split up into its spectral components, each at a different focus. The focus for violet light is always closer to the lens than the focus for red light, as depicted (with great exaggeration) in the sketch.

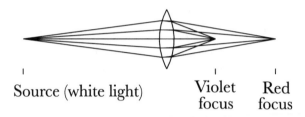

Source (white light)   Violet   Red
                              focus   focus

The injurious effects of dispersion can, however, be overcome to a considerable extent in one lens by combining it with a second lens. If the refractive power and dispersive tendency of its material are properly chosen, the auxiliary lens can be so shaped as to spread out the violet rays and collect the red rays, and so bring all the rays back to a common focus. The best combination is usually a convex lens made of less refractive glass, cemented to a concave lens of more refractive glass, as sketched here.

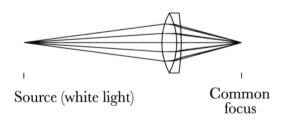

Source (white light)   Common
                               focus

A suitable combination of low- and high-refractivity materials is therefore indispensable in any optical instrument whose use is not restricted to monochromatic light (light of a single color). In addition, optical glass must possess exceptional transparency, durability, and homogeneity.

Some twenty years before commencing the experiments described in the present Series, Faraday had been drafted for a government-backed research project aimed at developing improved optical glass, primarily for navigational telescopes. As the most difficult manufacture, and the most serious deficiencies, were found in the existing highly refractive "flint" glass, Faraday and his associates after some time directed their efforts exclusively to the development of a new variety of high-refractivity glass that might improve upon the often-unsatisfactory "flint."

In practice, the desired high refractivity dictated that the new glass contain a high proportion of lead compounds in its makeup; this resulted in an unusually dense, that is, "heavy," material. The experiments were messy and laborious and would constitute a serious drain on Faraday's time and energy for four or five years in all; but at last a practical fabrication technique was developed and preliminary samples of *silicated lead borate glass*—distinguished for its unusually high refractive index, as well as for its extraordinary heaviness—were successfully produced. The photograph shows a few of Faraday's glass samples, which are still in the possession of the Royal Institution.

Faraday reported extensively on these experiments, but they failed to stimulate any significant large-scale improvements in glass manufacture. His hard-won specimens of "heavy" glass lay idle on the laboratory shelf for the next sixteen years, until the present investigations indicated a fresh need for a highly refractive medium. Then the "heavy" glass was recalled from ignominy to fulfill a role more consequential, perhaps, than its intended industrial exploitation might ever have become.

To see why a highly refractive glass was thought desirable for these investigations, we must look briefly at Faraday's experimental apparatus and rationale.

Magnetic rotation of a light ray

As Faraday states at the outset of the present Series, his initial aim was to find evidence of some kind of magnetic influence on light. He could not find any *direct* action; if any existed it was too weak to be detected. But glass and other optical media are obviously capable of influencing light—the more so, the higher their refractivity. Perhaps, therefore, a magnetic effect on light, too feeble to be observed in air, might be enhanced in a highly refractive medium. This conjecture led Faraday to call his old "heavy" glass specimens into service for the procedure described in the present Series.

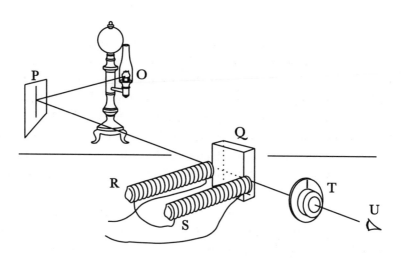

The sketch illustrates Faraday's apparatus. Light from lamp O is reflected (and polarized) at the glass plate P. It passes through Q, the highly refractive glass block, and thence through the Nicol eyepiece T. If the eyepiece is turned into the position of maximum transmission, the light proceeds to the observer at U. If it is then rotated 90°—that is, into the position of zero transmission—the light is of course blocked. But Faraday will find that the eyepiece position for extinguishing the beam is different when electromagnets R and S are energized than when they are inactive. Evidently, then, the directional orientation of the polarized beam is *rotated* clockwise or counter-clockwise by the magnetic action! This is exactly the sort of magneto-optical effect he had sought to exhibit.

But we are not to suppose that Faraday had anticipated the precise form of this effect, or that he knew in advance exactly what geometrical relation between light beam and magnet would suffice to produce it. Early in the present Series (paragraph 2148) he makes laconic reference to "many unproductive experiments" which had preceded the successful trials. In several of them he had directed the lines of magnetic force *across* the path of the light beam. It took considerable experimenting before he tried the present arrangement, in which the lines of magnetic force run *parallel* to the light beam.

A sketch from Faraday's *Diary* shows the details. The electromagnets are mounted so as to present opposite poles to one and the same side of the glass block. Although with this arrangement the magnetic lines intersecting the glass will be *curves*, not straight lines, they nevertheless run generally parallel to the light beam for much of their length. Try to visualize the course of the magnetic lines of force through the glass in Faraday's sketch.*

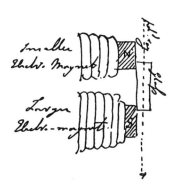

"New magnetic condition of matter"

That *magnetism* can exert an influence on *light* conforms perfectly to a belief in the fundamental interrelatedness of *all* natural powers. Faraday voices such a conviction—which while deeply characteristic was by no means unique to him—at the outset of the present Series; and writers have tended to discuss the "magneto-optical effect" exclusively from that perspective, which, to be sure, is a very important one.

But as the paper proceeds, Faraday will gradually change the focus of his attention from the *relation between magnetism and light* to the *condition of the optical material.* Whether the glass merely *amplifies* an action between magnetism and light, or whether it plays some more essential role, it must acquire *some* condition under magnetic action in order to render any effect whatsoever. Yet the glass surely does not

---

* The sketch shows a top view of the apparatus with horizontal light beam, not a vertical light beam as might at first appear.

magnetize like iron—it does not, for example, show any sign of having developed *poles.** It must, therefore, be in *a new kind of magnetic state.* Faraday names the new state "diamagnetic" in paragraph 2149 of the present Series; and it will become the principal topic of investigation in subsequent Series.

* Faraday wistfully points out (paragraph 2227) that if only things were otherwise, we should have the opportunity to study a transparent magnet!

EXPERIMENTAL RESEARCHES
IN
ELECTRICITY.

NINETEENTH SERIES.[1]

Received November 6,—Read November 20, 1845.

¶ i. *Action of magnets on light.*

2146. I HAVE long held an opinion, almost amounting to conviction, in common I believe with many other lovers of natural knowledge, that the various forms under which the forces of matter are made manifest have one common origin; or, in other words, are so directly related and mutually dependent, that they are convertible, as it were, one into another, and possess equivalents of power in their action.[3] In modern

---

[1] Philosophical Transactions, 1846, p. 1.

[2] The title of this paper has, I understand, led many to a misapprehension of its contents, and I therefore take the liberty of appending this explanatory note. Neither accepting nor rejecting the hypothesis of an æther, or the corpuscular, or any other view that may be entertained of the nature of light; and, as far as I can see, nothing being really known of a ray of light more than of a line of magnetic or electric force, or even of a line of gravitating force, except as it and they are manifest in and by substances; I believe that, in the experiments I describe in the paper, light has been magnetically affected, i. e. that that which is magnetic in the forces of matter has been affected, and in turn has affected that which is truly magnetic in the force of light: by the term magnetic I include here either of the peculiar exertions of the power of a magnet, whether it be that which is manifest in the magnetic or the diamagnetic class of bodies. The phrase "illumination of the lines of magnetic force" has been understood to imply that I had rendered them luminous. This was not within my thought. I intended to express that the line of magnetic force was illuminated as the earth is illuminated by the sun, or the spider's web illuminated by the astronomer's lamp. Employing a ray of light, we can tell, by the *eye*, the direction of the magnetic lines through a body; and by the alteration of the ray and its optical effect on the eye, can see the course of the lines just as we can see the course of a thread of glass, or any other transparent substance, rendered visible by the light: and this is what I meant by *illumination,* as the paper fully explains. —December 15, 1845. M. F.

[3] Experimental Researches, 57, 366, 376, 877, 961, 2071.

2146. *the various forms ... have one common origin*: What immense scope Faraday displays in his conviction of the *unity of powers*! This is not just another suspicion of reciprocity between two forces, such as opened the First Series (paragraph 3). Here, *all* the "forces of matter" have a common origin, a comprehensive relation, and may be expected to exhibit mutual convertibility.

times the proofs of their convertibility have been accumulated to a very considerable extent, and a commencement made of the determination of their equivalent forces.

2147. This strong persuasion extended to the powers of light, and led, on a former occasion, to many exertions, having for their object the discovery of the direct relation of light and electricity, and their mutual action in bodies subject jointly to their power;[4] but the results were negative and were afterwards confirmed, in that respect, by Wartmann.[5]

2148. These ineffectual exertions, and many others which were never published, could not remove my strong persuasion derived from philosophical considerations; and, therefore, I recently resumed the inquiry by experiment in a most strict and searching manner, and have at last succeeded in *magnetizing and electrifying a ray of light, and in illuminating a magnetic line of force.* These results, without entering into the detail of many unproductive experiments, I will describe as briefly and clearly as I can.

2149. But before I proceed to them, I will define the meaning I connect with certain terms which I shall have occasion to use:—thus, by *line of magnetic force,* or *magnetic line of force,* or *magnetic curve,* I mean that exercise of magnetic force which is exerted in the lines usually called magnetic curves, and which equally exist as passing from or to magnetic poles, or forming concentric circles round an electric current. By *line of electric force,* I mean the force exerted in the lines joining two bodies, acting on each other according to the principles of static electric induction (1161, &c.), which may also be either in curved or straight lines. By a *diamagnetic,* I mean a body through which lines of magnetic force are passing, and which does not by their action assume the usual magnetic state of iron or loadstone.

[4] Philosophical Transactions, 1834. Experimental Researches, 951–955.

[5] Archives de l'Electricité, ii. pp. 596–600.

2149. *line of force:* Note that according to the definitions here offered, a line of force is *an actual force,* not just the symbol or representative of a force.

diamagnetic: The term is surely patterned after *dielectric* in Series 11 (paragraph 1168, *note*). There, a dielectric was a substance "through or across which the electric forces are acting"; here, similarly, a diamagnetic is "a body through which lines of magnetic force are passing." It is in that sense that Faraday can speak, in paragraph 2152 below, of a glass block being "placed as a *diamagnetic* ... between the poles"—so placed, in other words, that magnetic lines of force will pass through it. But the definition goes further, to characterize (even if only negatively) the *internal condition* of the diamagnetic

2150. A ray of light issuing from an Argand lamp, was polarized in a horizontal plane by reflexion from a surface of glass, and the polarized ray passed through a Nichol's eye-piece revolving on a horizontal axis, so as to be easily examined by the latter. Between the polarizing mirror and the eye-piece two powerful electro-magnetic poles were arranged, being either the poles of a horse-shoe magnet, or the contrary poles of two cylinder magnets; they were separated from each other about 2 inches in the direction of the line of the ray, and so placed, that, if on the same side of the polarized ray, it might pass near them; or if on contrary sides, it might go between them, its direction being always parallel, or nearly so, to the magnetic lines of force (2149.). After that, any transparent substance placed between the two poles, would have passing through it, both the polarized ray and the magnetic lines of force at the same time and in the same direction.

2151. Sixteen years ago I published certain experiments made upon optical glass,[6] and described the formation and general characters of one variety of heavy glass, which, from its materials, was called silicated borate of lead. It was this glass which first gave me the discovery of the relation between light and magnetism, and it has power to illustrate it in a degree beyond that of any other body; for the sake of perspicuity I will first describe the phænomena as presented by this substance.

[6] Philosophical Transactions, 1830, p. 1. I cannot resist the occasion which is thus offered to me of mentioning the name of Mr. Anderson, who came to me as an assistant in the glass experiments, and has remained ever since in the Laboratory of the Royal Institution. He has assisted me in all the researches into which I have entered since that time, and to his care, steadiness, exactitude, and faithfulness in the performance of all that has been committed to his charge, I am much indebted. —M. F.

---

material: it "does not ... assume the usual magnetic state." Perhaps a better wording would be, "assumes a state different from the usual magnetic state." Otherwise a substance that was simply indifferent to magnetic power, developing *no particular condition*, would also have to be called a "diamagnetic." The present investigation will demonstrate the existence of materials that conform to the stronger wording. Whether there actually are any *magnetically indifferent* materials is not addressed in the present Series—but it will become a pressing question in the next Series.

2151. Faraday's "heavy glass" is described in the editor's introduction to the present Series. While not the only material capable of revealing an interaction between magnetism and light, its high refractivity turns out to show the effect with great prominence. That is important because when Faraday first begins the investigation he cannot know exactly what to look for; and too subtle an effect might easily escape his notice.

2152. A piece of this glass, about 2 inches square and 0.5 of an inch thick, having flat and polished edges, was placed as a *diamagnetic* (2149.) between the poles (not as yet magnetized by the electric current), so that the polarized ray should pass through its length; the glass acted as air, water, or any other indifferent substance would do; and if the eye-piece were previously turned into such a position that the polarized ray was extinguished, or rather the image produced by it rendered invisible, then the introduction of this glass made no alteration in that respect. In this state of circumstances the force of the electro-magnet was developed, by sending an electric current through its coils, and immediately the image of the lamp-flame became visible, and continued so as long as the arrangement continued magnetic. On stopping the electric current, and so causing the magnetic force to cease, the light instantly disappeared; these phænomena could be renewed at pleasure, at any instant of time, and upon any occasion, showing a perfect dependence of cause and effect.

2153. The voltaic current which I used upon this occasion, was that of five pair of Grove's construction, and the electromagnets were of such power that the poles would singly sustain a weight of from twenty-eight to fifty-six, or more, pounds. A person looking for the phænomenon for the first time would not be able to see it with a weak magnet.

2154. The character of the force thus impressed upon the diamagnetic is that of *rotation*; for when the image of the lamp-flame has thus been rendered visible, revolution of the eye-piece to the right or left, more or less, will cause its extinction; and the further motion of the eye-piece to the one side or other of this position will produce the reappearance of the light, and that with complementary tints, according as this further motion is to the right- or left-hand.

2152. *the force of the electro-magnet was developed … and immediately the image of the lamp-flame became visible…*: The correlation between visibility of the flame image and activation of the electromagnet certainly reveals a magnetic effect on light. But what *is* the effect, precisely? Did the polarized light become *un*-polarized, for example? If not, what *did* happen?

2153. *A person looking for the phænomenon for the first time would not be able to see it with a weak magnet.* Nor, as I mentioned in a comment to paragraph 2151, with a weakly refractive glass. The reasons are similar in both cases.

2154. *when the image of the lamp-flame has thus been rendered visible, revolution of the eye-piece to the right or left … will cause its extinction*: With the magnet turned *on*, the eyepiece extinguishes the polarized beam when turned to a *new* position, different from that which had previously extinguished the beam when the magnet was off. The light beam is not, therefore, depolarized by magnetic action; rather, it is *rotated* axially.

2155. When the pole nearest to the observer was a marked pole, *i. e.* the same as the north end of a magnetic needle, and the further pole was unmarked, the rotation of the ray was right-handed; for the eye-piece had to be turned to the right-hand, or clock fashion, to overtake the ray and restore the image to its first condition. When the poles were reversed, which was instantly done by changing the direction of the electric current, the rotation was changed also and became left-handed, the alteration being to an equal degree in extent as before. The direction was always the same for the same *line of magnetic force* (2149.).

*   *   *

2160. Magnetic lines, then, in passing through silicated borate of lead, and a great number of other substances (2173.), cause these bodies to act upon a polarized ray of light when the lines are parallel to the ray, or in proportion as they are parallel to it: if they are per-pendicular to the ray, they have no action upon it. They give the diamagnetic the power of rotating the ray; and the *law* of this action on light is, that if a magnetic line of force be *going from* a north pole, or *coming* from a south pole, along the path of a polarized ray coming to the observer, it will rotate that ray to the right-hand; or, that if such a line of force be coming from a north pole, or going from a south pole, it will rotate such a ray to the left-hand.

2161. If a cork or a cylinder of glass, representing the diamagnetic, be marked at its ends with the letters N and S, to repre-sent the poles of a magnet, the line joining these letters may be considered as a mag-netic line of force; and further, if a line be traced round the cylinder with arrow heads on it to represent direction, as in the figure, such a simple model, held up before the

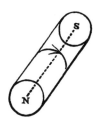

2160. *going from … coming from*: These directions are in relation to the *observer*; they do not express the inherent directionality of the lines of force (which is conventionally *from* a north pole *to* a south pole). It will be helpful to consult Faraday's sketch in paragraph 2161. Thus, if the north pole adjacent to the glass is *towards* the observer, and the south pole *away*, Faraday speaks of the line of force as "going from a north pole" and "coming from a south pole." In such a case the beam of light—which is of course directed towards the viewer—will be rotated to the viewer's right. If instead the poles are reversed, the line will be "coming from a north pole" and "going from a south pole"; and rotation of the light beam will be to the observer's left.

eye, will express the whole of the law, and give every position and consequence of direction resulting from it. If a watch be considered as the diamagnetic, the north pole of a magnet being imagined against the face, and a south pole against the back, then the motion of the hand will indicate the direction of rotation which a ray of light undergoes by magnetization.

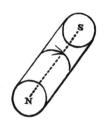

*   *   *

¶ iii. General considerations.

2221. Thus is established, I think for the first time,[7] a true, direct relation and dependence between light and the magnetic and electric forces; and thus a great addition made to the facts and considerations which tend to prove that all natural forces are tied together, and have one common origin (2146.). It is, no doubt, difficult in the present state of our knowledge to express our expectation in exact terms; and, though I have said that another of the powers of nature is, in these experiments, directly related to the rest, I ought, perhaps, rather to say that another form of the great power is distinctly and directly related to the other forms; or, that the great power manifested by particular phænomena in particular forms, is here further identified and recognized, by the direct relation of its form of light to its forms of electricity and magnetism.

2222. The relation existing between *polarized* light and magnetism and electricity, is even more interesting than if it had been shown to exist with common light only. It cannot but extend to common light;

---

[7] [I have omitted Faraday's footnote, and subsequent codicil, on whether the claim "for the first time" is accurate. —Editor]

2161. Notice that the "law" embodied in Faraday's handy model is formally identical to the *right-hand rule* identified in the First Series (paragraph 38, *note*). With the right hand grasping the glass around the dimension along which the light travels, let the thumb point in the direction of the magnetic lines of force from north pole to south pole. Then the fingers give the direction of rotation of the light beam *for glass and most materials*—but there are a few exceptions which were unknown to Faraday; see the comment on paragraph 2224 below.

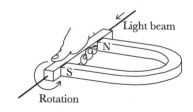

and, as it belongs to light made, in a certain respect, more precise in its character and properties by polarization, it collates and connects it with these powers, in that duality of character which they possess, and yields an opening, which before was wanting to us, for the appliance of these powers to the investigation of the nature of this and other radiant agencies.

2223. Referring to the conventional distinction before made (2149.), it may be again stated, that it is the magnetic lines of force *only* which are effectual on the rays of light, and they only (in appearance) when parallel to the ray of light, or as they tend to parallelism with it. As, in reference to matter not magnetic after the manner of iron, the phænomena of electric induction and electrolyzation show a vast superiority in the energy with which electric forces can act as compared to magnetic forces, so here, in another direction and in the peculiar and correspondent effects which belong to magnetic forces, they are shown, in turn, to possess great superiority, and to have their full equivalent of action on the same kind of matter.

2224. The magnetic forces do not act on the ray of light directly and without the intervention of matter, but through the mediation of the substance in which they and the ray have a simultaneous existence; the substances and the forces giving to and receiving from each other the power of acting on the light. This is shown by the non-action of a vacuum, of air or gases; and it is also further shown by the special degree in which different matters possess the property. That

---

2222. *made … more precise in its character*: Inasmuch as polarized light has a specific orientation it may be said to be "more precise in its character" than ordinary light, which represents all possible orientations about the beam axis.

appliance: that is, *application.*

2223. *it is the magnetic lines of force only which are effectual…*: In an earlier section here omitted, Faraday showed that *electric currents* are capable of producing the rotations—but *only* by means of their electromagnetic action. Thus, if current passes through a helix coiled longitudinally about the glass bar, a light beam directed through the bar will revolve in *the same direction* as that in which the current is passing. Of course this result is also consistent with the electro-magnetic right-hand rule, mentioned the comment to paragraph 2161 above.

in reference to matter not magnetic after the manner of iron, the phænomena of electric induction and electrolyzation show a vast superiority…: Faraday's tortuous statement notes the superior strength which, *in diamagnetic materials* (that is, materials "not magnetic after the manner of iron"), electrical phenomena generally exhibit in comparison to magnetic phenomena. Contrastingly, here is a phenomenon, specifically produced in diamagnetic material, for which only magnetic, not electric action, is effectual.

magnetic force acts upon the ray of light always with the same character of manner and in the same direction, independent of the different varieties of substance, or their states of solid or liquid, or their specific rotative force (2232.), shows that the magnetic force and the light have a direct relation: but that substances are necessary, and that these act in different degrees, shows that the magnetism and the light act on each other through the intervention of the matter.

2225. Recognizing or perceiving *matter* only by its powers, and knowing nothing of any imaginary nucleus, abstract from the idea of these powers, the phænomena described in this paper much strengthen my inclination to trust in the views I have on a former occasion advanced in reference to its nature.[8]

2226. It cannot be doubted that the magnetic forces act upon and affect the internal constitution of the diamagnetic, just as freely in the dark as when a ray of light is passing through it; though the phænomena produced by light seem, as yet, to present the only means of observing this constitution and the change. Further, any such change as this must belong to opake bodies, such as wood, stone, and

---

[8] Vol. ii. p. 284, or Philosophical Magazine, 1844, vol. xxiv. p. 136.

2224. *magnetic force acts upon the ray of light always with the same character of manner and in the same direction, independent of the different varieties of substance…*: Faraday describes experiments with many other media, besides the distinctive heavy glass originally tested, in omitted portions of this paper. Some show no detectable effects at all; the rest exhibit rotations in the same direction, through specifically characteristic angles. Faraday assumes that all materials will conform to this pattern, although a few exceptions will later be discovered later that exhibit rotations in the opposite direction. Nevertheless, the fact that there is found a degree of effect characteristic of each medium shows that the medium must play an essential role in the action. It must, therefore, be thrown into some specific condition when under magnetic influence. Faraday will draw this conclusion explicitly in paragraphs 2226 and 2227 below.

2225. *imaginary nucleus*: Faraday does not here refer to the atomic nucleus—a concept which was first proposed only in the early twentieth century. This is, instead, his openly dismissive term for the hypothetical *atom* itself! We earlier saw him express skepticism about the atomic idea in the Seventh Series (paragraph 869). In addition to other objections to the usual atomic theory, Faraday finds it philosophically incoherent to ground *natural powers* (weight, inertia, chemical—or electric—force) in a tiny little piece of *matter*. What can it possibly mean for matter to *possess* and *exercise* powers? What can matter possibly be, *apart* from those powers? Conventional atomism sidesteps those questions. The 1844 article cited in Faraday's note was published in the second volume of the Experimental Researches but, regrettably, it could not be included in the present selection.

metal; for as diamagnetics, there is no distinction between them and those which are transparent. The degree of transparency can at the utmost, in this respect, only make a distinction between the individuals of a class.

2227. If the magnetic forces had made these bodies magnets, we could, by light, have examined a transparent magnet; and that would have been a great help to our investigation of the forces of matter. But it does not make them magnets (2171.), and therefore the molecular condition of these bodies, when in the state described, must be specifically distinct from that of magnetized iron, or other such matter, and must be a *new magnetic condition*; and as the condition is a state of tension (manifested by its instantaneous return to the normal state when the magnetic induction is removed), so the *force* which the matter in this state possesses and its mode of action, must be to us a *new magnetic force* or *mode of action* of matter.

2228. For it is impossible, I think, to observe and see the action of magnetic forces, rising in intensity, upon a piece of heavy glass or a tube of water, without also perceiving that the latter acquire properties which are not only *new* to the substance, but are also in subjection to very definite and precise laws (2160. 2199.), and are equivalent in proportion to the magnetic forces producing them.

*   *   *

Royal Institution,
   Oct. 29, 1845.

Twentieth Series — Editor's Introduction

A new magnetic condition of matter

In the course of the Nineteenth Series Faraday's narrative underwent a dramatic shift of perspective. The magneto-optical rotations, which at the outset signified a *relation between magnetism and light*, quickly acquired fresh significance by indicating the existence of *a new magnetic condition* in the glass medium through which light rays and magnetic lines of force jointly passed. Indeed, in the course of that Series—or so I there suggested—this hitherto unsuspected species of magnetization actually replaced the magnetic-optic relation as the focus of Faraday's attention.

The present Series only substantiates that shift of emphasis. It opens with a retrospective reference to the Nineteenth Series in which Faraday ignores entirely the question of a direct relation between magnetism and light, attending instead to the "new magnetic condition of matter" exclusively. Faraday will call this new magnetic state the *diamagnetic* condition; its further investigation will occupy both the present Series and its sequel.

This is surely a remarkable transformation. For in the very opening sentence of the Nineteenth Series, Faraday had affirmed his belief that all the forces of matter are "directly related and mutually dependent." Nor, he indicated, was this a recent or tentative opinion with him but, on the contrary, a firm belief of long standing. His demonstration of a magnetic effect on light therefore represents a triumphant confirmation of that conviction. How, then, can he permit his attention to be so swept away from the magnetic and optical *powers*, to focus instead on the *material medium* in which they subsist? What is so compelling about the new "diamagnetic" condition of the medium as to raise it to such overarching significance?

Certainly the discovery of any new thing is of interest to science—all the more so when the novelty makes its appearance in a subject as familiar and seemingly well-explored as *magnetism*, which had been known for 2400 years and studied systematically for three centuries. But Faraday may have more particular reasons, and deeper ones, for viewing diamagnetism as a compelling discovery. These reasons derive from the mutual relation between magnetism and electricity. Since the Eleventh Series, at least, Faraday has understood that *all materials sustain a relation to electricity.* Materials in general *carry and pass onward* the electrical lines of force—in Faraday's terminology, all are *dielectrics.** Moreover each material transmits the lines of force with a characteristic degree of efficacy, which Faraday has called its *specific inductive capacity.*

The relation of materials to magnetism, on the other hand, has up to now looked very different. Among elementary substances only iron, nickel, and cobalt have exhibited the ability to become magnetic—iron predominating by far in this respect. Other materials have been seemingly passive and indifferent with respect to the magnetic force—permitting the magnetic lines of force to pass through but developing no detectable active relation with them. In nonmagnetic materials, there has been found nothing corresponding to the *specific inductive capacity* which those very same materials exhibit in their electrical relations.

Such asymmetry in the material relations of magnetism and electricity is disagreeable and unsatisfying in any case. But far more than a taste for symmetry and order is at stake. Remember that Faraday established the *mutual convertibility of electricity and magnetism* in the very first Series of these Experimental Researches: wherever the magnetic force in a material (or a space) changes, electrical force is developed; similarly, wherever electric current passes, magnetic force is developed. In any circumstance other than the strictest static condition, therefore, electricity and magnetism mutually imply one another. *How then can any material sustain a relation to electric force and not also to magnetic power?* And yet that must be the case for most materials, if indeed none but iron, nickel and cobalt are bearers of the magnetic condition.

But the discovery of diamagnetism extends the reach of the magnetic force beyond iron, nickel, and cobalt to *potentially all materials,* thereby matching the scope of the electric force. The material domains of magnetism and electricity being (at least in prospect) thus identical, they no longer militate against the First Series' unification of electricity and magnetism. Of course diamagnetism remains a novelty, its singular phenomena demanding thorough investigation and documentation; and without doubt it defies all existing theories to assimilate and explain it. But along with its imposing challenge, diamagnetism shows promise to *confirm and extend the fundamental unity* which Faraday has found in electricity and magnetism, and which—so his declaration in

---

* Every material, and vacuum too, is a *dielectric,* whether or not it is also a *conductor.* The two categories are independent, with one exception: a *perfect conductor*—if such exists—would terminate all electrical lines of force incident on it; it would not transmit any. A perfect conductor would not, therefore, be a dielectric in even the slightest degree.

the Nineteenth Series suggests—he hopes and expects to find among *all* the powers of nature.

Earlier in this section I asked why Faraday in the Nineteenth Series should have taken more interest in the diamagnetic state than in the direct relation between magnetism and light. It is a delicious irony that *electro-magnetic unity*, which is indirectly secured through the discovery of diamagnetism, is precisely an example of that *direct relation between natural powers* from which diamagnetism at first seemed to be a distraction!

A new magnetic dimension

But what is the specific character of diamagnetism? In the previous Series, Faraday could infer the *existence* of diamagnetism from the magnetic rotation of polarized light; yet the *nature* of this new magnetic condition—other than its not exhibiting "poles"—was far from evident. Now Faraday will document some diamagnetic actions that set diamagnetic materials apart from ordinary magnetic matter in a direct and unmistakable way, at the same time inviting interpretation as to their nature. But their most immediate consequence is to call attention to a neglected geometrical dimension in the arena of magnetic action.

Since the very opening of the Experimental Researches, Faraday has been attentive to the "magnetic curves," as revealed by magnetized needles or tiny slivers of iron. By definition, such indicators everywhere point in the direction of the curves; and Faraday has learned to interpret the *direction of the magnetic curves* as identical with the *direction of the magnetic power*. The curves associated with any magnet run, on the whole, from pole to pole; and when a magnet has *facing* poles, like the one sketched here, the shortest and straightest of those curves constitutes a polar *axis*. Faraday will define the straight line joining the magnetic poles as the *axial line*, and its direction as the *axial direction*.

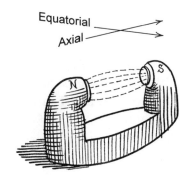

But a diamagnetic needle does *not* point in the direction of the magnetic curves. In the present Series Faraday will find that an elongated sample of diamagnetic material aligns itself not along, but *perpendicularly* to them! Is diamagnetism then another power, acting concurrently with the magnetic, but along the perpendicular dimension? Faraday will define the perpendicular bisector of the axial line as the *equatorial line*, and its direction as the *equatorial direction*. The sketch illustrates both directions; an implicit reference to the earth's polar axis and the terrestrial equator is evident.

A peculiar magnetic behavior of copper

Copper is not one of the magnetic metals. It is nevertheless capable of entering into important magnetic relations, as we saw in earlier Series; and under certain magnetic influences it performs some spectacular and rather enigmatic movements. Faraday will give an extended account of these phenomena beginning at paragraph 2309 below. Briefly, they are:

i. The "first turn": Copper is diamagnetic and points equatorially under a steady magnetic influence, just as bismuth does. But if a copper bar is placed midway between the axial and equatorial positions with respect to an electromagnet *not yet energized*, then when the electromagnet is switched on the bar shows a short-lived but unmistakable impulse towards the *axial* position! Faraday several times refers to this impulse as the "first turn."

The bar thereafter resists all attempts to push it in any direction. But the "feel" of this resistance is very different from, for example, the *restoring force* of a pendulum, which becomes stronger as the pendulum is displaced further from its equilibrium position. Instead, the copper's resistance resembles a viscous drag, which favors no particular position but opposes movement *per se*. The resistance persists, so long as the electromagnet remains energized.

ii. The "revulsion": With the magnet still energized, suppose the bar is restored to the same stationary position it had originally occupied (neither fully axial nor fully equatorial). If then the electromagnet is suddenly switched *off*, the bar exhibits another strong impulse to motion—this time towards the *equatorial* position. Faraday calls this impulse a "revulsion," since it suggests a sort of recoil from the original impulse.

Puzzling as these phenomena may be, the element of a "dragging" force is strongly reminiscent of the Arago rotations which, Faraday showed in the First Series, resulted from circulating electrical currents that were induced in the highly conductive Arago wheel by *changing* magnetic influences. Similar currents, Faraday will argue, underlie the peculiar gyrations observed when copper is present in the changing magnetic field of a powerful electromagnet.*

The role of *time* in electro-magnetism

When an electric battery is connected to a long wire or a coil—especially a coil containing an iron core—the current that passes does not develop its full value instantaneously but only gradually. Similarly, if

---

* These induced circulating currents are today called "eddy-currents." The term comes from water-lore, where an eddy is a pocket of circular motion in river or tidal flow.

the battery is suddenly removed the current does not instantly disappear but tends to persist, sometimes producing a momentary spark *after* the battery has been disconnected. Faraday investigated this phenomenon in the Ninth Series and showed that it was essentially a variation of *magneto-electric induction*, that production of electricity by magnetism which he had announced in the First Series. There, you will recall, Faraday showed (paragraphs 7 and 19) that when a current commenced in one wire its associated magnetic action induced a momentary current, *in the opposite direction*, in a parallel wire or coil. When the primary current ceased, the decaying magnetic force again induced a transient secondary current, now *in the same direction* as that of the diminishing primary current.

Clearly such interaction is not restricted to independent conductors but must occur also between the adjacent turns of a single coil. The diagram shows a two-turn coil connected to the poles of a battery. As battery current rises in each winding, an induced current having opposite direction forms in the adjacent winding.* The induced currents necessarily oppose the battery current—as a result, the total coil current rises more slowly than it otherwise would. Similarly, a *diminishing* battery current in each winding will induce current having the same direction in the adjacent winding. The induced currents reinforce the battery current, so that the total current falls more slowly than otherwise. These effects are greatly enhanced if an iron core is present. That is of course to be expected, since Faraday showed in the First Series that the tendency of a changing current to induce another current is a magnetic action, intensified in the presence of iron.

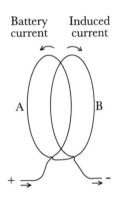

Thus the current through a coil or helix—or through any circuit whose parts experience mutual magnetic influence—will develop to its full value more slowly, and fall again to zero more gradually, than if such magnetic linkage were not present. This effect is generally termed *self-induction* today. In one of his splendid textbooks, Sylvanus P. Thompson summarizes the result, offering an example that is particularly appropriate for our study. He writes:

---

* For simplicity, the diagram represents *battery current only* in coil A, and *induced current only* in coil B; but since both windings belong to the same wire, each of them must in actuality carry both currents.

It requires time to magnetize an iron core. This is mainly due to the fact that a current, when first switched on, does not instantly attain its full strength, being retarded by the self-induced counter-electro-motive force... Faraday's large electromagnet at the Royal Institution takes about two seconds to attain its maximum strength.*

Faraday will describe this electromagnet—which is indeed "large," and which is still in the possession of the Royal Institution—in paragraph 2247 below.

* Written 1894.

TWENTIETH SERIES.[1]

§ 27. *On new magnetic actions, and on the magnetic condition of all matter.*[2] ¶ i. *Apparatus required.* ¶ ii. *Action of magnets on heavy glass.* ¶ iii. *Action of magnets on other substances acting magnetically on light.* ¶ iv. *Action of magnets on the metals generally.*

Received December 6,—Read December 18, 1845.

2243. THE contents of the last series of these Researches were, I think, sufficient to justify the statement, that a new magnetic condition (*i. e.* one new to our knowledge) had been impressed on matter by subjecting it to the action of magnetic and electric forces (2227.); which new condition was made manifest by the powers of action which the matter had acquired over light. The phænomena now to be described are altogether different in their nature; and they prove, not only a magnetic condition of the substances referred to unknown to us before, but also of many others, including a vast number of opake and metallic bodies, and perhaps all except the magnetic metals and their compounds: and they also, through that condition, present us with the means of undertaking the correlation of magnetic phænomena, and perhaps the construction of a theory of general magnetic action founded on simple fundamental principles.

2244. The whole matter is so new, and the phænomena so varied and general, that I must, with every desire to be brief, describe much which at last will be found to concentrate under simple principles of action. Still, in the present state of our knowledge, such is the only method by which I can make these principles and their results sufficiently manifest.

¶ i. *Apparatus required.*

2245. The effects to be described require magnetic apparatus of great power, and under perfect command. Both these points are obtained by the use of electro-magnets, which can be raised to a degree of force far beyond that of natural or steel magnets; and further, can be suddenly altogether deprived of power, or made energetic to the

---

[1] Philosophical Transactions, 1846, p. 21.

[2] [I have omitted a lengthy note of Faraday's, in which he recognizes an earlier paper by Becquerel. —Editor]

highest degree, without the slightest alteration of the arrangement, or of any other circumstance belonging to an experiment.

2246. One of the electro-magnets which I use is that already described under the term Woolwich helix (2192.). The soft iron core belonging to it is 28 inches in length and 2.5 inches in diameter. When thrown into action by ten pair of Grove's plates, either end will sustain one or two half-hundred weights hanging to it. The magnet can be placed either in the vertical or the horizontal position. The iron core is a cylinder with flat ends, but I have had a cone of iron made, 2 inches in diameter at the base and 1 inch in height, and this placed at the end of the core, forms a conical termination to it, when required.

2247. Another magnet which I have had made has the horseshoe form. The bar of iron is 46 inches in length and 3.75 inches in diameter, and is so bent that the extremities forming the poles are 6 inches from each other; 522 feet of copper wire 0.17 of an inch in diameter, and covered with tape, are wound round the two straight parts of the bar, forming two coils on these parts, each 16 inches in length, and composed of three layers of wire: the poles are, of course, 6 inches apart, the ends are planed true, and against these move two short bars of soft iron, 7 inches long and 2½ by 1 inch thick, which can be adjusted by screws, and held at any distance less than 6 inches from each other. The ends of these bars form the opposite poles of contrary name; the magnetic field between them can be made of greater or smaller extent, and the intensity of the lines of magnetic force be proportionately varied.

2247. *Another magnet...*:   This monster magnet, more powerful than any Faraday had used before, is still in the possession of the Royal Institution and is pictured here. Its iron core was fabricated from a link of ship's anchor chain.

magnetic field: Note Faraday's usage here; the magnetic field is a *place*, not an agent. Faraday ascribes "extent" to the field, but "intensity" to the *lines of magnetic force*. It is therefore the *lines of force* that sustain magnetic power, the field being only the space which those lines occupy. Today we are accustomed to a modern locution, in which "field" denotes a spatial disposition of power—or even the power itself. The phraseology is current not only in academic science but in popular culture (think of the "force fields" which abounded in classic science-fiction literature). Faraday is customarily credited with laying the ground for the modern view, for reasons we shall be better able to appreciate in later Series. But here the "magnetic field" is only a *place where magnetic events happen*—just as a baseball field is a place where baseball games happen.

2248. For the suspension of substances between and near the poles of these magnets, I occasionally used a glass jar, with a plate and sliding wire at the top. Six or eight lengths of cocoon silk being equally stretched, were made into one thread and attached, at the upper end, to the sliding rod, and at the lower end to a stirrup of paper, in which anything to be experimented on could be sustained.

2249. Another very useful mode of suspension was to attach one end of a fine thread, 6 feet long, to an adjustable arm near the ceiling of the room, and terminating at the lower end by a little ring of copper wire; any substance to be suspended could be held in a simple cradle of fine copper wire having 8 or 10 inches of the wire prolonged upward; this being bent into a hook at the superior extremity, gave the means of attachment to the ring. The height of the suspended substance could be varied at pleasure, by bending any part of the wire at the instant into the hook form. A glass cylinder placed between the magnetic poles was quite sufficient to keep the suspended substance free from any motion, due to the agitation of the air.

2250. It is necessary, before entering upon an experimental investigation with such an apparatus, to be aware of the effect of any magnetism which the bodies used may possess; the power of the apparatus to make manifest such magnetism is so great, that it is difficult on that account to find writing-paper fit for the stirrup above mentioned. Before therefore any experiments are instituted, it must be ascertained that the suspending apparatus employed does not point, *i. e.* does not take up a position parallel to the lines joining the magnetic poles, by virtue of the magnetic force. When copper suspensions are employed, a peculiar effect is produced (2309.), but when understood, as it will be hereafter, it does not interfere with the results of experiment. The wire should be fine, not magnetic as iron, and the form of the suspending cradle should not be elongated horizontally, but be round or square as to its general dimensions, in that direction.

2251. The substances to be experimented with should be carefully examined, and rejected if not found free from magnetism. Their state is easily ascertained; for, if magnetic, they will either be attracted to

---

2250. *writing-paper fit for the stirrup…*: It would be "fit" by being immune to magnetic influence. We seldom think of *paper* as having magnetic properties, but the enormous magnetic power his apparatus can develop reveals magnetic effects even in this material. It is important to ascertain and minimize such effects in advance, lest they distort the experimental results.

should not be elongated horizontally: Clearly, elongation would *magnify* any undesirable pointing tendency the suspending cradle might already have.

the one or the other pole of the great magnet, or else point between them. No examination by smaller magnets, or by a magnetic needle, is sufficient for this purpose.

2252. I shall have such frequent occasion to refer to two chief directions of position across the magnetic field, that to avoid periphrasis, I will here ask leave to use a term or two, conditionally. One of these directions is that from pole to pole, or along the line of magnetic force; I will call it the axial direction: the other is the direction perpendicular to this, and across the line of magnetic force; and for the time, and as respects the space between the poles, I will call it the equatorial direction. Other terms that I may use, I hope will explain themselves.

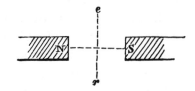

¶ ii. *Action of magnets on heavy glass.*

2253. The bar of silicated borate of lead, or heavy glass already described as the substance in which magnetic forces were first made effectually to bear on a ray of light (2152.), and which is 2 inches long, and about 0.5 of an inch wide and thick, was suspended centrally between the magnetic poles (2247.), and left until the effect of torsion was over. The magnet was then thrown into action by making contact at the voltaic battery: immediately the bar moved, turning round its point of suspension, into a position across the magnetic curve or line of force, and after a few vibrations took up its place of rest there. On being displaced by hand from this position, it returned to it, and this occurred many times in succession.

2251. *the great magnet*: That is, the magnet described in paragraph 2247 above. Since the actual experiment will be carried out under high magnetic intensities, the preliminary testing must be performed under equally high intensities.

2253. *silicated borate of lead*: This is the distinctive "heavy" glass, which figured so prominently in the magnetic rotation of light (Nineteenth Series). Note that in the present Series, Faraday is using a sample having a more elongated barlike shape.

left until the effect of torsion was over: If there is any twist in the suspension thread, the bar will swing back and forth after its release; but it will eventually come to rest in a position of equilibrium.

a position across the magnetic curve: That is, perpendicular to the lines of force—the position called "equatorial" above. If the axial position runs north-and-south, the equatorial position runs east-and-west.

it returned to it: The equatorial setting is *stable*, since the bar is clearly subject to a restoring force when slightly displaced from that position.

2254. Either end of the bar indifferently went to either side of the axial line. The determining circumstance was simply inclination of the bar one way or the other to the axial line, at the beginning of the experiment. If a particular or marked end of the bar were on one side of the magnetic, or axial line, when the magnet was rendered active that end went further outwards, until the bar had taken up the equatorial position.

2255. Neither did any change in the magnetism of the poles, by change in the direction of the electric current, cause any difference in this respect. The bar went by the shortest course to the equatorial position.

2256. The power which urged the bar into this position was so thoroughly under command, that if the bar were swinging it could easily be hastened in its course into this position, or arrested as it was passing from it, by seasonable contacts at the voltaic battery.

2257. There are two positions of equilibrium for the bar; one stable, the other unstable. When in the direction of the axis or magnetic line of force, the completion of the electric communication causes no change of place; but if it be the least oblique to this position, then the obliquity increases until the bar arrives at the equatorial position; or if the bar be originally in the equatorial position, then the magnetism causes no further changes, but retains it there (2298. 2299. 2384.).

2258. Here then we have a magnetic bar which points east and west, in relation to north and south poles, *i. e.* points perpendicularly to the lines of magnetic force.

2259. If the bar be adjusted so that its point of suspension, being in the axial line, is not equidistant from the poles, but near to one of them, then the magnetism again makes the bar take up a position perpendicular to the magnetic lines of force; either end of the bar being on the one side of the axial line, or the other, at pleasure. But at the same time there is another effect, for at the moment of completing the

---

2254. *Either end of the bar indifferently…*: Because a compass needle is stable in only *one* north-south orientation, it is possible to designate one end as "north-seeking." But the bar of heavy glass is equally stable in *either* of the two east-west alignments—clearly there is nothing like an *east-seeking end.*

2257. Equatorial orientation of the glass bar (perpendicular to the lines of force) is *stable.* The bar may also be retained in an axial position (parallel with the lines of force), but the balance is *unstable*—the slightest disturbance will destroy it. Here too, though Faraday does not say so, the bar shows no preference between the two possible axial orientations; both are equally unstable.

electric contact, the centre of gravity of the bar recedes from the pole and remains repelled from it as long as the magnet is retained excited. On allowing the magnetism to pass away, the bar returns to the place due to it by its gravity.

2260. Precisely the same effect takes place at the other pole of the magnet. Either of them is able to repel the bar, whatever its position may be, and at the same time the bar is made to assume a position, at right angles, to the line of magnetic force.

2261. If the bar be equidistant from the two poles, and in the axial line, then no repulsive effect is or can be observed.

2262. But preserving the point of suspension in the equatorial line, *i. e.* equidistant from the two poles, and removing it a little on one side or the other of the axial line (2252.), then another effect is brought forth. The bar points as before across the magnetic line of force, but at the same time it recedes from the axial line, increasing its distance from it, and this new position is retained as long as the magnetism continues, and is quitted with its cessation.

2263. Instead of two magnetic poles, a single pole may be used, and that either in a vertical or a horizontal position. The effects are in perfect accordance with those described above; for the bar, when near the pole, is repelled from it in the direction of the line of magnetic force, and at the same time it moves into a position perpendicular to the direction of the magnetic lines passing through it. When the

---

2259. *the centre of gravity … recedes from the pole*: In previous trials Faraday had suspended the glass bar midway between the magnet poles. Now when the suspension position is displaced towards one of the poles (remaining still along the axial line) he finds the same perpendicular "pointing" as before; but in addition the bar as a whole *recedes* from the nearer pole! Note that Faraday uses the word "repelled" here and in several subsequent paragraphs. Does he mean to suggest a mutual *pushing away* between pole and bar?

2262. *it recedes from the axial line*: Similarly if the suspension point is displaced in the *equatorial* direction (remaining still equidistant from the poles), the bar as a whole recedes from the axial line. Note that even if we thought the previous recession *from the pole* was a case of mutual pushing away (paragraph 2259), it would be difficult to hold the same interpretation here, since the *axial line* is not a body! How could there be *pushes* exerted between a *body* and a *geometrical line*?

2263. A "single pole" might be, for example, either end of the straight "Woolwich" helix. As described in paragraph 2246, its iron core is 2.5 inches in diameter.

perpendicular to the … magnetic lines…: Not *all* the lines of force that intersect the bar can be perpendicular to it, of course; but since the bar is only 2 inches

magnet is vertical (2246.) and the bar by its side, this action makes the bar a tangent to the curve of its surface.

2264. To produce these effects, of pointing across the magnetic curves, the form of the heavy glass must be long; a cube, or a fragment approaching roundness in form, will not point, but a long piece will. Two or three rounded pieces or cubes, placed side by side in a paper tray, so as to form an oblong accumulation, will also point.

2265. Portions, however, of any form, are *repelled*: so if two pieces be hung up at once in the axial line, one near each pole, they are repelled by their respective poles, and approach, seeming to attract each other. Or if two pieces be hung up in the equatorial line, one on each side of the axis, then they both recede from the axis, seeming to repel each other.

2266. From the little that has been said, it is evident that the bar presents in its motion a complicated result of the force exerted by the magnetic power over the heavy glass, and that, when cubes or spheres are employed, a much simpler indication of the effect may be obtained. Accordingly, when a cube was thus used with the two poles, the effect was repulsion or recession from either pole, and also recession from the magnetic axis on either side.

---

in length (paragraph 2253), compared to the pole diameter of 2.5 inches, when the bar is in a position "tangent to the curve" of a vertical cylindrical magnet, most of the lines that intersect it will be at least roughly perpendicular to it. The sketch depicts the "tangent" position.

2264. *a fragment approaching roundness in form, will not point, but a long piece will*: Thus it is the elongation itself, not an inherently directional axis in the glass, that responds to the magnetic direction. Note also that the blunt pieces "placed side by side" act as a single elongated body, even though they are merely linked together by their paper holder, not physically joined. This topic will come up again in paragraph 2283 below. In a magnificent later paper, Faraday will rely on an image of the magnetic lines of force to visualize *how* separate pieces of material can act magnetically as *one*. See "On the Physical Character of the Lines of Magnetic Force," which follows the 29th Series.

2265. The repulsion (perhaps it would be better to say "recession," as Faraday does in the next paragraph) does not depend on having a particular shape. Does that mean that *recession* is more fundamental than *pointing*? In the next paragraph Faraday seems to draw that conclusion.

2266. *repulsion or recession*: Faraday seems to be detaching the word "repulsion" from its ordinary intimations of *pushing away*, using it instead as a virtual synonym for "recession," which does not normally carry such connotations.

2267. So, the indicating particle would move, either along the magnetic curves, or across them; and it would do this either in one direction or the other; the only constant point being, that its tendency was to move from stronger to weaker places of magnetic force.

2268. This appeared much more simply in the case of a single magnetic pole, for then the tendency of the indicating cube or sphere was to move outwards, in the direction of the magnetic lines of force. The appearance was remarkably like a case of weak electric repulsion.

2269. The cause of the pointing of the bar, or any oblong arrangement of the heavy glass, is now evident. It is merely a result of the tendency of the particles to move outwards, or into the positions of weakest magnetic action. The joint exertion of the action of all the particles brings the mass into the position, which, by experiment, is found to belong to it.

2270. When one or two magnetic poles are active at once, the courses described by particles of heavy glass free to move, form a set of lines or curves, which I may have occasion hereafter to refer to; and as I have called air, glass, water, &c. diamagnetic (2149.), so I will distinguish these lines by the term *diamagnetic curves,* both in relation to, and contradistinction from, the lines called magnetic curves.

2271. When the bar of heavy glass is immersed in water, alcohol, or æther, contained in a vessel between the poles, all the preceding effects occur; the bar points and the cube recedes exactly in the same manner as in air.

2272. The effects equally occur in vessels of wood, stone, earth, copper, lead, silver, or any of those substances which belong to the diamagnetic class (2149.).

2267. *to move from stronger to weaker places of magnetic force*: This principle may be termed *migratory* in that it refers the glass body's movements to *local conditions* of magnetic power, specifically to the direction of its decrease. In contrast, actions of *pointing* and *recession* ordinarily imply *distant* centers, though Faraday has been steadily freeing those terms from such connotations.

The migration principle now shows itself as underlying all the variety of motions that have been observed. Note that in paragraphs 2253–2267, those movements presented themselves under a succession of reinterpretations, being viewed first as pointings, then repulsions (or recessions), and finally migrations. It is distinctly the character of Faraday's experimentation to bring forth such provisional interpretations repeatedly and successively—until the essential form of the phenomenon becomes evident at last.

2269. Faraday states explicitly what was only implied in paragraph 2267 above—that the fundamental form of the glass's movements are *migrations* in the direction from stronger to weaker magnetic conditions.

2271. *æther*: That is, *ether*—the volatile liquid used as a medical anaesthetic.

2273. I have obtained the same equatorial direction and motions of the heavy glass bar as those just described, but in a very feeble degree, by the use of a good common steel horse-shoe magnet (2157.). I have not obtained them by the use of the helices (2191. 2192.) without the iron cores.

2274. Here therefore we have magnetic repulsion without polarity, *i. e.* without reference to a particular pole of the magnet, for either pole will repel the substance, and both poles will repel it at once (2262.). The heavy glass, though subject to magnetic action, cannot be considered as magnetic, in the usual acceptation of that term, or as iron, nickel, cobalt, and their compounds. It presents to us, under these circumstances, a magnetic property new to our knowledge; and though the phænomena are very different in their nature and character to those presented by the action of the heavy glass on light (2152.), still they appear to be dependent on, or connected with, the same condition of the glass as made it then effective, and therefore, with those phænomena, prove the reality of this new condition.

¶ iii. *Action of magnets on other substances acting magnetically on light.*

2275. We may now pass from heavy glass to the examination of the other substances, which, when under the power of magnetic or electric forces, are able to affect and rotate a polarized ray (2173.), and may also easily extend the investigation to bodies which, from their irregularity of form, imperfect transparency, or actual opacity, could not be examined by a polarized ray, for here we have no difficulty in the application of the test to all such substances.

2276. The property of being thus repelled and affected by magnetic poles, was soon found not to be peculiar to heavy glass. Borate of lead, flint-glass, and crown-glass set in the same manner equatorially, and were repelled when near to the poles, though not to the same degree as the heavy glass.

2274. *repulsion without polarity*: The "repulsions" exhibited by heavy glass are very different from those displayed by iron, nickel, or cobalt—the magnetic substances known previously. Ordinary magnetic repulsions take place between two bodies, both "magnetized" in the sense of exhibiting north and south "poles." But the migratory motions of heavy glass are no longer to be seen as in reference to the *poles* of the external magnet, nor is the glass itself endowed with anything like "northness" or "southness." The glass displays an entirely new kind of magnetic condition, in which *poles* are absent.

2276. *Borate of lead, flint-glass, and crown-glass*: All are glasses having different compositions and optical properties. With the possible exception of borate of lead (see paragraph 2284), none is as heavy, or as highly refractive, as his "heavy glass," which bears the chemical name "*silicated* borate of lead" (paragraph 2253).

2277. Amongst substances which could not be subjected to the examination by light, phosphorus in the form of a cylinder presented the phænomena very well; I think as powerfully as heavy glass, if not more so. A cylinder of sulphur, and a long piece of thick India rubber, neither being magnetic after the ordinary fashion, were well directed and repelled.

2278. Crystalline bodies were equally obedient, whether taken from the single or double refracting class (2237.). Prisms of quartz, calcareous spar, nitre and sulphate of soda, all pointed well, and were repelled.

2279. I then proceeded to subject a great number of bodies, taken from every class, to the magnetic forces, and will, to illustrate the variety in the nature of the substances, give a comparatively short list of crystalline, amorphous, liquid and organic bodies below. When the bodies were fluids, I enclosed them in thin glass tubes. Flint-glass points equatorially, but if the tube be of very thin glass, this effect is found to be small when 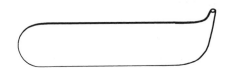 the tube is experimented with alone; afterwards, when it is filled with liquid and examined, the effect is such that there is no fear of mistaking that due to the glass for that of the fluid. The tubes must not be closed with cork, sealing-wax, or any ordinary substance taken at random, for these are generally magnetic (2285.). I have usually so shaped them in the making, and drawn them off at the neck, as to leave the aperture on one side, so that when filled with liquid they require no closing.

2280. Rock crystal.	Water.
Sulphate of lime.	Alcohol.
Sulphate of baryta.	Æther.
Sulphate of soda.	Nitric acid.
Sulphate of potassa.	Sulphuric acid.
Sulphate of magnesia.	Muriatic acid.
Alum.	Solutions of various
Muriate of ammonia.	alkaline and earthy
Chloride of lead.	salts
Chloride of sodium.	Glass.
Nitrate of potassa.	Litharge.
Nitrate of lead.	White arsenic.
Carbonate of soda.	Iodine.

2280. The table lists a variety of substances which all exhibit the distinctive equatorial pointing behavior. They are not tabulated in order of precedence, but Faraday offers a provisional ranking in paragraph 2284 below.

Iceland spar.
Acetate of lead.
Tartrate of potash and
    antimony.
Tartrate of potash and
    soda.
Tartaric acid.
Citric acid.
Olive oil.
Oil of turpentine.
Jet.
Caoutchouc.
Sugar.
Starch.
Gum-arabic.
Wood.
Ivory.

Phosphorus.
Sulphur.
Resin.
Spermaceti.
Caffeine.
Cinchonia.
Margaric acid.
Wax from shell-lac.
Sealing-wax.
Mutton, dried.
Beef, fresh.
Beef, dried.
Blood, fresh.
Blood, dried.
Leather.
Apple.
Bread.

2281. It is curious to see such a list as this of bodies presenting on a sudden this remarkable property, and it is strange to find a piece of wood, or beef, or apple, obedient to or repelled by a magnet. If a man could be suspended, with sufficient delicacy, after the manner of Dufay, and placed in the magnetic field, he would point equatorially; for all the substances of which he is formed, including the blood, possess this property.

2282. The setting equatorially depends upon the form of the body, and the diversity of form presented by the different substances in the list was very great; still the general result, that elongation in one direction was sufficient to make them take up an equatorial position, was established. It was not difficult to perceive that comparatively large masses would point as readily as small ones, because in larger masses more lines of magnetic force would bear in their action on the body,

---

2281. *the manner of Dufay*: Charles François Dufay (1698–1739) had himself hung up by silk threads in order to pursue investigations that involved electrical sparks and shocks, but not magnetism. Faraday seems to think that Dufay's silken suspension permitted him to *revolve*, but contemporary reports do not bear out that idea.

2282. *large masses would point as readily as small ones*: If so, larger masses must be acted upon by correspondingly greater turning forces. Faraday attributes these heightened forces to the fact that larger masses occupy greater volumes, which therefore intersect a greater number of lines of force. This implies a far-reaching principle—that *each line of force intersecting a body exerts a determinate share of the total force which that body experiences*. That principle, in turn, invokes the lines of force as *real structures or agents* in the material or in space.

and this was proved to be the case. Neither was it long before it evidently appeared that the form of a plate or a ring was quite as good as that of a cylinder or a prism; and in practice it was found that plates and flat rings of wood, spermaceti, sulphur, &c., if suspended in the right direction, took up the equatorial position very well. If a plate or ring of heavy glass could be floated in water, so as to be free to move in every direction, and were in that condition subject to magnetic forces diminishing in intensity, it would immediately set itself equatorially, and if its centre coincided with the axis of magnetic power, would remain there; but if its centre were out of this line, it would then, perhaps, gradually pass off from this axis in the plane of the equator, and go out from between the poles.

2283. I do not find that division of the substance has any distinct influence on the effects. A piece of Iceland spar was observed, as to the degree of force with which it set equatorially; it was then broken into six or eight fragments, put into a glass tube and tried again; as well as I could ascertain, the effect was the same. By a second operation, the calcareous spar was reduced into coarse particles; afterwards to a coarse powder, and ultimately to a fine powder: being examined as to the equatorial set each time, I could perceive no difference in the effect, until the very last, when I thought there might be a slight diminution of the tendency; but if so, it was almost insensible. I made the same experiment on silica with the same result, of no diminution of power. In reference to this point I may observe, that starch and other bodies in fine powder exhibited the effect very well.

2284. It would require very nice experiments and great care to ascertain the specific degree of this power of magnetic action possessed

---

2283. *the degree of force with which it set equatorially*: That is, the comparative strength of the *restoring force* which characterizes both equatorial and axial pointing; see paragraph 2253).

division of the substance has no evident effect either on the strength of diamagnetic pointing or its direction—as Faraday earlier noted in paragraph 2264 above. This suggests that a substance's ability to exhibit *as a whole* the diamagnetic condition does not depend on its being a whole *mechanically*. But how is that possible? By what means can mechanically separated pieces or particles of a body establish a *common relation* to magnetic, or any other, force? And if they can indeed sustain a common relation, then why has Faraday been unable to detect diamagnetism in *gases*?—for gases are, essentially, just very highly divided solids! (He will detect gas diamagnetism eventually; the report is in the 25th Series, not included here.)

2284. *experiments ... to ascertain the specific degree of this power*: Faraday does not explicitly state what phenomenon indicates this *specific degree*. But his phrase

by different bodies, and I have made very little progress in that part of the subject. Heavy glass stands above flint-glass, and the latter above plate-glass. Water is beneath all these, and I think alcohol is below water, and æther below alcohol. The borate of lead is I think as high as heavy glass, if not above it, and phosphorus is probably at the head of all the substances just named. I verified the equatorial set of phosphorus between the poles of a common magnet (2273.).

2285. I was much impressed by the fact that blood was not magnetic (2280.), nor any of the specimens tried of red muscular fibre of beef or mutton. This was the more striking, because, as will be seen hereafter, iron is *always* and in almost *all states* magnetic. But in respect to this point it may be observed, that the ordinary magnetic property of matter and this *new property* are in their effects opposed to each other; and that when this property is strong it may overcome a very slight degree of ordinary magnetic force, just as also a certain amount of the magnetic property may oppose and effectually hide the presence of this force (2422.). It is this circumstance which makes it so necessary to be careful in examining the magnetic condition of the bodies in the first instance (2250.). The following list of a few substances which were found slightly magnetic, will illustrate this point:—Paper, sealing-wax, china ink, Berlin porcelain, silkworm-gut, asbestos, fluor-spar, red lead, vermilion, peroxide of lead, sulphate of zinc, tourmaline, plumbago, shell-lac, charcoal. In some of these cases the magnetism was generally diffused through the body, in other cases it was limited to a particular part.

2286. Having arrived at this point, I may observe, that we can now have no difficulty in admitting that the phænomena abundantly establish the existence of a magnetic property in matter, new to our knowledge. Not the least interesting of the consequences that flow

---

"the degree of force with which it set equatorially" in the preceding paragraph suggests that he looks to the *strength or vigor* of the pointing actions, which would permit at least a rough ranking of different materials.

2285. *blood was not magnetic…*: This is a surprising result, since the red blood cells are known to contain iron; and iron is "*always* and in almost *all states* magnetic."

it may overcome… [it] may oppose and effectually hide…: Note how he here conceives ordinary magnetism and this "new property" as *independent and opposed tendencies*, capable of masking one another's effects. Is that only because blood, like many other materials, is a mixture—and may therefore include both magnetic and diamagnetic substances among its constituents? Or does he think that one and the same homogeneous substance might possess both magnetic and diamagnetic properties separately?

from it, is the manner in which it disposes of the assertion which has sometimes been made, that all bodies are magnetic. Those who hold this view, mean that all bodies are magnetic as iron is, and say that they point between the poles. The new facts give not a mere negative to this statement, but something beyond, namely, an affirmative as to the existence of forces in all ordinary bodies, directly the opposite of those existing in magnetic bodies, for whereas those practically produce attraction, these produce repulsion; those set a body in the axial direction, but these make it take up an equatorial position: and the facts, with regard to bodies generally, are exactly the reverse of those which the view quoted indicates.

¶ iv. *Action of magnets on metals generally.*

2287. The metals, as a class, stand amongst bodies having a high and distinct interest in relation both to magnetic and electric forces, and might at first well be expected to present some peculiar phænomena, in relation to the striking property found to be possessed in common by so large a number of substances, so varied in their general characters. As yet no distinction associated with conduction or non-conduction, transparent or opake, solid or liquid, crystalline or amorphous, whole or broken, has presented itself; whether the metals, distinct as they are as a class, would fall into the great generalization, or whether at last a separation would occur, was to me a point of the highest interest.

2288. That the metals, iron, nickel and cobalt, would stand in a distinct class, appeared almost undoubted; and it will be, I think, for the advantage of the inquiry, that I should consider them in a section apart by themselves. Further, if any other metals appeared to be magnetic, as these are, it would be right and expedient to include them in the same class.

2286. *Those who hold this view, mean that all bodies are magnetic as iron is, and say that they point between the poles*: Some investigators deny that a natural power, such as magnetism, can be fundamental to some materials but simply lacking in others; and therefore they insist that *all* bodies must be magnetic as iron is—though of course in different degrees. In this precise thesis they are mistaken; but the discovery of diamagnetism does at least vindicate their refusal to view nonmagnetic bodies as merely *lacking* a magnetic power. In the next Series (paragraph 2420) Faraday will suggest more explicitly that *diamagnetism*, though in certain respects *opposite* to magnetism, is more nearly allied to it than a mere indifference would be, and thus should be counted as a species of magnetic action.

2288. *iron, nickel, cobalt*: All exhibit the ordinary *magnetic* condition. Since that condition is specifically different from the "new" magnetic state, these metals ought to form a distinct class.

2289. My first point, therefore, was to examine the metals for any indication of ordinary magnetism. Such an examination cannot be carried on by magnets anything short in power of those to be used in the further investigation; and in proof of this point I found many specimens of the metals, which appeared to be perfectly free from magnetism when in the presence of a magnetic needle, or a strong horse-shoe magnet (2157.), that yet gave abundant indications when suspended near to one or both poles of the magnets described (2246.).

2290. My test of magnetism was this. If a bar of the metal to be examined, about 2 inches long, was suspended (2249.) in the magnetic field, and being at first oblique to the axial line was upon the supervention of the magnetic forces drawn into the axial position instead of being driven into the equatorial line, or remaining in some oblique direction, then I considered it magnetic. Or, if being near one magnetic pole, it was attracted by the pole, instead of being repelled, then I concluded it was magnetic. It is evident that the test is not strict, because, as before pointed out (2285.), a body may have a slight degree of magnetic force, and yet the power of the new property be so great as to neutralize or surpass it. In the first case, it might seem neither to have the one property nor the other; in the second case, it might appear free from magnetism, and possessing the special property in a *small* degree.

2291. I obtained the following metals, so that when examined as above, they did not appear to be magnetic; and in fact, if magnetic, were so to an amount so small as not to destroy the results of the other force, or to stop the progress of the inquiry.

Antimony.	Lead.
Bismuth.	Mercury.
Cadmium.	Silver.
Copper.	Tin.
Gold.	Zinc.

2292. The following metals were, and are as yet to me, magnetic, and therefore companions of iron, nickel and cobalt:—

2290. *My test of magnetisim*: The characteristic response of ordinary magnetic materials when presented to the pole of a magnet is to *approach* the pole and, if elongated, to point in the axial direction. In contrast, materials belonging to the new diamagnetic class *recede* from the pole or point equatorially.

so great as to neutralize or surpass it: If, however, a body possesses *both* forms of magnetism, he acknowledges that his test is incapable of distinguishing between them; it can only reveal whichever condition predominates in the body. Here again Faraday conceives the usual magnetism and the "new" condition as *independent and separate dispositions*. But in a later Series he will find reason to view them as *different degrees of a single characteristic*. And even that will not mark the end of these continual reinterpretations!

Platinum. | Titanium.
Palladium. |

2293. Whether all these metals are magnetic, in consequence of the presence of a little iron, nickel, or cobalt in them, or whether any of them are really so of themselves, I do not undertake to decide at present; nor do I mean to say that the metals of the former list are free. I have been much struck by the apparent freedom from iron of almost all the specimens of zinc, copper, antimony and bismuth, which I have examined; and it appears to me very likely that some metals, as arsenic, &c., may have much power in quelling and suppressing the magnetic properties of any portion of iron in them, whilst other metals, as silver or platinum, may have little or no power in this respect.

2294. Resuming the consideration of the influence exerted by the magnetic force over those metals which are not magnetic after the manner of iron (2291.), I may state that there are two sets of effects produced which require to be carefully distinguished. One of these depends upon induced magneto-electric currents, and shall be resumed hereafter (2309.). The other includes effects of the same nature as those produced with heavy glass and many other bodies (2276.).

2295. All the non-magnetic metals are subject to the magnetic power, and produce the same general effects as the large class of bodies already described. The force which they then manifest, they possess in different degrees. Antimony and bismuth show it well, and bismuth appears to be especially fitted for the purpose. It excels heavy glass, or borate of lead, and perhaps phosphorus; and a small bar or cylinder of it about 2 inches long, and from 0.25 to 0.5 of an inch in width, is as well fitted to show the various peculiar phænomena as anything I have yet submitted to examination.

2296. To speak accurately, the bismuth bar which I employed was 2 inches long, 0.33 of an inch wide, and 0.2 of an inch thick. When this bar was suspended in the magnetic field, between the two poles, and subject to the magnetic force, it pointed freely in the equatorial direction, as the heavy glass did (2253.), and if disturbed from that position, returned *freely* to it. This latter point, though perfectly in

---

2295. *All the non-magnetic metals are subject to the magnetic power*: This seemingly paradoxical statement simply means that none of the metals is *indifferent* to the magnetic power. Metals that are not "magnetic"—that is, in the manner of iron, nickel, and cobalt—exhibit the "new" condition first identified in heavy glass.

2296. *freely*: Faraday emphasizes this word because further on, copper and some other metals will exhibit motions that appear to be *impeded* in some way.

accordance with the former phænomena, is in such striking contrast with the phænomena presented by copper and some other of the metals (2309.), as to require particular notice here.

2297. The comparative sensibility of bismuth causes several movements to take place under various circumstances, which being complicated in their nature, require careful analysis and explanation. The chief of these, with their causes, I will proceed to point out.

2298. If the cylinder electro-magnet (2246.) be placed vertically so as to present one pole upwards, that pole will exist in the upper end of an iron cylinder, having a flat horizontal face 2½ inches in diameter. A small indicating sphere (2266.) of bismuth, hung over the centre of this face and close to it, does not move by the magnetism. If the ball be carried outwards, half-way, for instance, between the centre and the edge, the magnetism makes it move inwards, or towards the axis (prolonged) of the iron cylinder. If carried still further outwards, it still moves inwards under the influence of the magnetism, and such continues to be the case until it is placed just over the edge of the terminal face of the core, where it has no motion at all (here, by another arrangement of the experiment, it is known to tend in what is at present an upward direction from the core). If carried a little further outwards, the magnetism then makes the bismuth ball tend to go outwards or be repelled, and such continues to be the direction of the force in any further position, or down the side of the end of the core.

2299. In fact, the circular edge formed by the intersection of the end of the core with its sides, is virtually the apex of the magnetic pole, to a body placed like the bismuth ball close to it, and it is because the lines

---

2297. *sensibility*: That is, *sensitivity*, as also in previous Series.

2298. As Faraday will state in paragraph 2300, all the movements of the bismuth ball narrated in this and succeeding paragraphs, illustrate a general tendency to move from stronger to weaker regions of magnetic power. Near the center of the pole face the ball does not move at all; which would imply that this is the locally weakest position—or at least that the magnetic strength is essentially uniform there. However, the ball *recedes* in both radial directions from the pole edge, except when it is in a circular region just outside the circumference of the pole face. In the next paragraph, Faraday will explain how both these radial motions illustrate the same migratory tendency.

2299. *apex of the magnetic pole*: The sharp circular edge of the pole is a place of *convergence* of the lines, so that their concentration—and the total magnetic power experienced by a body—must diminish both inside and outside the circular edge ("the lines … diverge"). A conical pole-piece placed over the circular face eliminates the sharp edge. Note that except for the slightly renunciatory "as it were," Faraday is again using very physical language about the lines of force.

of magnetic force issuing from it diverge as it were, and weaken rapidly in all directions from it, that the ball also tends to pass in all directions either inwards or upwards, or outwards from it, and thus produces the motions described. These same effects do not in fact all occur when the ball, being taken to a greater distance from the iron, is placed in magnetic curves, having generally a simpler direction. In order to remove the effect of the edge, an iron cone was placed on the top of the core, converting the flat end into a cone, and then the indicating ball was urged to move upwards, only when over the apex of the cone, and upwards and outwards, as it was more or less on one side of it, being always repelled from the pole in that direction, which transferred it most rapidly from strong to weaker points of magnetic force.

2300. To return to the vertical flat pole: when a horizontal bar of bismuth was suspended concentrically and close to the pole, it could take up a position in any direction relative to the axis of the pole, having at the same time a tendency to move upwards or be repelled from it. If its point of suspension was a little excentric, the bar gradually turned, until it was parallel to a line joining its point of suspension with the prolonged axis of the pole, and the centre of gravity moved inwards. When its point of suspension was just outside the edge of the flat circular terminating face, and the bar formed a certain angle with a radial line joining the axis of the core and the point of suspension, then the movements of the bar were uncertain and wavering. If the angle with the radial line were less than that above, the bar would move into parallelism with the radius and go inwards: if the angle were greater, the bar would move until perpendicular to the radial line and go outwards. If the centre of the bar were still further out than in the last case, or down by the side of the core, the bar would always place itself perpendicular to the radius and go outwards. All these complications of motion are easily resolved into their simple elementary origin, if reference be had to the character of the circular angle bounding the end of the core; to the direction of the magnetic lines of force issuing from it and the other parts of the pole; to the position of the different parts of the bar in these lines; and the ruling principle that each particle tends to go by the nearest course from *strong* to *weaker* points of magnetic force.

2300. *the ruling principle*:   The general orientation taken by a bismuth bar suspended horizontally *above* the pole face is: (*i*) when within the circumference of the pole—*radial*; (*ii*) when well outside the circumference—*tangential*; (*iii*) when near the circumference—*intermediate and wavering*. Try to interpret each position according to the "ruling principle" that each portion of the bar endeavors to migrate from stronger to weaker places of magnetic force. Remember that the lines of force diverge in the *upward* (axial) direction, as well as radially.

2301. The bismuth points well, and is well repelled (2296.) when immersed in water, alcohol, æther, oil, mercury, &c., and also when enclosed within vessels of earth, glass, copper, lead, &c. (2272.), or when screens of 0.75 or 1 inch in thickness of bismuth, copper, or lead intervene. Even when a bismuth cube (2266.) was placed in an iron vessel 2½ inches in diameter and 0.17 of an inch in thickness, it was well and freely repelled by the magnetic pole.

2302. Whether the bismuth be in one piece or in very fine powder, appears to make no difference in the character or in the degree of its magnetic property (2283.).

2303. I made many experiments with masses and bars of bismuth suspended, or otherwise circumstanced, to ascertain whether two pieces had any mutual action on each other, either of attraction or repulsion, whilst jointly under the influence of the magnetic forces, but I could not find any indication of such mutual action: they appeared to be perfectly indifferent one to another, each tending only to go from stronger to weaker points of magnetic power.

2304. Bismuth, in very fine powder, was sprinkled upon paper, laid over the horizontal circular termination of the vertical pole (2246.). If the paper were tapped, the magnet not being excited, nothing particular occurred; but if the magnetic power were on, then the powder retreated in both directions, inwards and outwards, from a circular line just over the edge of the core, leaving the circle clear, and at the same time showing the tendency of the particles of bismuth in all directions from that line (2299.).

2305. When the pole was terminated by a cone (2246.) and the magnet not in action, paper with bismuth powder sprinkled over it being drawn over the point of the cone, gave no particular result; but when the magnetism was on, such an operation cleared the powder

---

2303. *mutual action*: Two bismuth pieces develop the distinctive "diamagnetic" condition when placed in a magnetic field. If the two pieces interact with one another, the patterns of their mutual action could reveal much about the character of their internal diagmagnetic condition. Although Faraday fails to detect any mutual attraction or repulsion here, he will find an apparent instance of it in the next Series.

2304. *if the magnetic power were on, then the powder retreated in both directions*: A dusting of bismuth powder exhibits at a glance what Faraday earlier had to infer piecemeal from the movements of a single bismuth ball (paragraphs 2298–2300). Here the bismuth particles retreat from the circular edge of the electromagnet's iron core—a place of presumed concentrated magnetic force. In the next paragraph he will confirm that interpretation.

from every point which came over the cone, so that a mark was traced or written out in clear lines running through the powder, and showing every place where the pole had passed.

2306. The bar of bismuth and a bar of antimony was found to set equatorially between the poles of the ordinary horse-shoe magnet.

2307. The following list may serve to give an idea of the apparent order of some metals, as regards their power of producing these new effects, but I cannot be sure that they are perfectly free from the magnetic metals. In addition to that, there are certain other effects produced by the action of magnetism on metals (2309.) which greatly interfere with the results due to the present property.

Bismuth.	Cadmium.
Antimony.	Mercury.
Zinc.	Silver.
Tin.	Copper.

2308. I have a vague impression that the repulsion of bismuth by a magnet has been observed and published several years ago. If so, it will appear that what must then have been considered as a peculiar and isolated effect, was the consequence of a general property, which is now shown to belong to all matter.[3]

[3] M. de la Rive has this day referred me to the Bibliothèque Universelle for 1829, tome xl. p. 82, where it will be found that the experiment spoken of above is due to la Baillif of Paris. M. la Baillif showed sixteen years ago that both bismuth and antimony repelled the magnetic needle. It is astonishing that such an experiment has remained so long without further results. I rejoice that I am able to insert this reference before the present series of these Researches goes to press. Those who read my papers will see here, as on many other occasions, the results of a memory which becomes continually weaker; I only hope that they will be excused, and that omissions and errors of that nature will be considered as involuntary. —M. F. Dec. 30, 1845.

2305. *every point … every place*: The tip of the conical pole piece is a place of high magnetic strength. Faraday mounts it vertically with point upwards. As he slides a horizontal sheet of paper over the pole, bismuth powder (previously sprinkled on the paper's upper surface) retreats from every point that passes above the tip, leaving a visible record of the paper's motion.

2307. As suggested in the comment to paragraph 2284, the "apparent order" of the listed materials is probably based on estimating the strength of their pointing actions. But the previous paragraph suggested a supplementary indication for highly-diamagnetic materials, namely, their ability to exhibit the distinctive "new" behavior even under ordinary levels of magnetic force. Bismuth is highest, copper is lowest, in equatorial pointing tendency. Presumably this means that, under equal magnetic action, bismuth develops the "new" (diamagnetic) condition in the greatest degree, copper develops it in the least degree.

2309. I now turn to the consideration of some peculiar phænomena which are presented by copper and several of the metals when they are subjected to the action of magnetic forces, and which so tend to mask effects of the kind already described, that if not known to the inquirer they would lead to much confusion and doubt. These I will first describe as to their appearances, and then proceed to consider their origin.

2310. If instead of a bar of bismuth (2296.) a bar of copper of the same size be suspended between the poles (2247.), and magnetic power be developed whilst the bar is in a position oblique to the axial and equatorial lines, the experimenter will perceive the bar to be affected, but this will not be manifest by any tendency of the bar to go to the equatorial line; on the contrary, it will advance towards the axial position as if it were magnetic. It will not, however, continue its course until in that position, but, unlike any effect produced by magnetism, will stop short, and making no vibration beyond or about a given point, will remain there coming at once to a dead rest: and this it will do even though the bar by the effect of torsion or momentum was previously moving with a force that would have caused it to make several gyrations. This effect is in striking contrast with that which occurs when antimony, bismuth, heavy glass, or other such bodies are employed, and it is equally removed from an ordinary magnetic effect.

2311. The position which the bar has taken up it retains with a considerable degree of tenacity, provided the magnetic force be continued. If pushed out of it, it does not return into it, but takes up its new position in the same manner, and holds it with the same stiffness; a push, however, which would make the bar spin round several times if no magnetism were present, will now not move it through more than 20° or 30°. This is not the case with bismuth or heavy glass; they vibrate freely in the magnetic field, and always return to the equatorial position.

Comment on **Faraday's footnote 3**: It would be difficult to stand unmoved by Faraday's candid and dignified acknowledgment of his failing memory—one of several symptoms that afflicted him at intervals throughout his life (see paragraph 863, *comment*).

2310–2317. In these paragraphs Faraday reports a variety of effects, very different from the "pointing" tendencies described heretofore. It may be helpful to state in advance that many of them involve a kind of *retarding* or *braking* tendency (the opposite of *"free"* in paragraph 2296 above), which persists so long as the magnetic power is continued.

2312. The position taken up by the bar may be any position. The bar is moved a little at the instant of superinducing the magnetism, but allowing and providing for that, it may be finally fixed in any position required. Even when swinging with considerable power by torsion or momentum, it may be caught and retained in any place the experimenter wishes.

2313. There are two positions in which the bar may be placed at the beginning of the experiment, from which the magnetism does not move it, the equatorial and the axial positions. When the bar is nearly midway between these, it is usually most strongly affected by the first action of the magnet, but the position of most effect varies with the form and dimensions of the magnetic poles and of the bar.

2314. If the centre of suspension of the bar be in the axial line, but near to one of the poles, these movements occur well, and are clear and distinct in their direction: if it be in the equatorial line, but on one side of the axial line, they are modified, but in a manner which will easily be understood hereafter.

2315. Having thus stated the effect of the supervention of the magnetic force, let us now remark what occurs at the moment of its cessation; for during its continuance there is no change. If, then, after the magnetism has been sustained for two or three seconds, the electric current be stopped, there is instantly a strong action on the bar, which has the appearance of a revulsion (for the bar returns upon the course which it took for a moment when the electric contact was made), but with such force, that whereas the advance might be perhaps 15° or 20°, the revulsion will cause the bar occasionally to move through two or three revolutions.

2316. Heavy glass or bismuth presents no such phænomena as this.

2317. If, whilst the bar is revolving from revulsion the electric current at the magnet be renewed, the bar instantly stops with the former appearances and results (2310.), and then upon removing the magnetic force is affected again, and, of course, now in a contrary direction to the former revulsion.

2312. *The position taken up by the bar may be any position*: The bar shows no preferred position, no "pointing," while the magnet acts. If the bar is in motion at the moment the magnet is energized, it will be quickly brought to rest, thereafter resisting subsequent displacement as described in the previous paragraph.

2313. The "first turn," which usually develops when the magnet is first energized, does *not* occur if initially the bar was at rest in either the axial or the equatorial position.

2315. *supervention*: Here, *commencement*; but the word generally indicates a takeover by something additional or extraneous.

*   *   *

2329. Without going much into the particular circumstances, I may say that the effect is fully explained by the electric currents induced in the copper mass. By reference to the Second Series of these Researches (160.),[4] it will be seen that when a globe, subject to the action of lines of magnetic force, is revolving on an axis perpendicular to these lines, an electric current runs round it in a plane parallel to the axis of rotation and to the magnetic lines, producing consequently a magnetic axis in the globe, at right angles to the magnetic curves of the inducing magnet. The magnetic poles of this axis therefore are in that direction which, in conjunction with the chief magnetic pole, tends to draw the globe back against the direction in which it is revolving. Thus, if a piece of copper be revolving before a north magnetic pole, so that the parts nearest the pole move towards the right-hand, then the right-hand side of that copper will have a south magnetic state, and the left-hand side a north magnetic state; and these states will tend to counteract the motion of the copper towards the right-hand: or if it revolve in the contrary direction, then the right-hand side will have a south magnetic state, and the left-hand side a north magnetic state. Whichever way,

[4] Philosophical Transactions, 1832, p. 168.

2329. *these states will tend to counteract the motion of the copper*: Faraday's globe experiment at paragraph 160 was omitted from these selections, but we can easily follow his present recounting of it. Consider the copper ball sketched here, spinning from left to right. Let the north pole of a magnet be in front of the paper, the south pole behind it. Then by the magneto-electric right-hand rule (paragraph 101), since the nearer parts of the globe are moving to the right, currents *i* will tend to pass *upwards* in the near hemisphere, *downwards* in the far hemisphere. Thus, current will *circulate* about the globe in planes that are parallel both to the axis of rotation and the polar axis of the magnet. This circulating current will, by the electro-magnetic right-hand rule, produce a north "pole" in the left-hand portion of the globe and a south "pole" in the right-hand portion.

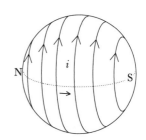

(The quotation marks indicate that these are not poles in the same sense as those possessed by iron magnets; Faraday will use the term "virtual polarity" in paragraph 2333 below.) Now as the globe turns from left to right, its motion tends to bring the left-hand portion, together with its virtual north pole, nearer to the fixed north pole—and bring the right-hand portion, with its virtual south pole, nearer to the fixed south pole. But since like poles *repel* one another, the interaction between the fixed and virtual poles will *oppose* the continued rotation of the globe. Note that this account invokes the poles as explanatory factors. In later Series, however, Faraday will jettison such appeals to magnetic "poles," in favor of a more fundamental explanatory principle.

therefore, the copper tends to revolve on its own axis, the instant it moves, a power is evolved in such a direction as tends to stop its motion and bring it to rest. Being at rest in reference to this direction of motion, then there is no residual or other effect which tends to disturb it, and it remains still.

*   *   *

2332. Before proceeding to the explanation of the other phenomena, it will be necessary to point out the fact generally understood and acknowledged, I believe, that *time* is required for the development of magnetism in an iron core by a current of electricity; and also for its fall back again when the current is stopped. One effect of the gradual rise in power was referred to in the last series of these Researches (2170.). This time is probably longer with iron not well annealed than with very good and perfectly annealed iron. The last portions of magnetism which a given current can develope in a certain core of iron, are also apparently acquired more slowly than the first portions; and these portions (or the condition of iron to which they are due) also appear to be lost more slowly than the other portions of the power. If electric contact be made for an instant only, the magnetism developed by the current disappears as instantly on the breaking of the current, as it appeared on its formation; but if contact be continued for three or four seconds, breaking the contact is by no means accompanied by a disappearance of the magnetism with equal rapidity.

2333. In order to trace the peculiar effect of the copper, and its cause, let us consider the condition of the horizontal bar (2310. 2313.) when in the equatorial position, between the two magnetic poles, or before a single pole; the point of suspension being in a line with the axis of the pole and its exciting wire helix. On sending an

---

2332. *time is required for the development of magnetism in an iron core by a current of electricity; and also for its fall back again when the current is stopped*: Stated this way, one might think that although the magnetism rises or falls gradually, the current starts and stops instantaneously. In fact the current, too, exhibits a certain retardation of its changes, which Faraday had already noted in the Ninth Series. See the discussion in the introduction to the present Series.

2333. *the horizontal bar ... in the equatorial position ... before a single pole*: In the sketch, a current-carrying helix with iron core develops a magnetic pole at each end. The copper bar is suspended adjacent to one of these poles, perpendicular to the pole's axis and therefore in the equatorial position.

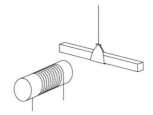

electric current through the helix, both it and the magnet it produces will conduce to the formation of currents in the copper bar in the contrary direction. This is shown from my former researches (26.), and may be proved, by placing a small or large wire helix-shaped (if it be desired) in the form of the bar, and carrying away the currents produced in it, by wires to a galvanometer at a distance. Such currents, being produced in the copper, only continue whilst the magnetism of the core is rising, and then cease (18. 39.), but *whilst* they continue, they give a virtual magnetic polarity to that face of the copper bar which is opposite to a certain pole, the polarity being the same in kind as the pole it faces. Thus on the side of the bar facing the north pole of the magnet, a north polarity will be developed; and on that side facing the south pole, a south polarity will be generated.

2334. It is easy to see that if the copper during this time were opposite only one pole, or being between two poles, were nearer to one than the other, this effect would cause its repulsion. Still, it cannot account for the whole amount of the repulsion observed alike with copper as with bismuth (2295.), because the currents are of but momentary duration, and the repulsion due to them would cease with them. They do, however, cause a brief repulsive effort, to which is chiefly due the first part of the peculiar effect.

---

current through the helix … currents in the copper bar in the contrary direction: While the electromagnetic power is rising to its full strength—but only while it is rising—current will be induced around the copper bar in a direction opposite to the current in the electromagnet's helix (paragraph 26). This creates in the bar surface a polarity similar to that of the pole which it faces. The electromagnetic right-hand rule shows why: Let two people grasp a broomstick with their right hands. The fingers of one hand represent current in the helix, while those of the other represent induced current in the bar. If the fingers point in opposite directions, the thumbs (representing north poles) will face either towards or away from each other.

by placing a small or large wire helix-shaped … in the form of the bar: That is, by substituting for the bar a coil, formed as though it had been wrapped around the bar *lengthwise*, in the manner sketched here. The induced current (and its direction) can then be detected directly by a galvanometer.

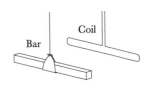

2334. *a brief repulsive effort*: As the inducing magnetism rises, the opposed similar magnetic poles produce a transient repulsion. This is in addition to, and not to be confused with, the ordinary diamagnetic recession from the pole which persists even after the electromagnet has developed full power. The transient repulsive impulse will cause *rotation* ("the peculiar effect") if the bar is oriented obliquely instead of equatorially—as the next paragraph will detail.

2335. For if the copper bar, instead of being parallel to the face of the magnetic pole, and therefore at right angles to the resultant of magnetic force, be inclined, forming, for instance, an angle of 45° with the face, then the induced currents will move generally in a plane corresponding more or less to that angle, nearly as they do in the examining helix (2333.), if it be inclined in the same manner. This throws the polar axis of the bar of copper on one side, so that the north polarity is not directly opposed to the north pole of the inducing magnet, and hence the action both of this and the other magnetic pole upon the two polarities of the copper will be to send it further round, or to place it edgeways to the poles, or with its breadth parallel to the magnetic resultant passing through it (2323.): the bar therefore receives an impulse, and the angle of it nearest to the magnet appears to be pulled up towards the magnet. This action of course stops the instant the magnetism of the helix core ceases to rise, and then the motion due to this cause ceases, and the copper is simply subject to the action before described (2295.). At the same time that this twist or small portion of a turn round the point of suspension occurs, the centre of gravity of the whole mass is repelled, and thus I believe all the actions up to this condition of things is accounted for.

2335. *the induced currents will move generally in a plane corresponding more or less to that angle*: That is, the currents will circulate around the *broad faces* of the oblique bar. This can be confirmed with the same elongated helix as in paragraph 2333, in the same oblique position as the bar. Since virtual poles are then formed on the broad faces of the bar, they will be repelled by the facing (and similar) fixed poles of the magnet. The immediate direction of this repulsion is towards the *axial* position (counter-clockwise in the sketch). As Faraday reported in paragraph 2310 above, "it will advance towards the axial position." But why do currents not also circulate about, and form virtual poles on, the *ends* of the bar as well as the broad

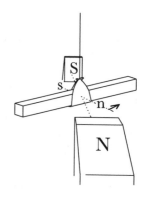

faces? Faraday does not explain that here. It will become evident as he develops the imagery of *magnetic conducting power*; see the introduction to the 26th Series.

This action of course stops the instant the magnetism of the helix core ceases to rise: Once the magnetic power of the core has attained its full value, the turning impulse disappears, and the characteristic *resistance* to further motion of the bar takes its place. As Faraday reported in paragraph 2310 above, "It will not … continue its course … [but] will stop short."

2336. Then comes the *revulsion* which occurs upon the cessation of the electric current, and the falling of the magnetism in the core. According to the law of magneto-electric induction, the disappearance of the magnetic force will induce brief currents in the copper bar (28.), but in the contrary direction to those induced in the first instance; and therefore the virtual magnetic pole belonging to the copper for the moment, which is nearest the north end of the electro-magnet, will be a south pole; and that which is furthest from the same pole of the magnet will be a north pole. Hence will arise an exertion of force on the bar tending to turn it round its centre of suspension in the contrary direction to that which occurred before, and hence the apparent revulsion; for the angle nearest the magnetic pole will recede from it, the broad face (2323.) or length (2315.) of the bar will come round and face towards the magnet, and an action the reverse in every respect of the first action will take place, except that whereas the motion was then only a few degrees, now it may extend to two or three revolutions.

2337. The cause of this difference is very obvious. In the first instance, the bar of copper was moving under influences powerfully tending to retard and stop it (2329); in the second case these influences are gone, and the bar revolves freely with a force proportionate to the power exerted by the magnet upon the currents induced by its own action.

2338. Even when the copper is of such form as not to give the oblique resultant of magnetic action from the currents induced in it, when, for instance, it is a cube or a sphere, still the effect of the action

---

2336. *the revulsion*: The previous paragraph described an initial impulse, associated with currents induced during the *rise* of magnetic power in the iron core. Corresponding currents will form in the opposite direction, when current to the helix is cut off and magnetic power in the iron core *falls*. Repulsion will be replaced by attraction; and a turning impulse, *opposite* to the direction of the initial impulse, will act—this is the "revulsion." The impulse vanishes when the magnetic power in the core decreases to zero (ceases to fall); but Faraday will explain in the next paragraph that with the external magnetic influence thus removed, there will then be no induced currents and hence no "obstruction" to continued motion of the bar. For that reason motion stemming from the revulsion "may extend to two or three revolutions."

2338. *such form as not to give the oblique resultant*: Now it is clear why the peculiar turning "effect" requires an *elongated* body. Everything depends on the conductor being *oblique* to the lines of force; but a sphere, or even a cube, is incapable of taking an "oblique" position. No matter how such a symmetrical body is placed, the currents induced in it will circulate in a plane *perpendicular* to the lines of force and therefore in such a body *repulsion alone* will be observed.

described above is evident (2325.). When a plate of copper about three-fourths of an inch in thickness, and weighing two pounds, was sustained upon some loose blocks of wood and placed about 0.1 of an inch from the face of the magnetic pole, it was repelled and held off a certain distance upon the making and continuing of electric contact at the battery; and when the battery current was stopped, it returned towards the pole; but the return was much more powerful than that due to gravity alone (as was ascertained by an experiment), the plate being at that moment actually *attracted*, as well as tending by gravitation towards the magnet, so that it gave a strong tap against it.

2339. Such is, I believe, the explanation of the peculiar phænomena presented by copper in the magnetic field; and the reason why they appear with this metal and not with bismuth or heavy glass, is almost certainly to be found in its high electro-conducting power, which permits the formation of currents in it by inductive forces, that cannot produce the same in a corresponding degree in bismuth, and of course not at all in heavy glass.

2340. Any ordinary magnetism due to metals by virtue of their inherent power, or the presence of small portions of the magnetic metals in them, must oppose the development of the results I have been describing; and hence metals not of absolute purity cannot be compared with each other in this respect. I have, nevertheless, observed the same phænomena in other metals; and as far as regards the sluggishness of rotatory motion, traced it even into bismuth. The following are the metals which have presented the phænomena in a greater or smaller degree:—

Copper.	Mercury.
Silver.	Platinum.
Gold.	Palladium.
Zinc.	Lead.
Cadmium.	Antimony.
Tin.	Bismuth.

2338, continued. *a plate of copper … was repelled and held off*: Faraday's mention of this particular effect is a little misleading, for since the repulsion *persists* it cannot be an example of the transient action he is chiefly describing. But the subsequent *attraction* when the electromagnet is switched off, which is both transient and superadded to the copper's weight, *does* illustrate the action under discussion.

2340. *greater or smaller degree*: Faraday does not say whether the metals listed are tabulated in order. Nevertheless copper, silver, and gold exhibit the "peculiar" phenomena to the highest degree—and they are also the best electrical conductors, which is precisely to be expected if Faraday's explanation by *induced currents* is correct.

2341. The accordance of these phænomena with the beautiful discovery of Arago,[5] with the results of the experiments of Herschel and Babbage,[6] and with my own former inquiries (81.),[7] is very evident. Whether the effect obtained by Ampère, with his copper cylinder and helix,[8] was of this nature, I cannot judge, inasmuch as the circumstances of the experiment and the energy of the apparatus are not sufficiently stated; but it probably may have been.

2342. As, because of other duties, three or four weeks may elapse before I shall be able to complete the verification of certain experiments and conclusions, I submit at once these results to the attention or the Royal Society, and will shortly embody the account of the action of magnets on magnetic metals, their action on gases and vapours, and the general considerations in another series of these Researches.

Royal Institution,
Nov. 27, 1845.

[5] Annales de Chimie, xxvii. 363; xxviii. 325; xxxii. 213. I am very glad to refer here to the Comptes Rendus of June 9, 1845, where it appears that it was M. Arago who first obtained his peculiar results by the use of electro- as well as common magnets.

[6] Philosophical Transactions, 1825, p. 467.

[7] Ibid. 1832, p. 146.

[8] Bibliothèque Universelle, xxi. p. 48.

2341. *the effect obtained by Ampère*: In paragraph 78 of the First Series Faraday had doubted that certain movements of a copper disk in a magnetic field, reportedly observed by Ampère, could really have taken place. Subsequently, in a long and respectful note appended to the Third Series, he suggested that the problem lay with an incomplete description of apparatus, and not in Ampère's observations.

Twenty-first Series — Editor's Introduction

A puzzle about gases and space

Up to now, air and the other gases have proved surprisingly indistinguishable, electrically, from vacuum or space. In the Eleventh Series, Faraday carried out measurements of specific inductive capacity on fifteen different gases under a range of different pressures. But, as he there reported (paragraph 1292),

> Notwithstanding the striking contrasts of all kinds which these gases present of property, of density, whether simple or compound, anions or cathions,* of high or low pressure, hot or cold, not the least difference in their capacity to favour or admit electrical induction through them could be perceived.

Faraday will undertake a new survey of gases and vacuum in the present Series, this time with respect to their magnetic or diamagnetic condition and degree. The gases appear to be as indistinguishable magnetically as they previously were electrically. But their magnetic uniformity is far more troubling than their equality of electrical condition. This is why.

According to Faraday's measurements and those of others, the electrical inductive capacities of solids and liquids occupy a considerable range, from glass at 1.76 times that of air,** to mica at about 8 times that of air, to pure water at 81.07 times that of air. But all these materials are *more* inductive than air—no solid or liquid exhibits a specific inductive capacity less than 1. This permits us to regard the inductivity of space or vacuum as a minimum, even indifferent, condition. It is fully in harmony with ordinary ideas that *absence of matter* should be incapable of doing anything, and that matter so highly rarefied as to be in the gaseous state might well exert no measurable effect.

But a comparable ranking of magnetic and diamagnetic materials places air, gases, and vacuum all together *between* the magnetic and the diamagnetic materials. Although just as before it is no surprise that highly rarefied gases might be indistinguishable from vacuum devoid of matter, it is no longer so obvious for the magnetic condition, as it had been for electrical inductivity, that vacuum represents a *minimum*

---

* Faraday defined anion and cathion at paragraph 665 in the Seventh Series.

** Faraday expressed the specific inductive capacity of each material in relation to *air* taken as standard—a practice that is still widely followed. The figures for glass are Faraday's from the Twelfth Series, but more recent measurements indicate values between 3 and 3.2 times that of air.

state—since it occupies an *intermediate* position in the magnetic ranking of materials. Hence the problem which will plague Faraday throughout this and subsequent Series: Does the intermediate position of vacuum constitute a magnetic *zero* between two absolute and opposite states? Or are magnetism and diamagnetism not to be seen as opposite powers, but as different gradations of one and the same magnetic power?—if so, space and the gases would have to be credited with a middling degree of the universal magnetic condition, exceeding some materials while falling short of others.

It might seem that the "two opposite classes" view is the only plausible one. After all, how can *absence of matter* exhibit a property of matter? If Faraday accepts that it can, he will be quite literally counting *space* as one of the materials! But the idea that magnetism and diamagnetism are simple opposites turns out to have its own difficulties. For one thing, it proves very hard to come up with a consistent *representation* of these supposedly opposite material conditions. Faraday will consider one such attempt, which grows out of the theory of active magnetic poles.

Magnetic induction according to the doctrine of *poles*

It is pretty common knowledge that one magnet will attract another magnet if their facing poles are opposite in kind, and will repel it if their facing poles are similar in kind. But how does a magnet attract a nail, or a paper clip, or any body *not already* a magnet?

For anyone who holds to a doctrine of *active poles* there can be only one answer. Since according to that doctrine it is the *poles* that bear the magnetic power, then if a body is attracted to the north pole of a magnet, its near surface *must contain a south pole*—and if that pole was not previously existent in the attracted body it must have been *raised up* in the body by some action of the approaching magnet.

You may recall that Faraday alluded at the very opening of the First Series to a similar action, by which an electrically charged body is able to attract bits of paper, foil, etc., that are not themselves already charged. In both cases the doctrine of poles holds that the approaching magnetized or electrified body must be capable of bringing into being, or *inducing*, a magnetic or electric pole of opposite quality in the facing surface of the attracted body. Thus a steel ball, previously unmagnetized, can be attracted by *either* the north pole *or* the south pole of a magnet—for whichever pole of the magnet it approaches, that pole induces an opposite magnetic pole in the ball's nearest surface, and attraction between the opposite poles draws the ball to the magnet. Such at least is the explanation of induced magnetism according to the doctrine of active poles.

Are there diamagnetic poles?

Now as we saw in the previous Series, diamagnetic bodies are not attracted to but *repelled from* the poles of a magnet. Understandably, then, some theorists hoped to interpret diamagnetic repulsion, too, as a case of induction—different only in raising up a similar, rather than an opposite, pole in the facing surface of the diamagnetic body. Thus if a small bismuth ball were brought near the north pole of a magnet, it was proposed that the bismuth would develop a north pole within its near surface—resulting in repulsion of the ball from the magnet. Faraday reports that he himself had toyed with such a view—even though the whole language and doctrine of "poles" or other centers of action is quite unappealing to him. But even within its own terms the hypothesis of "induced diamagnetic polarity" proves unsatisfactory and even inconsistent. Faraday pointed out its failings on several occasions. Here is a particularly neat paradox which he published about nine years after completing the present Series:*

Suppose an ordinary magnet M acts upon a piece of iron I, and that piece in turn upon a second iron body I'. According to the hypothesis of induced magnetic polarity, the bodies under induction, I and I', will develop poles as shown in the drawing. For convenience, we define the *direction of polarity* in a material as being from s to n; thus we shall say that bodies I and I' have developed polarity in the left-hand direction.

Again, suppose magnet M acts upon a piece of bismuth B and that piece in turn upon a second bismuth body B'. According to the hypothesis that diamagnetic bodies under induction develop a polarity the reverse of magnetic bodies, bars B and B' must develop poles as shown. Body B will be polarized in the *right-hand direction* (the reverse of inducing magnet M), and body B' will be polarized in the *left-hand direction* (the reverse of inducing body B).

Since bodies I and I' have polarity in the left-hand direction, the successive particles of iron in both bodies must individually be polarized

---

* The following account is adapted from Faraday's "On Some Points of Magnetic Philosophy," *Philosophical Magazine*, February 1855. Reprinted in *Experimental Researches*, vol. III, p. 528.

in the left-hand direction also—as the experiment of breaking out a piece from the middle of a bar-magnet (see the introduction to the First Series) confirms. Whether within the same body or between juxtaposed bodies, each iron particle induces upon the next a like polarity to its own—in this case a polarity in the left-hand direction. Adjacent iron particles present opposite faces to one another—as indeed the notion of induced magnetic polarity requires:

Now let us inquire as to the successive particles of bismuth in B and B'. The hypothesis of *induced reverse polarity* would seem to demand that neighboring particles of a diamagnetic substance present *similar* faces to one another; but in that case the characters (*n* or *s*) found at opposite ends of B would depend on the number of B particles from end to end—*same* if even, *opposite* if odd—an absurd notion! So within each diamagnetic body the bismuth particles too must all individually be assumed to have similar polarity to one another. They too will present opposite faces to one another, only the polarity in B will be reverse to the polarity in B':

Then each bismuth particle in B induces, upon its neighbor to the right, a like polarity to its own—namely, a polarity in the right-hand direction. However, the *last* particle in B has to induce an *opposite* polarity to its own—a polarity to the left—in the first particle of B'! But how can the interaction between adjacent bismuth particles be different between *neighboring particles of the same bar*, from what it is between *terminating particles of juxtaposed bars*? Why should a polarized bismuth particle induce *like* polarity upon one bismuth particle, but *opposite* polarity upon another bismuth particle?

Incongruities like this one must inevitably beset any polar theory of diamagnetism. Not all theorists were deterred, but the difficulties were enough to convince Faraday—who was never enthusiastic about "poles" anyway—that a wholly different approach was needed.

Differential action

A more direct challenge to the "two opposite classes" view arises with experiments which suggest that "magnetic" and "diamagnetic" denote *relations to the surrounding medium*, rather than absolute internal conditions of bodies. For example, iron sulphate belongs to the *magnetic* class of materials; iron sulphate solutions of any concentration exhibit axial attraction when tested with a magnet in air. And in the present Series Faraday shows that a tube of iron sulphate solution continues to be attracted axially when immersed in a less concentrated solution of the same material. But the same sample is urged equatorially—that is, it behaves like a *diamagnetic* material—when immersed in a more concentrated solution! It would seem, then, that materials cannot be labeled "magnetic" or "diamagnetic" of themselves, but only in relation to the surrounding medium. If so, however, the sample and the solution surrounding it must share some common characteristic, with respect to which they surpass or fall short of one another. Such relations are generally termed "differential."

As Faraday reminds us in the present Series (paragraph 2438), he came across a similarly-suggestive *electrical* example in the Twelfth Series when he suspended bits of silk thread in turpentine.* When he first described the phenomenon he either did not appreciate, or at any rate did not advertise, its "differential" significance; but now he reconsiders the experiment with a fresh interpretation.

When the electrodes are energized, the silk bits gather together to form a single chain. One might suppose that when the threads align themselves in this way, it is because the silk is being attracted by electric force and the

turpentine repelled. *But both silk and turpentine are dielectrics*; both materials ought therefore to be repelled from the axial line of electrical force. Nevertheless if they differ in their *degree* of facility in transmitting the electric force, and if silk is even slightly more efficacious than turpentine in that respect, then the overall transmission of force *will be maximized* when the silk occupies the regions of greatest concentration of lines of force—or, as Faraday expresses it (paragraph 2438), when "the bodies best fitted to carry on the force are drawn to the shortest line of action."

Thus Faraday ascribes this electrical phenomenon to a *difference in degree* among the materials, rather than to opposition between them. No doubt a comparable interpretation would lend a similar elegance and readability to the phenomena of magnetism and diamagnetism. And yet, in the present Series, Faraday holds back from taking that

---

* Faraday's description appears in paragraph 1350. I omitted it from the Twelfth Series selections but offered a brief account of the experiment in the introduction to that Series.

step. Or rather, it would be more accurate to say he *agonizes* over it, continually turning first towards, then away from that interpretation, and making no secret of the competing interests of the question. For the differential view, by positing *differences in degree* in place of *opposition between contraries*, inevitably nurtures that disconcerting implication for *space* I mentioned earlier—the proposition that *absence of matter* can sustain relations and participate in physical conditions.

§ 27. *On new magnetic actions, and on the magnetic condition of all matter—continued.* ¶ v. *Action of magnets on the magnetic metals and their compounds.* ¶ vi. *Action of magnets on air and gases.* ¶ vii. *General considerations.*

Received December 24, 1845,—Read January 8, 1846.

*   *   *

2362. A clear solution of the proto-sulphate of iron was prepared, in which one ounce of the liquid contained seventy-four grains of the hydrated crystals; a second solution was prepared containing one volume of the former and three volumes of water; a third solution was made of one volume of the stronger solution and fifteen volumes of water. These solutions I will distinguish as Nos. 1, 2, and 3; the proportions of crystals of sulphate of iron in them were respectively as 16, 4, and 1 per. cent. nearly. These numbers may, therefore, be taken as representing (generally only (2423.)) the strength of the magnetic part of the liquids.

2363. Tubes like that before described (2279.) were prepared and filled respectively with these solutions and then hermetically sealed, as little air as possible being left in them. Glasses of the solutions were also prepared, large enough to allow the tubes to move freely in them, and yet of such size and shape as would permit of their being placed

[1] Philosophical Transactions, 1846, p. 41.

2362. *16, 4, and 1 per. cent. nearly*: These figures conveniently express successive dilutions by factors of 4, but are only approximate. Faraday states that one "ounce of the liquid"—that is, one ounce of solution—contained 74 grains of the proto-sulphate crystals. Since one ounce (avoirdupois) equals 437.5 grains, the initial solution actually contains $74 \div 437.5$ or 16.91% proto-sulphate of iron, by weight.

2363. A *tube* of any fluid could be suspended horizontally within a *glass* of any other fluid. With the combination then placed between the magnet poles, the fluid's pointing behavior could be observed.

hermetically sealed: That is, impervious, impermeable. The legendary Hermes Trismegistus was credited with the invention of various alchemical techniques. According to tradition, these included a magic seal to make vessels airtight.

between the magnetic poles. In this manner the action of the magnetic forces upon the matter in the tubes could be examined and observed, both when the tubes were in diamagnetic media, as air, water; alcohol, &c., and also in magnetic media, either stronger or weaker in magnetic force, than the substances in the tubes.

2364. When these tubes were suspended in air between the poles, they all pointed axially or magnetically, as was to be expected; and with forces apparently proportionate to the strengths of the solutions. When they were immersed in alcohol or water, they also pointed in the same direction; the strongest solution very well, and also the second, but the weakest solution was feeble in its action, though very distinct in its character (2422.).

2365. When the tubes, immersed in the different ferruginous solutions, were acted upon, the results were very interesting. The tube No. 1 (the strongest magnetically), when in solution No. 1, had no tendency, under the influence of the magnetic power, to any particular position, but remained wherever it was placed. Being placed in solution No. 2, it pointed well axially, and in solution No. 3 it took the same direction, but with still more power.

2366. The tube No. 2, when in the solution No. 1, pointed equatorially, *i. e.* as heavy glass, bismuth, or a diamagnetic body generally, in air. In solution No. 2 it was indifferent, not pointing either way; and in solution No. 3 it pointed axially, or as a magnetic body. The tube No. 3, containing the weakest solution, pointed equatorially in solutions No. 1 and 2, and not at all in solution No. 3.

2367. Several other ferruginous solutions varying in strength were prepared, and, as a general and constant result, it was found that any tube pointed axially if the solution in it was stronger than the surrounding solution, and equatorially if the tube solution was the weaker of the two.

2368. The tubes were now suspended vertically, so that being in the different solutions they could be brought near to one of the magnetic poles, and employed in place of the indicating cube or sphere of bismuth, or heavy glass (2266.). The constant result was, that when the tube contained a stronger solution than that which surrounded it, it was attracted to the pole, but when its solution was the weaker of the two it was repelled. The latter phænomena were as to appearance in every respect the same as those presented in the repulsion of heavy glass, bismuth, or any other diamagnetic body in air.

2365. *the different ferruginous solutions*: That is, the four solutions with varying concentrations of proto-sulphate of iron.

2369. Having described these phænomena, I will defer their further consideration until I arrive at the last division of this paper, and proceed to certain results more especially belonging to the present part of these researches.

*   *   *

¶ vi. *Action of magnets on air and gases.*

2400. It was impossible to advance in an experimental investigation of the kind now described, without having the mind impressed with various theoretical views of the mode of action of the bodies producing the phænomena. In the passing consideration of these views, the apparently middle condition which *air* held between magnetic and diamagnetic substances was of the utmost interest, and led to many experiments upon its probable influence, which I will now proceed briefly to describe.

*   *   *

2408. In order to extend the experimental relations of air and gases, I proceeded to place substances of the diamagnetic class in them. Thus the bar of heavy glass (2253.) was suspended in a jar of air, and then the air about it more or less rarefied, but as before, in the case of the air-tube (2402.), alterations of this kind produced no effect. Whether the bar were in air at the ordinary pressure, or as rare as the pump could render it, it still pointed equatorially, and apparently always with the same degree of force.

2409. The bar of bismuth (2296.) was suspended in the jar and the same alteration in the density of the air made as before; but this caused no difference in the action of the bismuth, either in kind or degree. Carbonic acid and hydrogen gases were then introduced in succession

---

2369. *Having described these phænomena*: We may summarize the results of the foregoing experiments thus: Whether judged by its *pointing* or its *change of place*, the iron sulphate solution acts like a magnetic substance when surrounded by a less highly concentrated solution—but it acts like a *diamagnetic* material when surrounded by a more highly concentrated solution. Evidently substances are not to be classified as magnetic or diamagnetic inherently, but only in relation to the surrounding medium. Faraday states that he intends to defer consideration of this line of thought for a time; he will return to it again in paragraph 2436 below.

2400. *the apparently middle condition which air held between magnetic and diamagnetic substances was of the utmost interest…*: In paragraph 2432 below, Faraday will state the problem this "middle" ranking presents, not only for air but all the gases.

into the jar, and these also were employed in different degrees of rarefaction, but the results were the same; no change took place in the action on the bismuth.

2410. A bismuth cube was suspended in air and gases at ordinary pressure, and also rarefied as much as could be, and under these circumstances it was brought near the magnetic pole and its repulsion observed; its action was in all these cases precisely the same as in the atmosphere.

2411. The perpendicular copper bar (2323.) was suspended near the magnetic pole in *vacuo,* but its set, sluggish movements and revulsion were just the same as before in air (2324.).

*   *   *

2416. In every kind of trial, therefore, and in every form of experiment, the gases and vapours still occupy a medium position between the magnetic and the diamagnetic classes. Further, whatever the chemical or other properties of the substances, however different in their specific gravity, or however varied in their own degree of rarefaction, they all become alike in their magnetic relation, and apparently equivalent to a perfect vacuum. Bodies which are very marked as diamagnetic substances, immediately lose all traces of this character when they become vaporous (2415.). It would be exceedingly interesting to know whether a body from the magnetic class, as chloride of iron, would undergo the same change.

¶ vii. *General considerations.*

2417. Such are the facts which, in addition to those presented by the phænomena of light, establish a magnetic action or condition of matter new to our knowledge. Under this action, an elongated portion of such matter usually (2253. 2384.) places itself at right angles to the lines of magnetic force; this result may be resolved into the simpler one of repulsion of the matter by either magnetic pole.

2417. *Such are the facts…*: Faraday is referring not so much to the experiments just completed (paragraphs 2408–2416), as to his earlier investigations of the pointing and migrating behavior of diamagnetic materials. The question now before him is how to interpret that behavior. Is *pointing* the key to its understanding? Apparently not, for in the very next sentence he characterizes pointing as expressive of a more fundamental action, that of *repulsion.*

may be resolved into the simpler one of repulsion: Since he intends to identify a "simpler" action, it is a little surprising to see Faraday use the term "repulsion," which generally carries suggestions of pushes and pulls. But Faraday

The set of the elongated portion, or the repulsion of the whole mass, continues as long as the magnetic force is sustained, and ceases with its cessation.

2418. By the exertion of this new condition of force, the body moved may pass either *along* the magnetic lines or *across* them; and it may move along or across them in either or any direction. So that two portions of matter, simultaneously subject to this power, may be made to approach each other as if they were mutually attracted, or recede as if mutually repelled. All the phænomena resolve themselves into this, that a portion of such matter, when under magnetic action, tends to move from stronger to weaker places or points of force. When the substance is surrounded by lines of magnetic force of equal power on all sides, it does not tend to move, and is

---

has—or thinks he has—effectively freed the word of those associations in the preceding Series (paragraphs 2266–2269), and he now appears to use it as a virtual synonym for "recession." Nevertheless even "recession" may be thought to imply a *center from which* the body recedes. In the next paragraph, the motion will be reinterpreted again, this time as a *migration* from stronger to weaker zones of magnetic force—which effectively does away with any specially favored location or other distinguished center of action.

The set: here, the *orientation* of the bar.

2418. *the body moved may pass either along the magnetic lines or across them*: For example, a diamagnetic cube tends to recede from either of a pair of opposing magnetic poles, and in so doing moves *along* the lines of force. If placed midway between the poles, however, it drifts to either side of the axial line, thereby moving *across* the magnetic lines. (Faraday recounted both examples in paragraph 2266.)

a portion of such matter, when under magnetic action, tends to move from stronger to weaker places or points of force: The regions in which a magnet exerts its strongest influence are (*i*) *immediately adjacent to its poles*, and (*ii*) *along its axial line* (in the case of a horshoe magnet). Both *recession from a pole* and *displacement from the axial line*, therefore, represent migration from places of stronger to places of weaker magnetic action.

When the substance is surrounded by lines of magnetic force of equal power on all sides, it does not tend to move: The principle of *migration from stronger to weaker places of magnetic force* is the *essential principle* of diamagnetic movements. It applies not only to the overall displacement of a diamagnetic body but to *every increment* of its path: even the smallest movement must be such as to bring it to a place of weaker magnetic action than before. In a region of *strictly* uniform magnetic force, therefore, a resting diamagnetic would exhibit no tendency to move—since any initial increment of motion would leave the magnetic action on the body still unchanged. However Faraday will show in the 26th Series (paragraphs 2809–2810) that introduction of a new material into a previously uniform field destroys the uniformity; in certain cases this can produce a tendency towards displacement.

then in marked contradistinction with a linear current of electricity under the same circumstances.

2419. This condition and effect is new, not *only* as it respects the exertion of power by a magnet over bodies previously supposed to be indifferent to its influence, but is *new* as a magnetic action, presenting us with a second mode in which the magnetic power can exert its influence. These two modes are in the same general antithetical relation to each other as positive and negative in electricity, or as northness and southness in polarity, or as the lines of electric and magnetic force in magneto-electricity; and the diamagnetic phænomena are the more important, because they extend largely, and in a new direction, that character of duality which the magnetic force already, in a certain degree, was known to possess.

2420. All matter appears to be subject to the magnetic force as universally as it is to the gravitating, the electric and the chemical or cohesive forces; for that which is not affected by it in the manner of ordinary magnetic action, is affected in the manner I have now described; the matter possessing for the time the solid or fluid state. Hence substances appear to arrange themselves into two great divisions; the magnetic, and that which I have called the diamagnetic classes; and between these classes the contrast is so great and direct, though varying in degree, that where a substance from the one class

---

2418, continued. *in marked contradistinction with a linear current of electricity*: A straight current-carrying wire tends to move perpendicularly to the lines of magnetic force, *even when the field of force is uniform*. Strictly, though, this too is a case where the magnetic force loses its uniformity as soon as the object is introduced.

2420. *All matter appears to be subject to the magnetic force…*: That is, *all solid or liquid matter*, as the end of the sentence makes clear. Electricity affects all matter, since whatever is not a *conductor* is a *dielectric*. Gravitation similarly pertains to all matter. Previously, magnetism presented an anomaly by appearing to affect only iron, nickel, and cobalt; but now it is evident that any material not *magnetic* will be *diamagnetic*, so that magnetism, too, may take its place as a universal force—although the state of gases, and of course of the vacuum, remains questionable.

the contrast is so great and direct…: Nevertheless, the relation between magnetism and diamagnetism remains one of contrariety—as exemplified by the contrast between *attraction* and *repulsion*, or by the disparity between *lines at right angles to one another*. At the beginning of this paragraph, Faraday acknowledged a sense in which magnetism and diamagnetism share a universal connection—why then does he now emphasize their contrast? His reason may become evident at paragraph 2440 below.

will be attracted, a body from the other will be repelled; and where a bar of the one will assume a certain position, a bar of the other will acquire a position at right angles to it.

2421. As yet I have not found a single solid or fluid body, not being a mixture, that is perfectly neutral in relation to the two lists; *i. e.* that is neither attracted nor repelled in air. It would probably be important to the consideration of magnetic action, to know if there were any natural simple substance possessing this condition in the solid or fluid state. Of compound or mixed bodies there may be many; and as it may be important to the advancement of experimental investigation, I will describe the principles on which such a substance was prepared when required for use as a circumambient medium.

2422. It is manifest that the properties of magnetic and diamagnetic bodies are in opposition as respects their dynamic effects; and, therefore, that by a due mixture of bodies from each class, a substance having any intermediate degree of the property of either may be obtained. Protosulphate of iron belongs to the magnetic, and water to the diamagnetic class; and using these substances, I found it easy to make a solution which was neither attracted nor repelled, nor pointed when in air. Such a solution pointed axially when surrounded by water. If made somewhat weaker in respect of the iron, it would point axially in water but equatorially in air; and it could be made to pass more and more into the magnetic or the diamagnetic class by the addition of more sulphate of iron or more water.

2421. *As yet I have not found...*: Here is more indication that the magnetic and diamagnetic states are to be thought of as strict *opposites*. In the ranking of magnetic and diamagnetic substances, at least among solids and liquids, the *only* materials that hold a place equal to air and vacuum are *compound*. Their ingredients, therefore, might partake of magnetic and diamagnetic powers individually, and so neutralize one another's effects. The perceived neutrality of such materials would in that case represent a mutual cancellation or equilibrium of powers, rather than indifference or absence of relation.

a circumambient medium: here, a medium surrounding or encompassing the experimental substances.

2422. Faraday's artificially neutral mixture of magnetic and diamagnetic substances exhibits magnetic behavior indistinguishable from that of air or vacuum. It illustrates the thinking behind the "two class" view—that magnetic and diamagnetic materials represent two opposite classes exercising mutually antagonistic powers and capable of annihilating one another's effects.

2423. Thus a *fluid* medium was obtained, which, practically, as far as I could perceive, had every magnetic character and effect of a gas, and even of a vacuum; and as we possess both magnetic and diamagnetic glass (2354.), it is evidently possible to prepare a *solid* substance possessing the same neutral magnetic character.

2424. The endeavour to form a general list of substances in the present imperfect state of our knowledge would be very premature: the one below is given therefore only for the purpose of conveying an idea of the singular association under which bodies come in relation to magnetic force, and for the purpose of general reference hereafter:—

Iron.
Nickel.
Cobalt.
Manganese.
Palladium.
Crown-glass.
Platinum.
Osmium.
0° Air and vacuum.
Arsenic.
Æther.
Alcohol.
Gold.
Water.
Mercury.
Flint-glass.
Tin.
Heavy glass.
Antimony.
Phosphorus.
Bismuth.

2425. It is very interesting to observe that metals are the substances which stand at the extremities of the list, being of all bodies those which

---

2423. *practically, as far as I could perceive, had every magnetic character and effect of a gas, and even of a vacuum*: But it remains a daunting challenge to understand the neutrality of things that are clearly *not* composite. *Space*, for example, must be simple, homogeneous. And what about *elementary* gases like oxygen (not air, which is a mixture)—are they indeed neutral as space is, or are they merely highly diluted magnetics and diamagnetics? In short, can *matter* really be, in and of itself, neutral—indifferent?

2424. The list is given in order of precedence, iron exhibiting the behavior most strongly magnetic and bismuth the most strongly diamagnetic. Faraday does not specify by what experimental means he determined the order within each class; but an earlier table (paragraph 2399, omitted here) was based on the *vigor* as well as the direction of pointing for samples of various materials.

are most powerfully opposed to each other in their magnetic condition. It is also a very remarkable circumstance, that these differences and departures from the medium condition, are in the metals at the two extremes, iron and bismuth, associated with a small conducting power for electricity. At the same time the *contrast* between these metals, as to their fibrous and granular state, their malleable and brittle character, will press upon the mind whilst contemplating the possible condition of their molecules when subjected to magnetic force.

2426. In reference to the metals, as well as the diamagnetics not of that class (2286.), it is satisfactory to have such an answer to the opinion that all bodies are magnetic as iron, as does not consist in a mere negation of that which is affirmed, but in proofs that they are in a different and opposed state, and are able to counteract a very considerable degree of magnetic force (2448.).

*   *   *

2429. Theoretically, an explanation of the movements of the diamagnetic bodies, and all the dynamic phænomena consequent upon the actions of magnets on them, might be offered in the supposition that magnetic induction caused in them a contrary state to that which it produced in magnetic matter; *i. e.* that if a particle of each

---

2425. *molecules*: As in earlier Series, Faraday's "molecules" are literally *tiny portions* of the substance in question, not assemblies of elementary *atoms* as in the nomenclature of a later chemical atomism.

2426. *an answer ... as does not consist in a mere negation ... but in proofs*: Faraday is gratified that experiment does not merely refute a position but goes further to disclose *substantial knowledge* about the new magnetic state. He made the same point at paragraph 2286 in the previous Series, noting there too that the experiments accomplish more than mere refutation. As ever with Faraday, the proper use of experiment is not to win debates but to reveal nature's powers.

2429. *a contrary state*: If we view diamagnetic behavior as essentially *repulsive* where magnetic action would be *attractive*, we might suppose that diamagnetic bodies under induction develop poles like magnetic bodies, only reversed in quality. Faraday gives that approach a fair hearing in the next two paragraphs, notwithstanding his earlier reinterpretation of diamagnetism as *migration* (paragraph 2418) and his long-standing distaste for "poles". Ultimately, however, a polar view of diamagnetism must fail, as the introduction explains. Faraday's examination of the reverse polar theory of diamagnetism is not a digression, as readers sometimes think. The theory of reversed polarity presents the most substantial and concrete image so far of the "two opposing classes" view of magnetic materials—a view that is very much on Faraday's mind.

kind of matter were placed in the magnetic field both would become magnetic, and each would have its axis parallel to the resultant of magnetic force passing through it; but the particle of magnetic matter would have its north and south poles opposite, or facing towards the contrary poles of the inducing magnet, whereas with the diamagnetic particles the reverse would be the case; and hence would result approximation in the one substance, recession in the other.

2430. Upon Ampère's theory, this view would be equivalent to the supposition, that as currents are induced in iron and magnetics parallel to those existing in the inducing magnet or battery wire; so in bismuth, heavy glass and diamagnetic bodies, the currents induced are in the contrary direction. This would make the currents in diamagnetics the same in direction as those which are induced in dia-magnetic conductors at the *commencement* of the inducing current; and those in magnetic bodies the same as those produced at the *cessation* of the same inducing current. No difficulty would occur as respects non-conducting magnetic and diamagnetic substances, because the hypothetical currents are supposed to exist not in the mass, but round the particles of the matter.

2431. As far as experiment yet bears upon such a notion, we may observe, that the known inductive effects upon masses of magnetic and diamagnetic metals *are the same*. If a straight rod of iron be carried across magnetic lines of force, or if it, or a helix of iron rods or wire, be held near a magnet, as the power in it rises electric currents are induced, which move through the bars or helix in

---

2430. *the currents induced are in the contrary direction*: Ampère's theory (which refers the magnetic condition of a body to hypothetical molecular currents) could easily accommodate the supposition of reversed poles by simply choosing the direction of the currents—one direction for magnetic substances, the opposite for diamagnetic substances.

No difficulty would occur as respects non-conducting magnetic and diamagnetic substances: Since Ampère's hypothetical currents are supposed to circulate about individual particles rather than through whole bodies, one could without contradiction ascribe such particular currents even to nonconductive materials.

2431. The previous paragraph suggested a minor adjustment to Ampère's theory, whereby it could accommodate the "reversed pole" view of diamag-netism. Magneto-electric experiments might appear to rule out such adjustments, since the induced currents are *always in the same direction*, whether the material under induction is magnetic or diamagnetic. But these are *conduction* currents, passing through the whole mass of the material, whereas Ampère's theory pos-tulates miniature electric currents circulating about individual particles; so the experimental results are not necessarily applicable. Nevertheless, Ampère's

certain determinate directions (38. 114., &c.). If a bar or a helix of bismuth be employed under the same circumstances the currents are again induced, and precisely in the same direction as in the iron, so that here no difference occurs in the direction of the induced current, and not very much in its force, nothing like so much indeed as between the current induced in either of these metals and a metal taken from near the neutral point (2399.). Still there is this difference remaining between the conditions of the experiment and the hypothetical case; that in the former the induction is manifested by currents in the masses, whilst in the latter, *i. e.* in the special magnetic and diamagnetic effects, the currents, if they exist, are probably about the particles of the matter.

2432. The magnetic relation of aëriform bodies is exceedingly remarkable. That oxygen or nitrogen gas should stand in a position intermediate between the magnetic and diamagnetic classes; that it should occupy the place which *no* solid or liquid element can take;

---

approach does not recommend itself to Faraday. From Faraday's point of view, Ampère's microscopic currents are *utterly hypothetical* ("if they exist"); and insofar as they appear to be beyond the reach of experiment, quibbles about their possible direction are bound to leave Faraday unmoved. Besides, we saw in the previous Series (paragraph 2283) that he has other reasons to attribute diamagnetism to the body *as a whole*, not to its particles individually.

2432. Faraday returns to consider aeriform (gaseous) bodies, as in paragraph 2421 above. He notes that (*i*) they all occupy the *same* classificatory position; (*ii*) they share that position with *space*; and finally (*iii*) the shared position is an *intermediate* rather than an extreme one. What is "exceedingly remarkable" about each point? (*i*) Their sameness might appear remarkable since *as solids and liquids* the same materials differ widely— some highly magnetic, others highly diamagnetic—and if the magnetic or diamagnetic identity really belongs to a substance in virtue of its being that particular substance (iron, copper, water, etc.), then one would expect its magnetic or diamagnetic character to persist through mere changes of state. But in fact neither (*i*) their sameness, nor (*ii*) their coincidence in rank with space, is extraordinary. A commonsense explanation is ready to hand: Since gases are so highly rarefied, they actually do not contain very much more matter than space does, volume for volume—so with respect to their effects they might well be indistinguishable from space, and from one another, within the limits of our measurements. It is (*iii*) their intermediate position that is remarkable, for the reason suggested in the introduction. It would be no surprise for either space or highly rarefied matter to exhibit some physical characteristic in a *minimal* degree; but it is very difficult to see how matter in an extreme degree of rarefaction—not to mention a total absence of matter—could sustain a *moderate or middle degree* of such a characteristic.

that it should show no change in its relations by rarefaction to any possible degree, or even when the space it occupies passes into a vacuum; that it should be the same magnetically with any other gas or vapour; that it should not take its place at one end but in the very middle of the great series of bodies; and that all gases or vapours should be alike, from the rarest state of hydrogen to the densest state of carbonic acid, sulphurous acid, or æther vapour, are points so striking, as to persuade one at once that air must have a great and perhaps an active part to play in the physical and terrestrial arrangement of magnetic forces.

2433. At one time I looked to air and gases as the bodies which, allowing attenuation of their substance without addition, would permit of the observation of corresponding variations in their magnetic properties; but now all such power by rarefaction appears to be taken away; and though it is easy to prepare a liquid medium which shall act with other bodies as air does (2422.), still it is not truly in the same relation to them; neither does it allow of dilution, for to add water or any such substance is to add to the diamagnetic power of the liquid; and if it were possible to convert it into vapour and so dilute it by heat, it would pass into the class of gases and be magnetically undistinguishable from the rest.

2434. It is also very remarkable to observe the apparent disappearance of magnetic condition and effect when bodies assume the vaporous or gaseous state, comparing it at the same time with the similar relation to light; for as yet no gas or vapour has been made to show any magnetic influence over the polarized ray, even by the use of powers far more than enough to manifest such action freely in liquid and solid bodies.

2435. Whether the negative results obtained by the use of gases and vapours depend upon the smaller quantity of matter in a given

---

2433. *attenuation of their substance without addition*: That is, indefinite rarefaction, ultimately approaching the state of *vacuum* or *space*. If the magnetic behavior of gases is indeed only a reflection of their low density, then we should expect their magnetic properties to vary as that density is changed. But rarefaction does *not* perceptibly alter the magnetic behavior of gases (paragraphs 2408–2411). Notice Faraday's fetching images of rarefaction as a *dilution* of material in space, and of warming as a dilution in heat—the universal "solvent." He seems quite disappointed that the beautiful image does not, in fact, appear to hold! In subsequent Series, though, he will finally detect specific differences of diamagnetism in different gases, as well as in the same gas at different pressures.

2435. Here Faraday voices explicitly the question about gases and vapors: Does their magnetic similarity to space depend merely on the small quantity

volume, or whether they are direct consequences of the altered physical condition of the substance, is a point of very great importance to the theory of magnetism. I have imagined, in elucidation of the subject, an experiment with one of M. Cagniard de la Tour's æther tubes, but expect to find great difficulty in carrying it into execution, chiefly on account of the strength, and therefore the mass of the tube necessary to resist the expansion of the imprisoned heated æther.

2436. The remarkable condition of air and its relation to bodies taken from the magnetic and the diamagnetic classes, causes it to point equatorially in the former and axially in the latter. Or, if the experiment presents its results under the form of attraction and repulsion, the air moves as if repelled in a magnetic medium and attracted in a

---

of matter (per unit volume) they present? Or upon some specific factor pertaining to the *aeriform state* itself? Notice that another puzzle about gases is the apparent *discontinuity* (in magnetic properties) between the gaseous and the solid or liquid states. That problem would not exist if gases gradually *approached* the magnetic condition of vacuum—for vacuum is obviously the limit of continued rarefaction. On the other hand the puzzle about vacuum remains: to grasp how an *absence* of matter can possibly have properties that are *intermediate* among materials, rather than properties in a *null* degree.

one of M. Cagniard de la Tour's æther tubes: In experiments on ether made a few years before, Charles Cagniard de la Tour had discovered a combination of high pressure and low temperature at which the liquid and gaseous states appeared to merge into each other. For each substance a precise combination of temperature and pressure constitute what is now called its "critical point"; but Faraday called it the *disliquefying point* because, as he wrote to Dumas, at that point "the liquid becomes vapour without increase in bulk." The present relevance of this "disliquefying state" is clear: if there is no volumetric discontinuity between the liquid and gaseous states, then perhaps the troubling discontinuity in magnetic properties might similarly disappear! Recall that in the Twelfth Series, Faraday had noted that Cagniard's experiments bore a corresponding significance for the electrical properties of materials (paragraph 1336).

2436. *The remarkable condition of air ... causes it to point equatorially in [magnetic media] and axially in [diamagnetic media]*: Faraday does not identify the experiments to which he here refers, but he will fully document the magnetic behavior of air in the 25th Series. I will rehearse some of those findings in the introduction to the 26th Series.

if the experiment presents its results under the form of attraction and repulsion...: Not, "if we choose to see" but "if it presents"! The phrase appropriately expresses Faraday's characteristic readiness to let the phenomena *reveal themselves*, as I suggested in the introduction to the First Series.

medium from the diamagnetic class. Hence it seems as if the air were magnetic when compared with diamagnetic bodies, and of the latter class when compared to magnetic bodies.

2437. This result I have considered as explained by the assumption that bismuth and its congeners are absolutely repelled by the magnetic poles, and would, if there were nothing else concerned in the phænomena than the magnet and the bismuth, be equally repelled. So also with the iron and its similars, the attraction has been assumed as a direct result of the mutual action of them and the magnets; further, these actions have been admitted as sufficient to account for the pointing of the air both axially and equatorially, as also for its apparent attraction and repulsion; the effect in these cases being considered as due to the travelling of the air to those positions which the magnetic or diamagnetic bodies tended to leave.

2438. The effects with air are, however, in these results precisely the same as those which were obtained with the solutions of iron of various strength (2365.), where *all* the bodies belonged to the magnetic class, and where the effect was evidently due to the greater or smaller degree of magnetic power possessed by the solutions. A weak solution in a

---

Hence it seems as if the air were magnetic when compared with diamagnetic bodies, and [diamagnetic] when compared to magnetic bodies: With this statement he not only takes notice of the view whose deferral he announced in paragraph 2369, but goes beyond it. There he suggested only that the *pointing behavior* of a material was dependent on the surrounding medium; now he considers— though only tentatively—whether the magnetic and diamagnetic characters themselves, instead of constituting *inherent and opposed material natures*, might express only *differing relations* to that medium.

2437. *I have considered as explained… attraction has been assumed… actions have been admitted as sufficient…*: All these phrases recall explanations that *have been* acceptable up to now: If there are *two distinct classes* of magnetic material—one class absolutely attracted and the other absolutely repelled from magnetic poles or other strongly-magnetic regions—then it will be easy to explain the dual behavior of air—it simply fills up the space vacated by the magnetic or diamagnetic body and thus only *appears* to favor the equatorial or axial positions, respectively. But he will consider a different view in the next paragraph.

congeners: Things resembling or belonging to the same class as one another. Thus "bismuth and its congeners" refers to the whole class of diamagnetic substances. In the same way, "iron and its similars" refers to the class of ordinary magnetic materials.

2438. The dual behavior of air with respect to different surrounding media need not imply that those media belong to two opposing magnetic classes (the view recounted in the previous paragraph), since his iron sulphate solutions exhibited

stronger pointed equatorially and was repelled like a diamagnetic, not because it did not tend by attraction to an axial position, but because it tended to that position with less force than the matter around it; so the question will enter the mind, whether the diamagnetics, when in air, are repelled and tend to the equatorial position for any other reason, than that the air is more magnetic than they are, and tends to occupy the axial space. It is easy to perceive that if all bodies were magnetic in different degrees, forming one great series from end to end, with air in the middle of the series, the effects would take place as they do actually occur. Any body from the middle part of the series would point equatorially in the bodies above it and axially in those beneath it; for the matter which, like bismuth, goes from a strong to a weak point of action, may do so only because that substance, which is already at the place of weak action, tends to come to the place where the action is strong; just as in electrical induction the bodies best fitted to carry on the force are drawn into the shortest line of action. And so air in water, or even under mercury, is, or appears to be, drawn towards the magnetic pole.

---

the very same duality with respect to one another (paragraphs 2365 and 2368); and he showed that they are *all* magnetic, though in varying degrees (paragraph 2364). Perhaps, then, similarly, air *and all other materials* are magnetically active, air being so merely in a greater degree than the so-called "diamagnetic" bodies, and in a lesser degree than the bodies commonly classified as "magnetic."

not because it did not tend by attraction … but because it tended to that position with less force than the matter around it: When a weak solution of iron sulphate receded from the magnet's pole (paragraph 2365), both it and the stronger solution must have been drawn towards the axis. But the weaker solution may have been attracted *less* than was the surrounding, stronger solution; in which case the stronger solution would have displaced it from the strongly magnetic axial regions. Thus the solutions' behavior suggests the possibility that all materials, including gases, *belong to a single class* and are absolutely susceptible to magnetic force, but in unequal degrees. As Faraday points out, "if all bodies were magnetic in different degrees, forming one great series," the results would be just what we in fact observe.

Such actions are generally called *differential*, since they depend on differences of degree in a single attribute rather than on opposing powers such as attraction and repulsion. Faraday cites an example of differential action in electrical induction, where "the bodies best fitted to carry on the force are drawn into the shortest line of action." The introduction discusses this experiment, which comes from the Twelfth Series. The economy and interpretive power of this new differential view, which encompasses all matter under a *single* variable character, are striking. We might expect it to supplant all other interpretations—and yet the next paragraph will begin, "if this were the true view…". How then does this elegant and cogent "differential" image fall short?

2439. But if this were the true view, and air had such power amongst other bodies as to stand in the midst of them, then one would be led to expect that rarefaction of the air would affect its place, rendering it, perhaps, more diamagnetic, or at all events altering its situation in the list. If such were the case, bodies that set equatorially in it in one state of density, would, as it varied, change their position, and at last set axially: but this they do not do; and whether the rarefied air be compared with the magnetic or the diamagnetic class, or even with dense air, it keeps its place.

2440. Such a view also would make mere space magnetic, and precisely to the same degree as air and gases. Now though it may very well be, that space, air and gases, have the same general relation to magnetic force, it seems to me a great additional assumption to suppose that they are all absolutely magnetic, and in the midst of a series of bodies, rather than to suppose that they are in a normal or zero state. For the present, therefore, I incline to the former view, and

---

2439. *But if this were the true view*: If there were indeed but one universal magnetic class, then *air* must occupy a middle position within it. Moreover, air when sufficiently condensed or rarefied ought to take different positions with respect to other bodies; but it does not! Or at least, no such alterations have been observed. Oddly, Faraday seems to ignore the possibility that the effects are simply too minute to be detected with his present instruments. He previously acknowledged such a possibility where specific inductive capacity was concerned (paragraph 1293; paragraph 2432, *comment*). Why not here? But he will return to the question; see the introduction to the 26th Series.

2440. *Such a view would make mere space magnetic…*: Now he is acknowledging the real problem. If there is a single universal magnetic class, then space, as well as air and gases, must be "absolutely magnetic and in the midst of a series," inasmuch as space, air, and gases all occupy the same position on the magnetic scale—but how can *empty space* be magnetic? If, on the contrary, magnetic and diamagnetic materials constitute two separate classes, it will not be necessary to include *space* in either class. Thus Faraday is faced with a dual dilemma. Experiment offers strong support for the "single class" or differential view; but that view has frightful theoretical consequences for space. The alternative "two-class" view is unobjectionable in theory, but experiments fail to confirm it (though they do not refute it either).

all … in a normal or zero state: That is, a *true indifference*, an utter lack of relation to the magnetic force, as opposed to that *neutrality* which represents the mutual cancellation of opposite powers. If space, air, and gases belong to neither magnetic class, then they are to be accounted as indifferent to magnetic force. Faraday is evidently willing to countenance matter being indifferent or inactive—what is intolerable is that an *absence of matter* should be active!

For the present, therefore, I incline to the former view…: That is, he favors the "two class" view, at least provisionally. Perhaps he is fortified in that inclination by the *contrast* he insisted on earlier, in paragraph 2420—for the starker that

consequently to the opinion that diamagnetics have a specific action antithetically distinct from ordinary magnetic action, and have thus presented us with a magnetic property new to our knowledge.

2441. The amount of this power in diamagnetic substances seems to be very small, when estimated by its dynamic effect, but the motion which it can generate is perhaps not the most striking measure of its force; and it is probable that when its nature is more intimately known to us, other effects produced by it and other indicators and measurers of its powers, than those so imperfectly made known in this paper, will come to our knowledge; and perhaps even new classes of phænomena will serve to make it manifest and indicate its operation. It is very striking to observe the feeble condition of a helix when alone, and the astonishing force which, in giving and receiving, it manifests by association with a piece of soft iron. So also here we may hope for some analogous development of this element of power, so new as yet to our experience. It cannot for a moment be supposed, that, being given to natural bodies, it is either superfluous or insufficient, or unnecessary. It doubtless has its appointed office, and that one which relates to the whole mass of the globe; and it is probably because of its relation to the whole earth, that its amount is necessarily so small (so to speak) in the portions of matter which we handle and subject to experiment. And small as it is, how vastly greater is this force, even in dynamic results, than the mighty power of gravitation, for instance, which binds the whole universe together, when manifested by masses of matter of equal magnitude!

---

contrast, the easier it will be to see magnetic and diamagnetic materials as forming two separate classes. But Faraday probably continues to be influenced less by positive evidence here than by his abiding reluctance to attribute magnetic character to "mere space." Nevertheless, he will continue to struggle with the question; for a mere three paragraphs further on (paragraph 2443), he voices once more the opposite tenet—namely, that space cannot be *simply* indifferent to magnetic power! And by the 26th Series it will become clear that even the postulate of two separate classes cannot wholly avert implications of the materiality of space.

2441. *the astonishing force which [a helix] manifests by association with a piece of soft iron*: In particular, when the iron serves as the core of the helix.

vastly greater … than the mighty power of gravitation: There is an apparent asymmetry between the *strong* magnetic state iron is capable of developing and the *weak* diamagnetic degree of bismuth, at least as estimated by the mechanical forces involved. But relative weakness need not mean lack of importance in the natural scheme. For comparable masses of material, *gravity* is far weaker than the forces developed in diamagnetism; yet gravitation "binds the whole universe together," and no one could suppose that it plays an insignificant role in the natural order!

2442. With a full conviction that the uses of this power in nature will be developed hereafter, and that they will prove, as all other natural results of force do, not merely important but essential, I will venture a few hasty observations.

2443. Matter cannot thus be affected by the magnetic forces without being itself concerned in the phænomenon, and exerting in turn a due amount of influence upon the magnetic force. It requires mere observation to be satisfied that when a magnet is acting upon a piece of soft iron, the iron itself, by the condition which its particles assume, carries on the force to distant points, giving it direction and concentration in a manner most striking. So also here the condition which the particles of intervening diamagnetics acquire, may be the very condition which carries on and causes the transfer of force through them. In former papers (1161. &c.)[2] I proposed a theory of electrical induction founded on the action of contiguous particles with which I am now even more content than at the time of its proposition: and I then ventured to suggest that probably the lateral action of electrical currents which is equivalent to electro-dynamic or *magnetic* action, was also conveyed onwards in a similar manner (1663. 1710. 1729. 1735.). At that time I could discover no peculiar condition of the intervening or diamagnetic matter; but now that we are able to distinguish such an action, so *like* in its nature in bodies so *unlike* in theirs, and by that so

[2] Philosophical Transactions, 1835, Part 1.

2443. *It requires mere observation...*: In the Eleventh and Twelfth Series, Faraday carried out a reinterpretation of conventional electrical "action at a distance," presenting it instead as an action founded on the *internal inductive state of dielectric materials.* Similarly, he says, propagation of magnetic action must depend on an *internal condition of iron*—at least whenever there *is* any iron in the magnetic field. Notice that Faraday does not offer this suggestion as a routine analogy but as *evident to the sight* ("mere observation"). One such "observation" is the popular demonstration illustrated here: The large permanent magnet supports a train of small iron rods (nails or paper clips will do); each of which, in carrying the magnetic power onwards, becomes itself a magnet.

Perhaps, then, with diamagnetic materials also, "the condition which the particles ... acquire, may be the very condition which carries on and causes the transfer of force...." But as applied to diamagnetism the idea is a mere hypothesis; Faraday's experiments have not yet disclosed an *image* of diamagnetism. His work in the 26th Series, though, will permit him to extend the imagery of lines of force to display both magnetism and diamagnetism in their essential relation.

like in character to the manner in which the magnetic force pervades all kinds of bodies, being at the same time as universal in its presence as it is in its action; now that diamagnetics are shown not to be indifferent bodies, I feel still more confidence in repeating the same suggestion, and asking whether it may not be by the action of the contiguous or next succeeding particles that the magnetic force is carried onwards, and whether the peculiar condition acquired by diamagnetics when subject to magnetic action, is not that condition by which such propagation of the force is affected?

2444. Whichever view we take of solid and liquid substances, whether as forming two lists, or one great magnetic class (2424. 2437.), it will not, as far as I can perceive, affect the question. They are all subject to the influence of the magnetic lines of force passing through them, and the virtual difference in property and character between any two substances taken from different places in the list (2424.) will be the same; for it is the differential relation of the two which governs their mutual effects.

I feel still more confidence in repeating the same suggestion...: The suggestion, that is, that the magnetic and diamagnetic conditions are precisely to be identified with the respective capacities of those materials for *carrying on* the magnetic force. Note, however, that such an idea significantly undermines the system of two independent magnetic categories along with its convenient disposition of the vacuum (paragraph 2440)—since magnetism and diamagnetism would then be treated together under a single scheme. On the other hand, the unity and harmony of natural powers represents a powerful conviction, as Faraday declared at the outset of the 19th Series (paragraph 2146). Ultimately, Faraday's evolving vision of *congruity between agents and powers* is more important to him than evading the paradox of the vacuum. He will go far towards accepting a unified view of magnetism and diamagnetism in the next paragraph; and in the 26th Series he will put the vision forward with exceptional clarity—and follow out implications of the lines of force to far greater lengths than heretofore. But the question of *vacuum* remains a tormenting one, as we will see in paragraph 2445 below.

2444. *whether ... forming two lists, or one great magnetic class, it will not ... affect the question*: The question Faraday refers to is whether the "magnetic" and "diamagnetic" characters may be considered as *identified* with materials' respective capacities for carrying on magnetic force.

for it is the differential relation of the two which governs their mutual effects: With this statement he virtually gives up the "two list" view as far as observable effects are concerned. He would give it up for theoretical purposes also, it seems, were it not for the problem of vacuum—see the next paragraph. Let us note, and appreciate, the fierce tension between competing views that has been driving the last few paragraphs. Faraday is struggling over the significance of *space* and makes no secret of that struggle.

2445. It is that group which includes air, gases, vapours, and even a vacuum which presents any difficulty to the mind; but here there is such a wonderful change in the physical constitution of the bodies, and such high powers in some respects are retained by them, whilst others seem to vanish, that we might almost expect some peculiar condition to be assumed in regard to a power so universal as the magnetic force. Electric induction being an action through distance, is varied enough amongst solid and liquid bodies; but, when it comes to be exerted in air or gases, where it most manifestly exists, it is alike in amount in all (1292.); neither does it vary in degree in air however rare or dense it may be (1284.). Now magnetic action may be considered as a mere function of electric force, and if it should be found to correspond with the latter in this particular relation to air, gases, &c., it would not excite in my mind any surprise.

2446. In reference to the manner in which it is possible for electric force, either static or dynamic, to be transferred from particle to particle when they are at a distance from each other, or across a vacuum, I have nothing to add to what I have said before (1614, &c.). The supposition that such can take place, can present nothing startling

---

2445. It *is that group which includes air, gases, vapours, and even a vacuum which presents any difficulty to the mind*: Even if the new interpretation of magnetism and diamagnetism as *capacities to propagate magnetic force* is unexceptionable experimentally, it continues to present a theoretical problem for air, gases, vapors and vacuum because of their position in the middle of the magnetic list. Faraday will have to say that *vacuum too*—and matter so rarefied as to approach vacuum—carries on the magnetic force. That seems hardly less daunting than the difficulty we noted earlier (paragraphs 2432, 2435)—that *absence of matter* would have to exhibit magnetic properties in an intermediate degree. Nevertheless Faraday appears to find the present formulation somewhat more tolerable than the earlier one. We have, he reminds us, long ago accepted that static *electrical* induction takes place with equal facility through air, gases, and vacuum. To the extent that magnetism is produced by electricity, at least, we should not be surprised to discover a similar magnetic homogeneity among those same media.

2446. *The supposition … can present nothing startling*: It is hard to believe that Faraday really finds *nothing* troubling in the transmission of static electric induction, and heat, through the vacuum. But even if he is as comfortable as the phrase suggests, can Faraday simply *extend* the treatment of electrical induction to magnetism? Such might be the course of consistency and economy. But consistency and economy are virtues of *speech*. Faraday's science does not aim for an edifice of speech but for the *direct revelation of nature herself*. That will require an experimental disclosure; so far he still has not found one.

to the mind of those who have endeavoured to comprehend the radiation and the conduction of heat under one principle of action.

*   *   *

Royal Institution,
   Dec. 22, 1845.

Twenty-Sixth Series — Editor's Introduction

The relation between magnetism and diamagnetism

Do magnetism and diamagnetism represent *two opposed and mutually exclusive conditions*? Or are they simply *different degrees of a single condition*—so that no special significance attaches to any particular degree? The question arose with the discovery of diamagnetism in the Nineteenth Series; it has become urgent since the Twenty-First. We have seen Faraday question the relation of magnetic and diamagnetic materials to each other, and to *space*, several times already. The inquiry has taken a variety of forms; perhaps he has been struggling as much to figure out what the question is, as to determine its answer. Recall that in the 21st Series Faraday formulated the question specifically for gases. If all materials possess the same magnetic condition but in different degrees, then air and other gases must represent a roughly middle degree of that condition—a degree, however, that should *vary* with compression or rarefaction of the gas, just as the magnetic condition of an iron sulphate solution varied with the changing concentration of sulphate:

> 2439. But if ... air had such power amongst other bodies as to stand in the midst of them, then one would be led to expect that rarefaction of the air would affect its place, rendering it, perhaps, more diamagnetic, or at all events altering its situation in the list. If such were the case, bodies that set equatorially in it in one state of density, would, as it varied, change their position, and at last set axially: but this they do not do....

In the 21st Series Faraday was unable to detect any differences in the magnetic condition of gases under varying pressures—neither in the same gas nor even when comparing different gases with one another. But more delicate measurements in the 25th Series disclose unambiguous magnetic distinctions both among the gases themselves, and between the gases and *vacuum*—thus raising anew the possibility of a *single magnetic class embracing all matter*. The 25th Series is not included in these selections, but I should like to review some of its topics and the questions they bear upon before turning to the present Series, the Twenty-Sixth.

Some topics from Faraday's Twenty-Fifth Series

1. The differential magnetic balance

The sketch illustrates Faraday's basic experimental design. The glass tubes are identical but filled with different gases. They are suspended from horizontal bar AB, which is aligned in the equatorial direction and is also free to move in that direc-

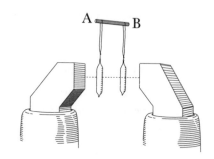

tion. With the electromagnet initially switched off, Faraday situates the tubes at equal distances on either side of the axial line.*

As Faraday recounted in paragraph 2290, magnetic substances tend to approach, diamagnetic substances tend to recede from, the polar axis. If then the gas-filled tubes are either equally magnetic or equally diamagnetic, their equal tendencies to approach or recede from the axial line when the electromagnet is energized will balance one another; and bar AB will not change position. But if they are magnetic or diamagnetic in unequal degrees—or if one is magnetic and the other diamagnetic—then the bar AB will be urged towards either A or B by a force representing the excess of one tendency over the other. When that excess is balanced by applying to the bar an external force just sufficient to retain the original position, the applied force serves as a measure of the relative difference, in magnetic or diamagnetic condition, of the two gases.**

Thus one may rank the gases in magnetic and diamagnetic order by making successive pairwise comparisons among them.*** And if evacuated tubes are included in these pairs, the rankings so obtained will also include *space*. With this new technique, the experiment that yielded a null result in the 21st Series receives fresh attention in the 25th. Faraday writes:

---

* In the actual experiments, Faraday employed specially-shaped pole pieces, not shown in the sketch, to insure a well-defined axial line.

** Recall that this was also the principle of the torsion balance used in the Eleventh Series; see paragraph 1180.

*** An advantage of this "differential" design is that the displacement tendencies of *the tubes themselves* balance each other and therefore need not be taken into account. Faraday explains that "being equal in nature and condition to each other, they tend to move with equal force when at equal distances, and at those distances compensate each other..." For similar reasons of symmetry the *air* or other surrounding medium may also be disregarded.

> A gas ... may be *rarefied* and *condensed* through a very extensive range, and the effect of this kind of change upon it ascertained independent of temperature or the presence of any other substance. Solids and liquids do not admit of these methods of examination, and do not therefore assist in the determination of the zero-point and of the true distinction of magnetic and diamagnetic bodies in the same manner that the gases do.
>
> It appeared to me that if a gaseous body were magnetic, then its magnetic properties ought to be diminished in proportion as it was rarefied, i.e. that equal volumes of such a gas at different pressures ought to be more magnetic, as they are denser; on the other hand, that if a gas were diamagnetic, rarefaction ought to diminish its diamagnetic character, until, when reduced to the condition of a vacuum, it should disappear.*

Using the differential balance he compares samples of different gases, as well as samples of the same gas at different densities, with one another. The results dramatically confirm what so many times previously seemed to have been refuted: *Gases do indeed alter their magnetic or diamagnetic strength when their densities are changed.* In one example, three samples of oxygen (at three different densities) are compared pairwise on the differential balance. Faraday finds that the denser sample of each pair consistently shows the stronger tendency towards the axial line. But *gases never alter their relation to vacuum.* When measured against an identical evacuated tube, a tube of any gas always approaches or always recedes from the axial line, whatever its density. When the gas-filled tube is fully exhausted it balances—but it never surpasses—the evacuated tube. Neither rarefaction nor condensation, therefore, can transform a magnetic gas into a diamagnetic one, or the reverse. The differential experiments thus offer at least a provisional answer to the question whether magnetism and diamagnetism are to be considered as one condition or two—they are *two.* By the same reasoning, the magnetic condition of space stands forth as being absolutely defined; Faraday lauds it as a "true zero" state (paragraph 2790). What about it is "true"?

2. The true zero

A "true zero" is the *absolute absence* of a specified condition. Opposed to a true zero is a condition of equilibrium or mutual cancellation between contrary powers. Such a composite neutrality is what Faraday produced earlier at paragraph 2422, by dissolving magnetic iron sulphate in diamagnetic water; his projected neutral glass (paragraph 2423) would have been similarly composite. A "true zero" state,

---

* Twenty-fifth Series, paragraphs 2778, 2779.

therefore, demands a medium that is not merely neutral, but *homogeneous and elementary*—so as to rule out any concealed opposition of powers. Vacuum is an obvious candidate—so were the *elementary* gases until the differential balance disclosed that most of them were not in fact neutral but magnetic or diamagnetic.

Notice therefore that if *air*, say, had yielded a different outcome in the differential experiments—if it had, when rarefied, actually passed from the magnetic to the diamagnetic condition—it would not thereby have implied a single all-embracing magnetic condition and controverted the absolute zero condition of space. For air is a mixture—of nitrogen and oxygen, mainly—and its magnetic condition may therefore be *composite*, the net result of magnetic and diamagnetic powers jointly. Under such circumstances, even a seemingly innocuous process like rarefaction might conceivably debilitate the predominating constituent more than the subordinate one; the result would be to transform the body from one magnetic character to the other. It would not then be permissible to conclude that magnetism and diamagnetism themselves were degrees of a single condition.

3. The magnetic characters of oxygen, nitrogen, and space

Air is a most important magnetic medium, whose properties must be accurately established. But since it is also a mixture, Faraday must separately investigate its constituents, nitrogen and oxygen. With the differential magnetic balance he finds in the 25th Series that oxygen is magnetic—more so, in fact, than any other gas tested. Nitrogen is, as far as he can determine, neutral.* And since nitrogen is elementary its neutral state is a *true* zero, not a composite. This is a very significant result; for it indicates that a material may be, absolutely and in itself, *indifferent* to magnetic power. I remarked in a comment to paragraph 2440 that Faraday seems prepared to acknowledge that matter might be indifferent or inactive—he only denies that an *absence of matter* could be active.

But it is not altogether satisfactory to contemplate a material that is *inactive* magnetically. Were there not very strong intimations in the 21st Series that magnetic power is *universal* among materials (paragraph 2420, *comment*)? If that is not in fact the case, and magnetic character is attributable only to certain materials, then what is the relation between the material and its magnetic or diamagnetic character? And in general—what is the relation between *any* material and the character or power we may attribute to it?**

* Is it then *obvious* that air, composed essentially of one magnetic and one neutral constituent, must therefore be magnetic?

** Recall that Faraday's electric eel (Fifteenth Series) presented a similar question.

Can there, in fact, *be* a "true zero" in the sense of *inactivity?* "Neutral" media clearly sustain some sort of magnetic relations—for they transmit magnetic forces. We cannot, therefore, view nitrogen and other "truly" neutral materials as simply devoid of magnetic power. But then—must it not follow?—if neutral materials are magnetically active we must equally attribute magnetic activity to *space!*

Faraday cannot escape this unwelcome consequence. Nevertheless it seems incomprehensible to him that an *absence of matter* can *act* in relation to magnetism. Although he acknowledges the *fact* that space can sustain a magnetic condition, he refuses to accept the equivalence of matter and space, which that fact would seem to imply:

> Before determining the place of zero amongst magnetic and diamagnetic bodies, we have to consider the true character and relation of *space* free from any material substance. Though one cannot procure a space perfectly free from matter, one can make a close approximation to it in a carefully prepared Torricellian vacuum. Perhaps it is hardly necessary for me to state, that I find both iron and bismuth in such vacua perfectly obedient to the magnet. From such experiments, and also from general observations and knowledge, it seems manifest that the lines of magnetic force (2149.) can traverse pure space, just as gravitating force does, and as static electrical forces do (1616.); and therefore space has a magnetic relation of its own, and one that we shall probably find hereafter to be of the utmost importance in natural phænomena. But this character of space is not of the same kind as that which, in relation to matter, we endeavour to express by the terms magnetic and diamagnetic. To confuse them together would be to confound space with matter, and to trouble all the conceptions by which we endeavour to understand and work out a progressively clearer view of the mode of action and the laws of natural forces. It would be as if in gravitation or electric forces (1613.), one were to confound the particles acting on each other with the space across which they are acting, and would, 1 think, shut the door to advancement. Mere space cannot act as matter acts....
>
> As space therefore comports itself independently of matter, and after another manner, the different varieties of matter must, in relation to their respective qualities, be considered amongst themselves. Those which produce no effect when added to space, appear to me to be neutral or to stand at zero. Those which bring with them an effect of one kind will be on the one side of zero, and those which produce an effect of the contrary kind will be on the other side of zero; by this division they constitute the two subdivisions of magnetic and diamagnetic bodies. The law which I formerly ventured to give (2267. 2418.), still expresses accurately their relations; for in an absolute vacuum or free space, a magnetic

body tends from weaker to stronger places of magnetic action, and a diamagnetic body under similar conditions from stronger to weaker places of action.*

Here for almost the only time in the Experimental Researches, Faraday's writing takes on a doctrinaire tone. Though compelled to acknowledge that space *has* a magnetic condition, he resolutely maintains that it cannot be the same kind of condition that matter has. His pronouncement that "mere space cannot act as matter acts" seems but a retreat into verbal orthodoxy. Why is he so insistent upon the separation of matter and space?

3. "Paramagnetism"

Although the magnetic balance experiments seem to answer the question "one magnetic class or two?" (page 443), recognition of *neutrality* as a condition of *activity* means that, in another sense, *all* materials are "magnetic"—they all respond to magnetic force, though indeed some of them respond to it in opposing ways. Further on in the 25th Series, therefore, Faraday proposes to convert the term "magnetic" into a general attribute, embracing two subclasses of materials. One subclass will contain the "diamagnetic" materials that have been under investigation since the Nineteenth Series. The other will contain the materials formerly called "magnetic"; but now that that term has been elevated to general status, Faraday proposes to call them *paramagnetic* substances instead.

> Now that the *true zero* is obtained, and the great variety of material substances satisfactorily divided into two general classes, it appears to me that we want another name for the magnetic class, that we may avoid confusion. The word *magnetic* ought to be general, and include *all* the phænomena and effects produced by the power. But then a word for the subdivision, opposed to the diamagnetic class, is necessary. As the language of this branch of science may soon require general and careful changes, I, assisted by a kind friend, have thought that a word not selected with particular care might be provisionally useful; and as the magnetism of iron, nickel and cobalt, when in the magnetic field, is like that of the earth as a whole, so that when rendered active they place themselves parallel to its axes or lines of magnetic force, I have supposed that they and their similars (including oxygen now) might be called paramagnetic bodies, giving the following division:

---

* Twenty-fifth Series, paragraphs 2787, 2789.

$$\text{Magnetic} \begin{cases} \text{Paramagnetic.} \\ \text{Diamagnetic.} \end{cases}$$

If the attempt to facilitate expression be not accepted, I hope it will be excused.*

The 25th Series thus proposes a genus/species system of classification, which in effect combines the previously irreconcilable one-class and two-class systems. But despite his own recommendation, Faraday does not hold very firmly to the system. In the Twenty-Sixth Series he will occasionally use the designation "magnetic" to refer to the specific condition of *permanently magnetized iron*, ignoring the term's proposed generic usage! This occurs even as he outlines a very important distinction between the temporary condition of paramagnetic bodies when brought under the influence of magnetic force, and the acquired or "permanent" condition exhibited by iron and allied substances. Later terminologists observed Faraday's distinction more faithfully than did Faraday himself, and the condition of permanently magnetized iron is now called *ferromagnetism.*

Magnetic conduction

Notwithstanding the highly important work of the 25th Series, it did not provide Faraday with an image of the essential nature of the paramagnetic and diamagnetic conditions. In the present Series, the Twenty-Sixth, he will adopt an image of *conduction* for provisional explication of the magnetic conditions. That image has already developed far beyond its first connotations of a circulating fluid and has grown to comprehend the *transmission of electric lines of force*, as well as that dynamic condition called *electric current.*** Applying it now to magnetism, Faraday identifies paramagnetic materials as *good* conductors of magnetic lines of force, diamagnetic materials as *poor* ones. As ever with Faraday, though, an image applied is an image changed. As he continues to explore the conduction image in the present Series, "good" and "poor" conduction will stand forth at last as actions of *gathering* and *dispersing* the lines of force, respectively.

Pointing

The conduction image bears immediate interpretive consequences for the action of *pointing,* as Faraday will note at paragraph 2811 below.

---

* Twenty-fifth Series, paragraph 2790.

** Unification of these two kinds of "conduction" was achieved in the 12th Series, where Faraday subsumed both *current conduction* and *insulation* under the theory of the *dielectric.*

When an elongated paramagnetic body aligns itself axially, the result is that *more force* will be carried on through the *better conductor* and for the *greatest possible distance*. Similarly, when an elongated *diamagnetic* body points equatorially the *least force* will be carried on through the *poorer conductor* and for the *shortest possible distance*. The drawings illustrate lines of force for paramagnetic axial pointing (sketch *a*) and diamagnetic equatorial pointing (sketch *b*).

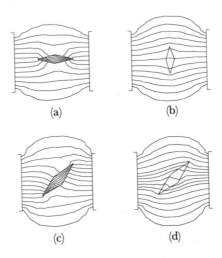

(a)   (b)

(c)   (d)

But in order to "point," bodies must be free to move. What if they are fixed in place instead? Sketches *c* and *d* show that in this case the lines of force alter their paths to achieve, as much as possible, the same condition as before: the greatest number of lines, transported for the greatest possible distance along the better (or the least possible distance along the poorer) conductor.

The conduction image thus provides Faraday with the germ of a more comprehensive view of "pointing," and of magnetic motions generally: We are to consider the whole system, consisting of paramagnets, diamagnets, and magnetic forces, as seeking *equilibrium*—which for magnetic systems is a condition of *optimal overall conduction*. Both the lines of force, and the material bodies (within the limits of their constraints), take up positions that best achieve that condition.

Note that this principle supports Faraday's tacit supposition in the Twentieth Series that induced currents circulate around the broad faces of the gyrating copper bar (paragraph 2335, *comment*). Since copper is diamagnetic, optimum conduction requires the magnetic lines to traverse the bar by the shortest possible path. This means entering and exiting through the broad faces, and it is about those faces, therefore, that the induced currents will circulate.

§ 32. *Magnetic conducting power.* ¶ i. *Magnetic conduction.*
¶ ii. *Conduction polarity.*

Received October 9,[2] — Read November 28, 1850.

¶ i. *Magnetic conduction.*

2797. THE remarkable results given in a former series of these Researches (2757. &c.) respecting the powerful tendency of certain gaseous substances to proceed either to or from the central line of magnetic force, according to their relation to other substances present at the same time, and yet the absence of all condensation or expansion of these bodies (2756.) which might be supposed to be consequent on such an amount of attractive or repulsive force as would be thought needful to produce this tendency and determination to particular places, have, upon consideration, led me to the idea, that if bodies possess different degrees of *conducting power* for magnetism, that difference may account for all the phænomena; and, further, that if such an idea be considered, it may assist in developing the nature of magnetic force. I shall therefore venture to think and speak freely on this matter for a while, for the purpose of drawing others into a consideration of the subject; though I run the risk, in doing so, of falling into error through imperfect experiments and reasoning. As yet, however, I only state the case hypothetically, and use the phrase *conducting power* as a general expression of the capability which bodies may possess of affecting the transmission of magnetic force; implying nothing as to how the process of conduction is carried on. Thus limited in sense, the phrase may be very useful, enabling us to take, for a time, a connected, consistent and general view of a large class of

[1] Philosophical Transactions, 1851, p. 29.

[2] Revised by the author and returned by him November 12, 1850.

2797. *the absence of all condensation or expansion of these bodies…:* Gases have no fixed volume but expand and contract according to their pressure and temperature. If a gaseous body is *attracted* to a region, therefore, should it not then become *concentrated* there? And yet no change in volume is observed when gases move into either stronger or weaker regions of magnetic force! Faraday will discuss the significance of this fact beginning at paragraph 2800 below.

developing: here, expounding, elucidating.

phænomena; may serve as a standard of meaning amongst them, and yet need not necessarily involve any error, inasmuch as whatever may be the principles and condition of conduction, the phænomena dependent on it must consist among themselves.

2798. If a medium having a certain conducting power occupy the magnetic field, and then a portion of another medium or substance be placed in the field having a greater conducting power, the latter will tend to draw up towards the place of greatest force, displacing the former. Such at least is the case with bodies that are freely magnetic, as iron, nickel, cobalt and their combinations (2357. 2363. 2367. &c.), and such a result is in analogy with the phænomena produced by electric induction. If a portion of still higher conducting power be brought into play, it will approach the axial line and displace that which had just gone there; so that a body having a certain amount of conducting power, will appear as if attracted in a medium of weaker power, and as if repelled in a medium of stronger power by this differential kind of action (2367. 2414.).

2799. At the same time that this idea of conduction will thus account for the place which a given substance would take up, as of oxygen in the axial line if in nitrogen, or of nitrogen at a distance if in oxygen, it also harmonizes with the fact, that there are no currents induced in a single gas occupying the magnetic field (2754.), for any one particle

---

2797 (continued). *must consist*: that is, must be consistent.

2798. *occupy the magnetic field*: Notice that here, as in paragraph 2247 in the Twentieth Series, a magnetic field is a *place that can be occupied*, not an agent or structure. But subsequently the "field" will acquire richer meanings.

the latter … displacing the former: But *why* should the medium having greater conducting power be able to displace the medium having lesser conducting power? At present he observes only that such seems to be the case among magnetic bodies; and he finds an analogy with electric induction (see the experiment with silk threads, discussed in the 21st Series' introduction). He will make a more serious attempt to *derive* displacement from inequality of conducting power beginning at paragraph 2803, below.

appear as if attracted…: According to this view, the poles of the magnet do not exert *pushes and pulls*. Rather, if a body is placed in a medium different from itself, and subjected to magnetic influence, *both it and the medium* will rearrange themselves in accordance with their relative conducting powers. While this may appear to involve attractions and repulsions, it is nevertheless *only* an appearance.

2799. *no currents induced in a single gas…*: Faraday is not referring to electric currents but to drift or circulation of the gas itself.

can then conduct as well as any other, and therefore will keep its place; and it also agrees, I think, with the unchangeability of volume (2750.).

2800. In reference to the latter point, we have to consider that the force which urges such a body as oxygen towards the middle of the field, is not a central force like gravitation, or the mutual attraction of a set of particles for each other; but an axial force, which, being very different in character in the direction of the axis and of the radii, may, and must produce its effect in a very different manner to a purely central force. That these differences exist, is manifest by the action of transparent bodies, when in the magnetic field, upon a ray of light; and also by the ordinary action of magnetic bodies: and hence, perhaps, the reason, that when oxygen is drawn into the middle of the field, in consequence of its conducting power, still its particles are not compressed together (2721.) by a force that otherwise would seem equal to that effect (2766.).

2801. So when two separate portions of oxygen or nitrogen are in the magnetic field, the one passes inwards and the other outwards, without any contraction or expansion of their relative volumes; and the result is differential, the two bodies being in *relation to and dependence on* each other, by being simultaneously related to the lines of magnetic force which pass conjointly through them both, or through them, and the medium in which they are conjointly immersed.

2802. I have already said, in reference to the transference onwards of magnetic force (2787.), that pure space or a vacuum permits that transference, independent of any function that can be considered as of the same nature as the conducting power of matter; and in a manner more analogous to that in which the lines of gravitating force, or of static electric force, pass across mere space. Then as respects those bodies which, like oxygen, facilitate the transmission of this power more or less, they class together as

---

2800. *not a central force...*: Central forces are directed towards a single point. But individual oxygen particles in the magnetic field are neither attracted to a common center, nor to one another; so it is clear that the forces acting on them are not central forces.

an axial force: A force directed along a line (an axis), but not towards a determinate center. This is by no means the first noncentral force we have come across; the magnetic rotation of light (Nineteenth Series) was an earlier example.

its particles are not compressed together: By contrast, if each particle were attracted to a common center, the particles would approach one another and the whole volume consequently diminish.

magnetic or paramagnetic substances (2790.); and those bodies, which, like olefiant gas or phosphorus, give more or less obstruction, may be arranged together as the diamagnetic class. Perhaps it is not correct to express both these qualities by the term *conduction*; but in the present state of the subject, and under the reservation already made (2797.), the phrase may I think be employed conveniently without introducing confusion.

2803. If such be a correct general view of the nature and differences of paramagnetic and diamagnetic substances, then the internal processes by which they perform their functions can hardly be the same, though they might be similar. Thus they *may* have circular electric currents in opposite directions, but their distinction can scarcely be supposed to depend upon the difference of force of currents in the *same* direction. If the view be correct also, though the results obtained when two bodies are simultaneously present in the magnetic field may be considered as differential (2770. 2768.) even though one of them be the general medium, yet the consequence of the presence of conducting power in matter renders a *single* body, when in space, subject to the magnetic force; and the result is, that when a paramagnetic substance is in a magnetic field of unequal force, it tends to proceed from weaker to stronger places of action, or is *attracted*; and when a diamagnetic body is similarly circumstanced, it tends to go from stronger to weaker places of action, or is *repelled* (2756.).

2804. Matter, when its powers are under consideration, may, as to its quantity, be considered either by weight or by volume. In the present

---

2802. *magnetic or paramagnetic substances*: As discussed in the introduction, Faraday introduced the term *paramagnetic* in the 25th Series to denote substances which, when placed under external magnetic influence, develop poles as iron does—but only temporarily, so long as the external magnetic power continues to be applied. Previously, such materials had been classified as "magnetic," along with iron, nickel, and cobalt. According to Faraday's present view, *iron, nickel, and cobalt*, together with all the *paramagnetic materials*, exhibit characteristically high degrees of conducting power, greatly facilitating the transmission of magnetic force. *Obstruction*—that is, low conducting power—is characteristic of *diamagnetic* materials.

olefiant gas: that is, ethylene.

2803. *circular electric currents*: A reference to Ampère's theory of magnetism (paragraph 3). If indeed *paramagnetism* is due to internal molecular currents having a certain direction, then currents in the *same* direction could not also account for *diamagnetism*. But currents in the *opposite* direction would imply merely *reversed poles*—a possibility Faraday contemplates here but soon (paragraphs 2818–2820) dismisses.

the result is, that … a paramagnetic substance … tends to proceed from weaker to stronger places of action…: Is this really an explanation? *How* does a body's conducting excellence result in its migration to places of stronger action?

case, where the effects produced have an immediate reference to mere space (2787. 2802.), it seems proper that the volume should be considered as the representation, and that in comparing one substance with another, equal volumes should be employed to give correct results. No other method could be used with the differential system of observation (2772. 2780.).

2805. Some experimental evidence, other than that of change of situation, of the existence of this conducting power, by differences in which, I am endeavouring to account for the peculiar characteristics of paramagnetic and diamagnetic bodies, may well be expected. This evidence exists; but as certain considerations connected with polarity preclude me from calling too freely upon iron, cobalt, or nickel (2832.) for illustrations, and as in other bodies which are paramagnetic, as well as in those that are diamagnetic, the effects are very weak, they will be better comprehended after some further general consideration of the subject (2843.).

2806. I will now endeavour to consider what the influence is which paramagnetic and diamagnetic bodies, viewed as conductors (2797.), exert upon the lines of force in a magnetic field. Any portion of space traversed by lines of magnetic power, may be taken as such a field, and there is probably no space without them. The condition of the field may vary in intensity of power from place to place, either along the lines or across them; but it will be better to assume for the present consideration a field of equal force throughout, and I have formerly

---

2804. *No other method [than comparing equal volumes] could be used with the differential system of observation*: The introduction to the present Series explains the operation of Faraday's differential balance. With unequal volumes, the action of the surrounding air would exert an undeterminable influence.

2805. *Some experimental evidence, other than that of change of situation ... may well be expected*: Since a link between a body's *conducting power* and its *change of situation* is the very interpretation these phenomena seem to present, Faraday must adduce evidence other than change of situation itself if that link is to be confirmed.

2806. *a field of equal force throughout*: Actually, a field in which the force is not only everywhere *equal* but also *same in direction*; today it would be called a *uniform field*. At paragraph 2463 Faraday reported how an approximately uniform field may be produced: "when flat-faced poles are used, though the lines of power are curved and vary in intensity at and towards the edges of the flat faces, yet there is a space at the middle of the magnetic field where they may be considered as parallel to the magnetic axes, and of equal force throughout." Moreover, the force becomes more nearly equal, and the lines more nearly parallel, as the distance between the pole faces is reduced.

described how this may, for a certain limited space, be produced (2465.). In such a field the power does not vary either along or across the lines, but the distinction of direction is as great and important as ever, and has been already marked and expressed by the term axial and equatorial, according as it is either parallel or transverse to the magnetic axis.

2807. When a paramagnetic conductor, as for instance, a sphere of oxygen, is introduced into such a magnetic field, considered previously as free from matter, it will cause a concentration of the lines of force on and through it, so that the space occupied by it transmits more magnetic power than before (fig. 1). If, on the other hand, a sphere of diamagnetic matter be placed in a similar field, it will cause a divergence or opening out of the lines in the equatorial direction (fig. 2); and less magnetic power will be transmitted through the space it occupies than if it were away.

2808. In this manner these two bodies will be found to affect, *first* the *direction* of the lines of force, not only within the space occupied by themselves, but also in the neighbouring space, into which the lines passing through them are prolonged; and this change in the course of the lines will be in the contrary direction for the two cases.

2809. *Secondly*, they will affect the *amount* of force in any particular part of the space within or near them; for as every section across the line of such a magnetic field must be definite in amount of force, and be in that respect the same as every other section, so it is impossible to cause a concentration within the sphere of oxygen (fig. 1) without causing also a simultaneous concentration in the parts axially situated as *a a* outside of it, and a corresponding diminution in the parts equatorially

---

2807. *concentration of the lines of force ... so that the space occupied by it transmits more magnetic power than before...*:   In keeping with the principle that each line represents a fixed increment of magnetic force (paragraph 1369, *comment*), concentration and separation of lines equate to intensification and diminution of force. Note that *conducting power*, the image with which the present Series opened, is now interpreted as the *gathering* (or the dispersal) *of lines of force.*

2808. *not only within ... but also in the neighbouring space...*:   As Figures 1 and 2 eloquently depict, the magnetic field is everywhere characterized by a fundamental *connectedness*; the lines of force are mutually related so that change anywhere means change everywhere.

2809. *every section across the line of such a magnetic field must be definite in amount of force...*:   Faraday already argued this for electric lines of force in the Twelfth Series. Together with the principle that each line of force represents everywhere a fixed, determinate quantity of force (paragraph 1369), it implies that *the total number of lines contained between any two lines is everywhere constant*—a consequence that will enjoy spectacular confirmation by the Moving Wire in the 28th Series.

Fig. 1.

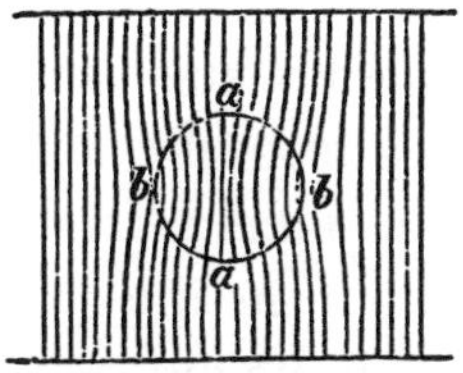

Fig. 2.

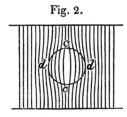

placed, *b b*. On the other hand, the diamagnetic body (fig. 2) will cause diminution of the magnetic force in the parts of space axially placed in respect of it, *c c*, and concentration in the near equatorial parts, *d d*. If the magnetic field be considered as limited in its extent by the walls of iron forming the faces of opposed poles (2465.), then even the distribution of the magnetism within the iron itself will be affected by the presence of the paramagnetic or diamagnetic bodies; and this will happen to a very large extent indeed, when, from among the paramagnetic class, such substances as iron, nickel or cobalt are selected.

2810. The influence of this disturbance of the forces upon the place and position of either a paramagnetic or a diamagnetic body placed

———————————————————————————————

even the distribution of the magnetism within the iron itself will be affected: The magnetic action becomes redistributed within the iron pole for the same reason that it suffers rearrangement throughout the magnetic field proper. The comment to paragraph 2810 will discuss the redistribution more fully.

2810. *The influence of this disturbance of the forces upon the place and position of either a paramagnetic or a diamagnetic body…*: As the body's presence changes the distribution of the lines of force, so in turn the new pattern of forces affects the position of the body. For example, one might think that in a *uniform* field a paramagnetic body would not move at all, since there is no "strong" region for it to move into, or "weak" region for it to move out of. But in fact, when paramagnetic bodies are placed in a uniform field, they invariably tend towards the nearest pole! How is that possible? As paragraph 2809 explained, the very ingression of a paramagnetic body into the field causes a concentration of lines not only inside the body but also in the axial regions outside it—even within the iron pole piece itself. And where the space between the body and the pole is smaller, the concentration will be even greater, as the sketch depicts. Since a paramagnetic body tends to move into regions of greater concentration, it will in this case move towards the pole.

S 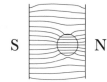 N

Once we understand the body's effect on the distribution of force, its motions can be "readily deduced" from the principle that paramagnetic bodies seek the stronger regions, and diamagnetic bodies the weaker. But that "principle" continues to leave several questions unanswered (paragraph 2803, *comment*). *Why* does a paramagnetic body move into stronger regions? And *how* does it acquire motion, since attraction at a distance is ruled out? The next paragraph begins to address these matters.

within the magnetic field, is readily deduced upon consideration and easily made manifest by experiment. A small sphere of iron placed within a field of equal magnetic power, bounded by the iron poles, has a position of unstable equilibrium, equidistant from the iron surfaces, and at such time a great concentration of force takes place through it, and at the iron faces opposite to it, and through the intervening axial spaces. If the sphere be on either side of the middle distance, it flies to the nearest iron surface, and then can determine the greatest amount of magnetic force to or upon the axial lines which pass through it.

2811. If the iron be a spheroid, then its greatest diameter points axially, whether it be in the position of unstable equilibrium, nearer to or in contact with the iron walls of the field. As the circumstances are now more favourable for the concentration of force in the axial line passing through the body than before, so this result can be produced by much weaker paramagnetics than iron, and I have no doubt could easily be produced by a vessel of oxygen or nitric oxide gas (2782. 2792.). It now becomes indeed a form, though not the best, of that experiment by which the magnetic condition of bodies is considered as most sensitively tested.

2812. The relative deficiency of power in diamagnetic bodies renders any attempt to obtain the converse phænomena to those of iron somewhat difficult; in order therefore to exalt the conditions, I used a saturated solution of protosulphate of iron in the magnetic field; by this means I strengthened the lines of power passing across it, without

---

2811. *circumstances are now more favourable for the concentration of force in the axial line...*: That is, when an elongated paramagnetic body aligns itself axially, *more force* will carried on through the *better conductor* and for a *greater distance* than before. As I suggested in the introduction to the present Series, this interpretation provides the germ of a more comprehensive view of the magnetic motions: Faraday will regard the whole ensemble of paramagnets, diamagnets, and magnetic forces as seeking a state of *equilibrium* which, for magnetic systems, is a condition of *optimal overall conduction*. In one of his last papers on the subject, the great essay *On the Physical Character of the Lines of Magnetic Force*, he will formulate this view explicitly.

2812. *in order therefore to exalt the conditions...*: Faraday introduces an elegant technique for intensifying the actions of weak diamagnets. Since diamagnetic behavior depends on the lessened concentration of lines of force within the test material in comparison to the surrounding medium, that behavior will be intensified if the lines can be concentrated even more highly *in the medium*. That is what Faraday accomplishes by replacing the original medium (air) with the highly paramagnetic iron sulphate solution. Note that this technique of substitution derives its rationale directly from the image of magnetic conducting power as a *tendency to gather (or to disperse) lines of force*, and would not make much sense apart from that image.

disturbing its equality in the parts employed, or introducing any error into the principle of the experiment, and then used bismuth as the diamagnetic body. A cylinder of this substance, suspended vertically, tended well towards the middle distance, finding its place of stable equilibrium in the spot where the paramagnetic body had unstable equilibrium. When the cylinder was suspended horizontally, then the direction it took was equatorial; and this effect also was very clear and distinct.

2813. These relative and reverse positions of paramagnetic and dia-magnetic bodies, in a field of equal magnetic force, accord well with their known relations to each other, and with the kind of action already laid down in principle (2807.) as that which they exert on the magnetic power to which they are subjected. One may retain them in the mind by conceiving that if a liquid sphere of a paramagnetic conductor were in the place of action, and then the magnetic force developed, it would change in form and be prolonged axially, becoming an oblong spheroid; whereas if such a sphere of diamagnetic matter were placed there, it would be extended in the equatorial direction and become an oblate spheroid.

2814. The *mutual action* of two portions of paramagnetic matter, when they are both in such a field of equal magnetic force, may be anticipated from the principles (2807. 2830.), or from the corresponding facts, which are generally known. Two spheres of iron, if retained in the same equatorial plane, repel each other strongly; but as they are allowed to depart out of that plane, they first lose their mutual repul-sive force and then attract each other, and that they do most powerfully when in an axial direction.

---

tended well towards the middle distance…: A bismuth body recedes from both poles and thus tends to situate itself midway between them.

2813. *the kind of action already laid down in principle*: The principle cited in paragraph 2807 is simply that of the gathering or dispersing of lines of force by para- and diamagnetic materials. But those actions do not, in themselves, imply anything about the motions or preferred positions of bodies in the presence of magnetic force. For that, additional principles are needed, as I suggested in comments to paragraphs 2810–2811 above.

2814. *a field of equal magnetic force*: that is, a *uniform magnetic field*, as in paragraph 2806. Faraday has reviewed the motions of solitary para- and dia-magnetic bodies in a uniform field; now he describes the mutual motions of a pair of bodies. Two spheres of iron recede from each other when placed parallel to the equatorial line; they approach each other when placed parallel to the axial line. Faraday merely reports the observed results here; he will interpret them more fully beginning at paragraph 2830 below.

2815. With diamagnetic bodies the mutual action is more difficult to determine, because of the comparative lowness of their condition. I therefore resorted to the expedient, before described, of using a saturated solution of protosulphate of iron as the medium occupying the field of equal magnetic force, and employing two cylinders of phosphorus, about an inch long and half an inch in diameter, as the diamagnetic bodies. One of these was suspended at the end of a lever, which was itself suspended by cocoon-silk, so as to have extremely free motion, and the adjustments were such, that when the phosphorus cylinder was in the middle of the magnetic field, it was free to move equatorially or across the lines of magnetic force; it however had *no tendency* to do so under the influence of the magnetic force. The other cylinder was attached to a copper wire handle, and could be placed in a fixed position on either side of the former cylinder; it was therefore adjusted close by the side of it, and the two retained together, until all disturbance from motion of the fluid or of the air had ceased; then the retaining body was removed, the two phosphorus cylinders still keeping their places; finally, the magnetic power was brought into action, and immediately the moveable cylinder separated slowly from the fixed one and passed to a distance. If brought back again whilst the magnet was active, when left at liberty it receded; but if restored to close vicinity, when the magnetic force was away, it retained that situation. The effect took place either in the one direction or the other, according as the fixed cylinder was on this or that side of the moving one; but the motion was in both cases across the lines of magnetic force, and was indeed mechanically and purposely limited to that direction by the mode of suspension. When two bismuth balls were placed, in respect of each other, in the direction of the magnetic axis, so that one might move, but

---

2815. *lowness*: that is, *weakness*. To observe interactions between diamagnets in a uniform field, Faraday selects phosphorus, one of the most highly diamagnetic materials, as the test material, and the highly paramagnetic iron sulphate solution as the surrounding medium, thus intensifying the effects in two ways.

One of these was suspended at the end of a lever, which was itself suspended by cocoon-silk...: The lever is suspended horizontally at its midpoint; it carries the cylinder at one end and a counterweight at the other, like the torsion-balance Faraday employed in the 25th Series (see the introduction); he actually calls it "the torsion balance" in paragraph 2817 below.

The effect took place either in the one direction or the other...: Despite the intensifying measures noted, the interaction between the diamagnets is very weak. Therefore Faraday pays special attention to the movable cylinder's initial tendency as the electromagnet is turned on, since this is more readily discerned. He finds that one diamagnet tends to recede from another diamagnet when they are lined up in the equatorial direction.

only in the direction of that axis, its place was not sensibly affected by the other; the tendency of the free one to go to the middle of the field (2812.) overpowered any other tendency that might really exist.

2816. Thus two diamagnetic bodies, when in the magnetic field, do truly affect each other; but the result is not opposed in its direction to that of paramagnetic bodies, being in both cases a separation of the substances from each other.

2817. The comparison of the action of para- and diamagnetic bodies on each other, was completed by using water as the medium in a field of equal magnetic force, and suspending a piece of phosphorus from the torsion balance. When the magnetic power was on, this phosphorus was repelled equatorially, as before, by another piece of phosphorus, but it was attracted by a tube filled with a saturated solution of protosulphate of iron; so paramagnetic and diamagnetic bodies attract each other equatorially in a mean medium, but each repels bodies of its own kind (2831.).

¶ ii. *Conduction polarity.*

2818. Having thus considered briefly the effects which the disturbance of the lines of force, by the presence of paramagnetic and diamagnetic bodies, is competent to produce (2807. &c.), I will ask attention to that which may be considered as their polarity: not wishing by the term to indicate any internal condition of the substances or their particles, but the condition of the mass as a whole, in respect of the state into which it is brought by its own disturbance of the lines of magnetic force; and that, both in regard to its condition with respect to other bodies similarly

---

2815 (continued). *the tendency of the free one to go to the middle of the field … overpowered any other…*: He tries to exhibit interaction between diamagnetic bismuth balls lined up *axially*. But any interaction between them proves to be insensibly weak, in comparison to the individual tendency of each ball to take up a central position in the axial direction.

2817. *a mean medium*: One having a magnetic rank *between* two test materials. In the present case the mean is water, which in paragraph 2424 was ranked above phosphorus, but below all magnetic materials, such as iron sulphate solution.

2818. *not … any internal condition of the substances or their particles…*: This in contrast to fluid theories, which posit internal concentrations of magnetic fluid within para- and diamagnetic bodies, or to Ampère's hypotheses, which attribute circulating electric currents to a body's individual particles.

…but the condition of the mass as a whole: That para- and diamagnetic action is to be attributed to the *whole*, was indicated by the absence of magnetic compression or expansion of gases—such as would be expected if the overall magnetic action were a summation of central forces acting on individual particles (paragraph 2800, *comment*).

affected; and also in regard to differences existing in different parts of its own mass. Such a condition concerns what may be called conduction polarity. Bodies in free space, when under magnetic action will possess it in its simplest condition; but bodies immersed in other media will also possess it under more complicated forms, and its amount may then be varied, being reversed or increased, or diminished to a very large extent.

2819. Taking the simplest case of paramagnetic polarity, or that presented in fig. 1 (2807.), it consists in a convergence of the lines of magnetic force on to two opposed parts of the body, which are to each other in the direction of the magnetic axis. The difference in character of the two poles at these parts is very great, being that which is due to the known difference of quality in the two oppo-

Fig. 1.

site directions of the line of magnetic force. Whether polar attraction or repulsion exists amongst paramagnetic bodies, when they present mere cases of conduction (as oxygen, for instance), is not yet certain (2827.), but it probably does; and if so, will doubtless be consistent with the attraction and repulsion of *magnets* having correspondent poles.

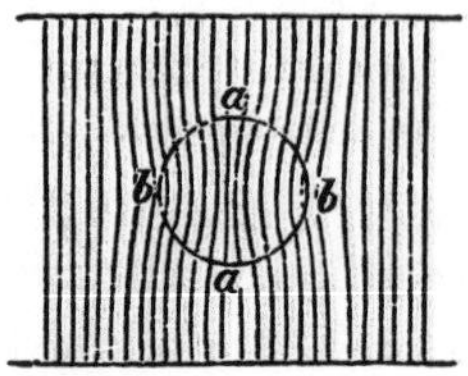

2820. When we consider the conduction polarity of a diamagnetic body, matters appear altogether different. It has not a polarity like that of a paramagnetic substance, or one the mere reverse (in name or direction of the lines of force) of such a substance, as I, Weber and others

---

2818 (continued). *conduction polarity*: Faraday is about to offer a new account of "polarity," founded on the image of *conduction* as a relative *gathering up* or *spreading out* of lines of force.

2819. *the simplest case of paramagnetic polarity*: In Faraday's new image, a "pole" is a place of *convergence of lines*. But if the new image is to be a competent replacement for the old one, it must adequately represent the essential phenomena on which the old image was based—though it need not interpret those phenomena in the same way. Now it is easy to discover, in the image of poles as convergences, that element of *oppositeness* which conventional thinking attributed to the poles. For suppose the lines in Figure 1 have a downward direction. Then they may be seen to *converge* at the upper pole *a* but to *diverge* at the lower pole *a*. But can the new image represent equally well the conventional appearances of *attraction and repulsion*? Experiments to be described in paragraphs 2826–2827 bear on that question; unfortunately they turn out to be inconclusive.

mere cases of conduction … magnets: Up to now, Faraday has not made any significant distinction between para- and diamagnetic *conductors*, and permanent *magnets*. He merely alludes to such a distinction here, but will explore it at length beginning at paragraph 2832 below.

2820. *conduction polarity of a diamagnetic body … altogether different*: If a "pole" is a place of convergence, a diamagnetic body has *no* poles. For refer to the regions

have at times assumed (2640.), but a state of its own altogether special. Its polarity consists of a divergence of the lines of power on to, or a convergence from the parts, which being opposite, are in the direction of the magnetic axis; so that these poles, having the *same* general and opposite relations to each other, which correspond to the differences in the poles of paramagnetic bodies, have still, under the circumstances, that striking contrast and difference from the polarity of the latter bodies which is given by convergence and divergence of the lines of force.

Fig. 3.

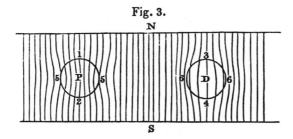

2821. Let fig. 3 represent a limited magnetic field with a paramagnetic body P, and a diamagnetic body D, in it, and let N and S represent the two walls of iron associated with the magnet (2465.) which form its boundary, we shall then be able to obtain a clear idea of the direction of the lines of magnetic force in the field. Now the two bodies, P and D, cannot be represented by supposing merely that they have the same polarities in opposite directions. The 1 polarity of P is importantly unlike the 3 polarity of D; but if D be considered as having the reverse polarities of P, then the one [1] polarity of P should be like the 4 polarity of D, whereas it is more unlike to that than to the 3 polarity of D, or even to its own 2 polarity.

---

marked *c, c* in Figure 2, which is reproduced here: the convergence belongs rather to the surrounding medium, than to the body! The widespread presumption that diamagnets develop poles merely *reversed* from those which permanently magnetizable materials develop—N where iron would exhibit S, S where iron would display N—is here altogether discarded. In the next paragraph Faraday will undertake a detailed exposition of para- and diamagnetism under the conduction image.

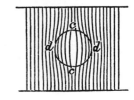

2821. *more unlike to [the 4 polarity of D] than to the 3 polarity ... or even to its own 2 polarity*: If D is thought to have poles of opposite character to P, then pole 1 should be the opposite of 3 and therefore the same as 4. Now proceeding from N to S, the lines at 1 converge as they enter P; but the lines at 4 converge as they *exit* D. This is difference enough to refute a supposed polarity in diamagnets that is simply opposite to that in paramagnets. But why does Faraday regard it as a *greater* difference than the dissimilarity of 1 and 3, or of 1 and 2?

2822. There are therefore two essential differences in the nature of the polarities dependent on conduction, the difference in the direction of the lines of force abutting on the polar surfaces, when the comparison is with a magnet reversed, and the difference of convergence and divergence of these lines, when compared with a magnet not reversed; and hence a diamagnetic body is not in that condition of polarity which may be represented by turning a paramagnetic body end for end, while it retains its magnetic state.

2823. Diamagnetic bodies in media more diamagnetic than themselves, would have the polar condition of paramagnetic bodies (2819.); and in like manner paramagnetic conductors in media more paramagnetic than themselves, would have the polarity of diamagnetic bodies.

2824. Besides these differences the bodies must have an equatorial condition, which, in the two classes of conductors, would be able to produce corresponding effects. The whole of the equatorial part of P (fig. 3) is alike in polar relation to the body P, or to the lines of force in the surrounding space; and there is a like correspondence in the equatorial parts of D, either to itself or to space; but these parts in P or in D differ in intensity of power one from the other, and both from the general intensity of the space. Such equatorial conditions must, I think, exist as a consequence of the definite character of any given section of the magnetic field (2809.).

2823. *Diamagnetic bodies in media more diamagnetic than themselves, would have the polar condition of paramagnetic bodies…:*   Earlier experiments abundantly confirmed, but could not explain, the fact. Now it becomes the evident and necessary consequence of the "convergence" image; for to have "the polar condition of paramagnetic bodies" is simply to disperse the lines *less*—that is, to be less diamagnetic—than the surrounding material.

Nevertheless Faraday's language suggests some reluctance to accept all the implications of the "convergence" image. He says that bodies less diamagnetic than the surrounding medium would *have the polar condition of* paramagnetic bodies—he does not say that they would *be* paramagnetic. Perhaps he still wishes to attribute a special status to *vacuum* (see the introduction), and to classify bodies as para- or diamagnetic only in relation to that medium. But the convergence image provides no support for such absolutism.

2824. *The whole of the equatorial part of P … is alike in polar relation to the body P…:*   Here Faraday's phraseology is more inventive than it is perspicuous; the line patterns depict relations that are not easy to express in words. He seems to discern a basic homogeneity among the equatorial regions of all systems, whether paramagnetic or diamagnetic—a degree of comparability that the axial regions fail to show.

such equatorial conditions must … exist as a consequence of the definite character of any given section: In order for lines of force to gather within a body, they must

Fig. 3.

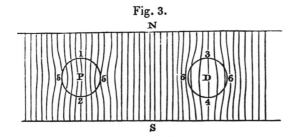

2825. Though the experimental results of these polarities are not absent, still they are not very evident or capable of being embodied in many striking forms; and that because of the extreme weakness of the forces brought into play, as compared with those larger forces exhibited in the mutual action of magnets. Hence it is, that the many attempts to show a polarity in bismuth have either failed, or other phænomena have been mistaken for those properly referable to such a cause. The highest, and therefore the most delicate, test of polarity we possess, is in the subjection of the polar body to the line of direction of magnetic forces of a very high degree, when developed around it; and hence it is, that the pointing of a substance between the poles of a powerful magnet is continually referred to for such a purpose. It would be, and is utterly in vain to look for any mutual action between the poles of two weak para-magnetic or diamagnetic conductors in many cases, when the action of these same poles is abundantly manifest in their relation to the almost infinitely stronger poles of a powerful horseshoe or electro-magnet.

---

be removed from adjacent (equatorial) regions outside the body. Similarly, in order for them to diverge within the body, they must become concentrated outside it. Since each line of force represents a fixed quantity of magnetic power, any concentration or dispersion of lines represents a corresponding intensification or diminution of force.

2825. *It would be … utterly in vain to look for any mutual action between the poles…*: The "polar" regions of an ordinary magnet may be easily located by causing another magnet to interact with it; for example, a compass needle will "point" to the poles of a bar magnet. According to the images of conduction polarity, bodies that have taken on the para- or diamagnetic condition should possess comparably distinguished regions—as 1, 2, 3, and 4 in Figure 3. It would be gratifying to be able to locate these regions by pointings or other interactions between para- and diamagnetic samples—and in fact that is no problem with strongly paramagnetic bodies; they interact just as ordinary magnets do. But weak paramagnets—and *all* diamagnets—interact so feebly as to render such tests hopelessly inconclusive. The most reliable indicator of the magnetic character of a body remains its interaction with a powerful conventional or electromagnet.

2826. I took a tube *a* (fig. 4), filled with a saturated solution of sulphate of cobalt, and suspended it between the poles of the great electro-magnet; it set readily and well. Another tube, *b*, was then filled with a saturated solution of sulphate of iron, and being associated with the S pole of the magnet, was brought near the cobalt tube in the manner shown, but not the slightest effect on the position of *a* was observable. The tube *b* was changed into the position *c*, to double any effect that might be present, but no trace of mutual action between the poles of *a* and *b* was visible (2819.).

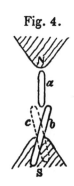

Fig. 4.

2827. To increase the effect, the magnetic solution tube was suspended in water, as a good diamagnetic medium, between flat-faced poles (fig. 5). It pointed well. Two bottles of saturated solution of sulphate of iron were placed at *d* and *e*, but they did not alter the position of *a*; being removed into the positions *f* and *g*, neither was any sensible alteration of the position of *a* produced. I made the same kind of experiment with an air-tube in water, in which case it points axially (2406.), with the same negative result. I do not mean to assert that there was absolutely no effect produced in these cases (2819.); but if any, it must have been inappreciably small, and shows how unfit such means are to compare with those which are supplied by the pointing of a body when under the influence of powerful magnets. If polarity cannot be found by these methods in paramagnetic bodies so strongly influential as saturated solution of iron, nickel or cobalt, it can hardly be expected

Fig. 5.

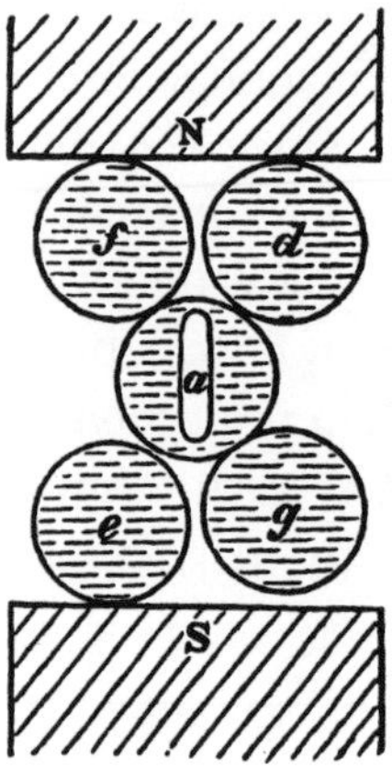

2826. These experiments show how weak is the interaction between paramagnetic bodies, in comparison to that between either body and the pole of the great magnet. How much more difficult, therefore, will it be to detect interaction between diamagnetic bodies!

it set readily and well: that is, it pointed axially.

The tube b was changed into the position c, to double any effect that might be present: If the tube in position *b* causes any displacement of *a*, then when in position *c* it should cause an equal displacement in the opposite direction. The movement of *a* from one side to the other would be double its initial displacement.

to manifest itself by analogous actions in the much weaker cases of diamagnetic substances.

2828. When a spherical paramagnetic conductor is placed midway in a field of equal magnetic force, it occupies a place of unstable equilibrium, from which, if it be displaced ever so little, it will continue to move until it has gained the iron boundary walls of the field (2465. 2810.); this is a consequence of its particular polar condition. If the sphere were free to change its form, it would elongate in the direction of the magnetic axis; or if it were a solid of an elongated form, it would point axially, both consequences of its polar condition (2811.).

2829. So also in the case of diamagnetic bodies, their peculiar condition of polarity is shown by corresponding facts, namely, by a spherical portion having its place of stable equilibrium in the middle of the magnetic field (2812.), by a fluid portion tending to expand equatorially and become an oblate spheroid (2813.), and by the equatorial pointing of an elongated portion (2812.). If pointed magnetic poles are used, then the effects are very much stronger, but are exactly the same in kind, and dependent upon the same causes and polar conditions.

2830. There are another set of effects produced, which are either the results of the axial polarity just referred to, or else may be considered as consequences of the condition of the equatorial parts of the conductors (2824.). Two balls of iron, in a field of equal force, if retained in a plane at right angles to the line of force, *i. e.* with their equatorial parts in juxtaposition, separate from each other with considerable power (2814.), and the probability is that two infinitely weaker bodies of the paramagnetic class would separate in like manner. Two portions of phosphorus, being a diamagnetic substance, have been found also to separate under the same circumstances (2815.).

2831. The motions here are of the same kind, whereas they might have been expected to be the reverse (2816.) of each other; still they

---

2830. *axial polarity ... condition of the equatorial parts*: Refer again to Figure 3, reproduced here, where regions 1, 2, 3, and 4 may be said to illustrate places of axial polarity; while regions 5 and 6 represent the equatorial parts.

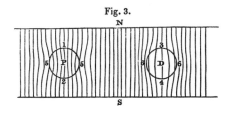

Fig. 3.

Fig. 6.

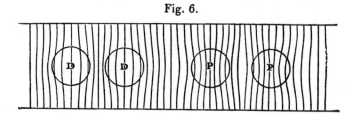

are perfectly consistent. The diamagnetics ought to separate, for the field is stronger in lines of magnetic force between them than on the outsides, as may easily be seen by considering the two spheres D D in fig. 6; and therefore this motion is consistent, and is in accordance generally with the opening or set equatorially, either of separate portions or of a continuous mass of such substances (2829.), in their tendency to go from stronger to weaker places of action. On the other hand, the two balls of iron, P P, have weaker lines of force between them than on the outside; and as their tendency is to pass from weaker to stronger places of action, they also separate to fulfil the requisite condition of equilibrium of forces. Finally, a paramagnetic and a diamagnetic body attract each other (2817.); and they ought to do so, for the diamagnetic body finds a place of weaker action towards the paramagnetic body, and the paramagnetic substance finds a place of stronger action in the vicinity of the diamagnetic body, D P, fig. 6.

2832. I have frequently spoken of iron in illustration of the action of paramagnetic conductors, and considered the polarity which it acquires as the same with that of these conductors; but I must now

---

2831. *the two balls ... PP, have weaker lines of force between them than on the outside*: The difference is not altogether evident in the diagram, but Faraday's point is clear: immediately outside a paramagnetic body the lines must be rarefied, since some of them will have been gathered into the interior of the body. Between two *adjacent* paramagnetic bodies that rarefaction will be magnified. For a similar reason "the diamagnetic body finds a place of weaker action towards the paramagnetic body, and the paramagnetic substance finds a place of stronger action in the vicinity of the diamagnetic body."

Faraday's clear account illustrates the impressive analytic power inherent in the new image of conducting power as an action of gathering (or dispersing) lines of force. Notice that if paramagnetism and diamagnetism are indeed two independent magnetic conditions (see the introduction), the conduction image will have to distinguish absolutely between paramagnetism as a "gathering" power, and diamagnetism as a "dispersing" power. On the other hand, the image easily accommodates an interpretation of "gathering" and "dispersing" as merely differing degrees of the same power to sustain and carry on lines of force. Faraday appears to take little notice of such intimations here; he will find them more compelling in later Series.

make clear a distinction, which exists in my mind, with regard to the polarity of a magnet, and the polarity, as I have called it, due to mere conduction. This distinction has an important influence in the case of iron. A permanent magnet has a polarity in itself, which is possessed also by its particles; and this polarity is essentially dependent upon the power which the magnet inherently possesses. It, as well as the power which produces it, is of such a nature, that we cannot conceive a mere space void of matter to possess either the one or the other, whatever form that space may be supposed to have, or however strong the lines of magnetic force passing across it. The polarity of a conductor is not necessarily of this kind; is not due to a determinate arrangement of the cause or source of the magnetic action, which in its turn overrules and determines the special direction of the lines of force (2807.); but is simply a consequence of the condensation or expansion of these lines of force, as the substance under consideration is more or less fitted to convey their influence onwards. It is evidently a very different thing to originate such lines of power and *determine their direction* on the one hand, and only to assist or retard their progression without any reference to their direction on the other. Speaking figuratively, the difference may be compared to that of a voltaic battery and the conducting wires, or substances, which connect its extremities. The stream of force passes through both, but it is the battery which originates it, and also determines its direction; the wire is only a better

---

2832. *the polarity of a magnet, and the polarity ... due to mere conduction*: Faraday has alluded before (paragraph 2819) to the distinction between that polarity which is the permanent attribute of a magnetized piece of iron, and the polarity exhibited by a magnetic conductor—whether para- or diamagnetic—while under the temporary action of an external magnetic force. Earlier, too, he attributed the para- and diamagnetic conditions to *whole bodies*, not to the individual particles of a body. But *permanent* magnetism, he now suggests, *does* belong to the particles. A consequence of this is that vacuum, having no "particles," cannot be *magnetized*—even though vacuum certainly possesses some degree of magnetic conducting power.

a very different thing to originate such lines ... on the one hand, and only to assist or retard their progression ... on the other: At stake is the distinction and the relation between a "source" and a "medium". That question was implicit in Faraday's study of the electric eel in the Fifteenth Series. As he attempts to articulate the distinction here, he continues to rely on metaphors: the *voltaic battery*, which "originates" the force, as distinguished from the *wires*, which assist or retard its progression. The imagery of convergence and divergence of lines of force has significantly clarified the relation between a *conductor* and the power it transmits. But Faraday still seeks to grasp the relation between a *source* and its own power, and how that power can belong to or be originated by it. In the present case, what is the relation between a *permanent magnet* and its own *lines of force*?

or worse conductor, however by variation of form or quality it may diffuse, condense, or vary the stream of power.

2833. If this distinction be admitted, we have to consider whether iron, when under the influence of lines of magnetic power, becomes a magnet and has its proper polarity, or is a mere paramagnetic conductor with conducting powers of the highest possible degree. In the first place, it would have the real polarity of the magnet, in the second only that which I assign to oxygen and other conducting bodies. To my mind the iron is a magnet. It can be raised as a *source* of lines of magnetic power to an extreme degree of energy in the electromagnet; and though, when very soft, it usually loses nearly all this power upon the cessation of the electric current, yet such is not the case if the mass of metal forms a continuous circuit or ring, for then it can retain the force for hours and weeks together, and is evidently for the time an original source of power independent of any voltaic current. Hence I think that the iron under the influence of lines of magnetic power becomes a magnet; and though it then has the same kind of polarity, as to direction, as a mere paramagnetic conductor, subject to the same lines of force, still with a great difference; for as the internal particles of iron become in a degree each a system producing magnetism, so their polarity is correlated and combined together into a polar whole, which, being infinitely more intense, may also be very different in the disposition of its force in different parts, to that equivalent to polarity, which a mere conductor possesses.

2833. *whether iron, when under the influence of lines of magnetic power, becomes a magnet*: When a sample of unmagnetized iron is exposed to magnetic action, does it act only as a highly effective *conductor*, or does it become an *originator* of magnetic power?

its proper polarity: that is, a polarity proper to itself, inherent, in contrast to the polarity of conduction which is relative to an external medium.

To my mind the iron is a magnet: Since hardened iron can be given a permanently polar condition, it must have acquired *possession* of that state and is therefore an originator of magnetic power. Soft iron does not long retain its acquired polarity and would therefore have to be regarded as a mere conductor—except that when formed into a ring it exhibits retentive power approaching that of hard iron! Hence Faraday thinks that *whenever* iron acts magnetically, it must be regarded as a *source* of magnetic power. Notice also his intriguing suggestion that the reason iron, nickel, and cobalt are so much more powerful than the merely paramagnetic materials is that, being *sources*, their particles are involved; and the combined and correlated powers of innumerable particles may well add up to an extraordinary degree of strength. Subsequent investigators have accepted Faraday's distinction and now refer to the permanently magnetized state of iron as *ferromagnetism*, as distinguished from the merely conductive paramagnetism and diamagnetism.

2834. It appears to me also as very probable, that when iron, nickel and cobalt, are heated up to the respective temperatures at which they lose their wonderful degree of power (2347.) and retain only so small a portion of it as to require the most sensible test to make it manifest (2343.), they then have passed into the condition of paramagnetic conductors, have lost all ability to acquire that state of internal polarity they could assume as magnets, and now have no other polarity than that which belongs to them as masses of paramagnetic matter (2819.). It is also probable that in many states of combination these metals may take up the mere conducting state; for instance, that whilst in the protoxide, iron may constitute a magnet, in the peroxide it is only a conductor; and in this respect it is not a little curious to find oxygen, which as a gas is a paramagnetic body (2782.), reducing iron down to, and indeed far below its own condition, weight for weight. In their various salts also and solutions, these metals may, in conjunction with the combined matter, be acting only as conductors.

2835. Perhaps I ought not to have called the condition of concentration or expansion of the lines of magnetic force in the bodies acting as conductors, a polarity; inasmuch as true magnetic polarity depends essentially and entirely on the *direction* of the line of force, and not on any mere compression or divergence of these lines. I have

---

2834. *have lost all ability to acquire that state of internal polarity*: He suspects that magnetized materials lose their (ferro)magnetic identity at high temperatures, becoming mere paramagnetic conductors. Iron may also lose its ferromagnetic power when it enters into chemical combination—especially with oxygen. What significance does Faraday find in these facts? Perhaps he means these examples, in which iron both *gains and loses* its ability to originate magnetic force, to serve as guideposts in thinking about the enigmatic relation between *source* and *power*.

2835. *true magnetic polarity depends essentially and entirely on the direction of the line of force, and not on any mere compression or divergence...*: By that criterion there would be "polarity" even in a uniform field! The idea seems strange here, but by the 28th Series Faraday will firmly identify *directionality*, and not convergence, as the essentially "polar" character.

Nevertheless neither we nor he should undervalue the imagery of "conduction polarity," to which the convergence and divergence of lines is fundamental. It represents a necessary reinterpretation of the conventional notion of "poles," whether depicted as active centers, accumulations of fluid, or fountains of power. The essential theme in all those images, once freed from its picturesque attire, is *concentration of force*—a theme that obtains perfect expression in the imagery of convergence of lines. The convergence image yields a beautiful rearticulation of the magnetic field; but Faraday well knows it cannot be the last such rearticulation. It has helped free us from the habitual canons of polar thinking, and our reward for obtaining that freedom is the prospect of one day becoming free from the convergence image itself.

done so only that I might point with the more facility to facts and views that have heretofore been associated with some supposed polarity in the bodies which, whether paramagnetic or diamagnetic, I have been considering as mere conductors, and I hope that no mistake of my meaning will arise in consequence. I have already asked for such liberty in the use of phrases (lines of force, conducting power, &c.) (2149. 2797.) as may, for the time, set me free from the bondage of preconceived notions; these are, for that very reason, exceedingly useful, provided they are for the time sufficiently restricted in their meaning, and do not admit of any hurtful looseness or inaccuracy in the representation of facts.

*   *   *

2870. If what is now often indifferently called magnetic force or intensity have its results distinguished as of two kinds, namely, those of *quantity* and those of *tension,* then we shall more readily comprehend this matter. At present a needle shows both these as magnetic force, making no distinction between them, yet they produce effects

---

2835, continued. *I hope that no mistake of my meaning will arise...*: Faraday's worries about the "conductor" image may recall similar concerns about the fluid image of electricity—which was sustained by another sort of conduction image. He hopes that readers will understand that he has no wish to resuscitate fluid thinking. Rather, having finally freed the idea of an *electric* conductor from fluid imagery, he hopes to do the same for *magnetic* "conducting power" too.

as may, for the time, set me free from the bondage of preconceived notions: According to this noble expression, liberty in *language* secures liberty in *thought*—but only provisionally! For lasting emancipation of our language, our thought, and our imagination we must look to *nature herself*—whose richly expressive and suggestive phenomena lead us to reflection and understanding.

2870. In this paragraph, which culminates a lengthy discussion of terrestrial atmospheric magnetic measurements, Faraday telescopes his argument to an uncharacteristically high degree. Perhaps it will be useful to consider it step by step:

If ... magnetic force or intensity have its results distinguished as ... those of quantity and those of tension: Faraday's earlier experiments with dielectrics distinguished *quantity* and *tension* as two aspects of electrical action (see the Eleventh Series' introduction, pages 224–227). Now he urges an analogous distinction within magnetic actions: we are to think of *magnetic quantity* and *magnetic tension* as related to each other through *magnetic conducting power,* as electric quantity (or *charge*) and electric tension in dielectrics are mutually related through *capacity for induction* (paragraph 1214, *comment*).

on it often in opposite directions; for as they increase or diminish they both affect the needle alike; but as it is assumed that the tension can change whilst the quantity remains the same, and the quantity can be altered, yet the tension remain unaffected, the result by the needle will then be uncertain. If the tension in a given region be increased by diminishing the conducting power, the needle will show *increased force*; if it be increased by an increase of magnetic power in the earth from some internal action, the needle will still show *increased force*, and will not distinguish the one effect from the other.

---

the tension can change whilst the quantity remains the same, and the quantity can be altered, yet the tension remain unaffected...:   Faraday's point is that magnetic quantity and intensity may vary independently. He will enumerate several cases.

If the tension ... be increased by diminishing the conducting power...:   This is the first case—increased tension and constant quantity. Faraday knows that a *dielectric* having less capacity for induction exhibits greater electric tension (while maintaining a specified quantity of induction) than a dielectric having more capacity (paragraph 1261, *comment*). Similarly, then, if a magnetic medium suffers diminution of magnetic conducting power it will develop greater magnetic tension in carrying on the same magnetic lines of force as before.

the needle will show increased force: That is, it will point more strongly when placed in the aforesaid medium. He already knows this as an experimental fact, since paramagnetic behavior became more pronounced in a less conductive medium (paragraph 2815). Faraday gives an interpretive explanation in an omitted passage: "A needle vibrates by gathering upon itself, because of its magnetic condition and polarity, a certain amount of the lines of force, which would otherwise traverse the space about it..." (paragraph 2868). Thus if the surrounding medium becomes less conductive, fewer of the lines of force will be retained in the medium and more will concentrate upon the needle, increasing the vigor of its response.

if it be increased by an increase of magnetic power in the earth...:   This is the second case—increased tension and increased quantity. If a dielectric's capacity for induction remains unaltered, any increase in the electric induction it supports requires a commensurate increase in the electric tension it has to sustain (paragraph 1372, *comment*). Similarly, if the earth should by some internal process become stronger as a magnet (the atmosphere remaining constant in magnetic conducting power), then more magnetic action would be exerted in every portion of the atmosphere and the atmosphere would sustain greater tension than before.

the needle will still show increased force, and will not distinguish the one effect from the other: It is well known that a needle will respond more strongly to a powerful magnet than to a weak one, other things being equal. Thus a magnetic needle behaves in the second case just as it does in the first case; the needle cannot, therefore, distinguish between constant and varying magnetic *quantity*. In the next case Faraday will examine its ability to represent magnetic *tension*.

If the quantity in a region be increased by increasing the conducting power, the needle will show no such increase; on the contrary, it will indicate *diminution* of force, because the tension is diminished; or if the quantity be diminished by diminishing the conducting power, it will show *increased* force. The force might even lose in quantity and gain in tension in such proportions that the needle should show no change; or it might gain in quantity and lose in tension, and the needle still be entirely indifferent to the whole result.

2871. If my view be correct, then the magnet is not, as at present applied, a perfect measure of the earth's magnetic force...

*   *   *

Royal Institution
September 14, 1850

2870, continued. *If the quantity in a region be increased by increasing the conducting power...*: This is the third case—increased quantity and constant tension (but see the next two comments). If the inductive capacity of a dielectric is increased, it will support a greater quantity of electric induction at the same tension as before (paragraph 1214, *comment*). The same relation will hold, analogously, for magnetic quantity, tension, and conducting power.

the needle ... will indicate diminution of force...: Again we know experimentally that when a paramagnetic body is immersed in a more highly conductive medium it responds less energetically to magnetic action. Or in terms of Faraday's earlier explanation, when the medium becomes more conductive, some lines of force that were previously carried on by the magnetic needle become rerouted through the medium instead; thus the needle's responses are correspondingly weakened.

...because the tension is diminished: This claim is confusing since, as we have just seen, the needle's diminished action does *not* presuppose reduced tension. Moreover, if the cases being considered are to illustrate how "the tension can change whilst the quantity remains the same, and the quantity can be altered, yet the tension remain unaffected," then the third case should be one of *constant*, not diminished, tension. It is certainly *possible* for magnetic quantity to increase while magnetic tension decreases—it requires only a sufficient increase in magnetic conducting power of the medium—but Faraday should not have asserted categorically that the third example constitutes such a case.

2871. *the magnet is not ... a perfect measure of the earth's magnetic force*: Since Faraday is discussing terrestrial magnetism, he mentions the earth's magnetic force specifically. But the magnetic needle's deficiencies limit its performance in *any* magnetic measurement.

Twenty-Eighth Series — Editor's Introduction

The Twenty-eighth Series begins a climactic stage in the Experimental Researches, in which the character and distribution of magnetic power will exhibit itself with unparalleled access and intelligibility. In a pair of breathtakingly beautiful exercises Faraday will demonstrate, first, how to *count* magnetic lines of force and, second, how to determine their distribution experimentally, not only in the regions surrounding a magnet but actually *inside* it! But of course those procedures are only as significant as the lines of force themselves. So before taking up any other topic, Faraday stops to review and justify his continuing employment of the lines of force as meaningful, accurate, and competent representatives of the magnetic power.

What is a line of force?

As he opens the Series Faraday acknowledges having made repeated appeal to "lines of force" without, up to now, stating clearly what they are, what formal properties they possess, and how literally we are to take them. Although it seems clear that Faraday himself is prepared to view them as highly concrete, he recognizes that some investigators may prefer to consider the lines of force solely in their representative capacity, as symbolic constructs that facilitate our thinking about magnetic and electric powers while making no claim to provide a literal delineation of those powers. Even in so limited a role, he urges, lines of force have many advantages over rival descriptive vocabularies, such as those which invoke "poles" and "fluids". Nevertheless, he warns, we should not overlook the far greater significance which lines of force may possess. For it is at least possible that in lines of force we are quite literally observing *powers made visible*—just as his "paradigm" for science (for so I characterized his opening footnote in the Nineteenth Series*) earlier prescribed.

We have seen already how many electrical and magnetic phenomena seem to imply a certain connectedness in things, a pattern of successive and progressive transmission of activity from the agent to the site of action. Such communication is a *physical* process; and if the lines of force do indeed exhibit it directly, they must themselves possess some sort of physical status. What, then, will lines of force be if they are "physical?" It is easier to say what they will *not* be. Lines of force will not be mathematical constructs, not hypothetical, not merely symbolic. They will be natural, corporeal beings—as are rocks, rivers, plants and animals, clouds, stars and planets.

---

* See also the editor's introduction to that Series.

What then are lines of force? For the present Faraday does not attempt a definitive answer. It is clear that he takes their possible physicality very seriously, but for now he merely advocates their more frequent employment as symbolic devices. He will take a stronger position, though, in a subsequent essay. Its title—*On the Physical Character of the Lines of Magnetic Force*—states clearly enough what that position will be.

The Moving Wire

"The Moving Wire" is Faraday's name for a device, or class of devices, which reveal magnetic lines of force by means of the currents developed in moving conductors. You will of course recall his experiments with moving conductors in the first two Series, long before he began to employ the Line of Force as an interpretive image. Now his return to that inquiry features a series of exercises that depend upon the image of the line of force for their very design. Here is one of them.

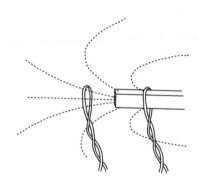

A wire loop or ring is connected to a galvanometer. When thrust over the end of a bar magnet it "cuts" lines of force—which emerge in all radial directions from the end of the bar. The galvanometer registers a forward deflection when the ring is placed over the magnet, a reverse deflection when it is removed. Thus each time the loop is placed over the magnet, it will cut the very same lines of force; and it will develop the same quantity of electricity in its circuit.

Now as we saw in earlier Series, a ballistic galvanometer will register this quantity of electricity, provided that the electricity fully completes its action before the (slow-moving) galvanometer needle has departed very far from its original position. For then the needle will be hurled to a maximum deflection, or "throw", proportional to the total electricity evolved. Suppose, then, the ring is placed successively once, twice, three times over the magnet, always well within the period of a single galvanometer throw. And let the galvanometer be connected during each placing of the loop, and disconnected during each withdrawal, so that only the forward currents pass through it. Faraday finds that the galvanometer deflections stand in the same proportion as the respective number of placements—that is, *in proportion to the number of lines of force cut*. He writes (paragraph 3086):

[W]hen the bend of the wires was formed into a loop, and that carried once over the pole of the [magnet], the galvanometer needle was deflected two degrees or more. The vibration of the needle was slow, and it was easy therefore to reiterate this action five or six times, or oftener, breaking and making contact with the galvanometer at right intervals, so as to combine the effect of like induced currents; and then a deflection of 10° or 15° on either side of zero could be readily obtained.

As a trial of proportionality, the results cited here are admittedly pretty rough: A figure of "two degrees or more," multiplied "five or six times" can only charitably be reckoned equal to "10° or 15°". But quantitative confirmation of the proportion is not Faraday's aim in this first exercise—a succession of increasingly more sophisticated measurements throughout this Series and the next will accomplish that. Rather, the loop procedure marks the beginning of a profound *transformation* which the Moving Wire is about to undergo.

When Faraday first announced *magneto-electric induction*, the evolution of currents in a moving wire was the very nucleus of the phenomenon, and a central focus of the overall investigation. But in the present Series Faraday reasons, from those very currents, to the *quantity* and *disposition* of the magnetic power that induced them. The Moving Wire is thus transformed from a *phenomenon* in its own right into an *instrument* by which other phenomena are themselves disclosed and interpreted.

We have seen this sort of metamorphosis before. Oersted's observation, that a compass needle was deflected in the vicinity of a current-carrying wire,* was initially a marvel and a wonder, itself the subject of extensive and vigorous investigation. But as the Experimental Researches begins, that effect has already been converted into a principle of measurement, habitually and routinely applied in the form of the *galvanometer*—that workhorse of the electrical laboratory.

The "counting" principle

The loop exercise, then, shows a rough proportionality between the *quantity of electricity evolved in*, and the *number of lines of force cut by*, the Moving Wire. As the Series progresses, Faraday will alternate between assuming that relation and confirming it more precisely. The proportionality becomes a powerful principle of *counting* because it permits him, from the quantities of electricity evolved in the circuit and

---

* Described in the editor's introduction to the First Series.

measured by the galvanometer, to calculate the relative number of magnetic lines of force that traverse any area the moving wire can be made to sweep out.*

The instrumental significance of this proportionality is great indeed. Nevertheless, Faraday is less interested in *proportional magnitudes* than in the *meaning of proportionality*. In his eyes the "counting" principle is much more than a new and powerful measurement technique—it reveals the germ of a new image of the magnet. For if a quantity of electrical action is strictly proportional to a quantity of magnetic lines, then *each line* may be accounted responsible for a determinate share of the total effect; and the magnetic lines of force are confirmed as *agents*, each embodying a determinate power.

The "counting" principle thus has interpretive as well as instrumental significance. Faraday will explore both of its aspects in experiments with a rotating magnet—a theme which, if you recall, marks another return to an early Series (paragraphs 218–220).

The rotating magnet

Faraday has an ingenious device which permits him to study a rotating magnet with the moving wire. His description of the apparatus is a model of clarity, and he recounts the individual exercises in full. Yet he touches upon their consequences with such delicacy that readers sometimes overlook their collective significance. It may be helpful, then, to summarize the experimental results in advance. I see them as threefold:

1. *Disposition of the lines.* The rotation experiments will establish the existence of lines of force *within* the magnet, identical in nature to those external to it, equal to them in number, and continuous with them in direction. In other words, the lines of force of a common magnet will be revealed as *closed curves*, neither arising nor terminating at the body of the magnet.

This is indeed a momentous result; for it removes at one stroke any visual imagery, at least, of "poles" or "magnetic fluids" or any supposed fountain of creation from which the lines of force might be imagined to pour out. In the absence of terminations, *any* point on the line of force is as strong a candidate for being the line's "origin" as any other point; so postulating a "pole" or a puddle of "fluid" at some location on the line becomes instantly futile.

---

* Recall that an independent *quantitative* measure is just what has been lacking for *electric* lines of force; see the comment to paragraph 1378 and the Twelfth Series' introduction.

2. *Force of the lines.* The experiments will establish that one line of force represents the same *quantity* of force (power, action) at every location—regardless of the line's angle, direction, or curvature; or its distance from any object; or its convergence towards or divergence from neighboring lines of force. When Faraday proposed a similar idea for electric lines in the Twelfth Series (paragraph 1369), he could offer it only as an interpretive rule. Now he has strict experimental demonstration of the principle, at least for magnetic lines.

3. *The role of iron and air in a magnet.* The iron interior of a magnet, and the air (or other exterior medium), sustain *equal power* since the same lines of force pass through both. They are moreover *equally essential* to the magnet, since a closed line of force cannot be sustained in one medium without passing through the other as well. Finally, the lines of force appear to have the same mode of existence in one medium as in the other. To use the language of the 26th Series, the media are merely unequal in conducting power—they differ in degree, not in kind. But this overthrows our conventional notion that the iron *owns* the lines of force, so to speak—that the iron *is* the magnet, while the surrounding air is only the incidental environment in which it happens to have been placed. Can we then, in any sense, continue to regard the magnetic power as "belonging" to the iron?

Considerations such as these make it harder and harder to distinguish "inner" from "outer" with respect to the magnet—or even to differentiate one magnetic medium from another. Eventually, I believe, they will cause Faraday to moderate somewhat that absolutism regarding matter and space he expressed in an earlier Series.* He will express such moderation overtly in later papers; but since the rotating magnet experiments of the present Series mark the real inauguration of that more temperate view, I call attention to it now.

---

* See the Twenty-sixth Series' introduction.

§ 34. *On lines of Magnetic Force; their definite character; and their distribution within a Magnet and through space.*

Received October 22,—Read November 27 and December 11, 1851.

3070. FROM my earliest experiments on the relation of electricity and magnetism (114. note), I have had to think and speak of lines of magnetic force as representations of the magnetic power; not merely in the points of quality and direction, but also in quantity. The necessity I was under of a more frequent use of the term in some recent researches (2149. &c.), has led me to believe that the time has arrived, when the idea conveyed by the phrase should be stated very clearly, and should also be carefully examined, that it may be ascertained how far it may be truly applied in representing magnetic conditions and phænomena; how far it may be useful in their elucidation; and, also, how far it may assist in leading the mind correctly on to further conceptions of the physical nature of the force, and the recognition of the possible effects, either new or old, which may be produced by it.

3071. A line of magnetic force may be defined as that line which is described by a very small magnetic needle, when it is so moved in either direction correspondent to its length, that the needle is constantly a tangent to the line of motion; or it is that line along which, if a transverse wire be moved in either direction, there is no tendency to the formation of any current in the wire, whilst if moved in any other direction there is such a tendency; or it is that line which coincides

---

[1] Philosophical Transactions, 1852, p. 1.

3070. *the term ... the phrase*: that is, "line of magnetic force".

3071. *A line of magnetic force may be defined...*: The *magnetic needle*, the *moving wire*, and *magnecrystallic alignment* (see below) represent three fundamental means by which to reveal the lines of magnetic force. Ordinary *iron filings*, a fourth medium for the lines' exhibition, constitute a mere variant of the first; inasmuch as each filing *is*, in effect, a magnetic needle. Soon, however, it will become clear that under certain circumstances the different means yield significantly different results; and Faraday will argue that the Moving Wire provides the best overall representation.

that line along which ... there is no tendency to the formation of any current...: While a magnetic needle indicates lines of force by pointing along them, the moving wire discloses their presence by *crossing* them. Thus if the moving wire develops no current when urged in a particular direction, it cannot be crossing any lines; hence its path must coincide with the lines themselves.

with the direction of the magnecrystallic axis of a crystal of bismuth, which is carried in either direction along it. The direction of these lines about and amongst magnets and electric currents, is easily represented and understood, in a general manner, by the ordinary use of iron filings.

3072. These lines have not merely a determinate direction, recognizable as above (3071.), but because they are related to a polar or antithetical power, have opposite qualities or conditions in opposite directions; these qualities, which have to be distinguished and identified, are made manifest to us, either by the position of the ends of the magnetic needle, or by the direction of the current induced in the moving wire.

3073. A point equally important to the definition of these lines is, that they represent a determinate and unchanging amount of force. Though, therefore, their forms, as they exist between two or more centres or sources of magnetic power, may vary very greatly, and also the space through which they may be traced, yet the sum of power contained in any one section of a given portion of the lines is exactly equal to the sum of power in any other section of the same lines, however altered in form, or however convergent or divergent they may be at the second place. The experimental proof of this character of the lines will be given hereafter (3109. &c.).

3074. Now it appears to me that these lines may be employed with great advantage to represent the nature, condition, direction and

---

3071, continued. *magnecrystallic axis:* In the 22nd Series, omitted from the present selection, Faraday investigated the tendency of certain crystals under magnetic influence to orient themselves in directions apparently unrelated to their para- or diamagnetic character. The axis of alignment proved to be related to the crystal structure, rather than to the bodily dimensions, of the sample; and this axis he termed "magnecrystallic."

3072. *opposite qualities or conditions in opposite directions:* that is, the qualities conventionally called "north" and "south."

3073. *these lines ... represent a determinate and unchanging amount of force:* In earlier Series Faraday repeatedly invoked this principle for both electric and magnetic lines of force. Up to now, though, it has served only as an interpretive rule—albeit a powerful and fruitful one. In the present Series Faraday will succeed in demonstrating it by direct experiment.

centres or sources of magnetic power: Faraday is permitting himself to use the conventional language of *poles*, even though he long ago satisfied himself that such active "centres" do not exist.

comparative amount of the magnetic forces; and that in many cases they have, to the physical reasoner at least, a superiority over that method which represents the forces as concentrated in centres of action, such as the poles of magnets or needles; or some other methods, as, for instance, that which considers north or south magnetisms as fluids diffused over the ends or amongst the particles of a bar. No doubt, any of these methods which does not assume too much, will, with a faithful application, give true results; and so they all ought to give the same results as far as they can respectively be applied. But some may, by their very nature, be applicable to a far greater extent, and give far more varied results, than others. For just as either geometry or analysis may be employed to solve correctly a particular problem, though one has far more power and capability, generally speaking, than the other; or just as either the idea of the reflexion of images, or that of the reverberation of sounds may be used to represent certain physical forces and conditions; so may the idea of the attractions and repulsions of centres, or that of the disposition of magnetic fluids, or that of lines of force, be applied in the consideration of magnetic phænomena. It is the occasional and more frequent use of the latter which I at present wish to advocate.

3075. I desire to restrict the meaning of the term *line of force*, so that it shall imply no more than the condition of the force in any given

---

3074. *the physical reasoner*: in contrast to the *mathematician*, for example. Today we would probably use the term "physicist." That title was gradually gaining acceptance when Faraday wrote, but he seems to have hated it and refused to use it.

No doubt, any of these methods which does not assume too much...: Considered only as aids to thinking, *all* the recognized magnetic images—whether *lines of force, poles or centers of power*, or *fluids*—are potentially valid representations of the phenomena. Any advantage one may have over another must, in practice, consist only in a *more comprehensive applicability* to a given situation. Faraday here seems surprisingly conciliatory towards notions, like "magnetic fluids," which he has conclusively rejected. Perhaps by setting an example of open-mindedness towards conventional and established views, he hopes to encourage in his readers a similar receptivity to new and revolutionary ones.

geometry or analysis: Cartesian (analytic) algebra is generally regarded as more powerful than Euclidean (synthetic) geometry. Nevertheless there are many individual problems to which both approaches are equally applicable, and some which yield more simply and swiftly to geometry than to algebra.

place, as to strength and direction; and not to include (at present) any idea of the nature of the physical cause of the phænomena; or be tied up with, or in any way dependent on, such an idea. Still, there is no impropriety in endeavouring to conceive the method in which the physical forces are either excited, or exist, or are transmitted; nor, when these by experiment and comparison are ascertained in any given degree, in representing them by any method which we adopt to represent the mere forces, provided no error is thereby introduced. On the contrary, when the natural truth and the conventional representation of it most closely agree, then are we most advanced in our knowledge. The emission and the æther theories present such cases in relation to light. The idea of a fluid or of two fluids is the same for electricity; and there the further idea of a current has been raised, which indeed has such hold on the mind as occasionally to embarrass the science as respects the true character of the physical agencies, and may be doing so, even now, to a degree which we at present little suspect. The same is the case with the idea of a magnetic fluid or fluids, or with the assumption of magnetic centres of action of which the resultants are at the poles. How the magnetic force is transferred through bodies or through space we know not:—whether the result is merely action at a distance, as in the case of gravity; or by

---

3075. *not to include (at present) any idea of the nature of the physical cause*: Strictly, the form of representation of a phenomenon should be independent of any hypotheses concerning its physical cause. Still, Faraday recognizes that different images are always more or less suggestive of different causal hypotheses. Towards the end of the present Series (paragraph 3175) he will acknowledge an inclination—though not a positive conviction—that lines of force also constitute the *physical* means by which magnetic force is transmitted.

when the natural truth and the conventional representation of it most closely agree, then are we most advanced in our knowledge...: Therefore we ought not to shrink from following out the many forms and images under which natural powers present themselves; for each of them holds promise for portraying its object in the most direct and fundamental guise ("the natural truth"). Faraday does not enlarge on what it means for truth and representation to "agree"; but *visual similarity* would harmonize perfectly with the aim of *making visible to the eye the powers of nature*, suggested earlier in the Nineteenth Series' introduction.

The emission and the æther theories: that is, the corpuscle and wave theories of light, respectively.

to embarrass the science...: Here, to perplex or confuse it.

action at a distance, as in the case of gravity: Although Newton himself professed to reject action-at-a-distance as a competent philosophical theory, the inverse-square law of gravitation appeared to many to convey precisely that idea; and conventional Newtonian mechanics became more and more inextricably associated with it.

some intermediate agency, as in the cases of light, heat, the electric current, and (as I believe) static electric action. The idea of magnetic fluids, as applied by some, or of magnetic centres of action, does not include that of the latter kind of transmission, but the idea of lines of force does. Nevertheless, because a particular method of representing the forces does not include such a mode of transmission, the latter is not therefore disproved; and that method of representation which harmonizes with it may be the most true to nature. The general conclusion of philosophers seems to be, that such cases are by far the most numerous, and for my own part, considering the relation of a vacuum to the magnetic force and the general character of magnetic phænomena external to the magnet, I am more inclined to the notion that in the transmission of the force there is such an action, external to the magnet, than that the effects are merely attraction and repulsion at a distance. Such an action may be a function of the æther; for it is not at all unlikely that, if there be an æther, it should have other uses than simply the conveyance of radiations (2591. 2787.). Perhaps when we are more clearly instructed in this matter, we shall see the source of the contradictions which are supposed to exist between the results of Coulomb, Harris and other philosophers, and find that they are not contradictions in reality, but mere differences in degree, dependent upon partial or imperfect views of the phænomena and their causes.

3076. Lines of magnetic force may be recognized, either by their action on a magnetic needle, or on a conducting body moving across them. Each of these actions may be employed also to indicate, either the direction of the line, or the force exerted at any given point in it, and this they do with advantages for the one method or the other under particular circumstances. The actions are however very different

---

Nevertheless, because a particular method of representing the forces does not include such a mode of transmission, the latter is not therefore disproved: If conventional discourse makes no allowance for a certain idea—as the theory of magnetic fluids cannot accommodate the idea of *propagation* of the magnetic action—that is no refutation of the idea. On the contrary, it may indicate deficiency and provinciality in the conventions!

not contradictions in reality, but mere differences in degree, dependent upon partial or imperfect views: Perhaps you have already noticed that Faraday hardly ever approaches an investigation forensically—as a trial between rigid and competing positions, each claiming the right. Rather, as here, questions are to be pursued incrementally, with the presumption that one's views are to undergo constant expansion and development.

in their nature. The needle shows its results by attractions and repulsions; the moving conductor or wire shows it by the production of a current of electricity. The latter is an effect entirely unlike that produced on the needle, and due to a different action of the forces; so that it gives a view and a result of properties of the lines of force, such as the attractions and repulsions of the needle could never show. For this and other reasons I propose to develope and apply the method by a moving conductor on the present occasion.

3077. The general principles of the development of an electric current in a wire moving under the influence of magnetic forces, were given on a former occasion, in the First and Second Series of these Researches (36. &c.); it will therefore be unnecessary to do more than to call attention, at this time, to the special character of its indications as compared to those of a magnetic needle, and to show how it becomes a peculiar and important addition to it, in the illustration of magnetic action.

3078. The moving wire produces its greatest effect and indication, not when passing from stronger to weaker places, or the reverse, but when moving in places of equal action, *i. e.* transversely across the lines of force (217.).

3079. It determines the direction of the polarity by an effect entirely independent of pointing or such like results of attraction or repulsion; *i. e.* by the direction of the electric current produced in it during the motion.[2]

---

[2] A natural standard of this polarity may be obtained, by referring to the lines of force of the earth, in the northern hemisphere, thus—if a person with arms extended move forward in these latitudes, then the direction of the electric current, which would tend to be produced in a wire represented by the arms, would be from the right hand through the arm and body towards the left.

3076. *The needle shows its results by attractions and repulsions*: or, if one wishes to avoid the language of pushes and pulls, by *alignments.*

3078. *moving in places of equal action, i. e. transversely...*: This is misleading. The moving wire does not develop a current *because* it moves in places of equal action, for it exhibits no effect when moving parallel to the lines of a uniform field, where *all* sites are "places of equal action." Moreover, a wire moving transversely in a nonuniform field will experience action that is only *approximately* constant— and even then only when its displacements are small. Nevertheless Faraday's main point, the contrast between the moving wire and magnetic needle, is clear.

3079. *It determines the direction of the polarity ... by the direction of the electric current produced*: If we know both the direction of motion of the wire, and the direction of current induced in it, we can distinguish the N- and S- directions of the lines of magnetic force; see Faraday's note to this paragraph.

   Comment on Faraday's note 2: His "natural standard" agrees with the magneto-electric right-hand rule (paragraph 101, *comment*). In English latitudes, earth's

3080. The principle can be applied to the examination of the forces *within* numerous solid bodies, as the metals, as well as outside in the air. It is not often embarrassed by the difference of the surrounding media, and can be used in fluids, gases or a vacuum with equal facility. Hence it can penetrate and be employed where the needle is forbidden; and in other cases where the needle might be resorted to, though greatly embarrassed by the media around it, the moving wire may be used with an immediate result (3142.).

3081. The method can even be applied with equal facility to the interior of a magnet (3116.), a place utterly inaccessible to the magnetic needle.

3082. The moving wire can be made to sum up or give the resultant at once of the magnetic action at many different places, *i. e.* the action due to an area or section of the lines of force, and so supply experimental comparisons which the needle could not give, except with very great labour, and then imperfectly. Whether the wire moves directly or obliquely across the lines of force, in one direction or another, it sums up, with the same accuracy in principle, the amount of the forces represented by the lines it has crossed (3113.).

3083. So a moving wire may be accepted as a correct philosophical indication of the presence of magnetic force. Illustrations of the capabilities already referred to, will arise and be pointed out in the

---

lines of force are directed obliquely downward. Thus with your right index finger pointing down, walk forward with thumb extended in front of you. Then the third finger (held perpendicular to both) will point from your right to your left—the direction in which current would tend to be induced by your forward motion. You can show similar agreement with the general law of paragraph 114.

3080. *embarrassed*: Here, *impaired in its function*—not quite the same sense as in paragraph 3075 above.

3081. *a place utterly inaccessible to the magnetic needle*: The problem with the magnetic needle is not only its lack of access to the magnet's interior. Faraday will explain in paragraph 3280 that even if the needle could be introduced into the interior of a bar magnet, it would give misleading indications—for it would point feebly or not at all, and thus appear to show a *disappearance* of power within the magnet. The moving wire, on the contrary, reveals a magnet's interior power in full measure, as he will show with a splendid example in paragraphs 3116–3117 below.

3082. *sum up … the magnetic action at many different places…*: Strictly speaking, it is the *galvanometer* rather than the *wire* that "sums up" the action when the wire moves. Faraday gives a full account of the galvanometer's performance in paragraphs 3104–3105 below.

3083. *a correct philosophical indication*: Perhaps today we would say "a valid scientific measure".

present paper; and though its sensibility does not as yet approach to that of the magnetic needle, still, there is no doubt that it may be very greatly increased. The diversity of its possible arrangements, and the great advantage of that diversity, is already very manifest to myself. Though both it and the needle depend for their results upon essential characters and qualities of the magnetic force, yet those which are influential, and, therefore indicated, in the one case, are very different from those which are active in the other; I mean, as far as we have been able as yet to refer directly the effects to essential characters: and this difference may, hereafter, enable the wire to give a new insight into the nature of the magnetic force; and so it may, finally, bear upon inquiries, such as whether magnetic polarity is axial or dependent upon transverse lateral conditions; whether the transmission of the force is after the manner of a vibration or current, or simply action at a distance; and the many other questions that arise in the minds of those who are pursuing this branch of knowledge.

3084. I will proceed to take the case of a simple bar magnet, employing it in illustration of what has been said respecting the lines of force and the moving conductor, and also for the purpose of ascertaining how these lines of force are disposed, both without and within the magnet itself, upon which they are dependent or to which

---

3083, continued. *sensibility*: that is, as often before, *sensitivity*.

yet those which are influential, and, therefore indicated ... are very different...: The moving wire and the magnetic needle respond to different aspects of the magnetic power. Each device therefore inevitably tends to single out a particular aspect as the *essential* magnetic character; for whether we realize it or not, every instrument already embodies images and presuppositions in its very design. But grasping essentials is the work not of instrumentation but of *intellect*. By interpreting instrumental responses critically, or by constructing new devices to bring out hitherto neglected aspects of phenomena, we may hope to bring forth continually more adequate representations of nature's powers, more reliably distinguishing between the essential and the incidental. Faraday proceeds to name some specific areas where this distinction has yet to be carried out satisfactorily:

whether magnetic polarity is axial or dependent upon ... lateral conditions: Faraday has pointed out before the apparent lateral relations of lines of force, whereby they maintain a rough paralleleity to one another. Is it an essential or a merely incidental aspect of the polar condition?

whether the transmission [is] a vibration or current, or simply action at a distance...: If the lines mark the *paths* of magnetic force, then what substantial mechanism propagates that force from place to place? Is there a vibratory medium—perhaps the aether? Is there a flow or transport of fluid—and what kind of fluid, if so? Or might even the old idea of *action at a distance* prove capable of rehabilitation?

Fig. 1.

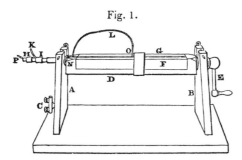

they belong. For this purpose the following apparatus was employed. Let fig. 1 represent a wooden stand, of which the base is a board 17.5 inches in length, and 6 inches in breadth, and 0.8 of an inch in thickness: these dimensions will serve as a scale for the other parts. A B are two wooden uprights; D is an axis of wood having two long depressions cut into it, for the purpose of carrying the two bar magnets F and G. The wood is not cut away quite across the axis, but is left in the middle, so that the magnets are about $^1/_{15}$th of an inch apart. From O towards the support A, it is removed, however, as low down as the axis of revolution, so as to form a notch between the two magnets when they are in their places; and by further removal of the wood, this notch is continued on to the end of the axis at P. This notch, or opening, is intended to receive a wire, which can be carried down the axis of rotation, and then passing out between the two magnets, anywhere between O and N, can be returned towards the end P on the outside. The magnets are so placed, that the central line of their compound system coincides with the axis of rotation; E being a handle by which rotation, when required, is given. H and I are two copper rings, slipping tightly on to the axis, by which communication is to be made between a wire adjusted so as to revolve with the magnets, and the fixed ends of wires proceeding from a galvanometer. Thus, let P L represent a covered wire; which being led along the bottom of the notch in the axis of the apparatus, and passing out at the equatorial parts of the magnets, returns into the notch again near N, and terminates at K. When the form of the wire loop is determined and given to it,

---

3084. *H and I are two copper rings*: These enable stationary wires from the galvanometer to make constant connection with a rotating wire. Such contacts would later be termed "slip rings" in electric motor and generator work.

let PL represent a covered wire: that is, an *insulated* wire. In Figure 1 it follows the path POLNK; Figures 4–9 below will clarify the wire's several alternate routes through and near the compound bar magnet.

Fig. 1.

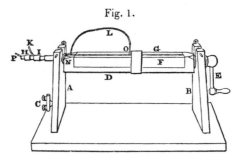

then a little piece of soft wood is placed between the wires in the notch at K, of such thickness, that when the ring I is put into its place, it shall press upon the upper wire, the piece of wood, and the lower wire, and keep all tightly fixed together, and at the same time leave the two wires effectually separated. The second ring, H, is then put into its place on the axis, and the introduction of a small wedge of wood, at the end of the axis, serves to press the end P into close and perfect contact with the ring H, and keep all in order. So the wire is free to revolve with the magnets, and the rings H and I are its virtual terminations. Two clips, as at C, hold the ends of the galvanometer wire (also of copper); and the latter are made to press against the rings by their elasticity, and give an effectual contact bearing, which generates no current, either by difference of nature or by friction, during the revolution of the axis.

3085. The two magnets are bars, each 12 inches long, 1 inch broad, and 0.4 of an inch thick. They weigh each 19 ounces, and are of such a strength as to lift each other end to end and no more. When the two are adjusted in their place, it is with the similar poles together, so that they shall act as one magnet, with a division down the middle: they are retained in their place by tying, or, at times, by a ring of copper which slips tightly over them and the axis.

3086. The galvanometer is a very delicate instrument made by Rhumkorff (2651.). It was placed about 6 feet from the magnet apparatus, and was not affected by any revolution of the latter. The wires, connecting it with the magnets, were of copper, 0.04 of an inch in diameter, and in their whole length about 25 feet. The length of the wire in the galvanometer I do not know; its diameter was $1/135$th of an inch.

3084, continued. *the rings H and I are its virtual terminations*: The K end of wire POLNK is joined to ring I while the P end passes through I without contacting it (since the wire is "covered") and is then connected to ring H.

3085. *they shall act as one magnet*: Faraday explains why the two parallel magnets would act as one in paragraph 3100 below.

The condition of the galvanometer, wires, and magnets, was such, that when the bend of the wires was formed into a loop, and that carried once over the pole of the united magnets, as from *a* to *b*, fig. 2, the galvanometer needle was deflected two degrees or more. The vibration of the needle was slow, and it was easy therefore to reiterate this action five or six times, or oftener, breaking and making contact with the galvanometer at right intervals, so as to combine the effect of like induced currents; and then a deflection of 10° or 15° on either side of zero could be readily obtained. The arrangement, therefore, was sufficiently sensible for first experiments; and though the resistance opposed by the thin long galvanometer wire to feeble currents was considerable, yet it would always be the same, and would not interfere with results, where the final effect was equal to 0°, nor in those where the consequences were shown, not by absolute measurement, but by comparative differences.

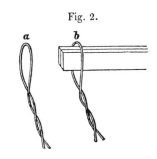

Fig. 2.

3087. The first practical result produced by the apparatus described, in respect of magneto-electric induction generally, is, that a piece of metal or conducting matter which moves across lines of magnetic force,

---

3086. *loop … carried once over the pole of the united magnets*: The wire, magnet, and galvanometer pertaining to Figure 2 are the same as for Figure 1, but he performs this exercise *before* fully assembling the rotary apparatus. Its purpose is to illustrate the sensitivity of the galvanometer connected as described.

breaking and making contact with the galvanometer at right intervals, so as to combine the effect of like induced currents: When the loop in Figure 2 is moved from *a* to *b* the galvanometer is deflected in one direction; it is deflected in the opposite direction when the loop returns from *b* to *a* (paragraph 256). As discussed in the introduction, Faraday keeps the galvanometer connections intact during the forward movements and breaks them during the return movements; thus the impulses experienced by the galvanometer needle will all be in the same direction and will propel the needle to a maximum displacement, proportional to their *cumulative* effect. This "ballistic" mode of the galvanometer is subject to several limitations; Faraday will describe them in paragraphs 3103–3105 below.

the resistance … would not interfere with results, where the final effect was *equal to 0°*: A long, thin wire in the galvanometer circuit cannot but further reduce the induced currents, which are feeble to begin with. But its influence on *equal* currents must be the same for all; so it will not interfere with tests for equality. Such is any test in which a succession of actions sums to *zero*—a *null experiment,* so called—for the opposite currents must have been equal in order to have canceled one another.

has, or tends to have, a current of electricity produced in it. A more restricted and precise expression of the full effect is the following. If a continuous circuit of conducting matter be traced out, or conceived of, either in a solid or fluid mass of metal or conducting matter, or in wires or bars of metal arranged in non-conducting matter or space; which being moved, crosses lines of magnetic force, or being still, is by the translation of a magnet crossed by such lines of force; and further, if, by inequality of angular motion, or by contrary motion of different parts of the circuit, or by inequality of the motion in the same direction, one part crosses either more or fewer lines than the other; then a current will exist round it, due to the differential relation of the two or more intersecting parts during the time of the motion: the direction of which current will be determined (with lines having a given direction of polarity) by the direction of the intersection, combined with the relative amount of the intersection in the two or more efficient and determining (or intersecting) parts of the circuit.

3088. Thus, if fig. 3 represent a magnetic pole N, and over it a circuit, formed of metal, which may be of any shape, and which is at first in the position c; then if that circuit be moved in one direction into the position 1; or in the contrary direction into position 2; or by a double direction of motion into position 3; or by translation into position 4; or into position 5; or any position between the first and these

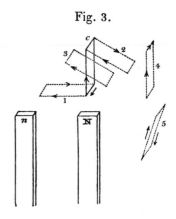

Fig. 3.

3087. *if … one part crosses either more or fewer lines than the other, then a current will exist around it*: Recall Faraday's discussion in paragraph 181: Two identical horizontal wires laid north and south will, during equal intervals of time, cut equal numbers of terrestrial magnetic curves (as the earth moves in its diurnal rotation) and so develop equal currents. If the wires are joined at the ends to form a closed path, the two currents necessarily oppose and cancel each other. But if one of the wires moves in relation to the other it will, during that interval, cut either more or fewer lines of force and develop a greater or smaller current. Since the currents oppose one another in the circuit, the greater current will prevail over the lesser and the excess will deflect a galvanometer.

the direction of which current will be determined…: To ascertain that direction, apply the right-hand rule (or its equivalent) to each moving portion of the wire. Depending on their relative directions in the circuit, the corresponding currents will combine or oppose. The sum or difference of these component currents will then give the quantity and direction of the overall current.

or any resembling them; or, if the first position *c* being retained, the pole move to, or towards, the position *n*; then, an electric current will be produced in the circuit, having in every case the same direction, being that which is marked in the figure by arrows. Reverse motions will give currents in the reverse direction (256. &c.).

3089. The general principles of the production of electrical currents by magnetic induction have been formerly given (27. &c.),[3] and the law of the direction of the current in relation to the lines of force, stated (114. 3079. note). But the full meaning of the above description can only be appreciated hereafter, when the experimental results, which supply a larger knowledge of the relations of the current to the *lines of force*, have been described.

3090. When *lines of force* are spoken of as crossing a conducting circuit (3087.), it must be considered as effected by the *translation* of a magnet. No mere rotation of a bar magnet on its axis, produces any induction effect on circuits exterior to it; for then, the conditions above described (3088.) are not fulfilled. The system of power about the magnet must not be considered as necessarily revolving with the magnet, any more than the rays of light which emanate from the sun are supposed to revolve with the sun. The magnet may even, in certain cases (3097.), be considered as revolving amongst its own forces, and producing a full electric effect, sensible at the galvanometer.

3091. In the first instance the wire was carried down the axis of the magnet to the middle distance, then led out at the equatorial part, and

[3] Philosophical Transactions, 1832, p.131, &c.

3089. *the relations of the current to the lines of force*: in particular, those relations involving *quantity*.

3090. *No mere rotation of a bar magnet on its axis, produces any induction effect on circuits exterior to it*: When a bar-magnet rotates about its axis, its lines of force remain *stationary*—and therefore are not "cut" by a stationary wire. Faraday showed this in the Second Series (paragraph 220), where he noted a "singular independence" between the magnetic power and the bar in which it resides.

3091. *the magnet*: that is, the pair of parallel magnets that "act as one" (paragraph 3085). In this and the following paragraphs Faraday is about to describe experiments in which now the wire, now the magnet, or both wire and magnet, are revolved; and in which the wire paths are also varied. For now, he merely reports the results; but he will explain their significance in later paragraphs.

Fig. 4

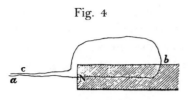

returned on the outside; fig. 4 will represent such a disposition. Supposing the magnet and wire to revolve once, it is evident that the wire *a* may be considered as passing in at the axis of the magnet, and returning from *b* across the lines of force external to the magnet, to the axis again at *c*; and that in one revolution, the wire from *b* to *c* has intersected once, all the lines of force emanating from the N end of the magnet. In other words, whatever course the wire may take from *b* to *c*, the whole system of lines belonging to the magnet has been *once* crossed by the wire. In order to have a correct notion of the relation of the result, we will suppose a person standing at the handle E, fig. 1 (3084.), and looking along the magnets, the magnets being fixed, and the wire

Fig. 1.

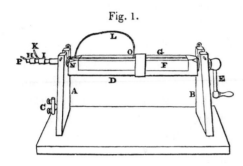

loop from *b* to *c* [fig. 4] turned over toward the left-hand into a horizontal plane; then, if that loop be moved over towards the right-hand, the magnet remaining stationary, it will be equivalent to a *direct* revolution (according to the hands of a watch or clock) of 180°, and will produce a feeble current in a given direction at the galvanometer. If it be carried back 180° in the reverse direction, it will produce a corresponding current in the reverse direction to the former. If the wire be held in a vertical, or any other plane, so that it may be considered as fixed, and the magnet

---

3091, continued. *in one revolution, the wire from b to c has intersected once, all the lines of force emanating from the N end of the magnet:* Faraday means us to imagine the course of lines of force in Figure 4. The sketch provided here shows a sample set of lines; the lines *not* shown will also be cut by wire *cb* as it revolves once about the magnet's axis.

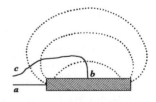

He finds that, whether the wire loop be rotated with the magnet held stationary, or the magnet rotated while the wire is held fixed, currents are produced. Furthermore, the current direction depends only on the *relative* motion of magnet and wire. These are both highly significant results, whose implications he will discuss in paragraphs 3106, 3116, and elsewhere below.

the wire loop from b to c turned over toward the left-hand ... towards the right-hand...: He turns the loop to the left from vertical to horizontal in order to permit a subsequent turn of 180° back to the right without striking the wooden supports.

be rotated through half a revolution, it will also produce a current; and if rotated in the contrary direction, will produce a contrary current; but as to the *direction* of the currents, that produced by the *direct* revolution of the wire is the same as that produced by the *reverse* revolution of the magnet; and that produced by the *reverse* revolution of the wire is the same as that produced by the *direct* revolution of the magnet. A more precise reference of the direction of the current to the particular pole employed, and the direction of the revolution of the wire or magnet, is not at present necessary; but if required is obtained at once by reference to fig. 3 (3088.), or to the general law (114. 3079. note).

3092. The magnet and loop being rotated together in either direction, no trace of an electric current was produced. In this case the effect, if any, could be greatly exalted, because the rotation could be continued for 10, 20, or any number of revolutions without derangement, and it was easy to make thirty revolutions or more within the time of the swing of the galvanometer needle in one direction. It was also easy, if any effect were produced, to accumulate it upon the galvanometer by reversing the rotation at the due time. But no amount of revolution of the magnet and wire together could produce any effect.

3093. The loop was then taken out of the axis of the magnet, but attached to it by a piece of pasteboard, so that all should be fixed together and revolve with the same angular velocity, fig. 5; but whatever the shape or disposition of the loop, whether large or small, near or distant, open or shut, in

Fig. 5.

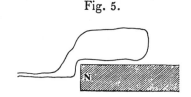

3092. *thirty revolutions … within the time of the swing of the galvanometer needle…*: In earlier Series, Faraday has used the "ballistic" galvanometer to measure single, brief discharges; but the present apparatus stands to generate multiple discharges if it produces any at all. In paragraphs 3103–3105 below, Faraday shows that the instrument will respond cumulatively to a succession of discharges, provided that their total duration is less—ideally, *much* less—than the *time of swing of the galvanometer needle.* Faraday therefore complies with that limitation here.

accumulate [the effect] upon the galvanometer by reversing the rotation at the due time: If joint rotation of magnet and wire produces even a feeble current, he should be able to build up large swings of the galvanometer needle by reversing the inducing rotation synchronously with the galvanometer's natural oscillations.

But no amount of revolution of the magnet and wire together could produce any effect: This result might seem obvious, since the relative motion of wire and magnet is *zero.* But remember that Faraday has reason to believe that the exterior lines of force are *not rotating* with the magnet (paragraph 3090); thus they are certainly being cut—and that ought to produce a current! Why then is no current detected? Faraday will give the answer at paragraph 3116.

one plane, or contorted into various planes; whatever the shape or condition, or place, provided it moved altogether with the magnet, no current was produced.

3094. Furthermore, when the loop was out of the magnets, and by expedients of arrangement, was retained immoveable, whilst the magnet revolved, no amount of rotation of the magnet (unaccompanied by translation of place) produced any degree of current through the loop.

3095. The loop of wire was then made of two parts; the portion *c*, fig. 6, on the outside of the magnet, was fixed at *b*, and the portion *a*, being a separate piece, was carried along the axis until it came in contact with the former at *d*; the revolution of one part was thus permitted either with or without the other, yet preserving always metallic contact and a complete circuit for the induced current. In this case, when the external wire and the magnet were fixed, no current was produced by any amount of revolution of the wire *a* on its axis. Neither was any current produced when the magnet and wire, *c d*, were revolved together, whether the wire *a* revolved with them or not. When the magnet was revolved without the external part of *c d*, or the latter revolved without the magnet, then currents were produced as before (3091.).

Fig. 6.

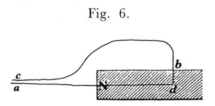

3093. *provided [the loop] moved altogether with the magnet, no current was produced*: If the lines of force about a rotating magnet really do remain stationary, it does not matter that the loop rotates *with the magnet*; what is important is that it rotates at all—for it will then be cutting fixed lines of force. Then why is no current produced? Faraday will discuss this configuration in paragraphs 3116–3117.

3094. Here the loop is wholly stationary; and if the lines of force are stationary too (for Faraday has determined that they do not rotate along with the magnet), then certainly no current will be induced.

3095. *no current was produced by … revolution of the wire a on its axis*: This might appear to be an obvious result, but it is not. For if there were lines of force emerging radially from wire *a*, they would be cut by its rotation and produce a current. Such lines can now, however, be ruled out.

Neither was any current produced when the magnet and wire, c d, were revolved together: In this case, *both* external part *c b* and internal part *b d* are revolving. Thus either both portions fail to produce a current, or they produce equal and opposite currents, canceling each other. The next variation reveals that the second of these alternatives is in fact the case.

When the magnet was revolved without the external part of c d [that is, without c b], or the latter revolved without the magnet, then currents were produced as before: In the

3096. The magnet was now included in the circuit, in the following manner. The wire *a*, fig. 7, was placed in metallic contact on both sides of the interval between the magnets at N (or the pole), and the part *c* was brought

Fig. 7.

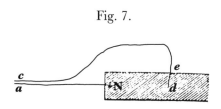

into contact with the centre at *d*. The result was in everything the same as when the wire *a* was continued up to *d, i. e.* no amount of revolution of the magnet and part *c* together could produce any electric current. When *c* was made to terminate at *e* or the equatorial part of the magnet, the result was precisely the same. Also, when *c* terminated at *e*, the part *a* of the wire was continued to the centre at *d*, and there the contact perfected, but the result was still the same. No difference, therefore, was produced, by the use between N and *d*, or *d* and *e*, of the parts of the magnet in place of an insulated copper wire, for the completion of the circuit in which the induced current was to travel. No rotation of the part *a* produced any effect, wherever it was made to terminate.

3097. In order to obtain the power of rotating the magnet without the external part of the wire, a copper ring was fixed round, and in contact with it at the equatorial part, and the wire *c*, fig. 8, made to bear by spring pressure against this ring, and also

Fig. 8.

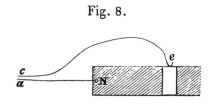

first case, *only* the interior portion *db* is revolving. Since currents are produced, it must be cutting lines! Where can these lines be, and in what direction must they run to be cut by *db*? In the second case only the exterior portion *cb* revolves. It is certainly cutting lines of force about the magnet, so naturally a current is produced—and it must, as stated previously, be equal and opposite to the currents induced in *db*. Faraday discusses this in paragraphs 3116–3118 below.

3096. From the results of paragraph 3095 it is clear that some process in the interior of the magnet is permitting a current to be induced there. Faraday thus focuses on the interior by replacing parts of the interior wire with the body of the magnet itself. First he eliminates the axial wire *Nd*, retaining the radial wire *de*.

3097. Now he eliminates both the axial *and* the radial portions of the interior wire.

a copper ring was fixed round…: Note this constitutes another slip ring, like the ones described earlier in paragraph 3084.

Fig. 1.

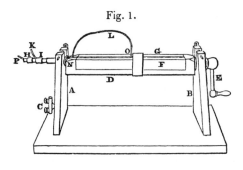

Fig. 8.

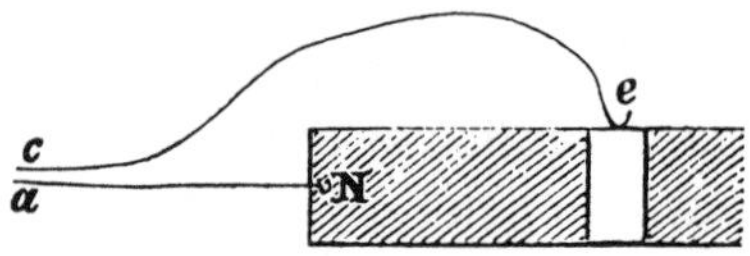

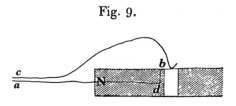

against the ring H on the axis, fig. 1 (3084.); the circuit was examined, and found complete. Now when the wire *c e* [fig. 8] was fixed and the magnet rotated, a current was produced, and that to the same amount for the same number of revolutions, whether the part of the wire *a* terminated at N, or was continued on to the centre of the magnet, or was insulated from the magnet and continued up to the copper ring *e*. When the wire, by expedients, which though rough were sufficient, was made to revolve whilst the magnet was still, currents in the contrary direction were produced, in accordance with the effect before described (3091.); and the results when the wire and magnet rotated together (3092.), show that these are in amount exactly equal to the former. When the inner and the outer wires were both motionless, and the magnet only revolved, a current in the full proportion was produced, and that, whether the axial wire *a* made contact at the pole of the magnet or in the centre.

3098. Another arrangement of the magnet and wires was of the following kind. A radial insulated wire was fixed in the middle of the

Fig. 9.

magnets, from the centre *d*, fig. 9, to the circumference *b*, being connected there with the equatorial ring (3097.); an axial wire touched this radial wire at the centre and passed out at the pole; the external part of the circuit, pressing on the ring at the equator, proceeded on the outside over the pole to form the communication as before. In the case where the magnet was revolved without the axial and the external wire, the full

---

3098. The arrangement here is comparable to that depicted in Figure 6 but improved by the slip ring, which permits continuous rotation.

and proper current was produced; the small wire, *d b*, being, however, the only part in which this current could be generated by the motion; for it replaced, under these circumstances, the body of the magnet employed on the former occasion (3097.).

3099. The external part of the wire instead of being carried back over that pole of the magnet at which the axial wire entered, was continued away over the other pole, and so round by a long circuit to the galvanometer; still the revolution of the magnet, under any of the described circumstances, produced exactly the same results as before. It will be evident by inspection of fig. 10, that, however the wires are carried away, the general result will, according to the assumed principles of action, be the same; for if *a* be the axial

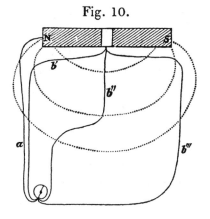

Fig. 10.

wire, and *b'*, *b''*, and *b'''* the equatorial wire, represented in three different positions, whatever magnetic lines of force pass across the latter wire in one position, will also pass across it in the other, or in *any* other position which can be given to it. The distance of the wire at the place of intersection with the lines of force, has been shown, by the experiments (3093.), to be unimportant.

3100. Whilst considering the condition of the forces of a magnet, it may be admitted, that the two magnets used in the experimental investigations described, act truly as one central magnet. We have only to conceive smaller similar magnets to be introduced to fill up the narrow space not occupied by the wire, and then the complete

---

the small wire, d b, being ... the only part in which this current could be generated: Again something about the interior condition of the magnet seems to focus on the little radial wire *d b*, which appears to be able to replace even *the whole body of the magnet*, as far as cutting lines of force is concerned. Faraday will offer an interpretation in paragraph 3118.

3099. *however the wires are carried away, the general result will ... be the same*: As Figure 10 depicts, *any* possible wire path (solid lines) from the revolving magnet's equator to the galvanometer will cut every line of force (dotted lines) *once* per revolution.

3100. *the two magnets ... act truly as one central magnet*: Faraday made this assertion in paragraph 3085; here now are arguments to support it.

magnet would be realized:—or it may be viewed as a magnet once perfect, which has had certain parts removed; and we know that neither of these changes would disturb the general disposition of the forces. In and around the bar magnet the forces are distributed in the simplest and most regular manner. Supposing the bar removed from other magnetic influences, then its power must be considered as extending to any distance, according to the recognized law; but, adopting the representative idea of *lines of force* (3074.), any wire or line proceeding from a point in the magnetic equator of the bar, over one of the poles, so as to pass through the magnetic axis, and so on to a point on the opposite side of the magnetic equator, must intersect *all* the lines in the plane through which it passes, whether its course be over the one pole or the other. So also a wire proceeding from the end of the magnet at the magnetic axis, to a point at the magnetic equator, must intersect curves equal to half those of a great plane, however small or great the length of the wire may be; and though by its tortuous course it may pass out of one plane into another on its way to the equator.

3101. Further, if such a wire as that last described be revolved once round the end of the magnet to which it is related, a slipping contact at the equator being permitted for the purpose, it will intersect *all* the lines of force during the revolution; and that, whether the polar contact is absolutely coincident with the magnetic axis, or is anywhere else at the end of the bar, provided it remain for the time unchanged. All this is true, though the magnet may be subject, by induction at a distance, to other magnets or bodies, and may be exerting part of its force on them, so as to make the distribution of its power very irregular as compared to the case of the independent bar (3084.), or may have an irregular or contorted shape, even up to

---

3101. Figure 10, reproduced here, will be helpful in imagining the different variations Faraday discusses in this paragraph. Both the *lines of force* and the *wire paths* may freely change their shape; but so long as their *terminations* remain as depicted, the wire intersects each line of force either *once*, or—equivalently—an *odd* number of times in each revolution. Moreover, the result will continue to hold even if the polar connection is displaced, so long as the polar and equatorial connections are on *opposite sides* of the axial line—that is, so long as "the loop of wire be made to pass over either pole."

Fig. 10.

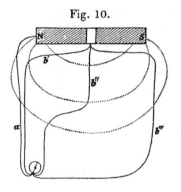

the horseshoe form. It is evident, indeed, that if a wire have one of its ends applied to *any* point on the surface of a magnet, and the other end to a point in the magnetic equator, and the latter be slipped once round the magnetic equator, and the loop of wire be made to pass over either pole, so as at last to resume its first position, it will in the course of its journey have intersected *once* every line of force belonging to the magnet.

3102. A wire from pole to pole which passes close to the equator, of course intersects half the external lines of force in a great plane, twice, in opposite directions as regards the polarity; and, therefore, when revolved round the magnet, has no electric current induced in it. If it do not touch at the equator, still, whatever lines it intersects, are twice intersected, and so the same equilibrium is preserved. If the magnet rotate under the wire, it acts the part of the central rotating wire already referred to (3095.); or if any course for the electric current other than a right line is assumed in it, that course is subject to the law of neutrality above stated, as will be seen by reference to the internal condition of the magnet itself (3117.). Hence the reason why no currents are produced, under any circumstances of motion, by the

---

3102. *A wire from pole to pole … intersects half the external lines of force in a great plane, twice…*: A wire from pole to *equator* intersects once (or an odd number of times) all the lines in its own plane *on one side of the axis*—that is to say, half the lines in its own plane. Therefore a wire from pole to *pole* must intersect each of the same lines one (or some odd number) additional time, for a total of two (or some even number) intersections.

…in opposite directions as regards the polarity: Consider a portion of wire parallel to the magnet's axis, sketched here; it will intersect the same line of force twice, as just explained. Now the direction of magnetic force will be *away* from the magnet at the intersection near the N pole, and it will be *towards* the magnet at the intersection near the S pole (in 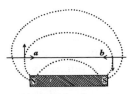 the sketch this is indicated by tangent arrows pointing *up* and *down*, respectively). But both portions of the wire move in the same direction—say, away from the viewer, in the sketch. Then by the *law* (paragraph 114) or the magneto-electric right-hand rule (paragraph 101, *comment*), the induced currents *a* and *b* will be in opposite directions and will cancel in the wire.

law of neutrality: that is, the mutual cancellation just described. Faraday already showed that straight, axial conducting paths within the magnet cannot be intersecting any *radial* lines of force (paragraph 3095). But what if the conducting paths within the magnet are curves? This (perhaps farfetched) possibility will have to await his further explication of the internal state of the magnet in paragraph 3117; nevertheless he assures us that when the situation is analyzed in light of that condition, the result will be the same cancellation as before.

application of such conducting circuits to the magnet. I may further observe, in reference to the intersection of the lines of force, that if a wire ring, a little larger in diameter than the magnet, be held edgeways at one of the poles, so that the lines of force there shall be in its plane, and be then turned 90° and carried over the pole to the equator (3088.), it will intersect *once* all the lines of the magnet, except the very few which will remain unintersected at the equator.

3103. Whilst endeavouring to establish experimentally the definite amount of the power represented by the *lines of force*, it is necessary to take certain precautions, or the results will be in error. For instance, ten revolutions of the wire about the magnet, or of the magnet within the fixed wire (3097.), ought to give a constant deflection at the galvanometer, and yet without any change in the position of the wire the results may at different times differ very much from each other; being at one time 9°, and at another only 4° or 5°. I found this to be due to difference of velocity within certain limits, and to be explained and guarded against as follows.

3104. If a wire move across lines of force slowly, a feeble electric current is produced in it, continuing for the time of the motion; if it move across the same lines quickly, a stronger current is produced for

---

3102, continued. *in reference to the intersection of the lines of force*: He means intersection of the lines *by a wire*—not by other lines! Lines of force cannot intersect one another—for what would be the direction of force at such an intersection, if it were possible?

turned 90° and carried over the pole…: Faraday imagines carrying a ring onto the magnet, as one places a ring on a finger. Compare Figure 2 in paragraph 3086 above. By initially turning through 90° the ring intersects all the lines contained between the axis and its circumference; the act of placing it on the equator intersects the remaining lines.

3103. *ten revolutions of the wire … ought to give a constant deflection at the galvanometer*: that is, the same number of revolutions should consistently give the same *throw* of the needle—for in ten revolutions the wire cuts the same lines of force ten times. In fact he finds that the *speed* of revolution has an effect, for reasons he will explain beginning in the next paragraph.

3104. Faraday gives a full account of the requirements necessary for proper *ballistic* use of the galvanometer. First, the discharges to be measured must be *short* compared to the time of swing of the needle; for a steady force applied to the needle will not continually increase its deflection, but only hold the needle at its maximum deflection for a longer interval. Faraday finds that with his particular instrument, equal discharges persisting for up to 75% of the (one-way) swing time consistently give the same throw of the needle; but more protracted discharges give inconsistent throws, which depend on their durations.

a shorter time. The effect of the current which deflects a galvanometer needle, is opposed by the action of the earth, which tends to return the needle to zero. A continuous weak current, therefore, cannot deflect it so far as a continuous stronger current. If the currents be limited in duration, the same effect will occur unless the time of the swing of the needle to one side be not considerably more than the time of either of the currents. If the time of the needle-swing be ten, and the time of ten quick rotations be six, then all the effect of the induced current is exerted in swinging the needle; but if the time of ten slow rotations be twelve or fifteen, then part of the current produced is not recognized by the extent of the vibration, but only by its holding the needle out awhile, at the extremity of a smaller arc of declination. Therefore, when quick and slow velocity was compared, and, indeed, in every case of comparative rotations of the wire and magnet, only that number of rotations was taken which could be well included within the time of the needle's journey to one side;—when the needle, therefore, was seen to travel on to its extreme distance after the rotation and the inducing current had ceased. If the needle began to return the instant the motion was over, such an experiment was rejected for purposes of comparison. When these precautions were attended to, and velocities of revolution taken, which occupied times from one-third to three-fourths of that required for the swing of the needle, then the same number of revolutions (ten) gave the same amount of deflection, namely, 9.5°, with my apparatus, though the time of revolution varied as 1 : 2, or even in a higher degree.

3105. Another cause of difference produced by varying velocity, is the diminution of the action of the current on the needle, as the angle which the latter forms with the convolutions of the coil increases. Hence a constant current produces more effect on the deflection of the needle for the first moments of time than afterwards. This effect, however, was scarcely sensible for swinging deflections of 9° or 10°, produced by currents which were over before the needle had moved through 4° or 5°.

3105. *more effect on the deflection of the needle for the first moments...*: Not only the consistency but the accuracy of the galvanometer is adversely affected if the discharge still continues after the needle has departed significantly from its initial position; for then the later stages of the discharge will contribute less to the needle's deflection than the earlier stages did.

Both considerations thus reduce to a single requirement: the discharges to be measured must complete their action before the galvanometer needle departs very far from its equilibrium position. In practice, this means a current duration that is small compared to the swing period of the galvanometer.

3106. It has already been shown, that it is a matter of indifference whether the wire revolve in one direction or the magnet in the other (3091.); and this is still further proved by the cases where the magnet and the wire revolve together (3092.); for then the currents which tend to form are exactly equal and opposed to each other, whatever the position of the wire may be. As the immobility of the needle is a point more easily ascertained than the extent of an arc, indicated only for a moment, and as the rotations of the magnet and wire conjointly can be made rapid and continuous, the proof in such cases is very satisfactory.

3107. Proceeding to experiment upon the effect of the *distance* of the wire *c*, fig. 11., from the magnet, the wire was made to vary, so that sometimes it was not more than 8 inches long (being of copper and 0.04 of an inch in diameter), and only half an inch from the magnet, whilst at other times it was 6 or 8 feet long and extended to a great distance. The deflection due to ten revolutions of the magnet was observed, and the average of several observations, for each position of the wire, taken: these were very close (with the precautions before described) for the same position; and the averages for different positions agreed perfectly together, being 9.5°. I endeavoured to repeat these experiments on distance by moving the wire and preserving the magnet stationary in the manner before described (3091.); they were not so striking because time would only allow of smaller deflections being obtained (3104.), but the same number of journeys through an arc of 180° gave the same deflection at the galvanometer, whether the course of the wire was close to the magnet or far off; and the deflection agreed with those obtained when the magnet was rotating and the wire at rest.

Fig. 11.

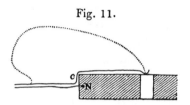

3108. As to *velocity* of motion; when the magnet was rotating and the wire placed at *different distances*, then ten revolutions of the magnet produced the same deflection of the needle, whether the motion was *quicker* or *slower*, and whatever the distance of the wire, provided the

---

3107. *the same deflection at the galvanometer, whether the course of the wire was close to the magnet or far off*: This result confirms what Faraday has maintained all along, that the power represented by the lines of force is independent of the *distance from the magnet* at which that power may be exerted.

these were very close…: Thus, so long as the durations fell within the permissible limit, wires both near and far from the magnet gave equal discharges for equal numbers of lines intersected.

precautions before described were attended to (3104.). That the same would be true if the wire were moving and the magnet still, is shown by this; that whatever the velocity with which the wire and magnet revolve together, and whatever their distance apart, they exactly neutralize and equal each other (3096.).

3109. From these results the following conclusions may be drawn. The *amount* of magnetic force, as shown by its effect in evolving electric currents, is determinate for the same lines of force, whatever the distance of the point or plane, at which their power is exerted, is from the magnet. Or it is the same in any two, or more, sections of the same lines of force, whatever their form or their distance from the seat of the power may be. This is shown by the results with the magnet and the wire, when both are in the circuit (3108.); and also by the wire loop revolving with the magnet (3092.); where the tendency of currents to form in the two parts oppose and exactly neutralize or compensate each other.

3110. In the latter case very varying sections outside of the magnet may be compared to each other; thus, the wire may be conceived of as passing (or be actually formed so as to intersect) lines of force near the pole, and then, being continued *along* a line of force until over the equator, may be directed so as to intersect the same lines of force in the contrary direction, and then return along a line of force to its commencement; and so two surface sections may be compared. It is manifest that every loop forming a complete circuit, which is in a great plane passing through the axis of the magnet, must have precisely the same lines of force passing into and passing out of it, though they may, so to say, be expanded in one part and compressed in another; or (speaking in the language of radiation) be more intense in one part

---

3109. In this paragraph Faraday summarizes what is meant by the "definite" character of the lines of force. Whatever may be their shape, their length, their distance from the magnet, or their degree of mutual approach or separation, a specified set of lines represents a *definite quantity of magnetic power.*

3110. *the language of radiation*: When we use a lens to focus sunlight onto a surface, we say that the *intensity* of the focused light increases as its rays are concentrated into smaller and smaller areas. Faraday proposes to apply this same language to the lines of force; thus as the lines become more or less "compressed" he describes them—together with their corresponding magnetic power—as exhibiting *greater or less intensity* at different locations in the magnetic field. In contrast to his Twelfth Series attempt to articulate electric "intensity" (paragraph 1370), the present image of line concentration makes no reference to *particles.*

and less intense in the other. It is also as manifest, that, if the loop be not in one plane, still, on making one complete revolution, either with or without the magnet, it will have intersected in its two opposite parts an exactly equal amount of lines of force. Hence the comparison of any one section of a given amount of lines of force with any other section is rendered, experimentally, very extensive.

3111. Such results prove, that, under the circumstances, there is no loss, or destruction, or evanescence, or latent state of the magnetic power by distance.

3112. Also that convergence or divergence of the lines of force causes no difference in their amount.

3113. That obliquity of intersection causes no difference. It is easy so to shape the loop (3110.), that it shall intersect the lines of force directly across at both places of intersection, or directly at one and obliquely at the other, or obliquely in any degree at both; and yet the result is always the same (3093.).

3114. It is also evident, by the results of the rotation of the wire and magnet (3097. 3106.), that when a wire is moving amongst equal lines (or in a field of equal magnetic force), and with an uniform motion, then the current of electricity produced is proportionate to the *time*; and also to the *velocity* of motion.

3111. *no loss, or destruction, or evanescence, or latent state of the magnetic power by distance*: We saw such results at paragraph 3107 above. Readers sometimes suppose that Faraday here denies what everyone knows—namely, that magnets commonly show stronger effects at small distances than at large. Not so! He means that the power represented by *a single line of force* remains constant throughout its length. Thus when the lines diverge, as they do from the ends of a bar magnet, a body will intersect more lines when it is nearby than when it is distant; and *that* is why nearer bodies experience stronger effects.

3112. This has the consequence, invoked several times already, that between any two lines, the same number of lines are everywhere contained.

3113. *obliquity of intersection*: Since wires of arbitrary shape evolve equal quantities of electricity in cutting the same lines of force, we may infer that the *angle* at which the wire cuts the lines plays no role in the results—since that angle is different in each part of each curve.

3114. *the current of electricity produced…*: Better, *the quantity of electricity discharged*; for it is to this quantity that the ballistic galvanometer swing is proportional. Faraday showed this in the Third Series (paragraphs 361–366).

…*is proportionate to the time; and also to the velocity of motion*: that is, proportional to the *distance* over which the moving wire travels, for distance is the product of velocity and time. This makes sense, since in a uniform field the lines are evenly distributed—so that in moving through a given distance,

3115. They also prove, generally, that the quantity of electricity thrown into a current is directly as the amount of curves intersected.

3116. In addition to these results, this method of investigation gives much insight into the internal condition of the magnet, and the manner in which the lines of force (which represent truly all that we are acquainted with of the peculiar action of the magnet) either terminate at its exterior, or at any assumed points, to be called poles; or are continued and disposed of within. For this purpose, let us consider the external loop (3093.) of fig. 5. When revolving with the magnet no current is produced, because the lines of force which are intersected on the one part, are again intersected in an opposing direction on the other (3110.).

Fig. 5.

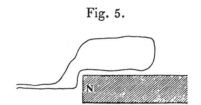

---

a wire will cut a proportional number of lines and produce a corresponding discharge, as is observed to be the case.

3115. Faraday finally asserts, as a formal conclusion, a relation of *proportionality* between the number of lines of force cut, and the quantity of electricity evolved in the cutting circuit. This relation was first exhibited, but only roughly, by the loop experiment (Fig. 2—paragraph 3086); subsequently it became a tacit assumption in several succeeding exercises. It will be thoroughly investigated and confirmed in the 29th Series.

Notice that in the stated proportionality between *quantity of electricity* and *number of curves intersected*, the *conductive material* plays no part, since it is the same for all the currents that are compared. This suggests a simpler interpretation of the old parallel-wire induction experiments Faraday discussed in the Second Series (paragraphs 201, 213). There, equal lengths of unequally-conducting materials were joined (along with a galvanometer) at their ends and moved together through the magnetic field. According to the present principle, the wires and the galvanometer together formed a single, common discharge path. Then since both wires must necessarily have intersected equal numbers of magnetic lines of force, equal quantities of electricity would have been "thrown into a current"—in opposite directions with respect to the discharge circuit. So the null result Faraday obtained is none other than the *expected* result—and the tortuous analysis suggested by paragraph 213 is superfluous.

3116. *this method of investigation gives much insight into ... the manner in which the lines of force ... either terminate at [the magnet's] exterior, or at ... poles; or are continued and disposed of within*: Do the lines of force terminate at the iron surface or penetrate it? If they enter the iron, what happens to them there? The Moving Wire will answer these questions! In Figure 5, a wholly external loop rotating through fixed lines of force produced no current, because it intersected each line of force *twice*—thus the two correspondingly induced currents canceled one another in the circuit.

But if one part of the loop be taken down the axis of the magnet, and the wire then pass out at the equator (3091.), still the same absence of effect is produced; and yet it is evident that, external to the magnet,

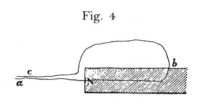

Fig. 4

every part of the wire passes through lines of force, which conspire together to produce a current; for all the external lines of force are then intersected by that wire in one revolution (3101.). We must therefore look to the part of the wire *within* the magnet, for a power equal to that capable of being exerted externally, and we find it in that small portion which represents a radius at the central and equatorial parts. When, in fact, the axial part of the wire was rotated it produced no effect (3095.); when the axial, the inner radial, and the external parts were revolved together, they produced no effect; when the external wire alone was revolved, *directly*, it produced a current (3091.); and when the internal radius wire alone (being insulated from the magnet) revolved, *directly*, it also produced a current (3095. 3098.) in the contrary direction to the former; and the two were exactly equal in power; for when both portions of the wire moved together *directly*, they perfectly compensated each other (3095.). This radius wire may be replaced by the magnet itself (3096. 3118.).

3116, continued. *But if one part of the loop be taken down the axis of the magnet, and the wire then pass out at the equator...*: Such is the arrangement in Figure 4, reproduced from paragraph 3091 which Faraday cites. A loop partly external and partly internal also produces no current, even though the exterior portion is certainly cutting each line of force once. It seems that here too, the wire must have cut each line twice.

We must therefore look to the part of the wire within the magnet, for a power equal to that capable of being exerted externally...: If the external part of the wire cut each line of force once, a second cut could only have been made by the *internal* portion of the wire—which shows that the external lines of force must continue on into the body of the magnet.

...and we find it in that small portion which represents a radius at the central and equatorial parts: Why the radius? Because rotation of the *axial* part of the interior wire is known to be ineffectual. Faraday showed this in paragraph 3095 with the arrangement depicted in Figure 6, reproduced here. In that configuration the axial part of the loop in Figure 4 was made a separate wire, capable of being rotated independently.

Fig. 6.

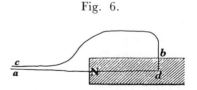

3117. So, by this test there exists lines of force within the magnet, of the same *nature* as those without. What is more, they are exactly equal in *amount* to those without. They have a relation in *direction* to those without; and in fact are continuations of them, absolutely unchanged in their nature, so far as the experimental test can be applied to them. Every line of force therefore, at whatever distance it may be taken from the magnet, must be considered as a closed circuit, passing in some part of its course through the magnet, and having an equal amount of force in every part of its course.

*   *   *

3117. *lines of force within the magnet ... of the same nature as those without*: for they evidently obey the same laws of magneto-electric induction as do the exterior lines.

...equal in amount to those without: because the interior and exterior portions of the wire must be cutting equal numbers of lines (or perhaps the *same* continuous lines!) in order to develop equal currents. And the currents *are* equal, for by their mutual cancellation they prove themselves to be both equal and opposite.

in direction ... are continuations of [the external lines]: that is, so that the external and internal lines together form *closed curves*. The internal lines of force must run axially, as shown in the sketch below, since we know they are cut by radial wire *d b* but not by axial wire *a d*. Furthermore, the current induced in external wire *c b* is known to be opposite to the current induced in radius *d b*. Then if the lines of force in the regions just outside of *b* cross the wire from left to right, the lines just inside of *b* must cross from right to left—opposite directions with respect to the *wire*, but a consistent direction with respect to circulation about the closed path. Moreover it is clear that if the internal and external lines of force are just continuations of one another, the total number of lines cut by external wire *c b* will be equal to the number cut by internal radius *d b*—as they must be, in order to produce currents that are equal as well as opposite. As the introduction to the present Series explains, this is a truly momentous result. It is probably the most dramatic example of what Faraday had stated earlier in paragraph 3081—that the moving wire "can even be applied ... to the interior of a magnet, a place utterly inaccessible to the magnetic needle."

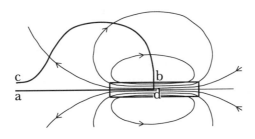

3120. In this striking disposition of the forces of a magnet, as exhibited by the moving wire, it exactly resembles an electromagnetic helix, both as to the direction of the lines of force in closed circuits, and in their equal sum within and without. No doubt, the magnet is the most heterogeneous in its nature, being composed, as we are well aware, of parts which differ much in the degree of their magnetic development; so much so, that some of the internal portions appear frequently to act as keepers or submagnets to the parts which are further from the centre, and so, for the time, to form complete circuits, or something equivalent to them, within. But these make no part of the resultant of force externally, and it is only that resultant which is sensible to us in any way; either by the action on a needle, or other magnets, or soft iron, or the moving wire. So also the power which is manifest *within* the magnet by its effect on the moving mass, is still only that same resultant; being equal to, and by polarity and other qualities, identical with it. No doubt, there are cases, as upon the approach of a keeper to the poles, or the approximation of other magnets, either in favourable or adverse positions, when more external force is developed, or it may be a portion apparently thrown inwards and so the external force diminished. But in these cases, that

---

3120. *it exactly resembles an electromagnetic helix*: Surely Faraday is not *surprised* at this similarity? Yet since the iron body of the magnet is so different from the surrounding air—and, moreover, so far from being homogeneous in itself—it might well occasion surprise to discover how similar, *magnetically*, the inside and the outside of an iron magnet are!

some of the internal portions … act as keepers or submagnets to [other internal parts]: The force that is sustained by a permanent magnet is generally carried on through the medium of *air*. But if the poles of a magnet are bridged by an iron bar or "keeper," not itself strongly magnetized, much of the force that would otherwise be propagated in the air becomes concentrated in the keeper instead. This is desirable when powerful magnets (particularly of the horse-shoe form) are kept in storage.

Now if the iron core of a magnet is inhomogeneous, so that some regions of the metal are less active magnetically than others, those less active regions may act as "keepers" in relation to the more powerful ones. What this means is that some of the magnetic force that would have appeared at the poles and in the external environs of the magnet will instead be concentrated in *internal regions of the core itself*, being thus effectively lost to the external field. But inasmuch as such internal "keepers" cause the number of external lines of force to diminish, they reduce the net number of internal lines accordingly—since they constitute closed loop paths whose opposite arcs mutually cancel. Therefore the principle that external and internal lines of force are equal in total quantity is preserved.

which remains externally existent, corresponds precisely to that which is the resultant internally; for when either the same, or contrary poles, of a powerful horseshoe magnet were placed within an inch and a half of the poles of the bar-magnets, prepared to rotate with the attached wires (3092.), as before described, still, upon their revolution, not the slightest action at the galvanometer was perceived; the forces within the magnet and those without perfectly compensating each other.

3121. The definite character of the forces of an invariable magnet, at whatever distance they are observed from the magnet, has been already insisted upon (3109.). How much more strikingly does that point come forth now, that, being able to observe within the magnet, we find the same definite character there; every section of the forces, whether within or without the magnet, being exactly of the same amount! The power of a magnet may therefore be easily represented by the effects of *any* section of its lines of force; and as the currents induced by two different magnets may easily be conducted through one wire, or be, in other ways, compared to each other, so facilities may thus arise for the establishment of a standard amongst magnets.

3122. On the other hand, the use of the idea of *lines of force*, which I recommend, to represent the true and real magnetic forces, makes it

---

3121. *every section of the forces, whether within or without the magnet, being exactly of the same amount*: When Faraday earlier proposed, and subsequently confirmed, that every section of the same lines represented the same quantity of force, he then spoke only of the *exterior* lines of force. That principle was already a striking and significant one; and now it is extended even into the *interior* of the magnet! In every way we are learning to rethink the relations between inside and outside. When our image of the magnet was founded on the difference between its material *body* and its air or vacuum *environment*, the temptation was strong to attribute the magnet's power essentially to its body. But if we now build our image of the magnet on *the pattern of its lines of force* we shall be compelled to view every magnet, not only as being a unity in itself, but as exercising its proper office in a vast and unified system of magnets and magnetic agents. Not only are *body and medium* intimately related, we find both *agents,* magnet and magnet, participating jointly in comparable relations which, extending through indefinite distances, attain even a cosmic wholeness.

a standard amongst magnets: that is, a standard of *equality*. Magnets are equal insofar as they sustain the same total number of lines of force. Thus two magnets might in principle be compared, by surveying them with a loop of wire in the manner indicated by Figure 2 (paragraph 3086). The respective galvanometer throws would disclose whether, and by what proportion, one magnet exceeds or falls short of another.

very desirable that we should find a unit of such force, if it can be attainable, by any experimental arrangement, just as one desires to have a unit for rays of light or heat. It does not seem to me improbable that further research will supply the means of establishing a standard of this kind. In the mean time, for the enlargement of the utility of the idea in relation to the magnetic force, and to indicate its conditions graphically, lines may be employed as representing these units in any given case. I have so employed them in former series of these Researches (2807. 2821. 2831. 2874. &c.), where the direction of the *line of force* is shown at once, and the relative amount of force, or of lines of force in a given space, indicated by their concentration or separation, *i. e.* by their number in that space. Such a use of unit lines involves, I believe, no error either in the direction of the polarity or in the amount of force indicated at any given spot included in the diagrams.

*   *   *

3154. Such a conclusion as that just arrived at, brings on the question of what is *magnetic polarity*, and how is it to be defined? For my

---

3122. *desirable that we should find a unit of such force*: This is far from a simple task. For (*i*) suppose we just choose a convenient magnet and designate *it* as the standard unit—comparing other magnets with it by the method envisaged in paragraph 3121. But magnets are notoriously unstable over time; so such a "standard" would be a misnomer. Then (*ii*) suppose we use the galvanometer and ring-shaped loop, as described in paragraph 3102, choosing some convenient angle of deflection as the unit. But the measurements would be dependent on that particular galvanometer, which (besides other reasons), since it contains *magnetic needles*, reduces again to the choice of a particular magnet as "standard." Or still another possibility: Since the galvanometer swing is proportional to the quantity of electricity discharged through it (paragraph 366), then (*iii*) if we can establish a unit of *electrical quantity*, then a unit of *magnetic* power might be defined in terms of electrical measurements. But then the task would simply shift from magnetism to electricity. By no means do these suggestions exhaust all possible or even all promising approaches. I only wish to show something of the magnitude of the project. It will fall to Faraday's successors; Maxwell, particularly, will play a decisive role.

3154. *Such a conclusion as that just arrived at*: Although Faraday refers to an omitted paragraph, the conclusion in question is practically the same as that which was voiced in paragraph 3122 above—that the relative amount of force in a given space is indicated by the convergence or divergence of lines within that space. The degree of convergence or divergence, in turn, is determined by the *conducting power* of the material through which the lines pass.

own part, I should understand the term to mean, the opposite and antithetical actions which are manifested at the opposite ends, or the opposite sides, of a limited (or unlimited) portion of a line of force (2835.). The line of dip of the earth, or a part of it, may again be referred to as the natural case; and a free needle above or below the part, or a wire moving across it (3076. 3079.), will give the direction of the polarity. If we refer to an entirely different and artificial source as the electro-magnetic helix, the same meaning and description will apply.

3155. If the term *polarity* have any meaning, which has reference to experimental facts and not to hypotheses only, beyond that included in the above description, I am not aware that it has ever been distinctly and clearly expressed. It may be so, for I dare not venture to say that I recollect all I have read, or even all the conclusions I myself have at different times come to. But if it neither have, nor should have, any

---

I should understand [magnetic polarity] to mean, the opposite and antithetical actions...: Note that Faraday has now passed completely beyond the earlier image of polarity as *convergence*, as indeed he already suggested in the 26th Series. The convergence image represented an essential refinement of the conventional action-at-a-distance concepts of "pole." It seized upon an aspect of "pole" that proved eminently suitable for guiding further experiments, at the same time transcending other, less benign elements in "polar" thinking— elements which (as Faraday will put it a few paragraphs hence) may have been functioning as "an obstruction to the advancement of truth and a defence of wrong assumptions and error." But now, by an evolution founded on those very experiments, the convergence image is itself to be refined and transcended.

3155. *If the term polarity have any meaning, which has reference to experimental facts and not to hypotheses only, beyond that included in the above description, I am not aware that it has ever been distinctly and clearly expressed*: Surely Faraday does himself an injustice here. His earlier image of polarity as *convergence of lines* has now been superseded, to be sure; but not because it lacked experimental foundation, or was not clearly and distinctly expressed!

I dare not venture to say that I recollect all I have read, or even all the conclusions I myself have at different times come to: This may refer to his recurring struggle with memory loss (paragraph 2308, *note*; see also paragraph 863). But perhaps more important, Faraday's candid admission implicitly rejects stereotypes of *completeness* and *perfection* in the investigative enterprise— evidently, that pursuit is compatible with omissions and deficiencies. Science, for Faraday, is not a commodity produced but a *life chosen*. As much as any life, therefore, it is at all times incomplete, tentative, continually retracing its paths even as it evolves new ones. It is not the activity of gods, as perhaps Aristotle taught; but it occasionally affords to mortals a vision of what must fall little short of the divine.

other meaning, then the question arises, is it correctly exhibited or indicated in every case by attractions and repulsions, *i. e.* by such like mutual actions of particular bodies on each other under the magnetic influence? A weak solution of protosulphate of iron, if surrounded by water, will, in the magnetic field, point axially; if in a stronger solution than itself, it will point equatorially (2357. 2366. 2422.). The same is true with stronger cases. We cannot doubt it would be true even up to iron, nickel, and cobalt, if we could render these bodies fluid in turn without altering their paramagnetic power, or if we had the command of magnets and of paramagnetic and diamagnetic media, stronger or weaker at pleasure. But in the case of the solutions, we cannot suppose that the weaker has one polarity in the stronger solution and another in the water. The lines of force across the magnetic field have the same general polarity in all the cases, and would be shown experimentally to have it, by the moving wire (3076.), though not by the attractions and repulsions.

3156. Here, therefore, we have a *difference* in the two modes of experimental indication; not merely as to the method, but as to the nature of the results, and the very principles which are concerned in their production. Hence the value I think of the moving wire as an investigator; for it leads us into inquiries which touch upon the very nature of the magnetic force. There is no doubt that the needle gives true experimental indications; but it is not so sure that we always interpret them correctly. To assume that pointing is always the direct effect of attractive and repulsive forces acting in couples (as in the cases in question, or as in bismuth crystals), is to shut out ideas, in relation to

---

3155, continued. *is [polarity] correctly exhibited or indicated in every case by attractions and repulsions...?*: An iconoclastic question which answers itself (in the negative) by the very condition of being asked. For according to the conventional notions, polarity would have been virtually *identified* with the display of attractions and repulsions!

we cannot suppose that the weaker [solution] has one polarity in the stronger solution and another in the water: But that is exactly what we should have to suppose if polarity were indicated by attractions and repulsions. By contrast, the moving wire would show the *same general direction* in the lines of force, however expanded or compressed in different media; and the image of polarity as directionality emphasizes this sameness, where the earlier image stressed the *contrast* between convergence and divergence of the lines. Perhaps as another consequence of the moving wire's suggestive revelations, Faraday will eventually give over some of his former absolutism about space and matter—for certainly the wire discloses no difference between space and other media beyond the relative differences of concentration or dispersion of magnetic lines.

magnetism, which are already applied in the theories of the nature of light and electricity; and the shutting out of such ideas *may be* an obstruction to the advancement of truth and a defence of wrong assumptions and error.

3157. What is the idea of polarity in a field of *equal force?* (whether it be occupied by air or by a mass of soft iron?). A magnetic needle, or an oblong piece of iron, would not show it in the air or elsewhere, except by disturbing the equal arrangement of the force and rendering it unequal; for on that the pointing of the needle or the iron, or the motions of either towards the walls of the magnetic field, if limited (2828.), would depend. A crystal of bismuth in showing this polarity by position (2464. 2839.), does it without much altering the distribution of the force, and the alteration which does take place is in the contrary direction to that effected by iron (2807.), for it expands the lines of force. It seems readily possible that a magnecrystal might exist, which, when in its stable position, should neither cause the convergence nor divergence of the lines of force within it. It need only be neutral in relation to space or any surrounding medium in that direction, and diamagnetic in its relation in the transverse direction, and the conditions would be fulfilled.

3158. But though an ordinary magnetic needle cannot show polarity in a field of equal force,[4] having no reference to it, and in fact ignoring such a condition of things, a moving wire makes it manifest instantly, and also shows the full amount of magnetic power to which such polarity belongs; and this it does without disturbing the distribution of the power, as far as we comprehend or understand distribution, when thinking of magnetic needles. At least such at present appears to me to be the case, from the consideration of the action of thin and thick wires (3141.) and wires of different substances (3153.).

[4] One could easily imagine hypothetically a needle that should do so.

3157. *What is the idea of polarity in a field of equal force?*: The very problem which cast doubt on the earlier *convergence* image of polarity (paragraph 2835), since there is obviously no convergence in a field traversed by parallel magnetic lines. Similarly Faraday will point out in paragraph 3174 below that there are no "centres of force" in the *circular magnetic field* surrounding an electric current. The twin examples of uniform and circular fields, therefore, show that *all* the earlier images of polarity—except the image of directionality—are deficient.

magnecrystal: as in paragraph 3071 above.

need only be neutral … in that direction, and diamagnetic … in the transverse direction: Such a crystal would, like the moving wire, reveal directionality in the uniform field without disturbing the uniformity, and would thereby lend additional support to *directionality* as the more fundamental (because more general) criterion.

*   *   *

3174. On bringing this paper to a close, I cannot refrain from again expressing my conviction of the truthfulness of the representation, which the idea of lines of force affords in regard to magnetic action. All the points which are experimentally established with regard to that action, *i. e.* all that is not hypothetical, appear to be well and truly represented by it. Whatever idea we employ to represent the power, ought ultimately to include electric forces, for the two are so related that one expression ought to serve for both. In this respect, the idea of lines of force appears to me to have advantages over the method of representing magnetic forces by centres of action. In a straight wire, for instance, carrying an electric current, it is apparently impossible to represent the magnetic forces by centres of action, whereas the lines of force simply and truly represent them. The study of these lines have, at different times, been greatly influential in leading me to various results, which I think prove their utility as well as fertility. Thus, the law of magneto-electric induction (114.); the earth's inductive action (149. 161. 171.); the relation of magnetism and light (2146. and note); diamagnetic action and its law (2243.), and magnecrystallic action (2454.), are cases of this kind: and a similar influence of them, over my mind, will be seen in the further instances of the polarity of diamagnetic bodies (2640.); the relation of magnetic curves and the evolved electric currents (243.); the explication of Arago's phænomenon (81.), and the distinction between that and ordinary magnetism (243. 245.); the relation of electric and magnetic forces (1709.); the views regarding magnetic conduction (2797.), and atmospheric magnetism (2847.). I have been so accustomed, indeed, to employ them, and especially in my last Researches, that I may, unwittingly, have become prejudiced in their favour, and ceased to be a clear-sighted judge. Still, I have always endeavoured to make experiment the test and controller of theory and opinion; but neither by that nor by close cross examination in principle, have I been made aware of any error involved in their use.

3175. Whilst writing this paper I perceive, that, in the late Series of these Researches, Nos. XXV. XXVI. XXVII., I have sometimes used the term *lines of force* so vaguely, as to leave the reader doubtful whether I intended it as a merely representative idea of the forces, or as the description of the path along which the power was continuously exerted. What I have said in the beginning of this paper (3075.) will render that matter clear. I have as yet found no reason to

---

3175. *What I have said in the beginning ... will render the matter clear*: that is, he affirms the representative role of the lines; and, while not positively asserting

wish any part of those papers altered, except these doubtful expressions: but that will be rectified if it be understood, that, wherever the expression *line of force* is taken simply to represent the disposition of the forces, it shall have the fullness of that meaning; but that wherever it may seem to represent the idea of the *physical mode* of transmission of the force, it expresses in that respect the opinion to which I incline at present. The opinion may be erroneous, and yet *all* that relates or refers to the disposition of the force will remain the same.

3176. The value of the moving wire or conductor, as an examiner of the magnetic forces, appears to me very great, because it touches the physics of the subject in a manner altogether different to the magnetic needle. It not only gives its indications upon a different principle and in a different manner, but in the mutual action of it and the source of power, it affects the power differently. The wire when quiescent does not sensibly disturb the arrangement of the force in the magnetic field; the needle when present does. When the wire is moving it does not sensibly disturb the forces external to it, unless perhaps in large masses, as in the discs (3163.), or when time is concerned (1730.), *i. e.* it does not disturb the disposition of the whole force, or the arrangement of the lines of force; a field of equal magnetic power is still equal to anything but the moving wire, whilst the wire moves across or through it. The moving wire also indicates quantity of force, independent of tension (2870.); it shows that the quantity within a magnet and that outside is the same, though the tension be very different. In addition to these advantageous points, the principle is available within magnets,

---

their physical existence, he favors that view as having at least a pragmatic value, if not the value of certain truth.

3176. *a field of equal magnetic power is still equal to anything but the moving wire, whilst the wire moves across or through it*: Even when the wire moves through it, a uniform field retains its equality by any standard *independent* of the wire. Note that this further supports the idea that lines of force have an *independent existence*, that they are not brought into being by the very device that indicates their presence.

The moving wire also indicates quantity of force, independent of tension … it shows that the quantity within a magnet and that outside is the same, though the tension be very different: Faraday attributed tension to magnetic power in the Twenty-sixth Series, where he suggested that lines of magnetic force must sustain greater tension in a poorer magnetic conductor than in a better one (paragraph 2870, *comment*). The tension inside and outside an iron magnet must be "very different" because iron and air differ so greatly in magnetic conducting power.

and paramagnetic and diamagnetic bodies, so as to have an application beyond that of the needle, and thus give experimental evidence, of a nature not otherwise attainable.

Royal Institution,
October 9, 1851.

Twenty-Ninth Series — Editor's Introduction

Single and multiple magnets

Faraday has been interested for some time in the various patterns of disposition of magnetic power. One case in particular, the lines of force of a common bar magnet, arises so often, and presents such a lucid and readable form, that Faraday finds in it a paradigm for other magnetic patterns. The magnetic lines of a solitary bar magnet are illustrated here, revealed by iron filings. In the present Series, Faraday will employ this distinctive 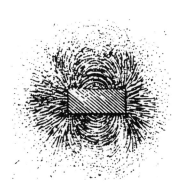 pattern as an interpretive tool, not only for single magnets but for configurations of multiple magnets. We will find a lovely example of the latter at paragraph 3231 below. There, Faraday traces the gradual and mutual separation of the two halves of a broken bar magnet. Stage by stage, he notes the progress from a *single complex pattern* to *two related patterns*, eventuating in *two virtually independent figures*—the familiarity and intelligibility of that culminating stage lending retrospective legibility to the earlier phases of relation and separation.

This "broken magnet" presentation, while one of the most engaging, is far from unique in the present Series. It is both preceded and followed by any number of similarly interpretive narratives. Indeed, in the 29th Series Faraday offers an unusually generous quota of *exhibitions* and *interpretations* of magnetic configurations. While the interpretation of natural powers is surely one of Faraday's most abiding philosophical aims, it was perhaps less overtly pursued in earlier Series, which often tended to focus on discoveries of new powers and effects. Now Faraday seems more openly intent on learning to see old effects with a new vision. While that has been a recurring theme in the Experimental Researches, it seems particularly prominent in this, the last of the collected Series.*

A voltaic analogy

Faraday draws on another interpretive source when reading the patterns of magnetic lines of force. As we saw in the Fifteenth Series, the *voltaic battery* presents, in the currents it produces in conductive

---

* Although Faraday published individual papers of a Thirtieth Series in various journals, they appeared subsequent to the third volume of collected *Experimental Researches in Electricity*. No fourth volume was issued.

environments, several suggestive analogies to the lines of force which encircle a magnet. Nor is that altogether surprising, since Faraday showed earlier that electric current is preceded by lines of induction; thus the current paths, too, are lines of force just as paths of magnetic influence are.

Faraday describes how the lines of force of two adjacent magnets may "coalesce," as he puts it, so that identical magnets lined up in a series will be found to deploy no more lines of force all together than any one of them exhibits when alone. This situation gains a perfect analogy when two or more voltaic batteries are connected, end to end, in a single circuit. For, he explains (paragraph 3232):

> If, as is well known, we separate a battery of 20 pair of plates into two batteries of 10 pair, or 4 batteries of 5 pair, each of the smaller batteries can supply as much dynamic electricity as the original battery, provided no sensible obstruction be thrown into the course of the lines, *i. e.* the path of the current.

But isn't this false? As anyone with a flashlight and a little wire can easily confirm, two batteries in line with one another ("in series") will light a bulb much brighter than a single battery will—how then can the single battery be said to supply "as much dynamic electricity" as the double? Nevertheless, Faraday is correct in his statement. The flashlight example is immaterial because the incandescent bulb violates Faraday's stipulation that "*no sensible obstruction* be thrown into ... the path of the current" (italics added). Indeed, it is natural for a modern reader to disregard that directive, for it would have disastrous consequences if acted upon today. To permit *no sensible obstruction* in the current path means to bridge the battery terminals with a perfect conductor—in other words, to *short-circuit* the battery.

Today no knowledgeable person would ever intentionally short-circuit a voltaic battery. Modern batteries would be destroyed, and might even explode under such treatment; indeed the very phrase "short circuit" has become practically synonymous with fault and danger. *But it was the ordinary mode of using voltaic batteries in Faraday's time.**

---

* Faraday's batteries were generally much bulkier than ours, in relation to the current they could produce, as well as being more open in construction; they could therefore better accommodate the heating and gas buildup that accompanies the discharge of all batteries. When exhausted they were not thought "dead," like our disposable batteries; all that was needed was to discard and replace the used electrolyte. When the latter was a liquid (as it usually was) that was easily done; in fact the original meaning of "charge," where batteries were concerned, was to *fill*—as one might charge a flagon with drink or a cannon with powder. Thus no stigma need attach to the rapid depletion of such a battery by demanding, through a "short circuit," its *maximum* current.

Consider then a single cell, such as the Bunsen design illustrated here. And suppose a *perfect conductor* is connected between the positive C (carbon) and negative Z (zinc) terminals. A current of definite and quite considerable strength will discharge through the conductor.* Ordinarily, the strength of that current would be limited by the conducting ability of the discharging path; but such cannot be the case here since we have supposed the discharge path to be perfect. The limiting factor or factors cannot, therefore, be external to the voltaic cell but must instead be *internal*; and we may infer that the cell—and, generally, every voltaic cell—has some *inherent* maximum discharge capability.

Now it is easy to see why several identical cells connected end to end—in series—will develop no more current *in a perfect conductor* than a single cell could

produce. For any current that develops in the series circuit must flow in *each* of the cells; but if the full force of each cell is expended in overcoming its own internal impediment, there will be no superfluity of force capable of discharging through other cells. Thus no cell, in a series of equals, ever serves as a path of discharge for any other cell.**

Such is not true, we may remark, for cells connected in *parallel*—that is, with similar terminals joined together. For although each cell still has an inherently limited discharge capacity, it affects only the current developed *in that cell*. When the individual currents combine in the common discharge path there is no further limitation (since we have assumed the path to be *perfect*); thus the conductor current will be the sum of the individual cell currents. Multiple cells in parallel,

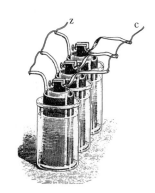

---

* Expressed in modern units, a single Bunsen cell could typically supply a short-circuit current of about 10–12 amperes. That is enough to develop a magnetic influence about equal to that of the earth's, at the center of a single conducting loop of 10 centimeters diameter. Alternatively, it would be sufficient to sustain a length of thin platinum wire at a bright red heat.

** Note that the same conclusion cannot be drawn if the cells are *unequal.*

then, combine their individual discharge capacities, as the same cells connected in series do not. Cells in parallel are capable of producing *any* amount of current (in a perfect conductor); it is only necessary to supply a sufficiently great number of them.

Lines of force and the theory of "poles"

The remarkably legible patterns of lines of force carry extraordinary suggestive powers, as we have abundantly seen. In Faraday's thinking they have served both as the instances leading to, and the primary expository representatives of, his powerful vision of natural powers interconnected and even unified in a continuous and shapely network of mutual action. On the other hand, lines of force are by no means the exclusive property of Faraday's thought nor do they rule out more conventional views. In particular, lines of force are *fully compatible* with the existence of isolated poles acting at a distance according to an inverse-square law.

Combative writers sometimes point out this broad compatibility as though it were somehow a reproach to Faraday. That is of course non-sense. Faraday, at least by now, knows perfectly well that the lines may be constructed by mathematical artifice—see paragraph 3238 below. The question is not whether lines of force exist *as symbols*—for that they do, on anyone's theory—but rather, what *particular patterns of lines*, experimentally obtained and interpreted, reveal about the modes of action and mutual relation of electric and magnetic powers.

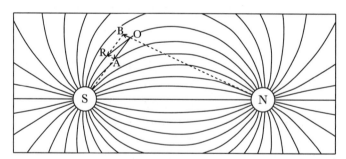

The drawing above shows how to construct the lines of force surrounding a pair of independent "poles", assuming only the inverse-square law and Newton's method of composition of forces.* Let N and S be the poles of a magnet, and O any point in the surrounding region. If the N-pole of a magnetic needle is placed at O it will be attracted by S along the line OA, and repelled by N along the line OB. By the inverse-square law, the force along OB will be inversely proportional to the square of NO, while the force along OA will be inversely proportional to the square of SO. Thus if, as here, the distance NO is

---

* See Newton's *Principia*, Corollary I to the Laws of Motion.

twice as great as distance SO, the force along OB will be one-quarter of the force along OA.

Newton's method of composition represents the magnitudes of these forces by the *lengths* OA and OB, respectively. Then, completing the parallelogram OARB, the diagonal OR will represent, in both magnitude and direction, the composite magnetic force exerted at O—its direction, in particular, is the direction in which the hypothetical magnetic needle would point. The construction may then be repeated at the point R, and at subsequent points towards both S and N, until a complete curve—and ultimately a set of curves like those shown—is completed.*

The construction is gratifying; and *if* a magnet has independent centers of action such as are represented by N and S, then certainly lines of force should be disposed about it in the very pattern that was constructed. But are they, in fact? There is, no doubt, much general resemblance between the synthetic pattern opposite and the results of actual experiment, as reproduced below in one of Faraday's figures. But as synthetic processes often do, the preceding construction misses some details that turn out to be disproportionately significant—a fact Faraday points out vigorously in paragraph 3238 below. Look, for example, at the lines near the central regions of the *real* bar-magnet; they do not appear to point toward the "poles" as in the synthetic pattern. And of course there will be a profound difference between the synthetic and the experimental representations of the magnet's

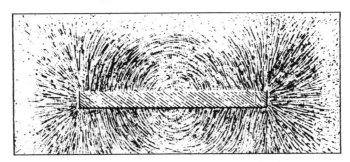

interior—where the moving wire has disclosed parallel continuations of the external lines of force, while the lines which the polar hypothesis constructs are presumed to begin and end at the hypothetical N and S poles, respectively.

---

* Note that this procedure treats each diagonal OR as a chord of the curve being sought. Strictly, however, each curve ought to be *tangent* to the diagonal OR at every point O. But as Newton shows in the *Principia*, Lemma VI, the error introduced by this approximation becomes ever smaller as OR becomes smaller; so for any desired accuracy it is only necessary to draw the parallelogram sufficiently small. Similarly, when actually tracing a line of force with a compass needle, one must use a sufficiently small needle.

"Ohm's Law"—or perhaps not

I stated Ohm's Law towards the end of the introduction to the Second Series, but only to dismiss it from further consideration. Not only was the Law in all likelihood unknown to Faraday at the time, but it implicitly drew upon the very fluid-flow imagery that he had consistently endeavored to avoid. Ohm's Law was, in short, *no part* of Faraday's vision of electric current—so I argued then.

However, an appreciative reference that appears early in the present Series might suggest that now—nearly twenty years later—Faraday has not only become acquainted with Ohm's Law but has learned to value it:

> 3180. The connexions for this galvanometer ... were all of copper rod or wire 0.2 of an inch in diameter; but even with wires of this thickness the extent of the conductors should not be made more than is necessary; for the increase from 6 to 8, 10 or 12 feet in length, makes a considerable difference at the galvanometer, when electric currents, low in intensity, are to be measured. *It is most beautiful to observe in such cases the application of Ohm's law of currents to the effects produced...* [italics added]

But though he invokes Ohm's Law, Faraday seems to have reduced its content to a most ordinary and common-sense relation, namely that *when resistance increases (or decreases), current decreases (or increases).* In fact, as I pointed out in the Second Series' introduction, Ohm's Law states far more than that. Specifically, it holds that the current in a circuit is inversely proportional to the resistance of the conducting path and directly proportional to the hypothetical "electromotive force" supposed to be active in the circuit. Stated symbolically,

$$Strength\ of\ current \propto \frac{electromotive\ force}{total\ electric\ resistance}$$

But in the passage quoted, Faraday mentions only the "considerable difference at the galvanometer" (indicating a *decrease in current*) when the circuit wires are lengthened (thereby increasing their resistance).* He does not mention having observed any *proportionality* in the relation between resistance and current. Finally, he makes no observations relating to *electromotive force*—that hypothetical *force which moves*

---

* That he indeed understands the *current decrease* to be the result of *resistance increase* is confirmed by his emphatic "even with wires of this thickness," since thicker wires would tend to keep resistance—or "obstruction" as he sometimes calls it—low.

*electricity about a circuit.** That is of course not surprising, since Faraday has long regarded the notion of electricity "moving" through its circuit as a mere figure of speech. Since the Eleventh and Twelfth Series, he has understood that what "moves" in a current is not a *substance* called electricity but a *wave* of tension and collapse. Thus an entire dimension of Ohm's Law—electromotive force—has to be virtually meaningless to Faraday.

But electromotive force is *essential* to Ohm's Law, which without that concept dwindles to a statement about resistance and current that is merely somewhat more precise than the common understanding— that *current decreases when resistance increases.* I do not believe that Faraday has, in any important sense, "observed the application of Ohm's law of currents." Or perhaps it would be more accurate to say that he has observed as much of Ohm's Law as makes sense to him. In any case, despite the allusion in paragraph 3180, Faraday's thinking seems no more hospitable to Ohmic tenets now than it was in the Second Series.

---

* In the Diary, too, Faraday confines his attention to the effect on galvanometer throw when wire length is changed.

TWENTY-NINTH SERIES.[1]

§ 35. *On the employment of the Induced Magneto-electric Current as a test and measure of Magnetic Forces.*

Received December 31, 1851,—Read March 25 and April 1, 1852.

3177. THE proposition which I have made to use the induced magneto-electric current as an experimental indication of the presence, direction and amount of magnetic forces (3074.), makes it requisite that I should also clearly demonstrate the principles and develope the practice necessary for such a purpose; and especially that I should prove that the amount of current induced is precisely proportionate to the amount of lines of magnetic force intersected by the moving wire, in which the electric current is generated and appears (3082. 3109.). The proof already given is, I think, sufficient for those who may repeat the experiments; but in order to accumulate evidence, as is indeed but proper in the first announcement of such a proposition, I proceeded to experiment with the magnetic power of the earth, which presents us with a field of action, not rapidly varying in force with the distance, as in the case of small magnets, but one which for a given place may be considered as uniform in power and direction; for if a room be cleared of all common magnets, then the terrestrial lines of magnetic force which pass through it, have one common direction, being that of the dip, as indicated by a free needle or other means, and are in every part in equal proportion or quantity, *i. e.* have equal power. Now the force being the same everywhere, the proportion of it to the current evolved in the moving wire is then perhaps more simply and directly determined, than in the case where, a small magnet being employed, the force rapidly changes in amount with the distance.

[1] Philosophical Transactions, 1852, p. 137.

3177. *the amount of current induced...*: Better, the *quantity of electricity* represented by the induced current; for it is that quantity which the galvanometer indicates (paragraph 366).

prove ... proof: He means experimental demonstration, not theoretical deduction.

¶ i. *Galvanometer.*

3178. For such experimental results as I now propose to give, I must refer to the galvanometer employed and the precautions requisite for its proper use. The instrument has been already described in principle (3123.), and a figure of the conductor which surrounds the needles, given. This conductor may be considered as a square copper bar, 0.2 of an inch in thickness, which passes twice round the plane of vibration of each of the needles forming the astatic combination, and then is continued outwards and terminates in two descending portions, which are intended to dip into cups of mercury. As both the needles are within the convolutions of this bar, an indicating bristle or fine wire of copper is fixed parallel to, and above them upon the same axis, and this, in travelling over the usual graduated circle, shows the place and the extent of vibration or swing of the needles below. The suspension is by cocoon silk, and in other respects the instrument is like a good ordinary galvanometer.

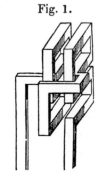

Fig. 1.

3179. It is highly important that the bar of copper about the needles should be perfectly clean. The vertical zero plane should, according to the construction, be midway between the two vertical coils of the bar, fig. 1; instead of which the needle at first pointed to the one side or the other, being evidently attracted by the upright portions of the bar. I at first feared that the copper was magnetic, but upon cleaning the surface carefully with fine sandpaper, I was able to remove this effect, due no doubt to iron communicated by handling or the use of tools, and

---

3178. *the needles forming the astatic combination*: The First Series' introduction described the dual-needle galvanometer on page 25. When the needles are equally and oppositely magnetized, the pair are jointly immune to extraneous magnetic influences and do not "point" as a single magnetic needle would do. Hence the name astatic—*without station*; the combination has no tendency towards a particular direction or alignment. But in paragraph 3181 Faraday will reveal that his galvanometer is not *perfectly* astatic.

the conductor which surrounds the needles: Faraday's Figure 1 depicts this conductor in glorious perspective, but the side view adapted here from the 28th Series will show its operation more clearly. Current carried by the conductor from A to B will run clockwise around the lower needle and counter-clockwise around the upper, thus producing opposite magnetic influences. But since the needles, too, are oppositely directed, both will be urged in the *same* direction.

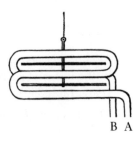

B A

the needle then stood truly in a plane equidistant from the two coils, when that plane corresponded with the magnetic meridian.

3180. The connexions for this galvanometer (3123. 3133.) were all of copper rod or wire 0.2 of an inch in diameter; but even with wires of this thickness the extent of the conductors should not be made more than is necessary; for the increase from 6 to 8, 10 or 12 feet in length, makes a considerable difference at the galvanometer, when electric currents, low in intensity, are to be measured. It is most beautiful to observe in such cases the application of Ohm's law of currents to the effects produced. When the connexions were extended to a distance, straight lengths of wire with dropping ends were provided, and these by dipping into cups of mercury completed the connexion and circuit. The cups consisted of cavities turned in flat pieces of wood. The ends of the connecting rods and of the galvanometer bar were first tinned, and then amalgamated; after which their contact with the mercury was both ready and certain. Even where connexion had to be made by contact of the solid substances, I found it very convenient and certain to tin and amalgamate the ends of the conductors, wiping off the excess of mercury. The surfaces thus prepared are always ready for a good and perfect contact.

3181. When the needle has taken up its position under the earth's influence, and the copper coil is adjusted to it, the needle ought to stand at true zero, and appears so to do. When that is really the case, equal forces applied in succession on opposite sides of the needle (by two contrary currents through the coil for instance) ought to deflect the needle *equally* on both sides, and they do so. But sometimes, when the needle appears to stand at zero, it may not be truly in the magnetic meridian; for a little torsion in the suspension thread, even though it be only 10° or 15° (for an indifferent needle), and quite insensible to the eye looking at the magnetic needle, does deflect it, and then the force which opposes the swing of the needle, and which stops and returns the needle towards zero (being due both to the torsion and the earth's force), is not equal on the two sides, and the consequence is, that the extent of swing in the two directions is not equal for equal powers, but is greater on one side than the other.

3180. *the application of Ohm's law of currents to the effects produced*: Faraday's laconic reference to Ohm's Law turns out to be less consequential than one might suppose; see the introduction to the present Series.

3181. *a little torsion in the suspension thread*: Though nominally "astatic," Faraday's galvanometer is not wholly immune to the earth's magnetic influence, and misalignment of the needle in the magnetic meridian might mask an inadvertent twist of the suspension fiber.

3182. I have not yet seen a galvanometer which has an adjustment for the torsion of the suspending filament. Also, there may be other causes, as the presence about a room, in its walls and other places, of unknown masses of iron, which may render the forces on opposite sides of the instrument zero unequal in a slight degree; for these reasons it is better to make *double observations*. All the phænomena we have to deal with, present effects in two contrary directions. If a loop pass over the pole of a magnet (3133.), it produces a swing in one direction; if it be taken away, the swing is in the other direction; if the rectangles and rings to be described (3192.) be rotated one way, they produce one current; if the contrary way, the other and contrary current is produced. I have therefore, always, in measuring the power of a pole or the effect of a revolving intersecting wire, made many observations in both directions, either alternately or irregularly; have then ascertained the average of those on the one side, and also on the other (which have differed in different cases from $\frac{1}{50}$th to $\frac{1}{300}$th part); and have then taken the mean of these averages as the expression of the power of the induced electric current, or of the magnetic forces inducing it.

3183. Care must be taken as to the position of the instrument and apparatus connected with it, in relation to a fire or sources of different temperatures, that parts which can generate thermo-currents may not become warmed or cooled in different degrees. The instrument is exceedingly sensible to thermo-electric currents; the accidental falling of a sun-beam upon one of two connecting mercury cups for a few moments disturbed the indications and rendered them useless for some time.

*   *   *

3185. It must be well understood, that, in all the observations made with this instrument, the *swing* is observed and counted as the effect produced, unless otherwise expressed. A constant current in an instrument will give a constant and continued deflection, but such is not the case here. The currents observed are for short periods, and they give,

---

3182. *a galvanometer which has an adjustment for the torsion of the suspending filament*: Such an adjustment might remedy the bias noted in the previous paragraph. Since that is not available, he prescribes a practice of *double measurements* for eliminating the error.

3185. *A constant current in an instrument will give a constant and continued deflection*: This is because a constant current will exert a constant magnetic influence, as discussed in the introduction to the First Series.

as it were, a blow or push to the needle, the effect of which, in swinging the needle, continues to increase the extent of the deflection long after the current is over. Nevertheless the extent of the swing is dependent on the electricity which passed in that brief current; and, as the experiments seem to indicate, is simply proportional to it, whether the electricity pass in a longer or a shorter time (3104.), and notwithstanding the comparative variability of the current in strength during the time of its continuance.

*   *   *

¶ ii. *Revolving Rectangles and Rings.*

3192. The form of moving wire which I have adopted for experiments with the magnetic forces of the earth (3177.), is either that of a rectangle or a ring. If a wire rectangle (fig. 3) be placed in a plane, perpendicular to the dip and then turned once round the axis Fig. 3. *a b*, the two parts *c d* and *e f* will twice intersect the lines of magnetic force within the area *c e d f*. In the first 180° of revolution the contrary direction in which the two parts *c d* and *e f* intersect those lines, will cause them to conspire in producing one current, tending to run round the rectangle (161.) in a given direction; in the following 180° of revolution they will combine in their effect to produce a contrary current; so that if the first current is from *d* by *c e* and *f* to *d* again, the second will be from *d* by *f e* and *c* to *d*. If the rectangle, instead of being closed, be open at *b*, and the ends there produced be connected with a commutator, which changes sides when the rectangle comes into the plane perpendicular to the dip, *i. e.*

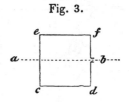

Fig. 3.

---

the extent of the swing is dependent on the electricity which passed in that brief current: More precisely, it depends on the *quantity* of electricity which passed.

3192. I have omitted a note at the head of this section, in which Faraday comments on some related experiments made by Nobili.

the contrary direction in which the two parts c d and e f intersect those lines, will cause them to conspire in producing one current…: Since the loop rotates about axis *a b*, the segments *c d* and *e f* traverse the lines of force in opposite directions. If then the current induced in *c d* is from *c* to *d*, the current induced in *e f* will be from *f* to *e*. Both currents have the same direction, as regards circulation about the loop.

a commutator, which changes sides when the rectangle comes into the plane…: To commute (Latin: *commutare*) is to change or, as here, to *exchange*. Faraday will explain the commutator's operation in the next paragraph.

at every half revolution, then these successive currents can be gathered up and sent on to the galvanometer to be measured. The parts *ce* and *df* of the rectangle may be looked upon simply as conductors; for as they do not in their motion intersect any of the lines of force, so they do not tend to produce any current.

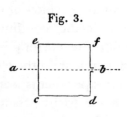

Fig. 3.

3193. The apparatus which carries these rectangles, and is also the commutator for changing the induced currents, consists of two uprights, fixed on a wooden stand, and carrying above a wooden horizontal axle, one end of which is furnished with a handle, whilst the other projects,

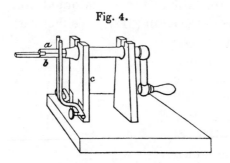

Fig. 4.

and is shaped as in fig. 4. It may there be seen, that two semi-cylindrical plates of copper *a b* are fixed on the axle, forming a cylinder round it, except that they do not touch each other at their edges, which therefore leave two lines of separation on opposite sides of the axle. Two strong copper rods, 0.2 of an inch in diameter, are fixed to the lower part of the upright *c*, terminating there in sockets with screws for the purpose of receiving the ends of the rods proceeding from the galvanometer cups (3180.); in the other direction the rods rise up parallel to each other, and being perfectly straight, press strongly against the curved plates of the commutator on opposite sides; the consequence is, that, whenever in the rotation of the axle, the lines of separation between the commutator plates arrive at and pass the horizontal plane, their contact with these bearing rods is changed, and consequently the direction of the current proceeding

---

3193. *whenever … the lines of separation between the commutator plates arrive at and pass the horizontal plane, their contact with these bearing rods is changed…*: If the connection between rotating loop and galvanometer had been *continuous* (as with the "slip rings" of paragraph 3084), the current would have reversed every half revolution, for an overall null deflection of the galvanometer. But here, the commutator transposes the galvanometer connections at the same moment as the induced current reverses; consequently, the successive discharges will sustain and accumulate their effects on the galvanometer.

The commutator exchanges contacts when the division between its plates is *horizontal*; ideally, that exchange should coincide with the loop becoming *perpendicular to the lines of force*, for it is in the perpendicular position that the loop's cutting wires *reverse their motion* across the lines. Faraday will discuss adjustment of the loop in the next paragraph.

from these plates to the rods, and so on to the galvanometer, is changed also. The other or outer ends of the commutator plates are tinned, for the purpose of being connected by soldering to the ends of any rectangle or ring which is to be subjected to experiment.

3194. The rectangle itself is tied on to a slight wooden cross (fig. 5), which has a socket on one arm that slides on to and over the part of the wooden axle projecting beyond the com-
mutator plates, so that it shall revolve with the axle. A small copper rod forms a continuation of that part of the frame which occupies the place of axle, and the end of this rod enters into a hole in a separate upright, serving to support and steady the rectangle and its frame. The frames are of two or three sizes, so as to receive rectangles of 12 inches in the side, or even larger, up to 36 inches square. The rectangle is adjusted in its place, so that it shall be in the horizontal plane when the division between the commutator plates is in the same plane, and then its extremities are soldered to the two commutator plates, one to each. It is now evident, that when dealing with the lines of force of the earth, or any other lines, the axle has only to be turned until the upright copper rods touch on each side at the separation of the commutator plates, and then the instrument adjusted in position, so that the plane of the ring or rectangle is perpendicular to the direction of the lines of force which are to be examined, and then any revolution of the commutator and intersecting wire will produce the maximum current which such wire and such magnetic force can produce. The lines of terrestrial magnetic force are inclined at an angle of 69° to the horizontal plane.

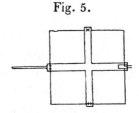

Fig. 5.

3194. *The rectangle itself is tied on to a slight wooden cross … which has a socket on one arm*: The large square in Figure 5 denotes the wire rectangle; the socket is shown as a hole drilled into the right-hand extremity of the wooden frame.

The rectangle … shall be in the horizontal plane when the division between the commutator plates is in the same plane…: So adjusted, the commutator exchanges connections when the rectangular loop is parallel to the base of the device.

the instrument adjusted … so that the plane of the ring or rectangle is perpendicular to the direction of the lines of force…: Ideally, the instrument need only be *tilted* so that its base, previously horizontal, is perpendicular to the lines. Then commutation would also take place when the loop is perpendicular to the terrestrial lines of force, as is desired.

The lines of terrestrial magnetic force are inclined at an angle of 69° to the horizontal: Therefore the proper angle at which to tilt the instrument would be 21° from the horizontal—the difference between 90° and 69°. In the next sentence, however, Faraday will explain why he neglects this step for the present.

As, however, only comparative results were required, the instrument was, in all the ensuing experiments, placed in the horizontal plane, with the axis of rotation perpendicular to the plane of the magnetic meridian; under which circumstances no cause of error or variation was introduced into the results. As no extra magnet was employed, the commutator was placed within 3 feet of the galvanometer, so that two pieces of copper wire 3 feet long and 0.2 of an inch in thickness,

sufficed to complete the communication. One end of each of these dipped into the galvanometer mercury cups, the other ends were tinned, amalgamated, introduced into the sockets of the commutator rods (3193.), and secured by the pinching screw (fig. 4).

Fig. 4.

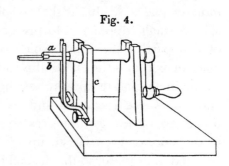

3195. When a given length of wire is to be disposed of in the form best suited to produce the maximum effect, then the circumstances to be considered are contrary for the case of a loop to be employed with a small magnet (39. 3184.), and a rectangle or other formed loop to be employed with the lines of terrestrial force. In the case of the small magnet, *all* the lines of force belonging to it are inclosed by the loop;

As ... only comparative results were required, the instrument was ... placed in the horizontal plane...: With a magnetic dip of 69°, the commutator should ideally reverse when the plane of the loop is 21° from the horizontal. Placed on a horizontal workbench, the apparatus actually reverses when the loop is horizontal, so that current will be *diminished* during 21° of each half-rotation, instead of being augmented by the same amount. But the percent of attenuation will be the same for any loop; so it does not prevent a comparison of different loops with one another—which is what Faraday intends.

As no extra magnet was employed...: Since no magnets are present, he can locate the galvanometer within 3 feet of the loop. Had he used an auxiliary magnet to augment the earth's effect on the loop, its influence might have affected the galvanometer, requiring him to place it at a greater distance.

3195. *In the case of the small magnet,* all *the lines of force belonging to it are inclosed by the loop...*: Thus if two loops A, B encircle a magnet, they both intersect the same number of lines of force; so that the additional wire used in forming the larger loop accomplishes nothing. But if that wire is instead made into a loop of multiple turns, as C, the same lines will be cut a corresponding number of times, and the total current will increase proportionally to the total length of wire in C. But among the

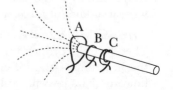

and if the wire is so long that it can be formed into a loop of two or more convolutions, and yet pass over the pole, then twice or many times the electricity will be evolved that a single loop can produce (36.). In the case of the earth's force, the contrary result is true; for as in circles, squares, similar rectangles, &c. the areas inclosed are as the squares of the periphery, and the lines of force intersected are as the areas, it is much better to arrange a given wire in one simple circuit than in two or more convolutions. Twelve feet of wire in one square intersects in one revolution the lines of force passing through an area of nine square feet, whilst if arranged in a triple circuit, about a square of one foot area, it will only intersect the lines due to that area; and it is thrice as advantageous to intersect the lines within nine square feet once, as it is to intersect those of one square foot three times.

3196. A square was prepared, containing 4 feet in length of copper wire 0.05 of an inch in diameter; it inclosed one square foot of area, and was mounted on the commutator and connected in the manner already described (3194.). Six revolutions of it produced a swing deflection of 14° or 15°, and twelve quick revolutions were possible within the required time (3104.). The results of *quick* and *slow* revolutions were first compared. Six slow revolutions gave as the average of several experiments 15.5° swing. Six moderate revolutions gave also an average of 15.5°; six quick revolutions gave an average of 15.66°. At another time twelve moderate revolutions gave an average of 28.75°, and twelve quick revolutions gave an average of 31.33° swing. As before explained (3186.), the probable reason why the quick revolutions gave a larger result than the moderate or slow revolutions is, that in slow time the later revolutions are performed at a period when the needle is so far from parallel with the copper coil of the galvanometer, that the

---

earth's lines of force, which extend far beyond any practical loop, a larger loop A would intersect more lines than a smaller loop B, proportional to its *area*—that is, to the *square* of its circumference.

thrice as advantageous to intersect the lines within nine square feet once, as it is to intersect those of one square foot three times: A square 1 foot on a side consumes 4 feet of wire and encloses 1 square foot; so that if encircled three times it will consume 12 feet of wire and produce three times as much current as a single turn of 1 square foot. But a square 3 feet on a side also consumes 12 feet of wire; and it encloses nine square feet, which will produce *nine* times as much current as a single turn of 1 square foot. Nine is "thrice as advantageous" as three.

3196. *the probable reason why the quick revolutions gave a larger result than the moderate or slow revolutions...*: Faraday cites an omitted paragraph. Refer instead to his substantially identical account at paragraph 3104 in the previous Series.

impulses due to them are less effectually exerted. Hence a small or moderate number of revolutions and a quick motion is best. The difference in the extreme case is less than might have been expected, and shows that there is no practical objection in this respect to the method proposed of experimenting with the lines of magnetic force.

3197. In order to obtain for the present an expression of the power of the earth's magnetic force by this rectangle, observations were made on both sides of zero, as already recommended (3182.). Nine moderately quick direct revolutions (*i. e.* as the hands of the clock) gave as the average of many experiments 23.87°, and nine reverse revolutions gave 23.37°; the mean of these is 23.62° for the nine revolutions of the rectangle, and therefore 2.624° per revolution. Now the six quick revolutions (3196.) gave 15.66°, which is 2.61° per revolution, and the twelve quick revolutions gave 31.33°, which is also 2.61° per revolution; and these results of 2.624°, 2.61°, and 2.61°, are very much in accordance, and give great confidence in this method of investigating magnetic forces.

3198. A rectangle was prepared of the same length (4 feet) of the same wire, but the sides were respectively 8 and 16 inches (fig. 6), so that when revolving the intersecting parts should be only 8 inches in length instead of 12. The area of the rectangle was necessarily 128 square inches instead of 144. This rectangle showed the same difference of quick and slow rotations as before (3196.).

Fig. 6.   **Fig. 7.**

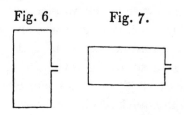

When nine direct revolutions were made, the result was 20.87° swing. Nine reverse revolutions gave an average of 20.25° swing; the mean is 20.56°, or 2.284° per revolution.

3197. *In order to obtain … an expression of the power of the earth's magnetic force…*: As he earlier gauged the relative power of a permanent magnet by a moving loop (paragraph 3086), so here he assesses the strength of the earth's magnetic influence by the same principle. First, using the 1-foot square loop (144 square inches), he obtains galvanometer deflections of about 2.6° per rotation.

these results … are very much in accordance, and give great confidence in this method: He obtains essentially the same deflection per revolution whether making nine moderate turns, six rapid turns, or twelve rapid turns—as he should, since each revolution cuts the same number of lines of force.

3198. He now prepares two loops, each having the same 4-foot circumference as before, but formed into 8-inch by 16-inch rectangles enclosing areas of only 128 square inches. One is rotated parallel to the 8-inch sides and yields an average deflection of 2.284° per revolution. The second is rotated parallel to the 16-inch sides, yielding 2.34° per revolution.

A third rectangle was prepared of the same length and kind of wire, the sides of which were respectively 8 and 16 inches long (fig. 7), but now so revolved that the intersecting parts were 16 inches, or twice as long as before; the area of the rectangle remained the same, *i. e.* 128 inches. The like effect of slow and quick revolutions appeared as in the former cases (3196. 3198.). Nine direct revolutions gave as the average effect 20.75°; and nine reverse revolutions produced 21.375°; the mean is 21.06°, or 2.34° per revolution.

3199. Now 2.34° is so near to 2.284°, that they may in the present state of the investigation be considered the same. The little difference that is evident, was, I suspect, occasioned by centrifugal power throwing out the middle of the longer intersecting parts during the revolution. The coincidence of the numbers shows, that the variation in the arrangement of the rectangle and in the length of the parts of the wires intersecting the lines of magnetic force, have had no influence in altering the result, which, being dependent alone on the number of lines of force intersected, is the same for both; for the area of the rectangles is the same. This is still further shown by comparing the results with those obtained with the square. The area in that case was 144 square inches, and the effect per revolution 2.61°. With the long rectangles the area is 128 square inches, and the mean of the two results is 2.312° per revolution. Now 144 square inches is to 128 square inches as 2.61° is to 2.32°; a result so near to 2.312° that it may be here considered as the same; proving that the electric current induced is directly as the lines of magnetic force intersected by the moving wire.

3200. It may also be perceived that no difference is produced when the lines of force are chiefly disposed in the direction of the motion of the wire, or else, chiefly in the direction of the length of the wire;

3199. The two rectangles produce very nearly equal deflections per revolution, as they should since they enclose equal areas and therefore intersect equal numbers of lines. The rectangle of Figure 7 produces a slightly larger deflection, which he explains as follows: Since this loop has its longer dimension parallel to the axis, its outer conductors span a longer distance between their radial supports and are therefore more susceptible to being bent outward by centrifugal force. This would slightly increase both the loop area and the number of lines intersected, thereby increasing the galvanometer deflection.

Prior to publication, Faraday added a note to this paragraph applying a correction to the galvanometer deflections. These, it had been pointed out to him, depart increasingly from proportionality at increasing angles. The correction is small and only corroborates Faraday's original inferences; I have therefore omitted it here and elsewhere.

i. e. no alterations are occasioned by variations in the *velocity* of the motion, or of the length of the wire, provided the amount of lines of magnetic force intersected remains the same.

3201. Having a square on the frame 12 inches in the side but consisting of copper wire 0.1 of an inch in thickness, I obtained the average result of many observations for one, two, three, four and five revolutions of the wire.

> One revolution gave   7°   equal to 7°   per revolution.
> Two revolutions gave   13.875 equal to 6.937 per revolution.
> Three revolutions gave 21.075 equal to 7.025 per revolution.
> Four revolutions gave   28.637 equal to 7.159 per revolution.
> Five revolutions gave   37.637 equal to 7.527 per revolution.

These results are exceedingly close upon each other, especially for the first 30°, and confirm several of the conclusions before drawn (3189. 3199.) as to the indications of the instrument, the amount of the curves, &c.

3202. At another time I compared the effect of equable revolutions with other revolutions very irregular in their rates, the motion being sometimes even backwards and continually differing in degree by fits and starts, yet always so that within the proper time a certain number of revolutions should have been completed. The rectangle was of wire 0.2 of an inch thick; the mean of many experiments, which were closely alike in their results, gave for two smooth, equable revolutions, 17.5°, and also for two irregular uncertain revolutions the same amount of 17.5°.

*   *   *

§ 36. *On the amount and general disposition* of *the Forces of a Magnet when associated with other magnets.*

3215. Prior to further progress in the experimental development by a moving wire of the disposition of the lines of magnetic force

---

3201. *These results are exceedingly close upon each other, especially for the first 30°...*: This remark could be misleading if it seems to suggest that readings are more consistent *because* they are smaller than 30°. While it is true, as I mentioned in the previous comment, that galvanometer error increases at larger angles, that error would account for less than half the discrepancy found here. The major consideration is probably the one Faraday suspected in paragraph 3199—distortion of the wire by centrifugal force, with consequent increase in loop area.

pertaining to a magnet, or of the physical nature of this power and its possible mode of action at a distance, it became quite essential to know what change, if any, took place in the amount of force possessed by a perfect magnet, when subjected to other magnets in favourable or adverse positions; and how the forces combined together, or were disposed of, *i. e.* generally, and in relation to the principle already asserted and I think proved, that the power is in every case definite under those different conditions. The representation of the magnetic power by *lines of force* (3074.), and the employment of the moving wire as a test of the force (3076.), will I think assist much in this investigation.

3216. For such a purpose an ordinary magnet is a very irregular and imperfect source of power. It not only, when magnetized to a given degree, is apt by slight circumstances to have its magnetic power diminished or exalted, in a manner which may be considered for the time, permanent; but if placed in adverse or favourable relations to other magnets, frequently admits of a considerable temporary diminution or increase of its power externally, which change disappears as soon as it is removed from the neighbourhood of the dominant magnet. These changes produce corresponding effects upon the moving wire, and they render any magnet subject to them unfit for investigation in relation to definite power. Unchangeable magnets are, therefore, required, and these are best obtained, as is well known, by selecting good steel for the bars, and then making them exceedingly hard; I therefore procured some plates of thin steel twelve inches long

---

3215. *its possible mode of action at a distance...*:   Surely Faraday has been continually arguing *against* "action at a distance"; why does he call it "possible" here? But in fact, Faraday has never denied that magnets have an influence in places that are distant from their iron substance. What he rejects is magnetism's (or electricity's) alleged ability to act distantly *independent of a medium*. The *fact* of action is clear to all; Faraday wishes to inquire into the physical *mode* by which the action is exerted.

what change, if any, took place in the amount of force possessed by a perfect magnet, when subjected to other magnets in favourable or adverse positions...:   Previously, hard steel magnets have been treated as virtually immutable. Now with the aid of the moving wire, that assumption can be put to precise experimental test.

3216. *making them exceedingly hard*:   Steel becomes very hard and highly retentive of magnetism if, after being heated to a bright red temperature, it is suddenly cooled by immersion in water, oil, or mercury. But that process, which is called *quenching*, leaves the metal in a brittle condition. The brittleness can be alleviated by subsequent *tempering*, a process of gentle reheating; but magnets generally show greatest retentive power when they are simply left in their quenched state.

and one inch broad, and making them as hard as I could, afterwards magnetized them very carefully and regularly, by two powerful steel bar-magnets, shook them together in different and adverse positions for a little while, and then examined the direction of the forces by iron filings. Small cracks and irregularities were in this way detected in several of them; but two which were very regular in the disposition of their forces were selected for further experiment, and may be distinguished as the subjected magnets D and E.

3217. These two magnets were examined by the moving loop precisely in the manner before described (3138.), *i. e.* by passing the loop over one of the poles, observing the swing, removing it, and again observing the swing and taking an average of many results; the process was performed over both poles at different times. The loop contained 7.25 inches in length of copper wire 0.1 of an inch in diameter, and was of course employed in all the following comparative experiments; the distance of the loop and magnets from the galvanometer was 9 feet. For one passage over the pole either on or off, *i. e.* for one intersection of the lines of force of the magnet D, the galvanometer deflection was 8.36°. For one intersection of the lines of force of the other bar E, the deflection was 8.78°. The two bars were then placed side by side with like poles together, and afterwards used as one magnet; their conjoined power was 16.3°, being only 0.84° less than the sum of the powers of the two when estimated separately. This indicates that the component magnets do affect, and in this position reduce, each other somewhat; but it also shows how small the effect is as compared with ordinary magnets (3222.).

3218. The compound magnet DE (3217.) was now subjected to the close action of another magnet, sometimes under adverse, and at other times under favourable conditions; and was examined by the loop as to the sum of its power (not the direction) under these circumstances. For this purpose it was fixed, and another magnet A brought near, and at times in contact with it, in the positions indicated by the figure 8; the loop in each case being applied many times to DE, that a correct average of its power might be procured. The dominant

---

3216, continued. *shook them together in different and adverse positions ... Small cracks and irregularities were in this way detected*: Under moderate shocks such as Faraday describes, excessively brittle samples develop visible cracks; or they may show irregularities in their magnetic patterns which indicate invisible cracks. In either case they are disqualified.

3218. *the positions indicated by the figure 8...*: The compound magnet DE was not labeled in Faraday's figure. I have supplied identification.

Fig. 8.

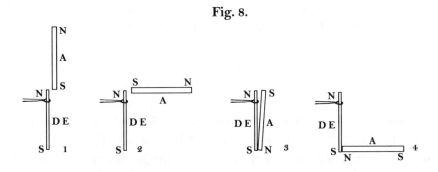

magnet A was much the stronger of the two, having the power indicated by a swing deflection of 25.74°.

3219. When the relative position of the magnets was as at 1, then the power of D E was 16.37°; when as at 2, the power was 16.4°; when as at 3, it was 18.75°; and when as at 4, it was 17.18°. All these positions are such as would tend to raise, by induction, the power of the magnet D E, and they do raise it above its first value, which was 16.3°; but it is seen at once how little the first and second positions elevate it, and even the third, which presents the most favourable conditions, only increases the power 2.45°, which falls again in the fourth position.

3220. Then the dominant magnet A was placed in the same positions, but with the ends reversed, so as to exert an adverse or depressing influence; and now the results with D E were as follows :—

Position 1	15.87°
Position 2	15.68
Position 3	15.37
Position 4	16.06

All these are a little below the original force of DE, or 16.3°, as they ought to be, and show how slightly this hard bar-magnet is affected.

3219. *All these positions are such as would tend to raise, by induction, the power of the magnet DE...*: In position 1, for example, the S-pole of magnet A would tend to induce an opposite N quality in the adjacent end of DE; such an induced characteristic would augment the power exerted at the N-pole that already exists there. But "to raise the power of the magnet" means intensifying *both* poles of DE. Can "induction" also exalt the S-pole, which is not even adjacent to magnet A?

3220. *All these are a little below the original force of DE ... as they ought to be...*: As a favorably aligned magnet A intensified magnet DE in the previous paragraph, so a reversed magnet depresses it. A similar explanation applies— with the same question about how "induction" is capable of affecting the distant pole.

3221. A soft iron bar, now applied in the first, second and third positions instead of the magnet A, raised DE to the following values respectively, 16.24°, 16.43°, and 18°.

3222. When an ordinary bar-magnet was employed instead of the hard magnet DE, great changes took place. Thus a bar B, corresponding to bar A in size and general character, was employed in place of the hard magnet. Alone, B had a power of 14.83°, but when associated adversely with A, as in position 3 (3218.), its power fell to 7.87°, being reduced nearly one-half. This loss was chiefly due to a coercion internally, and not to a permanent destruction of the state of magnet B; for when A was removed, B rose again to 13.06°. When B was laid for a few moments favourably on A and then removed, it was found that the latter had been raised to a permanent external action of 15.25°.

3223. A very hard steel bar 6 inches long, 0.5 broad and 0.1 in thickness, given to me by Dr. Scoresby, was magnetized and then found, by the use of the loop, to have a value at my galvanometer of 6.88° (3189.). It was submitted in position 2 to a compound bar-magnet like DE, having a power of 11.73°, or almost twice its own force, but

---

3221. *A soft iron bar ... raised DE to the following values...*: Even if we accept that a magnet can raise the power of DE by induction, why would the application of a *nonmagnetized* bar affect the power of DE? Note that the first of the cited values actually represents a *lowering*, not a raising of power; Faraday ignores this. Nevertheless, the third value is hardly less, and the second is even greater, than those resulting when magnet A was placed in the corresponding positions (paragraph 3219); thus a non-magnetized bar has here affected DE as much or more than a magnet did! Does the idea of *conduction of lines of force* help to explain this seeming paradox?

3222. *a coercion internally, [not] a permanent destruction...*: Coercion, here roughly synonymous with *domination*, often implies the presence of a continuing *opposition* which can be expected to reassert itself when the coercing force is removed. Thus pulling your chair away from the table is not coercion, because the chair stays put afterwards, with no further force required. But *lifting* the chair is coercion; for the moment you cease your upward effort the chair will fall back down again. In the Twelfth Series Faraday used the related term "coercitive force" in an electrical context, where it had a similar connotation of working against some opposing action (paragraph 1323). Here, as there, we should not confuse Faraday's expression with the hypothetical "coercive force" some magnetic theorists ascribed to hard steel.

the latter had been raised to a permanent external action: "the latter" here must denote ordinary magnet B, not magnet A. Thus ordinary (not *hard*) magnets can be affected either temporarily or permanently.

Fig. 8.

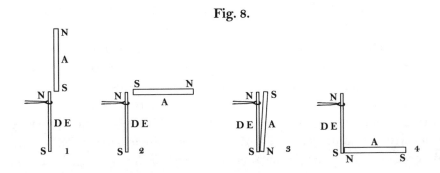

whether in the adverse or the favourable position, its power was not sensibly altered. When submitted in like manner to a 12-inch bar-magnet having a force of 40.21°, it was raised to 7.53°, or lowered to 5.87°, but here the dominant magnet had nearly six times the power of the one affected.

3224. The variability of soft steel magnets, both in respect of their *absolute* degree of excitation or charge, and also of the disposition of the force externally and internally, when their degree of excitation may for the time be considered as the same, is made very manifest by this mode of examination; and the results agree well with our former knowledge in this respect. It is equally manifest, that hard and invariable magnets are requisite for a correct and close investigation of the disposition and characters of the magnetic force. A common soft bar-magnet may be considered as an assemblage of hard and soft parts, disposed in a manner utterly uncertain; of which some parts take a much higher charge than others, and change less under the influence of external magnets; whilst, because of the presence of other parts within, acting as the keeper or submagnet, they may seem to undergo far greater changes than they really do. Hence the value of these hard and comparatively unchangeable magnets which Scoresby describes.

3225. From these and such results, it appears to me, that with perfect, unchangeable magnets, and using the term *line of force* as a mere representant of the force as before defined (3071. 3072.), the following useful conclusions may be drawn.

3224. *because of the presence of other parts within, acting as keeper or submagnet, they may seem to undergo far greater changes than they really do:* In paragraph 3120 Faraday compared inhomogeneities in the iron of a magnet to tiny "keepers" which concentrate magnetic force within the core, thereby rendering some of a magnet's power effectively lost to the external field. Any variability in this action will produce disproportionate changes in the amount of force disposed externally, even when the overall magnetic force is relatively constant.

3226. Lines of force of different magnets in favourable positions to each other coalesce.

3227. There is no increase of the total force of the lines by this coalescence; the section between the two associated poles gives the same sum of power as that of the section of the lines of the invariable magnet when it is alone (3217.). Under these circumstances there is, I think, no doubt that the external and internal forces of the same magnet have the same relation and are equivalent to each other, as was determined in a former part of these Researches (3117.); and that therefore the equatorial section, which represents the sum of forces or lines of forces passing through the magnet, remains also unchanged (3232.).

3228. In this case the analogy with two or more voltaic batteries associated end to end in one circuit is perfect. Probably some effect, correspondent to *intensity* in the case of the batteries, will be found to exist amongst the magnets.

3229. The increase of power upon a magnetic needle, or piece of soft iron placed between two opposite, favourable poles, is caused by concentration upon it of the lines which before were diffused, and not by the addition of the power represented by the lines of force of one

---

3226. *Lines ... coalesce*: Magnet A in itself conducts a number of lines proportional to 25.74 (paragraph 3218), while magnet DE conducts a number proportional to 16.3 (paragraph 3217). Then with the magnets placed as shown (position 1 of Figure 8), if their respective lines of force retained their individuality we would expect to find magnet DE, say, conducting not only its own lines, but most of magnet A's lines as well—because DE is a better conductor for those lines than is air. If so, examination with the moving wire would produce a deflection of as much as 42° (that is, the *sum* of the separate deflections) at DE. But in fact DE is found to retain its proper power, *nearly unchanged*: there are virtually *no additional lines* passing through DE. (Faraday does not bother to survey magnet A as it is not "hard" and would therefore be expected to change in any case.) Then either all of A's lines are passing through the *air*—hardly likely given the proximity of A and DE—or else the bulk of A's lines must have become *united* with an equal number of DE's lines, so that the unified lines pass through both magnets.

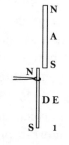

3227. *There is no increase of the total force of the lines by this coalescence*: The total number of lines passing through "hard" magnet DE, or through the space immediately adjacent to it, is essentially the same as when DE was alone. (As before, no such claim can be sustained for magnet A, as it is not "hard".)

3228. *two or more voltaic batteries associated end to end in one circuit...*: See the discussion, in the editor's introduction, of batteries connected "in series."

pole to that of the lines of force of the other. There is no more power represented by all the lines of force than before; and a line of force is not more powerful because it coalesces with a line of force of another magnet. In this respect the analogy with the voltaic pile is also perfect.

3230. A line of magnetic force being considered as a closed circuit (3117.), passes in its course through *both* the magnets, which are for the time placed so as to act on each other favourably, *i. e.* whose lines coincide and coalesce. Coalescence is not the addition of one line of force to another *in power*, but their union in one common circuit.

3231. A line of force may pass through many magnets before its circuit is complete; and these many magnets coincide as a case with that of a single magnet. If a thin bar-magnet 12 inches long be examined by filings (3235.), it will be found to present the well-known beautiful system of forces, perfectly simple in its arrangement. If it be broken in half, without being separated, and again examined, the manner in which, from the destruction of the continuity, the transmission of the force at the equator is interfered with, and many of the lines, which before were within are made to appear externally there, is at once evident, Plate III. fig. 6. Of those lines, which thus become external, some return back to the pole which is nearest to the new place, at which the lines issue into the air, making their circuit through only one of the halves of the magnet; whilst others

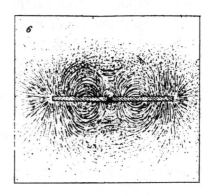

proceed onward by paths more or less curved into the second half of the magnet, keeping generally the direction or polarity which they had whilst within the magnet, and complete their circuit through the two.

3230. *Coalescence is not the addition of one line of force to another in power, but their union in one common circuit.* A lovely clarification of "coalescence." Note that if lines of force do indeed coalesce without increase or decrease of power, each line of force must necessarily be equivalent in power to any other line of force.

3231. *the well-known beautiful system of forces*: That is, a line pattern similar to the one shown here, reproduced from Figure 1 of Plate III, on page 553. It will acquire a memorable name in Faraday's essay *On the Physical Character of the Lines of Magnetic Force.* This paragraph also refers to Figures 6 and 7 from Plate III; the plate is printed in full on pages 553 and 554 below.

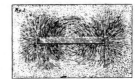

Gradually separating the two halves, and continuing to examine the course of the lines of force, it is beautiful to observe how more and more of the lines which issue from the two new terminations, turn back to the original extremities of the bar, fig. 7, and how the portion which makes a common circuit through the two halves diminishes,

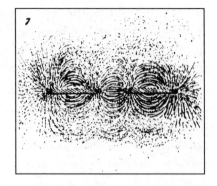

until the halves are entirely removed from each other's influence, and then become two separate and independent magnets. The same process may be repeated until there are many magnets in place of one.

3232. All this time the amount of lines of force is the same if the fragments of the bar preserve their full state of magnetism; *i. e.* the sum of lines of force in the equator of *either* of the new magnets is equal to the sum of lines of force in the equator of the original unbroken bar. I took a steel bar 12 inches long, 1 inch broad and 0.05 of an inch thick, made it very hard, and magnetized it to saturation by the use of soft iron cores and a helix; its power was 6.9°. I broke it into two pieces nearly in the middle, and found the power of these respectively 5.94° and 5.89°; indicating a fall not more than was to be expected considering the saturated state of the original magnet. When these halves were placed side by side, with like poles together as a compound magnet, they had a joint power of 11.06°, which, though it shows a mutual quelling influence, is not much below the sum of their powers

---

3231, contined. *how the portion which makes a common circuit through the two halves diminishes, until the halves ... become two separate and independent magnets*: Figures 6 and 7 show how, as the magnet halves are separated, the pattern of lines between them spreads out over a larger and larger space. When that spreading becomes sufficiently extensive, a line exiting from, say, the N-end of one half can be expected to *return to the S-end of that same half*, instead of crossing to the S-end of the adjacent half. When the lines return for the most part to their own halves, the two will have essentially become "separate and independent magnets."

3232. *a fall not more than was to be expected considering the saturated state of the original magnet*: Faraday has not previously introduced the idea of "saturation" of magnetic material; he will discuss it at length in his paper *On the Physical Character of the Lines of Magnetic Force*. For now, it must remain a question why, upon separating the two halves, they should exhibit *any* drop in magnetic power at all.

it shows a mutual quelling influence: That is, a mutual degradation of power. Such quelling is to be expected, since parallel magnets arranged with like poles

ascertained separately. All this is in perfect harmony with the voltaic battery, where lines of dynamic electric force are concerned. If, as is well known, we separate a battery of 20 pair of plates into two batteries of 10 pair, or 4 batteries of 5 pair, each of the smaller batteries can supply as much dynamic electricity as the original battery, provided no sensible obstruction be thrown into the course of the lines, *i. e.* the path of the current.

3233. When magnets are placed in an adverse position, as neither could add power to the other in the former case, so now each retains its own power; and the lines of magnetic force represent this condition accurately. Two magnets placed end to end with like poles together are in this relation; so also are they if placed with like poles together side by side. In the latter case the two acting as one compound magnet, give a system of lines of force equal to the sum of the two separately (3232.), minus the portion which, as in imperfect magnets, is either directed inwards by the softer parts or ceases to be excited altogether.

§ 37. *Delineation of Lines of Magnetic Force by iron filings.*

3234. It would be a voluntary and unnecessary abandonment of most valuable aid, if an experimentalist, who chooses to consider

---

juxtaposed are in an adverse position to one another. Faraday will identify some other "adverse" cases in the next paragraph.

each of the smaller batteries can supply as much dynamic electricity as the original battery: That is, each supplies a *current* equal to that supplied by the original battery under the stated conditions—as indicated, for example, by equal deflections of a compass needle. But how can a battery of only 5 pairs of plates equal the effect of 20 pairs? The question is discussed in the introduction.

3233. *Two magnets placed end to end with like poles together ... so also if placed with like poles together side by side*: Line patterns of magnets placed end-to-end are illustrated here. Faraday shows the side-by-side case in Plate III Figure 9, on page 553.

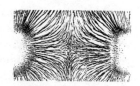

either directed inwards by the softer parts or ceases to be excited altogether: Two magnets placed parallel with their like poles adjacent are in mutually "adverse" positions. Each, then, may be expected to develop less than its full power—as earlier experiments (paragraphs 3220, 3222) showed, but did not explain. Faraday characterizes this reduction as a partial cessation of "excitation," which he will explain more fully in *On the Physical Character of the Lines of Magnetic Force.* Additionally, some lines of force may succeed in passing *between* the magnets instead of coursing around the regions exterior to both of them. Such lines will be lost to the exterior of the compound pair and correspond to lines "directed inwards by the softer parts" of imperfectly homogeneous magnets—as Faraday discussed in the previous Series (paragraph 3120).

magnetic power as represented by lines of magnetic force, were to deny himself the use of iron filings. By their employment he may make many conditions of the power, even in complicated casess, visible to the eye at once; may trace the varying direction of the lines of force and determine the relative polarity; may observe in which direction the power is increasing or diminishing; and in complex systems may determine the neutral points or places where there is neither polarity nor power, even when they occur in the midst of powerful magnets. By their use probable results may be seen at once, and many a valuable suggestion gained for future leading experiments.

3235. Nothing is simpler than to lay a magnet upon a table, place a flat piece of paper over it, and then sprinkling iron filings on the paper, to observe the forms they assume. Nevertheless, to obtain the best and most generally useful results, a few particular instructions may be desirable. The table on which the magnet is laid should be quite horizontal and steady. Means should be taken, by the use of thin boards or laths, or otherwise, to block up round the magnet, so that the paper which is laid over it should be level. The paper should be without any cockle or bend, and perfectly flat, that the filings may be free to assume the position which the magnet tends to give them. I have found well-made cartridge or thin drawing-paper good for the

---

3234. *visible to the eye at once*: The phrase recalls Faraday's "paradigm" of scientific explanation, so called in the introduction to the Nineteenth Series.

the varying direction … the relative polarity: The patterns obviously reveal varying direction—that is, curvature—in the lines. Moreover, from Faraday's revised view of polarity (paragraphs 3154–3157) it follows that two surfaces that are joined by a line of force must have opposite polarities to one another.

in which direction the power is increasing or diminishing: Power increases—more strictly, force becomes more intense—as one proceeds towards places of greater concentration of lines; power diminishes in the direction of their greater dispersion. Faraday suggested this interpretation of line concentration in (paragraph 3110), where he also noted its optical origins.

the neutral points: These are points occupied by *no lines of force whatever*, since if a line passes through a point, force must by definition be exerted there. Neutral points cannot, strictly, be displayed by iron filings, since a mere absence of filings does not prove anything. But neutral points can often be inferred from the pattern as a whole. For example, there would appear to be a neutral point between the two facing similar poles illustrated in the comment to paragraph 3233, as evidenced by a tendency of all the lines to veer away from that point.

3235. *cockle*: Here, an unwanted wrinkle or pucker.

cartridge … paper: A strong paper suitable for rough drawings; originally used in making gunpowder cartridges. During Faraday's youthful apprenticeship as

purpose. It should not be too smooth in ordinary cases, or the filings, when slightly agitated, move too freely towards the magnet. With very weak or distant magnets I have found silvered paper sometimes useful. The filings should be clean, *i. e.* free from much dirt or oxide; the latter forms the lines but does not give good delineations. The filings should be distributed over the paper by means of a sieve more or less fine, their quantity being partly a matter of taste. It is to be remembered, however, that the filings disturb in some degree the conditions of the magnetic power where they are present, and that in the case of small magnets, as needles, a large proportion of them should be avoided. Large and also fine filings are equally useful in turn, when the object is to preserve the forms obtained. For the distribution of the latter it is better to use a fine sieve with the ordinary filings than to separate the filings first: a better distribution on the paper is obtained. The filings being sifted evenly on the paper, the latter should be tapped very lightly by a small piece of wood, as a pen-holder; the taps being applied wherever the particles are not sufficiently arranged. The taps must be perpendicularly downwards, not obliquely, so that the particles, whilst they have the liberty of motion for an instant, are not driven out of their places, and the paper should be held down firmly at one corner, so as not to shift right or left; the lines are instantly formed, especially with fine filings.

3236. The designs thus obtained may be fixed in the following manner, and then form very valuable records of the disposition of the forces in any given case. By turning up two corners of the paper on which the filings rest, they may be used as handles to raise the paper upwards from the magnet, to be deposited on a flat board or other plane surface. A solution of one part of gum in three or four of water

---

a bookbinder (see the editor's introduction to the Eleventh Series), he would have known cartridge paper as a bookbinding material.

silvered paper: A paper-backed metal foil, used even today for gift wrapping. Its smooth surface would be advantageous when the magnetic forces are weak.

the filings disturb ... the magnetic power where they are present: Because their combined bulk, especially where densely scattered, may concentrate the lines in the same way, if not to the same extent, as a solid iron bar would do. Note that the very tendency of magnetic materials to point or align themselves in the magnetic field occurs precisely because they deflect some of the lines of force onto themselves; thus they necessarily alter the magnetic landscape in the very act of revealing it. The moving wire does not have this drawback.

3236. *one part of gum in three or four of water*: *Gum* (or *gum arabic*) is a water-soluble substance exuded by the tree *Acacia arabica* as well as other species. It was commonly used in the manufacture of adhesives such as mucilage.

having been prepared, a coat of this is to be applied equably by a broad camel-hair pencil, to a piece of cartridge paper, so as to make it fairly wet, but not to float it, and after wafting it through the air once or twice to break the bubbles, it is to be laid carefully over the filings, then covered with ten or twelve folds of equable soft paper, a board placed over the paper, and a half-hundred weight on the board for thirty or forty seconds. Or else, and for large designs it is a better process, whilst the papers are held so that they shall not shift on each other, the hand should be applied so as to rub with moderate pressure over all the surface equably and in one direction. If; after that, the paper be taken up, all the filings will be found to adhere to it with very little injury to the forms of the lines delineated; and when dry they are firmly fixed. If a little solution of the red ferroprussiate of potassa and a small proportion of tartaric acid be added to the gum-water, a yellow tint is given to the paper, which is not unpleasant; but besides that, prussian blue is formed under every particle of iron; and then when the filings are purposely or otherwise displaced, the design still remains recorded. When the designs are to be preserved in blue only, the gum may be dispensed with and the red ferroprussiate solution only be used.

3237. It must be well understood that these forms give no indication by their appearance of the relative strength of the magnetic force at different places, inasmuch as the appearance of the lines depends greatly upon the quantity of filings and the amount of tapping; but the direction and forms of the lines are well given, and these indicate, in a considerable degree, the direction in which the forces increase and diminish.

3236, continued. *make it fairly wet, but not to float it*: Here *float* bears its antique meaning, to *flood.* That usage survives in parts of New England, where one may still be said to "float" a cranberry bog.

red ferroprussiate of potassa: That is, potassium ferricyanide.

prussian blue: Properly, ferric ferrocyanide; although other blue pigments were sometimes called "prussian blue," inaccurately.

3237. *these forms give no indication … of the relative strength of the magnetic force at different places*: Strictly, iron filings do not reveal lines of force, since there is no way to be sure that successive particles belong to "the same" line. If we cannot trace individual lines, we cannot accurately assess their number and relative concentration. Therefore, though we can often identify the *places* of greater and lesser concentration of force from the general appearance of the patterns formed, we are unable to ascertain from those patterns the *relative strength* of the magnetic force at different locations.

3238. Plate III. fig. 1, shows the forms assumed about a bar-magnet. On using a little electro-magnet and varying the strength of the current passed through it, I could not find that a variation in the strength of the magnet produced any alteration in the forms of the lines of force external to it. Fig. 2 shows the lines over a pole, and fig. 3 those between contrary poles. The latter accord with the magnetic curves, as determined and described by Dr. Roget and others, with the

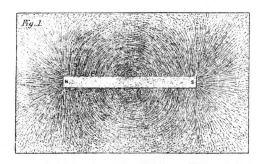

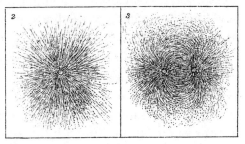

assumption of the poles as centres of force. The difference between them and those belonging to a continuous magnet, shown in fig. 1, is evident.

3238. *I could not find that a variation in the strength of the magnet produced any alteration in the forms of the lines of force*: A variation in the magnet's strength *must*, according to the moving wire, increase or decrease the concentration of lines of force in and around the electromagnet. But the pattern of iron filings is capable of revealing only shape, not density, of line distribution. Faraday finds that a change in density does not affect the shape; here then is another indication that iron filings, for all their advantages, are ill-suited for determining quantitative measures of magnetic power.

Figure 2 shows the lines over a pole...: To make this pattern, a cylindrical magnet was placed vertically beneath the paper.

...and figure 3 those between contrary poles: Two magnets, with contrary poles facing upward, were mounted vertically beneath the paper.

The latter [that is, those in Figure 3] *accord with the magnetic curves, as determined and described by Dr. Roget and others, with the assumption of the poles as centres of force*: Roget and other partisans of the reality of magnetic "poles" are obliged to regard Figure 3 as the paradigm representation of a bar magnet. But Figure 3 is significantly different from the pattern of magnetic curves of an actual bar magnet, as Faraday points out in the next sentence.

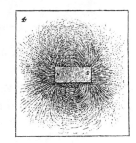

The difference between them and those belonging to ... fig. 1, is evident: A difference is indeed evident; but since the bar magnet of Figure 1 is so long, a more telling comparison might be made to the short magnet shown here in Faraday's Figure 4, which he will discuss on the next page.

Figs. 4, 5 show the lines produced by short magnets. In the latter case the magnet was a steel disc about one inch in diameter and 0.05 in thickness. Fig. 6 shows the result when a bar-magnet is broken in half, but not separated. Fig. 7 shows the development of the lines externally at the two new ends as the halves are more and more separated (3231.). Figs. 8, 9 and 10 present the results, with the two halves or new magnets in different positions. Figs. 11, 12, 13 and 14 show the results with disc magnets. Fig. 15 shows the condition of a system of magnetic forces when it is inclosed by a larger one, and is contrary to it. Fig. 16 shows the coalescence of the lines of force (3226.) when the magnets are so placed that the polarities are in accordance.

3239. Fig. 17 exhibits the lines of force round a vertical wire carrying a current of electricity. Whether the wire was thick or thin appeared to make no difference as to the intensity of the forces, the current remaining the same. Fig. 18 represents the lines round two like currents when within mutual influence. Fig. 19 shows the result when a third current is introduced in the contrary direction. Fig. 20 presents the transition to a helix of three convolutions. Fig. 21 indicates the direction of the lines within and outside the end of a cylindrical helix, on a plane through its axis. Fig. 22 presents the effect when a very small soft iron core is within the helix.

*   *   *

Royal Institution,
December 20, 1851.

3238, continued. *Figs. 4, 5 show the lines produced by short magnets*: These and the remaining figures Faraday mentions belong to Plate III, which is reproduced in full on pages 553 and 554.

a system of magnetic forces … inclosed by a larger one, and … contrary to it: Faraday will find this to be a particularly significant case at paragraph 3275 in *On the Physical Character of the Lines of Magnetic Force,* below.

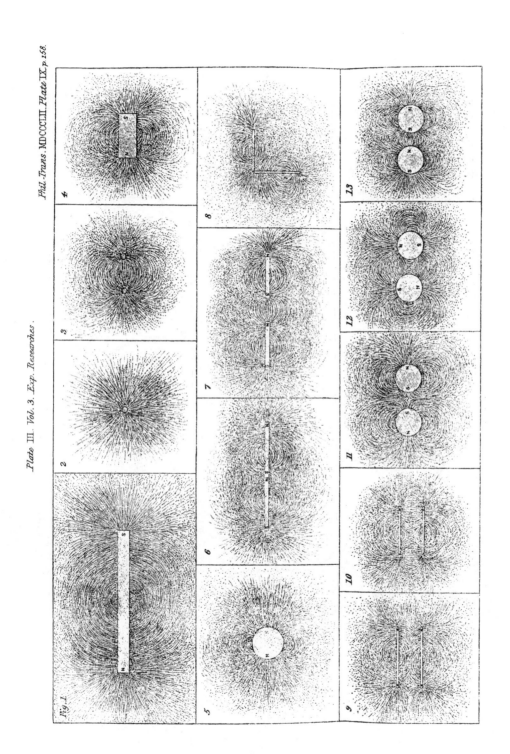

Plate III. Vol. 3. Exp. Researches.

Phil. Trans. MDCCCLII. Plate IX. p. 158.

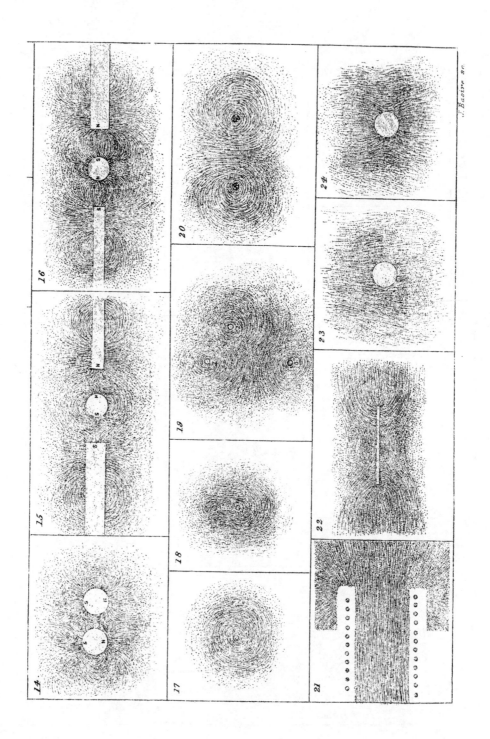

J. Basire sc.

On the Physical Character
of the Lines of Magnetic Force

Editor's Introduction

Faraday's paper *On the Physical Character of the Lines of Magnetic Force*
appeared in the Philosophical Magazine for June 1852.* This marked
a change of venue from the Philosophical Transactions, which had
carried his previous twenty-nine Series of Experimental Researches.
Faraday explains at the outset that the present paper deals with
questions so fundamental and consequential, yet so far exceeding our
capacity to answer in a rigorous or determinate way, as to make it
unsuitable for a journal of exact scientific results. Faraday's epithet for
questions of this sort and scope is "speculative." But thought is not
"speculative" simply because it goes beyond the established domain of
standard science. Judging by the contents of the present essay, speculative reasoning also involves a lively element of playfulness. By
contrast, what Faraday calls "the *strict* line of reasoning" (paragraph
3243) is *committed thought*, deriving consequences from established and
admitted premises.

In his remarks Faraday implies that "speculation" and "strict"
reasoning are to be carefully distinguished. But such a duality seems
alien to Faraday's actual practice. As we have seen for ourselves in
several preceding Series, the most prosaic of experiments may lead
naturally and inescapably to the largest of questions. The devastating
conundrum concerning a possible materiality of space, for example,
grew out of a simple tabulation of the relative magnetic and diamagnetic order of various materials. Despite his own distinction, I think
Faraday's scientific enterprise embraces both "strict" investigation and
imaginative "speculation" *as a unity*. The interpretation of nature
must indeed rest upon experience, since only through experience can
we be said to be in the presence of nature. On the other hand
experimentation—that deliberate exercise of the senses, aided by

---

* It should not be confused with his similarly-titled article, *On the Physical Lines of
Magnetic Force*, which was also published in June 1852, but in the Royal Institution
Proceedings.

appropriate artifice—is finally *for the sake of* interpretation of nature. If it never launches us into that risky and transcending activity, we must expect always to experience a baffling disconnection between the natural world and our life and work within it.

As though to affirm his continuity of purpose, Faraday has numbered the paragraphs of *On the Physical Character* consecutively with those of the Twenty-ninth Series. That numerical succession declares the present paper, with its appreciative and inventive rhythms, to be the legitimate and accredited successor to the patrimony of methods, devices, themes, questions, and investigative styles which Faraday's previous research has continually originated and refined.

What is "physical" character?

Repeatedly in the Experimental Researches we have been led to question the status of the lines of force. Are they merely figurative contrivances, like the graphs and vectors of mathematical physics? Or are they *real beings*, the bearers of natural powers? Faraday has looked with ever increasing assurance to the second view, but only in the present essay will he champion that interpretation outright.

To be physical is to be *in action*—and Faraday will call our attention to the many ways in which magnetic lines of force show themselves as being indeed at work and in action: the definite power associated with them, their ability to occupy place, and (most of all) their readily legible configurations, which imply relations of connectedness and mutual equilibrium. In these and other ways, the lines of force invite us to respond to them just as we would to any other natural objects. Faraday will argue that magnetic lines, in particular, are real structures physically present in all the materials through which they run, present even in so-called empty space.

A secondary element in Faraday's reasoning—though in the present essay he offers it first—is analogical. He finds in magnetism some highly suggestive analogies both with lines of static *electric induction* and with lines of *electric current*.* Correspondence with current is particularly significant. Since no one doubts that current electricity is a physical entity—for it undeniably represents the exercise of a force or power—a pattern of analogies with electric current may betoken a similarly physical character in magnetic action too.

---

* Or, in the phrase Faraday here prefers, "electricity in a dynamic condition." In comparison to "electric current," the new phrase is less burdened with fluid-flow imagery, expressing only that the electricity is *in action* in some sense not illustrated by static induction.

The evolving interpretations of lines of force

At several earlier places in the Experimental Researches I pointed out a growing substantiality in Faraday's allusions to the lines of force, and an increasingly prominent explanatory role for them in his narrative. The very title of the present essay stands as a kind of culmination of that growth. The former "magnetic curves" were initially little more than imaginative schemata, comprising the successive orientations of iron filings or small compass needles. Now we find them on the threshold of recognition as legitimate physical objects!

In earlier Series, Faraday interpreted the pointing and migratory tendencies of magnetic and diamagnetic materials in terms of their relative gathering power for lines of force. In the 29th Series, though, he began to suggest some alternative principles which represent the *lines of force themselves*, not steel needles or iron bars or cylinders, as the primary agents. Among them are the following: (*i*) magnetic lines of force tend to shorten themselves, (*ii*) parallel lines oppositely directed tend to approach each other, while (*iii*) parallel lines in the same direction recede from each other. Such principles certainly bear on the physical interpretation of lines of force. It may be helpful to look in advance at some of the examples he offers in this essay.

1. Pointing of the magnetic needle

The sketch maps the line patterns about a paramagnetic needle in air, held at an angle to the magnetic axis. Its pointing tendency illustrates principle (*i*), since it is evident that lines traversing the needle longitudinally will be shortest when the paramagnetic needle takes up the axial position. (Note that if the needle were diamagnetic, lines of force would run through it *transversely*. Can you show that the lines of force will be shortened if the needle takes up an *equatorial* position in that case?)

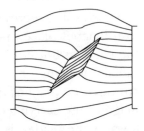

2. Parallel magnets

The illustration is from Faraday's Plate III for the 29th Series. Iron filings reveal the lines of force about parallel, like-directed magnets. Lines emerging from like poles veer away from each other. In the equatorial regions, parallel curves joining the poles of one magnet expand and recede from each other, even as similar curves from the other magnet show a comparable expansive tendency. It is the example Faraday cites to illustrate principle (*iii*).

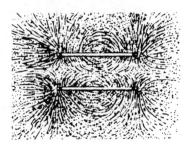

3. Parallel currents

The drawing shows parallel and similarly-directed currents directed downwards through the page. Lines of force encircle the conductors.

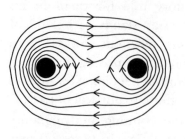

Notice that since both currents have the same direction, the center point between them is necessarily ringed by line segments having opposite directions, producing a presumptive null point there. Ampère showed experimentally that wires carrying such currents tend to approach each other; and it is easy to interpret their mutual approach in accordance with principle (*i*) above, that magnetic lines tend to shorten. Indeed the drawing almost suggests looped elastic bands bringing the conductors together. The mutual attraction also illustrates principle (*ii*), that oppositely-directed lines of force tend to approach each other. But on either principle it is the lines of force which drag the material bodies about, just as much as the other way around!

4. Electro-magnetic rotations

Faraday's first major independent discovery was the "rotation" which a current-carrying wire would perform about a magnet, or a magnet about the current. The Second Series' introduction reviewed it in terms of the right-hand rule; but Faraday's new principles based on the lines of force would now appear to provide a deeper explanation.

The drawing represents a cylindrical iron magnet placed perpendicular to the page, its lines of force converging radially towards one end. Likewise a conductor placed alongside the magnet, and carrying current directed towards the reader, is encircled by its own lines of force running counter-clockwise in accordance with the right-hand rule (this will be so 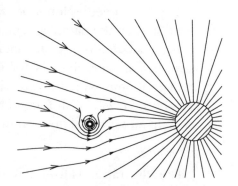 whether a magnet is present or not; see the Second Series' introduction). You can see that any tendency of the lines of force to shorten—principle (*i*) above—would have the effect of driving the conductor clockwise around the magnet—or of driving the magnet around the conductor, if it is the latter that is anchored in place. The same result is deducible from principles (*ii*) and (*iii*); for if like-directed portions of lines of force recede while opposing segments approach each other, both actions will drive the wire in a clockwise direction.

The new principles differ among themselves, but all represent the lines of force as carrying more explanatory power than the compass needles or iron filings that first made them manifest. That Faraday can explicitly contemplate the lines of force as physical structures culminates a long process of responding to them more and more seriously as objects of thought. Indeed, the highest objects of thought must be *beings*.

The sphondyloid

In the 29th Series the Moving Wire showed the line of force to be not only the bearer of the magnet's exterior action, but indicative of its interior condition as well. The line of force was shown to be unchanged in its nature, whether we view that part of it which resides within the iron bar, or its continuation outward into the surrounding places. Power resides in both phases of the line of force, equally. It is inaccurate, therefore, to identify the iron exclusively as "the" magnet. The outer medium, no less than the iron, is *essential* to the magnet and defines what Faraday calls the system or atmosphere of power (paragraphs 3271, 3275).

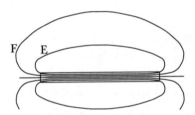

Lines of force E and F

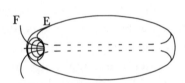

"Sphondyloid" developed from E

The sketch shows magnetic curves E, F about a straight bar magnet; similar curves will of course lie also in every other plane that contains the bar's axis. Thus the family of closed magnetic curves filling and surrounding a bar magnet constitutes an atmosphere whose shape is that of a solid of rotation—really a nest of surfaces of rotation—about the bar. This interesting shape Faraday calls "sphondyloid," from the Greek word for *beetle*.*

The sphondyloid form is characteristic and exemplary. As Faraday will show in the present paper, even those magnetic systems which do not display the sphondyloid shape have forms which can be viewed as distortions and transformations of it. The atmosphere about a spherical magnet, for example, comprises lines of force whose angles of

---

* Other sources cite a similar word denoting one of the cervical vertebrae. I hope you can make out suggestions of both meanings in the drawing. My own preference is the beetle.

refraction are rather gentler than those of a hard, well-charged bar magnet; the shape of the atmosphere is somewhat stubby by comparison, but in it the characteristic sphondyloid structure can readily be perceived. The sphondyloid eminently portrays the magnet's action as an integrally shaped and quantitatively definite physical structure, and thus it contributes substantially to the physical vision of magnetic lines of force.

Gymnotus revisited

As Faraday closed the Fifteenth Series he expressed above all a desire to experiment with Gymnotus' ability to convert nervous force into electrical activity. Experiments on the fish might well clarify the relation between other agents and the powers they exercise—magnets, electrified bodies, light sources—since animate and inanimate bodies are in many respects commensurable and are therefore capable of serving as interpretive images for one another. Indeed, in the Fifteenth Series Faraday drew upon both the imagery of the voltaic cell and the imagery of the bar magnet, in his efforts to represent Gymnotus—there borrowing some of the apparatus's geometrical legibility to interpret the fish's peculiar electrical activity.

Nevertheless it is something of a surprise, fourteen years later, to find Faraday returning to contemplate the electric fish in the course of the present essay on magnetic power. The moments of recollection pass quickly and make up but a tiny part of the essay. Still, that they occur at all is a powerful indication of Faraday's synoptic conception of natural philosophy. The Experimental Researches is not a path of ever-forward progress, but a field of continuing cultivation and re-evaluation of facts, images, and theories, each one capable of lending prospective or retrospective significance to the others. For that reason, please forgive me if I devote more attention to the Gymnotus episode than its brief appearance in the essay might appear to warrant.

Faraday's recollection of Gymnotus in the present paper involves a startling reversal of his earlier narrative protocol, for he invokes both the animal and the voltaic battery as explanatory images. By doing so, Faraday now places the *fish* prior in explanatory order to the *magnet*! At paragraph 3276 below he writes:

> The magnet, with its surrounding sphondyloid of power, may be considered as analogous in its condition to a voltaic battery immersed in water or any other electrolyte; or to a gymnotus or torpedo, at the moment when these creatures, at their own will, fill the surrounding fluid with lines of electric force.

In the Fifteenth Series he represented the *Fish* as *Magnet*. Now he represents the *Magnet* as *Fish*. How does the electric fish, which formerly had been interpreted *by* the magnet, come now to be the interpreter *of* the magnet?

When Faraday calls upon Gymnotus in the present essay, he does so primarily to represent the magnet's quantitatively and geometrically definite distribution of power. True, Faraday had revealed the *definite quantity* of magnetic action through the phenomena of the Moving Wire (paragraph 3109); but it was his early studies of the voltaic cell, and more especially the Gymnotus mapping exercises of the Fifteenth Series, which had yielded the first intimations of a power that *fills up* its medium, and whose exterior action bears an *essential relation* to the interior condition of the agent.

In order to convey his vision of the *magnetic* lines of force in the present paper, Faraday will describe typical methods for making visible the lines of *electric force* about an immersed voltaic battery (paragraph 3276). *These procedures are virtual recapitulations of earlier Gymnotus exercises.* For example, he describes how the lines of electric force may be probed with the galvanometer; for if its leads are dipped into the conducting fluid the instrument will show deflection when the line joining its collector ends is parallel or oblique to the lines of electric force, but no deflection when at right angles to those lines. It is evident that this technique rehearses the earlier Gymnotus mapping, both with hands and with the disk collectors (paragraphs 1775–1781). He describes also an electrochemical direction-indicator for lines of force, which recalls the role of electro-decomposition in establishing the direction of Gymnotus's discharge (paragraph 1763). Each of these exercises draws on earlier imagery from the Gymnotus mapping experiments to articulate the physical occupation, by external action, of the ambient medium.

Another element in the recollection is Faraday's appreciation of shape and proportion in magnetic systems; for it is clear that variations in form of the magnet correspond to the earlier coiling configurations of Gymnotus's body. Faraday will devote five full pages of the present essay to a lovingly detailed exposition of various changes in external disposition of magnetic power that result when a bar magnet is bent, stretched, or squeezed out of its original proportions. All the differently shaped "atmospheres" of magnetic lines of force become "readable" as derivatives and variants of the standard sphondyloid shape.

But the Gymnotus had bent and distorted *itself* in the course of its habitual movements fourteen years earlier, and in its natural predatory activity it presented *itself* in manifold electrical relations to other animals in the surrounding medium. Gymnotus's habitual behavior thus had occasioned the direct display of much the same topology for the animal that artifice and more ingressive experimentation now make evident for the magnet! The animal's habitual action was at the same time a heuristic, self-interpretive action. In the present paper Faraday will reflect (paragraph 3282):

> When, therefore, a magnet, in place of being a bar, is made into a horseshoe form, we see at once that the lines of force and the sphondyloids are greatly distorted or removed from their former regularity; that a line of maximum force from pole to pole grows up as the horseshoe form is more completely given; that the power gathers in, or accumulates about this line, just because the badly conducting medium, *i. e.* the space or air between the poles, is shortened. A bent voltaic battery in its surrounding medium, or a gymnotus curved at the moment of its peculiar action, present exactly the like results.

The close relation between shape of external action and shape of the body proper can be read more surely in the magnet, in part because Gymnotus called that vision forth for itself fourteen years before.

"Nature's school"

Faraday's implicit and explicit recollections of Gymnotus thus confirm, albeit retrospectively, the intimations of intelligibility and readability in animal powers he had responded to in the Fifteenth Series. To be sure, the earlier image of Gymnotus falls far short of the present vision of the magnet in comprehensiveness and depth. The magnet especially benefits from a view of its *interior* that the Moving Wire makes possible, while no comparable interior view could be secured for the electric fish. Nevertheless Gymnotus may be credited with having presented a more accessible starting point for Faraday's ultimate vision than the magnet itself could provide. The "promise" it showed then might now, therefore, be described as instructional or even pedagogical. Gymnotus's contribution to the elucidation of the magnet consisted not in data, perhaps not even in concepts. It provided rather a concrete object which both instigated and served as the practice ground for a kind of thinking which, we now see, the magnet ultimately demands. The Gymnotus in his tub has turned out to be a *school for interpretation.* Or, if not a school in its own right, Gymnotus surely qualifies as a constituent tutorial within—to use Faraday's own phrase—nature's school.* The brief reversal of images in *On the Physical Character* looks back upon a long period of schooling for the image of the magnet.

---

* Faraday used this expression in the Christmas Lectures of 1851-52. As David Gooding points out in a beautiful and instructive essay which takes the phrase as its title, Faraday did not simply learn from experiments; he also had to learn *how to learn from experiments.* See Gooding, David and Frank A. J. L. James, eds., *Faraday Rediscovered. Essays on the Life and Work of Michael Faraday, 1791–1867.* New York: American Institute of Physics, 1989.

On the Physical Character of the Lines of Magnetic Force.[1]

NOTE.—The following paper contains so much of a speculative and hypothetical nature, that I have thought it more fitted for the pages of the Philosophical Magazine than those of the Philosophical Transactions. Still it is so connected with, and dependent upon former researches, that I have continued the system and series of paragraph numbers from them to it. I beg, therefore, to inform the reader, that those in the body of the text refer chiefly to papers already published, or ordered for publication, in the Philosophical Transactions; and that they are not quite essential to him in the reading of the present paper, unless he is led to a serious consideration of its contents. The paper, as is evident, follows Series xxviii. and xxix., now printing in the Philosophical Transactions, and depends much for its experimental support upon the more strict results and conclusions contained in them.

3243. I have recently been engaged in describing and defining the lines of magnetic force (3070.), *i. e.* those lines which are indicated in a general manner by the disposition of iron filings or small magnetic needles, around or between magnets; and I have shown, I hope satisfactorily, how these lines may be taken as exact representants of the magnetic power, both as to disposition and amount; also how they may be recognized by a moving wire in a manner altogether different in principle from the indications given by a magnetic needle, and in numerous cases with great and peculiar advantages. The definition then given had no reference to the physical nature of the force at the place of action, and will apply with equal accuracy whatever that may be; and this being very thoroughly understood, I am now about to leave the strict line of reasoning for a time, and enter upon a few speculations respecting the physical character of the lines of force, and the manner in which they may be supposed to be continued through space. We are obliged to enter into such speculations with regard to numerous natural powers, and, indeed, that of gravity is the only instance where they are apparently shut out.

[1] Philosophical Magazine for June 1852.

3243. *the manner in which they may be supposed to be continued through space*: The question of transmission applies to all magnetic conductors, but it is especially urgent for *space*, where there is an absence of ponderable matter.

gravity is the only instance where they are apparently shut out: Faraday will explain in paragraph 3245 why gravitational lines would appear to have no "physical" properties.

3244. It is not to be supposed for a moment that speculations of this kind are useless, or necessarily hurtful, in natural philosophy. They should ever be held as doubtful, and liable to error and to change; but they are wonderful aids in the hands of the experimentalist and mathematician. For not only are they useful in rendering the vague idea more clear for the time, giving it something like a definite shape, that it may be submitted to experiment and calculation; but they lead on, by deduction and correction, to the discovery of new phænomena, and so cause an increase and advance of real physical truth, which, unlike the hypothesis that led to it, becomes fundamental knowledge not subject to change. Who is not aware of the remarkable progress in the development of the nature of light and radiation in modern times, and the extent to which that progress has been aided by the hypotheses both of emission and undulation? Such considerations form my excuse for entering now and then upon speculations; but though I value them highly when cautiously advanced, I consider it as an essential character of a sound mind to hold them in doubt; scarcely giving them the character of opinions, but esteeming them merely as probabilities and possibilities, and making a very broad distinction between them and the facts and laws of nature.

3245. In the numerous cases of force acting at a distance, the philosopher has gradually learned that it is by no means sufficient to rest satisfied with the mere fact, and has therefore directed his attention to the manner in which the force is transmitted across the intervening space; and even when he can learn nothing sure of the manner, he is still able to make clear distinctions in different cases, by what may be called the affections of the lines of power; and thus, by these and other means, to make distinctions in the nature of the lines of force of different kinds of power as compared with each other, and therefore between the powers to which they belong. In the action of gravity, for instance, the line of force is a straight line as far as we can test it by the resultant phænomena. It cannot be deflected, or even affected, in its course. Neither is the action in one line at all influenced, either in direction or

---

3244. *speculations*: Faraday's penchant for—and cautious attitude toward—"speculation" is discussed in the introduction to the present paper.

hypotheses ... of emission and undulation: that is, the corpuscle and wave theories of light. Only one of them can be true—and both may be false. Yet they have proved fruitful guides in both theoretical and practical optics.

3245. *the mere fact*: that is, of action at a distance. Newton formulated the inverse-square law for gravity but *framed no hypotheses* concerning its cause. Such punctilious restraint may befit the mathematician; it is not the part of the philosopher, whose very work is to search out causes and essential characters.

amount, by a like action in another line; *i.e.* one particle gravitating toward another particle has exactly the same amount of force in the same direction, whether it gravitates to that one alone or towards myriads of other like particles, exerting in the latter case upon each one of them a force equal to that which it can exert upon the single one when alone: the results of course can combine, but the direction and amount of force between any two given particles remain unchanged. So gravity presents us with the simplest case of attraction; and appearing to have no relation to any physical process by which the power of the particles is carried on between them, seems to be a pure case of attraction or action at a distance, and offers therefore the simplest type of the cases which may be like it in that respect. My object is to consider how far magnetism is such an action at a distance; or how far it may partake of the nature of other powers, the lines of which depend, for the communication of force, upon intermediate physical agencies (3075.).

3246. There is one question in relation to gravity, which, if we could ascertain or touch it, would greatly enlighten us. It is, whether gravitation requires *time*. If it did, it would show undeniably that a physical agency existed in the course of the line of force. It seems equally impossible to prove or disprove this point; since there is no capability of suspending, changing, or annihilating the power (gravity), or annihilating the matter in which the power resides.

3247. When we turn to radiation phænomena, then we obtain the highest proof, that though nothing ponderable passes, yet the lines of force have a physical existence independent, in a manner, of the body

---

gravity … appearing to have no relation to any physical process…: As Faraday has just recounted, lines of gravitational force are everywhere straight, never curved or deflected; moreover the gravitating force between any two masses is unaffected by the presence or absence of other gravitating matter in the vicinity. Such *insusceptibility to external influence* suggests that gravitational lines of force are mere mathematical concepts, not physical entities.

3246. *time … would show undeniably that a physical agency existed…*: If the gravitational lines involved *time* they could not be purely geometrical. With the irreversibility of time there arises history—and nature.

3247. *radiation phænomena … lines of force…*: Note Faraday regards even *light beams* as examples of "lines of force"; but that is not really surprising since Faraday (in common with many 19th-century thinkers) has never limited his conception of "force" to mechanical pushes and pulls. In an 1858 addendum to his paper On the Conservation of Force, he will explain: "What I mean by the word "force," is the *cause* of a physical action; the source or sources of all possible changes amongst the particles or materials of the universe." (*Experimental Researches in Chemistry and Physics*, page 460.)

radiating, or of the body receiving the rays. They may be turned aside in their course, and then deviate from a straight into a bent or a curved line. They may be affected in their nature so as to be turned on their axis, or else to have different properties impressed on different sides. Their sum of power is limited; so that if the force, as it issues from its source, is directed on to or determined upon a given set of particles, or in a given direction, it cannot be in any degree directed upon other particles, or into another direction, without being proportionately removed from the first. The lines have no dependence upon a second or reacting body, as in gravitation; and they require time for their propagation. In all these things they are in marked contrast with the lines of gravitating force.

3248. When we turn to the electric force, we are presented with a very remarkable general condition intermediate between the conditions of the two former cases. The power (and its lines) here requires the *presence* of two or more acting particles or masses, as in the case of gravity; and cannot exist with one only, as in the case of light. But though two particles are requisite, they must be in an *antithetical* condition in respect of each other, and not, as in the case of gravity,

---

3247, continued. *They may be … turned on their axis*: Faraday showed in the Nineteenth Series that magnetic influence can rotate the plane of polarization of a light beam. And in general, the susceptibility of light beams to incidental influence suggests their *physical* existence.

Their sum of power is limited: An opaque body held in front of a lighted surface *casts a shadow*, showing that light which falls on one surface cannot illuminate additional area unless it is withdrawn from the first—the total illumination is redistributed but not increased. The illumination provided by a light source is thus *limited*, as any physical entity must be. In contrast, a gravitating body exerts the same influence on one particle, whether or not a second particle is present (paragraph 3245)—gravity casts no shadows! So in the presence of an *ever-increasing quantity* of other gravitating matter however distant, a gravitating body would exert an *ever-increasing total* of gravitational action. Faraday regards such limitlessness as acceptable in a mathematical hypothesis, absurd in a physical power.

3248. *the electric force [is] intermediate between the … two former cases*: A line of electric force terminates in two oppositely-electrified surfaces or particles. In this it resembles the line of gravitational force, which also terminates in two particles or surfaces; but it differs from light, which requires only a single terminus—the illuminating source. In another respect the electric line of force differs from gravity, for there is nothing like *positive* and *negative* in gravitating matter, while it resembles the light beam in having "limited" power—any influence that an electrified particle exerts on one body reduces the influence it exerts on another body, simultaneously present (paragraph 3247).

alike in relation to the force. The power is now dual; there it was simple. Requiring two or more particles like gravity, it is unlike gravity in that the power is limited. One electro-particle cannot affect a second, third and fourth, as much as it does the first; to act upon the latter its power must be proportionately removed from the former, and this limitation appears to exist as a necessity in the dual character of the force; for the two states, or places, or directions of force must be equal to each other.

3249. With the electric force we have both the static and dynamic state. I use these words merely as names, without pretending to have a clear notion of the physical condition which they seem meaningly to imply. Whether there are two fluids or one, or any fluid of electricity, or such a thing as may be rightly called a current, I do not know; still there are well-established electric conditions and effects which the words *static, dynamic,* and *current* are generally employed to express; and with this reservation they express them as well as any other. The lines of force of the *static* condition of electricity are present in all cases of induction. They terminate at the surfaces of the conductors under induction, or at the particles of non-conductors, which, being electrified, are in that condition. They are subject to inflection in their course (1215. 1230.), and may be compressed or rarefied by bodies of different inductive capacities (1252. 1277.); but they are in those cases affected by the intervening matter; and it is not certain how the line of electric force would exist in relation to a perfect vacuum, *i. e.* whether it would be a straight line, as that of gravity is assumed to be, or curved in such a manner as to show something like physical existence separate from the mere distant actions of the surfaces or particles bounding or terminating the induction. No condition of *quality* or *polarity* has as yet been discovered in the line of static electric force; nor has any relation of *time* been established in respect of it.

3249. *static and dynamic ... I use these words merely as names...:* They should not be taken to imply literal rest and motion. But certainly some kind of *doing* is to be associated with electricity in the dynamic state: production of heat, light, etc.

The lines of force of the static *condition of electricity ... terminate at the surfaces...:* Faraday thinks that *static* electric lines of force necessarily terminate. In later years, Maxwell will show how electric lines of force can be both *static* and *circular* when they surround a changing magnetic environment in the absence of conductors (in particular, when the magnetic intensity changes at a uniform rate); but Faraday knows circular electric lines only as *induced currents*—and current is electric force in the *dynamic* condition.

No condition of quality or polarity has as yet been discovered in the line of static electric force: Faraday does not mean to deny "polarity" in the sense he defined in the Eleventh Series (paragraph 1304). He means, rather, that electric lines of force cannot be polarized like a light beam, as he makes clear at paragraph 3252 below.

3250. The lines of force of dynamic electricity are either limited in their extent, as in the lowering by discharge, or otherwise[,] of the inductive condition of static electricity; or endless and continuous, as closed curves in the case of a voltaic circuit. Being definite in their amount for a given source, they can still be expanded, contracted, and deflected almost to any extent, according to the nature and size of the media through which they pass, and to which they have a direct relation. It is probable that matter is always essentially present; but the hypothetical æther may perhaps be admitted here as well as elsewhere. No condition of quality or polarity has as yet been recognized in them. In respect of *time*, it has been found, in the case of a Leyden discharge, that time is necessary even with the best conductors; indeed there is reason to think it is as necessary there as in the cases dependent on bad conducting media, as, for instance, in the lightning flash.

3251. Three great distinctions at least may be taken among these cases of the exertion of force at a distance; that of gravitation, where propagation of the force by physical lines through the intermediate space is not supposed to exist; that of radiation, where the propagation does exist, and where the propagating line or ray, once produced, has existence independent either of its source, or termination; and that of electricity, where the propagating process has intermediate existence, like a ray, but at the same time depends upon both extremities of the line of force, or upon conditions (as in the connected voltaic pile) equivalent to such extremities. Magnetic action at a distance has to be compared with these. It may be unlike any of them; for who shall say we are aware of all the physical methods or forms under which force is communicated? It has been assumed, however, by some, to be a pure case of force at a distance, and so like that of gravity; whilst others have considered it as better represented by the idea of streams of power. The question at present appears to be, whether the lines of magnetic

---

3250. *The lines of force of dynamic electricity are either limited in their extent, … or endless and continuous:* Lines of dynamic electricity are either *continuous currents*—in which case they constitute closed curves—or they represent the increasing or decreasing *induction* between two surfaces—and are therefore terminated by those surfaces. An example of the latter would be the *sudden discharge* of a Leyden jar—or, as Faraday puts it, "the lowering by discharge, or otherwise[,] of the inductive condition of static electricity."

No condition of quality or polarity has as yet been recognized in them: "Polarity" is as in the previous paragraph; he clarifies "quality" in paragraph 3252 below.

time is necessary even with the best conductors: Wheatstone's revolving mirror showed this; see the Twelfth Series' introduction.

force have or have not a physical existence; and if they have, whether such physical existence has a static or dynamic form (3075. 3156. 3172. 3173.).

3252. The lines of magnetic force have not as yet been affected in their *qualities, i. e.* nothing analogous to the polarization of a ray of light or heat has been impressed on them. A relation between them and the rays of light when polarized has been discovered (2146.);[2] but it is not of such a nature as to give proof as yet, either that the lines of magnetic force have a separate existence, or that they have not; though I think the facts are in favour of the former supposition. The investigation is an open one, and very important.

3253. No relation of *time* to the lines of magnetic force has as yet been discovered. That iron requires *time* for its magnetization is well known. Plucker says the same is the case for bismuth, but I have not been able to obtain the effect showing this result. If that were the case, then mere space with its æther ought to have a similar relation, for it comes between bismuth and iron (2787.); and such a result would go far to show that the lines of magnetic force have a separate physical existence. At present such results as we have cannot be accepted as in any degree proving the point of *time*; though if that point were proved, they would most probably come under it. It may be as well to state, that in the case also of the moving wire or conductor (125. 3076.), time is required.[3] There seems no hope of touching the investigation by any method like those we are able to apply to a ray of light, or to the current of the Leyden discharge; but the mere statement of the problem may help towards its solution.

3254. If an action in *curved* lines or directions could be proved to exist in the case of the lines of magnetic force, it would also prove their physical existence external to the magnet on which they might depend; just as the same proof applies in the ease of static electric induction.[4] But the simple disposition of the lines, as they are shown by iron particles, cannot as yet be brought in proof of such a curvature,

[2] Philosophical Transactions, 1846, p. 1.

[3] Experimental Researches, 8vo edition, vol. ii. pp. 191, 195.

[4] Philosophical Transactions, 1838, p. 16.

3252. *their* qualities, *i. e.* ... *analogous to the polarization of a ray of light*: This explains the phrase "quality or polarity" in paragraphs 3249–3250 above.

3253. *any method like those we are able to apply to a ray of light, or to the current of the Leyden discharge*: For example, Wheatstone's revolving mirror, as also in paragraph 3250 above.

because they may be dependent upon the presence of these particles and their mutual action on each other and the magnets; and it is possible that attractions and repulsions in right lines might produce the same arrangement. The results therefore obtained by the moving wire (3076. 3176.),[5] are more likely to supply data fitted to elucidate this point, when they are extended, and the true magnetic relation of the moving wire to the space which it occupies is finally ascertained.

3255. The *amount* of the lines of magnetic force, or the force which they represent, is clearly limited, and therefore quite unlike the force of gravity in that respect (3245); and this is true, even though the force of a magnet in free space must be conceived of as extending to incalculable distances. This limitation in amount of force appears to be intimately dependent upon the dual nature of the power, and is accompanied by a displacement or removability of it from one object to another, utterly unlike anything which occurs in gravitation. The lines of force abutting on one end or pole of a magnet may be changed in their direction almost at pleasure (3238.), though the original seats of their further parts may otherwise remain the same. For, by bringing fresh terminals of power into presence, a new disposition of the force upon them may be occasioned; but though these may be made, either in part or entirely, to receive the external power, and thus alter its direction, no change in the amount of the force is thus produced. And this is the case in strict experiments, whether the new bodies introduced are soft iron or magnets (3218. 3223.).[6] In this respect,

[5] Ibid. 1852.

[6] Philosophical Transactions, 1852.

3254. *they may be dependent upon the presence of these particles*…: Curves exhibited by iron filings or compass needles are inconclusive, since the very presence of an indicating material alters the disposition of the forces to be indicated.

attractions and repulsions in right lines might produce the same arrangement: Curved resultants may arise from straight lines of action by *composition of forces* (see the introduction to the 29th Series). But the *moving wire* is not subject to Newtonian attraction or repulsion, neither does its presence disturb the disposition of the forces; so the curves it discloses escape all objection.

3255. *This limitation in amount of force appears to be intimately dependent upon the dual nature of the power … a displacement or removability*: We saw in paragraph 3247 that the clearest sign of a force's "limitation in amount" is its removability from one terminus to another; so that insofar as lines of force representing "dual" powers are *terminated*, the dependence Faraday speaks of will follow. But remember that not all lines of a dual power need be terminated—the lines corresponding to induced electric currents are *endless and continuous*, as Faraday acknowledged in paragraph 3250 above.

therefore, the lines of magnetic force and of electric force agree. Results of this kind are well shown in some recent experiments on the effect of iron, when passing by a copper wire in the magnetic field of a horseshoe magnet (3129. 3130.), and also by the action of iron and magnets on each other (3218. 3223.).

3256. It is evident, I think, that the experimental data are as yet insufficient for a full comparison of the various lines of power. They do not enable us to conclude, with much assurance, whether the magnetic lines of force are analogous to those of gravitation, or direct actions at a distance; or whether, having a physical existence, they are more like in their nature to those of electric induction or the electric current. I incline at present to the latter view more than to the former, and will proceed to certain considerations bearing on the question, with a view to the further and future elucidation of the subject.

———————

3257. I think I have understood that the mathematical expression of the laws of magnetic action at a distance is the same as that of the laws of static electric actions; and it has been assumed at times that the supposition of north and south magnetisms, spread over the poles or respective ends of a magnet, would account for all its external actions on other magnets or bodies. In either the static or dynamic view, or in any other view like them, the exertion of the magnetic forces outwards, at the poles or ends of the magnet, must be an essential condition. Then, with a given bar-magnet, can these forces exist without a mutual relation of the two, or else a relation to contrary magnetic forces of equal amount originating in other sources? I do not believe they can; because, as I have shown in recent researches, the sum of the lines of force is equal for any section across them taken anywhere externally between the poles (3109.). Besides that, there are many other experimental facts which show the relation and connexion of the

———————

3257. *the mathematical expression of the laws of magnetic action at a distance is the same as that of the laws of static electric actions*: That is, both may be expressed as inversely proportional to the square of the corresponding distance. But such an expression conveys no more than the "mere fact"—and this, Faraday reminded us in paragraph 3245, is not sufficient for philosophical understanding.

Then … can these forces exist without a mutual relation of the two, or else a relation to contrary magnetic forces of equal amount originating in other sources? I do not believe they can…: Recall that Faraday asked the same question, and formed the same opinion, concerning positive and negative electrification—see the Eleventh Series.

forces at one pole to those at the other;[7] and there is also the analogy with static electrical induction, where the one electricity cannot exist without relation to, equality with, and dependence on the other. Every dual power appears subject to this law as a law of necessity. If the opposite magnetic forces could be independent of each other, then it is evident that *a charge* with one magnetism only is possible; but such a possibility is negatived by every known experiment and fact.

3258. But supposing this necessary relation, which constitutes polarity, to exist, then how is it sustained or permitted in the case of an independent bar-magnet situated in free space? It appears to me, that the outer forces at the poles can only have relation to each other by *curved* lines of force through the surrounding space; and I cannot conceive curved lines of force without the conditions of a physical existence in that intermediate space. If they exist, it is not by a succession of particles, as in the case of static electric induction (1215. 1231.), but by the condition of space free from such material particles. A magnet placed in the middle of the best vacuum we can produce, and whether that vacuum be formed in a space previously occupied by

[7] The manner in which a large powerful magnet deranges, overpowers, and even inverts the magnetism of a smaller magnet, when it is brought near it in different directions without touching it, presents a number of such cases.

3258. *how is it sustained ... in the case of an independent bar-magnet situated in free space?* Faraday finally begins to confront again the long-averted question about *space* (paragraph 2435, *comment*; paragraphs 2445 and 2446).

the poles can only have relation to each other by curved lines of force through the surrounding space: The lines must be *curved* since there are no straight exterior paths between the bar ends.

I cannot conceive curved lines of force without the conditions of a physical existence: He previously argued (paragraph 1166) that any curvature in the lines of force implies participation of the medium—in particular the *contiguous particles of the medium*—in sustaining that force. But, as he will point out in the next two sentences, lines of force can be delineated in *vacuum*, where there is no material. Then either *space itself* must play the participatory role formerly ascribed to a material medium, or the lines must require no medium but be the bearers of the action in their own right. Either case will constitute a "condition of physical existence," and Faraday will not always distinguish between them.

not by a succession of particles ... but by the condition of space free from such material particles: With this acknowledgment he finally abandons the conventional doctrine that *matter* is the only physical agent. The demonstrable efficacy of magnets even in "the best vacuum we can produce" has at last persuaded him that *space has a magnetic condition*. If there are no material particles in space, therefore, that only shows such particles to be inessential for physicality. Contrary to his earlier declaration (cited in the 26th Series' introduction), it seems that mere space *can* act as matter acts! In another article, which Faraday

paramagnetic or diamagnetic bodies, acts as well upon a needle as if it were surrounded by air, water or glass; and therefore these lines exist in such a vacuum as well as where there is matter.

*　　*　　*

3265. The well-known relation of the electric and magnetic forces may be thus stated. Let two rings, in planes at right angles to each other, represent them, as in Plate IV fig. 1. If a current of electricity be sent round the ring E in the direction marked, then lines of magnetic force will be produced, correspondent to the polarity indicated by a supposed magnetic needle placed at NS, or in any other part of the ring M to which such a needle may be supposed to be shifted. As these rings represent the lines of electro-dynamic force and of magnetic force respectively, they will serve for a standard of comparison. I have elsewhere called the electric current, or the line of electrodynamic force, "an axis of power having contrary forces exactly equal in amount in contrary directions" (517.). The line of magnetic force may be described in *precisely the same terms*; and these two axes of power, considered as right lines, are perpendicular to each other; with this additional condition, which determines their mutual direction, that they are separated by a right line perpendicular to both.

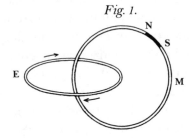

Fig. 1.

---

published in the same year (and with nearly the same title) as the present paper, he states outright: "Experimentally mere space is magnetic…"

3265. *The well-known relation of the electric and magnetic forces…*:　To see that Faraday's Figure 1 is consistent with the electromagnetic right-hand rule, let the ring E be grasped in the right hand, with thumb pointing in the direction of the arrow. Then the fingers will point downward through the center of ring E—that is, in the direction counter-clockwise about ring M. As the drawing shows, that is also the direction in which N, the north-seeking pole of a magnetic needle, will point.

The line of magnetic force may be described in precisely the same terms:　Faraday's Figure 1 draws an implicit analogy between electric current and magnetic force—both are represented as *closed curves*, suggesting that electric and magnetic lines of force may represent *comparable configurations of power*. Here Faraday makes the comparison explicit. He previously characterized the electric current as *an axis of power having contrary forces … in contrary directions*; now he extends that representation to the magnetic line of force as well. Such analogies between electric and magnetic force are significant because they suggest the physical substantiality of magnetic lines. Faraday will outline the argument at paragraph 3269 below.

The meaning of the words above, when applied to the electric current, is precise, and does not imply that the forces are contrary because they are in reverse directions, but are *contrary in nature*; the turning one round, end for end, would not at all make it resemble the other; a consideration which may have influence with those who admit electric fluids, and endeavour to decide whether there are one or two electricities.

3266. When these two axes of power are compared, they have some remarkable correspondences, especially in relation to their position at right angles to each other. As a physical fact, Ampère[8] and Davy[9] have shown, that an electric current tends to elongate itself; and, so far, that may be considered as marking a character of the *electric* axis of power. When a free magnetic needle near the end of a bar-magnet first points and then tends to approach it, I see in the action a character of the contrary kind in the *magnetic* axis of power; for the lines of magnetic force, which, according to my recent researches, are common to the magnet and the needle (3230.), are shortened, first by the motion of the needle when it points, and again by the action which causes the needle to approach the magnet. I think I may say, that all the other actions of a magnet upon magnets, or soft iron, or other paramagnetic and diamagnetic bodies, are in harmony with the same effect and conclusions.

3267. Again:—like electric currents, or lines of force, or axes of power, when placed side by side, attract each other. This is well known and well seen, when wires carrying such currents are placed parallel to each

[8] *Ann. de Chim.* 1822, vol. xxi p 47.

[9] Phil. Trans. 1823, p. 153.

3265, continued. *turning one round, end for end, would not at all make it resemble the other*: Faraday asserts—but on what grounds?—that if a positive current, say, were reversed, it would not become a negative current. He thus denies what proponents of a single electric "fluid" would have to assert.

3266. *an electric current tends to elongate itself*: Davy led electric currents into and out of a mercury bath through insulated vertical conductors. The liquid appeared to accumulate and the surface rise up immediately above the conductors' submerged ends; he viewed this as "elongation" of the current. Davy's interpretation is far from conclusive, but Faraday appears to accept and even welcome it.

the lines of magnetic force … are shortened: The introduction reviews some specific examples. The contractile tendency of magnetic lines appears to contrast with the lengthening tendency of electric currents. But the contrast masks an underlying unity, as Faraday will point out in paragraph 3268 below.

like electric currents, or lines of force … when placed side by side, attract each other…: I discuss one example in the introduction.

other. But like magnetic axes of power or lines of force repel each other: the parallel case to that of the electric currents is given, by placing two magnetic needles side by side with like poles in the same direction; and by the use of iron filings, numerous pictorial representations (3234.) of the same general result may be obtained.

3268. Now these effects are not merely *contrasts* continued through two or more different relations, but they are contrasts which *coincide* when the position of the two axes of power at right angles to each other are considered (1659. 3265.). The tendency to *elongate* in the electric current, and the tendency to *lateral* separation of the magnetic lines of force which surround that current, are both tendencies in the same direction, though they seem like contrasts, when the two axes are considered out of their relation of mutual position; and this, with other considerations to be immediately referred to, probably points to the intimate physical relation, and it may be, to the oneness of condition of that which is apparently two powers or forms of power, electric and magnetic. In that case many other relations, of which the following are some forms, will merge in the same result. Thus, unlike magnetic lines, when end on, repel each other, as when similar poles are face to face; and unlike electric currents, if placed in the same relation, stop each other; or if raised in intensity, when thus made static, repel each other.

But like magnetic axes of power or lines of force repel each other: The introduction gives an example from Faraday showing lines of force among parallel magnets.

3268. *they are contrasts which coincide when the position of the two axes of power at right angles to each other are considered*: Consider again Figure 1, reproduced here. The

drawing depicts a single magnetic curve M, but in actuality there must be a flock of such curves arrayed about the circumference of current E. If then a segment of E were straightened out, the circular magnetic lines surrounding that portion would become parallel to one another, like the rings on an earthworm's body when it lies straight. Then if the straightened portion of E were lengthened, the rings M surrounding it would be distributed over a longer

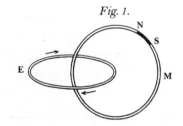

Fig. 1.

axis—equivalent to mutual recession. Conversely, increasing the separation of the magnetic curves M would require elongation of the current path E. Thus, the tendencies of electric currents to elongate (paragraph 3266) and magnetic lines to separate are harmonious rather than contrary to one another. This again represents a fundamental affinity between electric and magnetic forces, whose significance for the present essay Faraday will articulate in the next paragraph.

unlike magnetic lines, when end on, repel each other: that is, they do not join up or coalesce. Similarly *"unlike electric currents … stop each other"*—which is of course figurative speech. It would be clearer to say that they cancel or neutralize one other.

Like electric currents or lines of force, when end on to each other, coalesce; like magnetic lines of force similarly placed do so too (3266. 3295.). Like electric currents, end to end, do not add their sums; but whilst there is no change in quantity, the intensity is increased. Like magnetic lines of force, similarly placed do not increase each other, for the power then also remains the same (3218.): perhaps some effect correspondent to the gain of intensity in the former case may be produced, but there is none as yet distinctly recognised. Like electric currents, side by side, add their quantities together; a case supplied either by uniting several batteries by their like ends, or comparing a large plate battery with a small one. Like magnetic lines of force do the same (3232.).

3269. The mutual relation of the magnetic lines of force and the electric axis of power has been known ever since the time of Œrsted and Ampère. This, with such considerations as I have endeavoured to advance, enables us to form a guess or judgement, with a certain degree of probability, respecting the nature of the lines of magnetic force. I incline to the opinion that they have a physical existence correspondent to that of their analogue, the electric lines; and having that notion, am further carried on to consider whether they have a probable dynamic condition, analogous to that of the electric axis to which they are so closely and, perhaps, inevitably related, in which case the idea of magnetic currents would arise; or whether they consist in a state of tension (of the æther?) round the electric axis, and may therefore be considered as static in their nature. Again and again the idea of an *electrotonic* state (60. 1114. 1661. 1729. 1733.) has been forced on my mind; such a state would coincide and become identified with that which would then constitute the physical lines of magnetic force. Another consideration tends in the same direction. I formerly remarked that the magnetic equivalent to *static* electricity was not known; for if the undeveloped state of electric force correspond to the like undeveloped condition of magnetic force, and if the electric

---

3268, continued. *Like electric currents, end to end...*: This odd phrase may be illustrated by voltaic cells in series, discharging through a common conductor. The current produced by all the cells is identical with the current through a single cell; but the tension (intensity) of the whole is the *sum* of the individual cell tensions. I discussed the series configuration of voltaic cells in the introduction to the 29th Series.

Like electric currents, side by side...: Exemplified by cells discharging in parallel, also described in the 29th Series introduction. The combined tension (intensity) of the whole is identical to that of a single cell; but current developed by all the cells is the sum of the individual cell currents.

current or axis of electric power correspond to the lines of magnetic force or axis of magnetic power, then there is no known magnetic condition which corresponds to the static state of the electric power (1734.). Now assuming that the physical lines of magnetic force are currents, it is very unlikely that such a link should be naturally absent; more unlikely, I think, than that the magnetic condition should depend upon a state of tension; the more especially as under the latter supposition, the lines of magnetic power would have a physical existence as positively as in the former case, and the curved condition of the lines, which seems to me such a necessary admission, according to the natural facts, would become a possibility.

3270. The considerations which arise during the contemplation of the phænomena and laws that are made manifest in the mutual action of magnets, currents of electricity, and *moving conductors* (3084. &c.), are, I think, altogether in favour of the physical existence of the lines of magnetic force. When only a single magnet is employed in such cases, and the use of iron or paramagnetic bodies is dismissed, then there is no effect of attraction or repulsion or any ordinary magnetic result produced. The phænomena may all very fairly be looked upon as purely electrical, for they are such in character; and if they coincide with magnetic actions (which is no doubt the case), it is probably because the two actions are one. But being considered as electrical actions, they convey a different idea of the condition of the field where they occur, to that involved in the thought of magnetic action at a distance. When a copper wire is placed in the neighbourhood of a bar-magnet, it does not, as far as we are aware (by the evidence of a magnetic needle or other means), disturb in the least degree the disposition of the magnetic forces, either in itself or in surrounding space. When it is moved across the lines of force, a current of electricity is developed in

---

3270. *The phænomena may ... be looked upon as purely electrical*: Between a single magnet and a (copper) wire there can be no magnetic interaction in the sense of attractions or repulsions. The *only* discernible effect is the electric current which motion produces in the wire.

a different idea of the condition of the field ... to that involved in the thought of magnetic action at a distance: The idea of magnetic action at a distance portrays the intervening space or medium as indifferent, unaffected by the action between detached bodies. But the electrical view of the phenomena emphasizes the dynamism, the *doing*, which distinguishes electric current. The surrounding space or medium thereby comes to be seen as sustaining a condition of *readiness to act*, which the motion of the wire fulfills. The conception of the magnetic field as *filled with the action* is the "different idea" Faraday has in mind. He will carry the image even further by introducing a locution of "equivalence" a few sentences later.

it, or tends to be developed; and there is every reason to believe, that if we could employ a perfect conductor, and obtain a perfect result, it would be the full equivalent to the force, electric or magnetic, which is exerted in the place occupied by the conductor. But, as I have elsewhere observed (3172.), this current, having its full and equivalent relation to the magnetic force, can hardly be conceived of as having its entire foundation in the mere fact of motion. The motion of an external body, otherwise physically indifferent, and having no relation to the magnet, could not beget a physical relation such as that which the moving wire presents. There must, I think, be a previous state, a state of tension or a static state, as regards the wire, which, when motion is superadded, produces the dynamic state or current of electricity. This state is sufficient to constitute and give a physical existence to the lines of magnetic force, and permit the occurrence of curvature or its equivalent external relation of poles, and also the various other conditions, which I conceive are incompatible with mere action at a distance, and which yet do exist amongst magnetic phænomena.

3271. All the phænomena of the moving wire seem to me to show the physical existence of an atmosphere of power about a magnet, which, as the power is antithetical, and marked in its direction by the lines of magnetic force, may be considered as disposed in sphondyloids,

---

3270, continued. *it would be the full equivalent to the force … which is exerted in the place occupied by the conductor*: Faraday suggests that the current induced in a moving wire does not merely *signify* the magnetic action that induced it; rather it is in some sense *equivalent* to that action. Deflection of the galvanometer is more than a symptom of the magnetic condition of the field but represents a portion of that action, converted into the form of an electric current. Indeed, instruments in general are not mere "artificial observers"; they must be in a relation of *doing*, that is, *conversion of forces*, with the objects or conditions they measure. Faraday voiced a similar interpretation of *equivalence* previously; see paragraph 855 in the Seventh Series.

hardly be conceived of as having its entire foundation in the mere fact of motion: Certainly, in order to develop induced current, the Moving Wire has to be *moving*. But the induced current and the antecedent magnetic action are equivalent regardless of the speed or even the *path* of the motion (so long as the same lines of force are cut). The motion would thus appear to be only incidental to induction—perhaps triggering the release of some antecedent state of tension whose discharge takes the form of induced current. Thus once again, Faraday's thoughts turn to the suspected *electrotonic state* (paragraphs 60, 71, 3269).

3271. *the power is antithetical*: that is, polar—it manifests opposite characters ("north" and "south" in the case of magnetism) in opposite directions.

sphondyloids: From Greek *sphondulê*, a kind of beetle; alternatively *sphondulos*, vertebra. The term and the shape it names are discussed in the introduction.

determined by the lines, or rather shells of force.[10] As the wire inter-sects the lines within a given sphondyloid external to the magnet, a current of electricity is generated, and that current is definite and the same for any or every intersection of the given sphondyloid. At the same time, whether the wire be quiescent or in motion, it does not cause derangement, or expansion, or contraction of the lines of force; the state of the power in the neighbouring or other parts of the sphondyloid remaining sensibly the same (3176.).

3272. The old experiment of a wire when carrying an electric current[11] moving round a magnetic pole, or of a current being produced in the same wire when it is carried per force round the same pole (114.), shows the electrical dependence of the magnet and the wire, both when the current is employed from the first, and when it is generated by the motion. It coincides in principle with the results already quoted, and it includes, experimentally, all currents of electricity, whatever the medium in which they occur, even up to that due to the discharge of the Leyden jar or that between the electrodes of the voltaic battery. I think it also indicates the state of magnetic or electric tension in the surrounding space, not only when that space is occupied by metal or a wire, but also by air and other bodies; for whatever be the state in one case, it is probably general and therefore common to all (3173.).

[10] The lines of magnetic force have been already defined (3071.). They have also been traced, as I think, and shown to be closed curves passing in one part of their course through the magnet to which they belong, and in the other part through space (3117.). If, in the case of a straight bar-magnet, any one of these lines, E, be considered as revolving round the axis of the magnet, it will describe a surface; and as the line itself is a closed curve, the surface will form a tube round the axis and inclose a solid form. Another line of force, F, will produce a similar result. The sphondyloid body may be either that contained by the surface of revolution of E, or that contained between the two surfaces of E and F, and which, for the sake of brevity, I have (by the advice of a kind friend) called simply the *sphondyloid*. The parts of the solid described, which are within and without the magnet, are in power equivalent to each other. When it is needful to speak of them separately, they are easily distinguished as the inner and outer sphondyloids; the surface of the magnet being then part of the bounding surface.

[11] Experimental Researches, 8vo edition, vol. ii. p. 127.

Comment on Faraday's note 10. See the introduction, page 559, for a sketch showing specimen lines of force E and F. Faraday did not supply a drawing to accompany his note.

3272. *The old experiment of a wire … moving round*: Faraday refers to his sensa-tional discovery of "electro-magnetic rotations," described in the Second Series' introduction. Thirty years later, it is "the old experiment." But it will demand a new interpretation—see the introduction to the present paper.

Fig. 2.

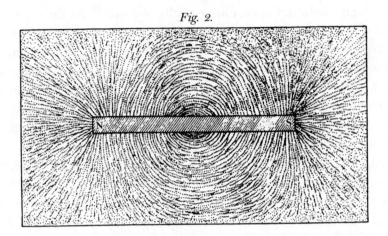

3273. I will now venture for a time to assume the physical existence of the external lines of magnetic force, for the purpose of considering how the idea will accord with the general phænomena of magnetism. The magnet is evidently the sustaining power, and in respect of its internal condition or that of its particles, there is no idea put forth to represent it which at all approaches in probability and beauty to that of Ampère (1659.). Its analogy with the helix is wonderful; nevertheless there is, as yet, a striking experimental distinction between them; for whereas an unchangeable magnet can never raise up a piece of soft iron to a state more than equal to its own, as measured by the moving wire (3219.), a helix carrying a current can develope in an iron core magnetic lines of force, of a hundred or more times as much power as that possessed by itself, when measured by the same means. In every point of view, therefore, the magnet deserves the utmost exertions of the philosopher for the development of its nature, both as a magnet and also as a source of electricity, that we may become acquainted with the great law under which the apparent anomaly may disappear, and by which all these various phænomena presented to us shall become *one*.

3274. The physical lines of force, in passing out of the magnet into space, present a great variety of conditions as to form (3238.). At times their refraction is very sudden, leaving the magnet at right, or obtuse, or acute angles, as in the case of a hard well-charged bar-magnet, fig. 2;

3274. *At times their refraction is very sudden, leaving the magnet at right, or obtuse, or acute angles, as in … fig.2*: Faraday showed that lines of force within the bar run generally parallel to its axis (paragraphs 3116–3117), but lines emerging from the bar's sides must evidently have curved laterally, away from the axis. Within a long, narrow bar, however, a comparatively small departure from the axial direction suffices. Lines will therefore undergo most refraction outside,

in other cases the change of form of
the line in passing from the magnet
into space is more gradual, as in the
circular plate or globe-magnet, figs.
3, 4, 5. Here the form of the magnet
as the source of the lines has much to
do with the result; but I think the
condition and relation of the
surrounding medium has an essential and evident influence, in a
manner I will endeavour to point out
presently. Again, this refraction of
the lines is affected by the relative
difference of the nature of the
magnet and the medium or space
around it; as the difference is greater,
and therefore the transition is more
sudden, so the line of force is more
instantaneously bent. In the case of
the earth, both the nature of its
substance and also its form, tend to
make the refractions of the line of
force at its surface very gradual; and
accordingly the line of dip does not
sensibly vary under ordinary circumstances, at the same place, whether it
be observed upon the surface or
above or below it.

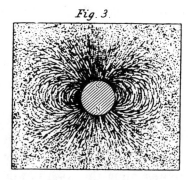

Fig. 3.

Fig. 4.

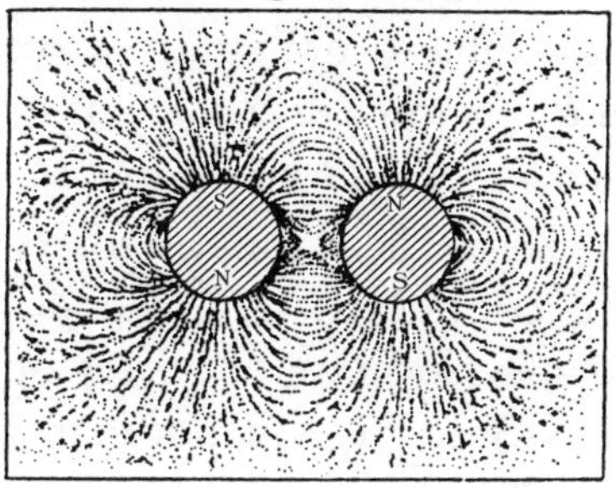

Fig. 5.

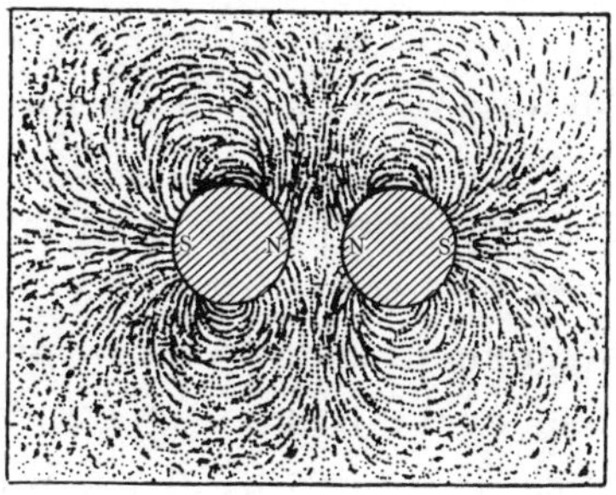

3275. Though the physical lines of force of a magnet may, and must
be considered as extending to infinite distance around it as long as the

---

not inside, the bar. In Figure 2, then, lines exiting from the polar faces will
have refracted very little, while those emerging laterally near the polar ends
will have refracted through nearly right angles as they pass from the iron into
the air. Lines closer to the central regions of the bar will have bent through
larger, even obtuse, angles.

in other cases ... more gradual, as in ... figs. 3, 4, 5: Does Faraday think there
is significant bending of lines *within* the circular magnets? He nowhere offers
his reasons, if so. Yet if not, it is hard to see why he regards the refraction as
"more gradual" in these cases, for the same range of exit angles, from zero to
right and obtuse, can be seen in these figures too. Perhaps he means that in
the circular magnet, *proportionally more* of the lines bend through acute angles,
and fewer through right and obtuse angles, than in the long bar magnet.

magnet is absolutely alone (3110.), yet they may be condensed and compressed into a very small local space, by the influence of other systems of magnetic power. This is indicated by fig. 6. I have no doubt, after the experimental results given in Series XXVIII. respecting definite magnetic action (3109.), that the sphondyloid representing the total power, which in the experiment that supplied the figure had a sectional area of not two square inches in surface, would have equal power upon the moving wire, with that infinite sphondyloid which would exist if the small magnet were in free space.

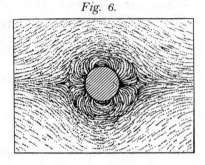

Fig. 6.

3276. The magnet, with its surrounding sphondyloid of power, may be considered as analogous in its condition to a voltaic battery immersed in water or any other electrolyte; or to a gymnotus (1773. 1784.) or torpedo, at the moment when these creatures, at their own will, fill the surrounding fluid with lines of electric force. I think the analogy with the voltaic battery so placed, is closer than with any case of *static* electric induction, because in the former instance the physical lines of electric force may be traced both through the battery and its surrounding medium, for they form continuous curves like those I have imagined within and without the magnet. The direction of these lines of electric force may be traced, experimentally, many ways. A magnetic needle freely suspended in the fluid will show them in and near to the battery, by standing at right angles to the course of the lines. Two wires from a galvanometer will show them; for if the line

---

3275. *would have equal power … with that infinite sphondyloid…*:   Thus the organized magnetic pattern represents an integral power which can be spread out or confined, just as other physical entities can be expanded or compressed.

3276. *analogous in its condition to a … gymnotus*: Here is the comparison of *magnet to electric fish* discussed in the introduction. It complements Faraday's comparison of *fish to magnet* in the Fifteenth Series.

physical lines of electric force may be traced both through the battery and its surrounding medium: Faraday refers to *dynamic* lines of electric force, that is, lines of *current*.

A magnetic needle … will show them … by standing at right angles to the course of the lines: As discussed in the First Series' introduction, a magnetic needle endeavors to point perpendicularly to the direction of an adjacent electric current.

Two wires from a galvanometer will show them…: The galvanometer deflects when the wires are immersed in the conducting medium; but the deflection dwindles to zero when the line joining the wires is turned perpendicularly to

joining the two ends in the fluid be at right angles to the lines of electric force (or the currents), there will be no action at the galvanometer; but if oblique or parallel to these lines, there will be deflection. A plate, or wire, or ball of metal in the fluid will show the direction, provided any electrolytic action can go on against it, as when a little acetate of lead is present in the medium, for then the electrolysis will be a maximum in the direction of the current or line of force, and nothing at all in the direction at right angles to it. The same ball will disturb and inflect the lines of electric force in the surrounding fluid, just as I have considered the case to be with paramagnetic bodies amongst magnetic lines of force (2806. 2821. 2874.). No one I think will doubt that as long as the battery is in the fluid, and has its extremities in communication by the fluid, lines of electric force having a physical existence occur in every part of it, and the fluid surrounding it.

3277. I conceive that when a magnet is in free space, there is such a medium (magnetically speaking) around it. That a vacuum has its own magnetic relations of attraction and repulsion is manifest from former experimental results (2787.); and these place the vacuum in relation to material bodies, not at either extremity of the list, but in the *midst* of them, as, for instance, between gold and platina (2399.), having other bodies on either side of it. What that surrounding magnetic medium, deprived of all material substance, may be, I cannot tell, perhaps the æther. I incline to consider this outer medium as *essential* to the magnet; that it is that which relates the external polarities to each other by curved lines of power; and that these must be so related as a matter of necessity. Just as in the case of the battery above, there is no line of force, either in or out of the battery, if this relation be cut off by removing or intercepting the conducting medium;—or in that of static electric induction, which is impossible until this related state be allowed (1169.);[12] so I conceive,

[12] Philosophical Magazine, March 1843; or Experimental Researches, 8vo, vol. ii. p. 279.

---

the line of current. Similarly, employing either their submerged hands or the disk collectors, Faraday and his colleagues were able to deduce the direction of electric lines of force in Gymnotus's tub (paragraphs 1775–1781).

3277. *I incline to consider this outer medium as essential to the magnet*: By contrast, the idea of action at a distance depreciates the regions surrounding an "active" center of force as mere passive territory having no role in determining or transmitting the action (paragraph 3270). In thus reinterpreting the magnet, Faraday implicitly reinterprets our general categories of agent and power—and begins to break down the distinction between them.

that without this external mutually related condition of the poles, or a related condition of them to other poles sustained and rendered possible in like manner, a magnet could not exist; an absolute north-ness or southness, or an unrelated northness or southness, being as impossible as an absolute or an unrelated state of positive or negative electricity (1178.).

3278. In this view of a magnet, the medium or space around it is as essential as the magnet itself; being a part of the true and complete magnetic system. There are numerous experimental results which show us that the relation of the lines to the surrounding space can be varied by occupying it with different substances; just as the relation of a ray of light to the space through which it passes can be varied by the presence of different bodies made to occupy that space, or as the lines of electric force are affected by the media through which either induction or conduction takes place. This variation in regard to the magnetic power may be considered as depending upon the aptitude which the surrounding space has to effect the mutual relation of the two external polarities, or to carry onwards the physical line of force; and I have on a former occasion in some degree considered it and its consequences, using the phrase *magnetic conduction* to represent the physical effect (2797.) produced by the presence either of paramagnetic or diamagnetic bodies.

3279. When, for instance, a piece of cold iron (3129.) or nickel (3240.) is introduced into the magnetic field, previously occupied by air or being even mere space, there is a concentration of lines of force on to it, and more power is transmitted through the space thus occupied than if the paramagnetic body were not there. The lines of force there-fore converge on to or diverge from it, giving what I have called conduction polarity (2818.); and this is the whole effect produced as regards the amount of the power; for not the slightest addition to, or diminution of, that external to the magnet is made (3218. 3223.). A new disposition of the force arises; for some passes now where it did not pass before, being removed from places where it was previously transmitted. Supposing that the magnet was inclosed in a surrounding solid mass of iron, then the effect of its superior conducting power would be to cause a great contraction inwards of the sphere of external action, and of the various sphondyloids, which we may suppose to be identified in different

---

3278. *In this view of a magnet, the medium or space around it is as essential as the magnet itself*: Thus it is a misleading figure of speech to identify the *iron* as "the" magnet. The iron is merely the tangible part of the magnetic system—its *body*, as Faraday will call it in the next paragraph and in paragraph 3284 below.

parts of it. A magnetic needle, if it could be introduced into the iron medium, would indicate extreme diminution, if not apparent annihilation, of the external power of this magnet; but the moving wire would show that it was there present to its full extent (3152. 3162.) in a very concentrated condition, just as it shows it in the very body of a magnet (3116.); and the power within the magnet, it being a hard and perfect one, would remain the same.

3280. The reason why a magnetic needle would fail as a correct indicator of the amount of power present in a given space is, that when perfect, it, because of the necessary condition of hardness, cannot carry on through its mass more lines of force than it can excite (3223.). But because of the coalescence of like lines of force end on (3226.), such a needle, when surrounded by a bad magnetic conductor, determines on to itself many of the lines which would otherwise pass elsewhere, has a high magnetic polarity, and is affected in proportion; every experiment, as far as I can perceive, tending to show that the attractions and repulsions are merely consequences of the tendency which the lines of physical magnetic force have to shorten themselves (3266.). So when the magnetic needle is surrounded by a medium gradually increasing in conducting power, it seems to show less and less force in its neighbourhood, though in reality the force is increasing there more and more. We can

---

3279. *A magnetic needle, if it could be introduced into the iron medium, would indicate extreme diminution…*: A needle having greater conducting power than its surrounding medium will *point* axially (paragraphs 2811, 2828), since in gathering up lines of force it re-routes some of them from the medium through itself. But if the medium is iron, equal in conducting power to the needle, few or none of its lines would be transferred to the needle; and consequently the pointing tendency would decline or disappear. Note that by this account the number of lines passing through the needle *varies* as the surrounding medium is varied.

the power within the magnet … would remain the same: In the 29th Series, Faraday was able to effect only small changes in the strength of a hard bar-magnet, and then only through the influence of other, more powerful magnets. The presence of mere unmagnetized iron, therefore, should cause no significant increase or decrease.

3280. *it … cannot carry on through its mass more lines of force than it can excite*: Faraday showed in the 29th Series that a hard, well-charged magnet is essentially *constant* in power—that is, in the number of lines it supports.

the magnetic needle … seems to show less and less force in its neighbourhood, though in reality the force is increasing there more and more: A comment to the previous paragraph recounted how this would be true for a needle formed of soft iron. But here Faraday says he is talking about a "hard" needle—which sustains a constant number of lines, as we have already seen. How can its pointing tendency then vary, if the number of lines passing through it does not vary?

easily conceive a very hard and feebly charged magnetic needle surrounded by a medium, as soft iron, better than itself in conducting power, *i. e.* carrying on by conduction more lines of force than the needle could determine or carry on by its state of charge (3298.). In that case I conceive it would, if free to move, point feebly in the iron, because of the coalescence of the lines of force, but would be repelled bodily from the chief magnet, in analogy with the action on a diamagnetic body. As I have before stated, the principle of the moving wire can be applied successfully in those cases where that of the magnetic needle fails (3155.).

3281. If other paramagnetic bodies than iron be considered in their relation to the surrounding space, then their effects may be assumed as proportionate to the conducting power. If the surrounding medium were hard steel, the continuation of the sphondyloid of power would be much less than with iron; and the effects, in respect of the magnetic needle, would occur in a limited degree. If a solution of protosulphate of iron were used, the effect would occur in a very much less degree. If a solution were prepared and adjusted so as to have no paramagnetic or diamagnetic relation (2422.), it would be the same to the lines of force as free space. If a diamagnetic body were employed, as water, glass, bismuth or phosphorus, the extent of action of the sphondyloids would expand (3279.); and a magnetic needle would appear to increase in intensity of action, though placed in a region having a smaller amount of magnetic force passing across it than before (3155.). Whether in any of these cases, even in that of iron, the body acting as a conductor has a state induced upon its particles for the time like that of a magnet in the corresponding state, is a question which I put upon a former occasion (2833.); but I leave its full investigation and decision for a future time.

3281. *If other paramagnetic bodies than iron be considered ... their effects may be assumed as proportionate to the conducting power*: By "their effects" Faraday means the diminution of a magnetic needle's pointing tendency when surrounded by various paramagnetic media; the greater that decline, the greater we may assume the magnetic conducting power of the medium to be. But perhaps this remark should be restricted to soft iron needles; see the previous comment.

If the surrounding medium were hard steel... If a solution of protosulphate of iron were used...: The materials iron, hard steel, iron protosulphate solution, are named in order of decreasing magnetic conducting power. Thus an iron needle surrounded by each of these materials in succession would point with increasing vigor.

3282. The circumstances dependent upon the shape and size of magnets appear to accord singularly well with the view I am putting forth of the action of the surrounding medium. If there be a function in that medium equivalent to conduction, involving differences of conduction in different cases, that of necessity implies also reaction or resistance. The differences could not exist without. The analogous case is presented to us in every part by the electric force. When, therefore, a magnet, in place of being a bar, is made into a horseshoe form, we see at once that the lines of force and the sphondyloids are greatly distorted or removed from their former regularity; that a line of maximum force from pole to pole grows up as the horseshoe form is more completely given; that the power gathers in, or accumulates about this line, just because the badly conducting medium, *i. e.* the space or air between the poles, is shortened. A bent voltaic battery in its surrounding medium (3276.), or a gymnotus curved at the moment of its peculiar action (1785.), present exactly the like result.

3283. The manner in which the keeper or sub-magnet, when in place, reduces the power of the magnet in the space or air around, is evident. It is the substitution of an excellent conductor for a poor one; far more of the power of the magnet is transmitted through it than through the same space before, and less, therefore, in other places. If a horseshoe magnet be charged to saturation with its keeper on, and its power be then ascertained, removing the keeper will cause the power to fall. This will be (according to the hypothesis) because the iron keeper could, by its conduction, sustain higher external conditions of the magnetic force, and therefore the *magnet* could take up and sustain a higher condition of charge.

3282. *The circumstances dependent upon the shape and size of magnets appear to accord singularly well with the view I am putting forth...*: In the following paragraphs Faraday will bring forth a parade of examples showing a close correlation between changes in form of the iron "body" with changes in the external disposition of a magnet's power. Note that before introducing these sometimes playful cases he is moved to recall the gymnotus "coiling incident" in the 15th Series.

3283. *the keeper ... reduces the power of the magnet in the space or air around*: By concentrating the lines of force within itself, an iron keeper effectively withdraws them from neighboring regions (paragraph 3120, *comment*). But from another viewpoint, by substituting itself for the poorly conductive *air* the keeper raises the conductivity of the whole system and thereby enables it to receive a stronger initial magnetization.

The case passes into that of a steel ring magnet, which being mag-netized, shows no external signs of power, because the lines of force of one part are continued on by every other part of the ring; and yet when broken exhibits strong polarity and external action, because then the lines, which, being determined at a given point, were before carried on through the continuous magnet, have now to be carried on and continued through the surrounding space.

3284. These results, again, pass into the fact, easily verified partially, that if soft iron surround a magnet, being in contact with its poles, that magnet may receive a much higher charge than it can take, being surrounded with a lower paramagnetic substance, as air: also another fact, that when masses of soft iron are at the ends of a magnet, the latter can receive and keep a higher charge than without them; for these masses carry on the physical lines of force, and deliver them to a body of surrounding space; which is either widened, and therefore increased in the direction across the lines of force, or shortened in that direction parallel to them, or both; and both are circumstances which facilitate the conduction from pole to pole, and the relation of the external lines to the lines of force *within* the magnet. In the same way the armature of a natural loadstone is useful. All these effects and expedients accord with the view, that the space or medium external to the magnet is as important to its existence as the body of the magnet itself.

3283, continued. *The case passes into that of a steel ring magnet…*: A continuous steel ring might be thought of as a magnet and keeper so perfectly integrated, no part of the power passes into the air! Such rings when magnetized have been observed to "keep" their power for many years.

Faraday will choose this phrase, "passes into," several times again in succeeding paragraphs. Phenomena that "pass into" one another disclose a commonality, an affinity, among themselves. Experiments that disclose such passages can effect fundamental transformations in our experience and in our representative imagery. Perhaps the essential scientific activity, insofar as it is directed to *knowing* nature (not just tabulating facts and statistics about it), is the cultivation of such leading and metamorphosing strains within natural phenomena themselves. Recall Faraday's earlier anticipation in paragraph 3273, that "all these various phænomena presented to us shall become *one.*"

3284. *masses of soft iron*: Affixed to the ends of a long, narrow magnet, they afford the lines of force a more gradual transition from their high concentration within the magnet body to their dispersion throughout the surrounding air.

the armature of a natural loadstone: A pair of iron caps covering the polar regions of an "armed" loadstone (paragraph 56). Their usefulness was long recognized; now Faraday identifies the principle of their operation.

the body of the magnet: That is, the *iron*—the palpable but by no means sole and exclusive constituent of the magnet.

3285. Magnets, whether large or small, may be supersaturated, and then they fall in power when left to themselves; quickly at first if strongly supersaturated, and more slowly afterwards. This, upon the hypothesis, would be accounted for by considering the surrounding medium as unable, by its feeble magneto-conducting power, to sustain the higher state of charge. If the conducting power were increased sufficiently, then the magnet would not be supersaturated, and its power would not fall. Thus, if a magnet were surrounded by iron, it might easily be made to assume and retain a state of charge, which, if the iron were suddenly replaced by air, would instantly fall. Indeed, magnets can only be supersaturated by placing them for the time under the dominion of other sources of magnetic power, or of other more favourable surrounding media than that in which they manifest themselves as supersaturated.

3286. The well-known result, that small bar-magnets are far stronger in proportion to their size than larger similar magnets, harmonizes and *sustains* that view of the action of the external medium which has now been given. A sewing-needle can be magnetized far more strongly than a bar twelve inches long and an inch in diameter; and the reason under the view taken is, that the excited system in the magnet (correspondent to the voltaic battery in the analogy quoted (3276.)) is better sustained by the necessary conjoint action of the surrounding medium in the case of the small magnet. For as the imperfect magneto-conducting power of that medium (or the consequent state of tension into which it is thrown) acts back upon the magnet (3282.), so the smaller the sum of exciting force in the centre of the magnetic sphondyloids, the better able will the surrounding medium be to do its part in sustaining the resultant of

---

3285. *supersaturated*: That is, magnetized to a higher degree than can be permanently retained. Supersaturation regularly results when a magnet is charged with its keeper in place, as described in paragraph 3283. As one nineteenth-century textbook relates, "A horseshoe-shaped steel magnet will support a greater weight immediately after being magnetized than it will do after its armature [keeper] has been once removed from its poles."

considering the surrounding medium as unable, by its feeble magneto-conducting power, to sustain the higher state of charge: As mentioned earlier, the highly conductive keeper "keeps" by taking the place of poorly-conducting air.

If the conducting power were increased sufficiently, then the magnet would not be supersaturated, and its power would not fall: Note that Faraday means increase in conducting power of the *medium*, not of the *iron*. Thus the customary phraseology is truly misleading—it is not the *magnet* (that is, the iron) but the *medium* that is "saturated" or "supersaturated"!

force. It is very manifest, that if the twelve-inch bar be conceived of as subdivided into sewing-needles, and these be separated from each other, the whole amount of exciting force acts upon, and is carried onwards in closed magnetic curves, by a very much larger amount of external surrounding medium than when they are all accumulated in the single bar.

3287. The results which have been observed in the relation of *length* and *thickness* of a bar-magnet, harmonize with the view of the office of the external medium now urged. If we take a small, well-proportioned, saturated magnet, as a sewing-needle; alone, it has, as just stated, such relation to the surrounding space as to have its high condition sustained; if we place a second like magnetic needle by the side of the first, the surrounding space of the two is scarcely enlarged, it is not at all improved in conducting character, and yet it has to sustain double the internal exciting magnetic force exerted when there was one needle only (3232.); this must react back upon the magnets, and cause

---

3286. *if the twelve-inch bar be conceived of as subdivided into sewing-needles...*: Faraday will employ this analytical image repeatedly in order to apprehend the correlation between a magnet's shape and the power it can acquire and retain. In the present instance, imagine the large magnet pared away into sewing-needles, each retaining a proportionate share of the magnetic power belonging to the whole bar. Faraday sees *at a glance* that the needles will enjoy "a very much larger amount of external surrounding medium" when separated than when packed together as a rod, and that they can therefore be more strongly magnetized individually than they had been collectively. But readers who find such an inference less than "very manifest" will appreciate Faraday's next two paragraphs.

exciting force: That is, the power whereby a magnet *originates* lines of force (paragraph 3298). Of the various efforts to explain this unknown power Faraday has found none convincing, not even Ampère's theory of circulating molecular currents. But he continues to regard as promising the analogy between magnetic lines of force and lines of electric current; in that analogy the "exciting force" of a magnet corresponds to the voltaic cell's ability to maintain a condition of electric tension between its terminals.

3287. *if we place a second like magnetic needle by the side of the first...*: If the second needle were to sustain the same number of lines as the first, then a virtually fixed expanse of external medium would have to carry on a double quantity of lines of force; and similarly for additional needles placed alongside the first two. But a medium having restricted volume and imperfect conducting power cannot continually increase the number of lines it carries. Its limitations must then constrain the whole system ("react back upon the magnets, and cause a reduction of their power"). Thus a needle that expands in thickness, with no extension in length, will be unable to retain a proportionally heightened degree of magnetization.

a reduction of their power. The addition of a third needle repeats the effect; and if we conceive that successive needles are added until the bundle is an inch thick, we have a result which will illustrate the effect of a thickness too large, and disproportionate to the length.

3288. On the other hand, if we assume two such needles similarly placed in a right line at a distance from each other, each has its surrounding system of curves occupying a certain amount of space; if brought together by unlike poles, they form a magnet of double the length; the external lines of force coalesce (3226.), those at the faces of contact nearly disappear; those which proceed from the extreme poles coalesce externally, and form one large outer system of force, the lines of which have a greater length than the corresponding lines of either of the two original needles. Still, by the supposition that the magnets are perfectly hard and invariable, the exciting force within remains, or tends to remain the same (3227.) in quantity, there is nothing to increase it. The increase in length, therefore, of the external circuit, which acts as a resisting medium upon the internal action, will tend to diminish the force of the whole system. Such would be the case if a voltaic battery surrounded by distilled water, as the analogous illustration (3276.), could be elongated in the water, and so its poles be removed further apart; and though in the case of magnets previously charged, some effect equivalent to intensity of excitement may be produced by conjoining several together end on, yet the diminished sustentation of power externally appears to follow as a consequence of the increased distance

---

3288. *two such needles similarly placed in a right line ... form a magnet of double the length...*: Since the lines of force native to the needles coalesce, the exciting force of the two needles arranged serially is no greater than that of a single needle. But Faraday makes an implicit simplifying assumption—that the lines of force exit *primarily from the extreme ends* of the combination. On this supposition, the external medium through which the lines must pass will be virtually doubled in length; and similarly for additional needles placed in line with the first two. But a constant exciting force cannot sustain the same number of lines of force through increasingly longer paths in the poorly conducting external medium; the total number of lines must therefore decrease. Thus a needle that increases in length, with no expansion in thickness, may actually suffer a *decrease* in the degree of magnetization it can retain.

Perhaps Faraday's speedy inference in paragraph 3286 above will now be more evident; since he has shown that if a needle grows in cross-section without a proportional increase in length—or in length without a proportional increase in section, it will suffer a relative or even an absolute decrease in its ability to retain magnetization. A bar 12 inches long and one inch in diameter is very far from the proportions of a typical sewing needle. The proportions may be represented by $12 \div 1^2$ or 12 for the bar; for a needle 2 inches long and .03 inches in diameter the corresponding quotient is $2 \div .03^2$ or 2222.

of the extreme poles, or external, mutually dependent parts. Static electric induction also supplies a correspondent and illustrative case.

3289. The usual case in which the influence of length and thickness becomes evident, is not, however, always or often that of the juxtaposition of magnets already as highly charged as they can be, but rather that of a bar about to be charged. If two bars alike in steel, hardness, &c., one an inch long and the tenth of an inch in diameter, and the other of the same length but five-tenths of an inch in diameter, be magnetized to supersaturation, the latter, though it contains twenty-five times the steel of the former, will not retain twenty-five times the power, for the reason already given (3287.); the surrounding medium not being able to sustain external lines of force to that amount. But if a third bar, two inches long and also five-tenths in diameter, be magnetized at the same time, it can receive much more power than the second one. A natural reason for this presents itself by the hypothesis; for the limitation of power in the two cases is not in the magnets themselves, but in the external medium. The shorter magnet has contact and connexion with that medium by a certain amount of surface; and just what power the medium outside that surface can support, the magnet will retain. Make the magnet as long again, and there is far more contact and relation with the surrounding medium than before;

3289. *juxtaposition of magnets already as highly charged as they can be...*: That is, several bars are individually magnetized to saturation and then combined into a whole—the procedure considered in the previous two paragraphs.

rather ... a bar about to be charged: That is, several unmagnetized bars are combined, then magnetized as a whole. The usual experimental situation resembles this second case, not the somewhat fictional procedure previously discussed. But for bars that differ only in their thickness, the experimental results agree with his conclusion in paragraph 3287—namely, that the thicker bar does not retain a proportionally greater power than the thinner bar.

a third [longer] bar ... can receive much more power than the second [shorter] one: Thus, for bars that differ only in their length, the experimental results do *not* agree with his previous conclusion. In paragraph 3288 Faraday had argued (under a simplifying assumption) that a lengthened needle would generally suffer a decrease in power. But experimentally, a longer bar exhibits an *increase* in the power it can retain, though, to be sure, not a proportional one. To accommodate the actual results, therefore, Faraday will give up the earlier assumption.

The shorter magnet has ... a certain amount of surface... Make that magnet as long again, and there is far more contact and relation with the surrounding medium than before: Previously, lines of force were assumed to exit only at the extreme ends of the serial combination. Lengthening such an idealized magnet, therefore, would do nothing to increase the exit area. But in a real magnet, a considerable fraction of the lines of force pass into the air through the *sides*, as his own

Fig. 2.

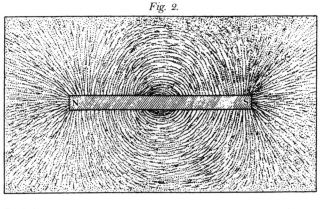

Fig. 3.   Fig. 4.   Fig. 5.

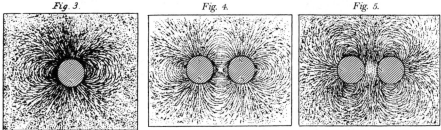

and therefore the power which the magnet can retain is greater. If there were such limited points of resulting action in the magnet as is often understood by the word *poles,* then such a result could hardly be the case, on my view of the physical actions. But such poles do not exist. Every part of the surface of the magnet, so to say, is pouring forth externally lines of magnetic force, as may be seen in figs. 2, 3, 4, 5 (3274.). The larger

---

experiments have shown. Clearly one consequence of this is that the lengths of exterior paths do not increase in strict proportion with the length of the bar— as had been previously supposed. Moreover, lengthening the bar provides more avenues of exit for the lines, which tends to *ease* their passage through the air and so increase the power which the longer magnet can retain. Thus a real magnet represents a balance between the tendency of longer external paths to reduce, and wider avenues of passage to increase, the overall power. In general, therefore, a longer magnet will accept a greater charge than a shorter one of the same diameter—but not a proportionally greater one.

If there were such limited points of resulting action in the magnet as is often understood by the word poles, then such a result could hardly be the case: For if lines of force originated in poles, they would exit predominantly from the ends, thereby approximating the earlier assumption; and the result of magnetizing a lengthened bar would be that deduced in paragraph 3288.

But such poles do not exist: This certain declaration, and the beautiful image of the magnet's surface "pouring forth externally lines of magnetic force," which immediately follows it, stand nearly at the midpoint of the paragraph— as indeed they represent a pivotal point in its content too.

the magnet, to a certain extent, and the larger the amount of external conducting medium in contact with it, the more freely is this transmission made. If the second magnet, being an inch long, be conceived to be charged to its full amount, and then, whilst in free space, could have half an inch of iron added to its length at each end, we see and know that many of the lines of force originally issuing from that part of its surface still left in contact with the air at the equatorial part, would now move internally towards the ends, and issue at a part of the soft iron surface; indicating the manner in which the tension would be relieved by this better conducting medium at the ends, and by increased surface of contact with the surrounding bad conductor of air or space. The thick, short magnet could evidently excite and carry on physical lines of magnetic power far more numerous than those which the space about it can receive and convey from pole to pole; and the increase in the length of the magnet may go on advantageously, until the increasing sum of power, sustainable by the increasing medium in the circuit, is equal to that which the magnet can sustain or transmit internally; for all the lines of power, wherever they issue from the magnet, have to pass through its equator; and in this way the equator or thickness of the magnet becomes related to its length. So the advantageous increase in length of the bar is limited by the increasing resistance within, and especially at the equator of the bar; and the increase in breadth, by the increasing resistance (for increasing powers) of the external surrounding medium (3287.).

*   *   *

3292. When, in place of considering the medium external to a magnet, as homogeneous or equal in magnetic power, we make it variable in different parts, then the effects in it appear to me still to be in perfect accordance with the notion of physical lines of magnetic force, which, being present externally, are definite in direction and amount. The series of substances at our command which affect the surrounding space in this respect, do not present a great choice of successive steps; but having iron, nickel and cobalt, very high as paramagnetic bodies, we then possess hard steel, as very far beneath them; next, perhaps, oxides of iron, and so on by solutions of the magnetic metals to oxygen, water, glass, bismuth and phosphorus, in the diamagnetic direction. Taking the magnetic force of the earth as supplying the source of power, and placing a globe of iron or nickel in the air, we see by the pointing of a small magnetic needle (or in another case, by the use of iron filings (3240.)), the deflected course of the lines of force as they enter into and pass out of the sphere, consequent upon the conducting power of the paramagnetic body. These have been

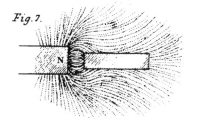

Fig. 7.

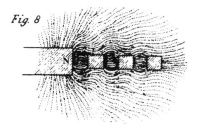

Fig. 8

described in their forms in another place (3238.). If we take a large bar-magnet, and place a piece of soft iron, about half the width of the magnet, and three or four times as long as it is wide, end on to, and about its own width from one pole, and covering that with paper, then observe the forms of the lines of force by iron filings; it will be seen how beautifully those issuing from the magnet converge, by fine inflections, on to the iron, entering by a comparatively small surface, and how they pass out in far more diffuse streams by a much larger surface at the further part of the bar, fig. 7. If we take several pieces of iron, cubes for instance, then the lines of force which are altogether outside of them, may be seen undergoing successive undulations in contrary directions, fig. 8. Yet in all these cases of the globe, bar and cubes, I, at least, am satisfied that a section across the same lines of force in any part of their course, however or whichever way deflected, would yield the same amount of effect (3109. 3218.); at the same time this effect of deflection is not only consistent with, but absolutely suggests the idea of a physical line of force.

3293. Then the manner in which the power disappears in such cases to an ordinary magnetic needle is perfectly consistent. A little needle held by the side of the soft bar described above (3292.), indicates much less magnetic power than if the iron were away. If held in a hole made in the iron, it is almost indifferent to the magnet; yet what power remains shows that the lines through the air in the hole are in the same general direction as those through the neighbouring iron. These effects are perfectly well known, no doubt; and my object is only to show that they are consistent with, and support the idea of external media having magnetic conducting power. But these apparent destructions of power, and even far more anomalous cases (2868. 3155.), are fully accounted for by the hypothesis; and the force absolutely unaffected in amount is found, experimentally, by the moving wire. I have had occasion before to refer to the modification of the magnetic force (in relation to the magnetic needle), where, its absolute quantity being the same, it passes across better or worse conductors, and I have temporarily used the words *quantity* and *intensity* (2866. 2868. 2870.). I would, however, rather not attempt to limit or define these or such like terms now, however much they may be wanted, but wait until what is at present little more than suggestion, may have been canvassed, and if true in itself; may have received assurance from the opinions or testimony of others.

3294. The association of magnet with magnet, and all the effects then produced (3218.), are in harmony, as far as I can perceive, with the idea of a physical line of magnetic force. If the magnets are all free to move, they set to each other, and then tend to approach; the great result being, that the lines from all the sources tend to coalesce, to pass through the best conductors, and to contract in length. When there are several magnets in presence and in restrained conditions, the lines of force, which they present by filings, are most varied and beautiful (3238.); but all are easily read and understood by the principles I have set forth. As the power is definite in amount, its removability from place to place, according to the changing disposition of the magnets, or the introduction of better or worse conductors into the surrounding media, becomes a perfectly simple result.

3295. As magnets may be looked upon as the habitations of bundles of lines of force, they probably show us the tendencies of the physical lines of force where they occur in the space around; just as electric currents, when conducted by solid wires, or when passing, as the Leyden or the voltaic spark, through air or a vacuum, are alike in their essential relations. In that case, the repulsion of magnets when placed side by side, indicates the lateral tendency of separation of lines of magnetic force (3267.). The effect, however, must be considered in relation to the simultaneous gathering up of the terrestrial lines of force in the surrounding space upon each magnet, and also the tendency of each magnet to secure its own independent external medium. The effect coincides with, and passes into that of the lateral repulsion of balls of iron in a previously equal magnetic field (2814.);

3294. *all are easily read and understood by the principles I have set forth*: With this golden phrase Faraday reminds us that the essential motive in natural philosophy is to recognize the phenomena as *legible*. The highest philosophical precepts are not laws (like Newton's Laws of Motion) but guides to cognition and interpretation. The application of laws to the phenomena reveals regularity; the pursuit of interpretation yields intelligibility.

3295. *habitations of bundles of lines of force*: Viewed as "habitations," magnets simply make visible the locations and mutual relations of the lines of force that occupy them—no different in that respect from iron filings. Such a view, though, passes over their still unresolved role as *originators* of lines.

which again, by a consideration of the action in two directions, *i. e.* parallel to and across the magnetic axis, links the phænomena of separation with those of attraction.

3296. When speaking of magnets, in illustration of the question under consideration, I mean magnets perfect in their kind, *i. e.* such as are very hard and hold their charge, so that there shall be neither internal reaction of discharge or development (3224.), nor any external change, except what may depend upon such absolute and permanent loss of exciting power as is consequent upon an over-ruling change of the external relations. Heterogeneous magnets, which might allow of irregular variations of power, are out of present consideration.

3297. With regard to the great point under consideration, it is simply, whether the lines of magnetic force have a *physical existence* or not? Such a point may be investigated, perhaps even satisfactorily, without our being able to go into the further questions of how they account for magnetic attraction and repulsion, or even by what condition of space, æther or matter, these lines consist. If the extremities of a straight bar-magnet, or if the polarities of a circular plate of steel (3274.), are in magnetic relation to each other externally (3257.), then I think the existence of *curved* lines of magnetic force must be conceded (3258. 3263.);[13] and if that be granted, then I think that the physical nature of the lines must be granted also. If the external relation of the poles or polarity is denied, then, as it appears to me, the internal relation must be denied also; and with it a vast number of old and new facts (3070. &c.) will be left without either theory, hypothesis, or even a vague supposition to explain them.

[13] See for a case of curved lines the inclosed and compressed system of forces belonging to the central circular magnet, fig. 6 (3275.).

---

links the phænomena of separation with those of attraction: In the first stages of magnetic interpretation, attraction and repulsion had been "linked" as the supposed complementary principles of magnetic action. In the present, more fully evolved vision, magnetic action is seen as reflecting continuous and extensive relations of tension and equilibrium; attraction and repulsion are merely two of many manifestations of those rich and fertile relations.

3296. *there shall be neither internal reaction of discharge or development*: That is, neither a decrease nor an increase of overall magnetic power. As Faraday showed in the 29th Series, hard well-charged magnets can meet this criterion pretty nearly; although no magnet is truly "perfect" in the sense described.

3298. Perhaps both magnetic attraction and repulsion, in all forms and cases, resolve themselves into the differential action (2757.) of the magnets and substances which occupy space, and modify its magnetic power. A magnet first originates lines of magnetic force; and then, if present with another magnet, offers in one position a very free conduction of the new lines, like a para-magnetic body; or if restrained in the contrary position, resists their passage, and resembles a highly diamagnetic substance. So, then, a source of magnetic lines being present, and also magnets or other bodies affecting and varying the conducting power of space, those bodies which can convey onwards the most force, may tend, by differential actions, with the others present, to take up the position in which they can do so the most freely, whether it is by pointing or by approximation; the best conductor passing to the place of strongest action (2757.), whilst the worst retreats from it, and so the effects both of attraction and repulsion be produced. The tendency of the lines of magnetic force to shorten (3266. 3280.) would be consistent with such a notion. The result would occur whether the physical lines of force were supposed to consist in a dynamic or a static state (3269.).

3299. Having applied the term *line of magnetic force* to an abstract idea, which I believe represents accurately the nature, condition, direction, and comparative amount of the magnetic forces, without reference to any physical condition of the force, I have now applied the term *physical line of force* to include the further idea of their physical nature. The first set of lines I *affirm* upon the evidence of strict experiment (3071. &c.). The second set of lines I advocate, chiefly with a view of stating the question of their existence; and though I should not have raised the argument unless I had thought it both important, and likely to be answered ultimately in the affirmative, I still hold the opinion with some hesitation, with as much, indeed, as accompanies any conclusion I endeavour to draw respecting points in the very depths of

---

3298. *A magnet first originates lines of magnetic force...*: Faraday has not been able to see very far into this act of origination. He has, indeed, alluded to an "exciting force" (paragraphs 3286–3289, 3296); but this only gives an alternate label to the problem without further elucidating it. What does it mean to "originate" lines of force; and what sort of relationship between force and matter does it imply? More generally, what is the relation between agents and the powers they exercise? Faraday has been asking this question continually since his researches with the electric eel (Fifteenth Series).

approximation: That is, approach.

science, as, for instance, regarding one, two, or no electric fluids; or the real nature of a ray of light, or the nature of attraction, even that of gravity itself; or the general nature of matter.

Royal Institution, March 6, 1852.

3299. *the real nature of a ray of light, … of attraction, … of gravity itself; or the general nature of matter*: Typically, as Faraday concludes this searching essay he has the very largest questions in view!

Bibliography

Agassi, Joseph. *Faraday as a Natural Philosopher*. Chicago and London: University of Chicago Press, 1971.

Atkinson, E, trans. *Ganot's Physics. Elementary Treatise on Physics, Experimental and Applied, for the use of Colleges and Schools*. Twelfth edition. New York: William Wood and Co., 1886.

Bence Jones, Henry. *The Life and Letters of Faraday*. 2 vols. London and Philadelphia: Longmans, Green, 1870.

Bowers, Brian. *Sir Charles Wheatstone FRS 1802–1875*. London: Her Majesty's Stationery Office, 1975.

Cantor, Geoffrey. *Michael Faraday, Sandemanian and Scientist: A Study of Science and Religion in the Nineteenth Century*. London: Macmillan and New York: St. Martin's Press, 1991.

Cantor, Geoffrey, David Gooding, Frank A. J. L. James. *Michael Faraday*. Atlantic Highlands, New Jersey: Humanities Press, 1996.

Davy, Humphry. *Elements of Chemical Philosophy*. London: J. Johnson & Co., 1812.

Faraday, Michael. *Faraday's Diary*. Thomas Martin (ed.), 8 vols., London: G. Bell and Sons, 1932–1936.

Faraday, Michael. *Chemical Manipulations*. London: W. Phillips, 1827.

Faraday, Michael. *Experimental Researches in Chemistry and Physics*. London: Richard Taylor and William Francis, 1859. Reprinted London: Taylor and Francis, 1991.

Faraday, Michael. *Experimental Researches in Electricity*. 3 vols. London: Richard and John Edward Taylor, 1839, 1844; Taylor and Francis, 1855. Reprinted Santa Fe: Green Lion Press, 2000.

Faraday, Michael. *On the Various Forces of Nature and their Relations to Each Other*. William Crookes (ed.), reprinted New York: The Viking Press, 1960.

Faraday, Michael. *The Chemical History of a Candle*. William Crookes (ed.), London: Griffin, Bohn, 1861. Reprinted New York: Thomas Y. Crowell Company, 1957.

Fisher, Howard J. "The Great Electrical Philosopher" in *The College*, **31**. Annapolis: St. John's College, 1979.

Fisher, Howard J. "The Body Electric" in *The St. John's Review*, **41**,1. Annapolis: St. John's College, 1992.

Fisher, Howard J. "Faraday's Two Voices" in *Physis, Revista Internazionale di Storia della Scienza*. **29**,1 (N.S.). Firenze: Leo S. Olschki Editore, 1992.

Gooding, David. *Experiment and the Making of Meaning*. Dordrecht and Boston: Kluwer Academic Publications, 1990.

Gooding, David and Frank A. J. L. James (eds.), *Faraday Rediscovered. Essays on the Life and Work of Michael Faraday, 1791–1867.* New York: American Institute of Physics, 1989.

Heilbron, J. L., *Elements of Early Modern Physics.* Berkeley, Los Angeles, London: University of California Press, 1982.

James, Frank A. J. L. (ed.), *The Correspondence of Michael Faraday,* vols. 1–4 (vols. 5–6 forthcoming). London: Institution of Electrical Engineers, 1991–1999.

King, Ronald, *Michael Faraday of the Royal Institution.* London: The Royal Institution of Great Britain, 1973.

King, Ronald, *Humphry Davy.* London: The Royal Institution of Great Britain, 1978.

Marcet, Jane, *Conversations in Chemistry.* New Haven: Sidney's Press, 1809.

Maxwell, James Clerk, *Treatise on Electricity and Magnetism,* 2 vols., 3rd edn., J. J. Thomson (ed.). Oxford: The Clarendon Press, 1891. Reprinted New York: Dover, 1956.

Ostwald, Wilhelm, *Electrochemistry: History and Theory.* N.P. Date (tr.), 2 vols., Smithsonian Institution and National Science Foundation, 1980.

Ostwald, Wilhelm, *Outlines of General Chemistry.* 3rd English edn., W. W. Taylor (tr.), Macmillan, 1912.

Poynting, J. H. and Sir J. J. Thomson, *A Text-Book of Physics.* Vol. IV, Parts 1 and 2, Electricity and Magnetism. London: Charles Griffin and Co., Ltd., 1914.

Ruben, Samuel, *The Founders of Electrochemistry.* Philadelphia: Dorrance, 1975. Reprinted La Salle, Illinois: Open Court, 1983.

Simpson, Thomas King, "Faraday's Thought on Electromagnetism" in *The College,* **22**, 2. Annapolis: St. John's College, July 1970.

Simpson, Thomas King, *A Critical Study of Maxwell's Dynamic Theory of the Electromagnetic Field in the Treatise on Electricity and Magnetism.* Diss., The Johns Hopkins University, 1968.

Simpson, Thomas King. *Maxwell on the Electromagnetic Field. A Guided Study.* Masterworks of Discovery Series. New Brunswick: Rutgers University Press, 1997.

Thomas, John Meurig, *Michael Faraday and the Royal Institution. The Genius of Man and Place.* Bristol and Philadelphia: Institute of Physics Publishing, 1991. Reprinted 1997.

Thompson, Silvanus P., *Elementary Lessons in Electricity and Magnetism.* London: Macmillan and Co., 1908; 7th edn., 1915. New York: The Macmillan Company, 1924.

Thompson, Silvanus P., *Michael Faraday: His Life and Work.* New York: Macmillan, 1898.

Tricker, R. A. R., *Early Electrodynamics; the First Law of Circulation.* London and New York: Pergamon Press, 1965.

Tweney, Ryan, with David Gooding (eds.), *Michael Farady's 'Chemical Notes, Hints, Suggestions and Objects of Pursuit' of 1822.* London: Institution of Electrical Engineers, History of Technology Series, No. 17, 1991.

Tyndall, John, *Faraday as a Discoverer.* London: Longmans, Green, 1868.

Webster, John W., M.D., *A Manual of Chemistry on the Basis of Professor Brande's.* 2nd ed. Boston: Richardson and Lord, 1829.

Whittaker, E. T., *A History of the Theories of Aether and Electricity.* 2 vols. London and New York: T. Nelson, 1951–1953. Reprinted (in one volume) New York: Dover Publications, Inc., 1989.

Williams, L. Pearce, *Michael Faraday.* New York: Basic Books, Inc., 1965.

Williams, L. Pearce, R. FitzGerald and O. Stallybrass (eds.), *The Selected Correspondence of Michael Faraday.* 2 vols. Cambridge, 1971.

Index

Note: The index entries have been chosen to suit the aims and contents of this Guide. Readers may also wish to consult Faraday's extensive index provided in the complete *Experimental Researches in Electricity* (reprinted Green Lion Press, 2000).

Biographical Note

photo by Guill

Howard J. Fisher

Howard Fisher teaches in the Great Books program at St. John's College in Annapolis. With a background in physics and philosophy, he has been studying Faraday in both laboratory and library for over twenty-five years.

Carnet